CHROMANS AND TOCOPHEROLS

This is the Thirty-Sixth Volume in the Series

THE CHEMISTRY OF HETEROCYCLIC COMPOUNDS

THE CHEMISTRY OF HETEROCYCLIC COMPOUNDS

A SERIES OF MONOGRAPHS

ARNOLD WEISSBERGER and EDWARD C. TAYLOR

Editors

An Interscience® Publication

Library of Congress Cataloging in Publication Data:

Main entry under title:
Chromans and tocopherols.

(The Chemistry of heterocyclic compounds; v. 36)
"An Interscience publication."
Includes indexes.
1. Vitamin E. 2. Chroman. I. Ellis, Gwynn Pennant. II. Lockhart, Ian M.

QP772.T6C47 547.7′4 80-16902
ISBN 0-471-03038-4

Printed in the United States of America

10 9 8 7 6 5 4 3 2 1

CHROMANS AND TOCOPHEROLS

Edited by

G. P. Ellis

DEPARTMENT OF CHEMISTRY
UNIVERSITY OF WALES INSTITUTE OF
SCIENCE AND TECHNOLOGY
CARDIFF, UNITED KINGDOM

I. M. Lockhart

BOC LIMITED SPECIAL GASES, LONDON, UNITED KINGDOM

AN INTERSCIENCE® PUBLICATION

JOHN WILEY & SONS

NEW YORK . CHICHESTER . BRISBANE . TORONTO

The Chemistry of Heterocyclic Compounds

The chemistry of heterocyclic compounds is one of the most complex branches of organic chemistry. It is equally interesting for its theoretical implications, for the diversity of its synthetic procedures, and for the physiological and industrial significance of heterocyclic compounds.

A field of such importance and intrinsic difficulty should be made as readily accessible as possible, and the lack of a modern detailed and comprehensive presentation of heterocyclic chemistry is therefore keenly felt. It is the intention of the present series to fill this gap by expert presentations of the various branches of heterocyclic chemistry. The subdivisions have been designed to cover the field in its entirety by monographs which reflect the importance and the interrelations of the various compounds, and accommodate the specific interests of the authors.

In order to continue to make heterocyclic chemistry as readily accessible as possible new editions are planned for those areas where the respective volumes in the first edition have become obsolete by overwhelming progress. If, however, the changes are not too great so that the first editions can be brought up-to-date by supplementary volumes, supplements to the respective volumes will be published in the first edition.

Arnold Weissberger

Research Laboratories
Eastman Kodak Company
Rochester, New York

Edward C. Taylor

Princeton University
Princeton, New Jersey

Contributors

G. P. Ellis, *Department of Chemistry, University of Wales Institute of Science and Technology, Cardiff* CF1 3NU, *United Kingdom*

R. Livingstone, *Department of Chemical Sciences, The Polytechnic, Huddersfield* HD1 3DH, *United Kingdom*

I. M. Lockhart, *BOC Limited Special Gases, London,* SW19 3UF, *United Kingdom*

R. M. Parkhurst, *Life Sciences Division, Stanford Research Institute International, Menlo Park, California 94025*

W. A. Skinner, *Life Sciences Division, Stanford Research Institute International, Menlo Park, California 94025*

S. Smolinski, *Institute of Chemistry, Jagiellonian University, Krakow, 30-060, Poland*

Preface

Volume 31 in this series made the first incursion into the extensive literature of the benzopyrans. In order to keep the volume to a manageable size, it was limited to a discussion of chromenes, chromones, and chromanones and chromanols having the functional group in the heterocyclic ring. The present volume extends the coverage to the chromans and includes the extensive field of the tocopherols as well as reviewing the chemistry of other chromanols having the hydroxyl group in the homocyclic nucleus. Spirochromans are also included, as are benzopyrans with an oxo substituent in a fully or partially saturated aromatic ring. The precise scope of the book is detailed in Chapter I. Flavanones, flavans, flavones, and the corresponding 3-aryl analogs are excluded, as are the coumarins and their derivatives. It is hoped that the preparation of these two volumes on the chemistry of the benzopyrans will provide a spur to further contributions to the series.

We are especially grateful to the various contributors whose cooperation and effort has made this volume possible. The General Editors of the series and the staff of John Wiley & Sons have been particularly helpful at all stages.

The willingness of BOC Limited to allow one of us (I. M. L.) to undertake this work is gratefully acknowledged. The patience and understanding of our families played a vital supportive role while the work was in progress.

G. P. Ellis
I. M. Lockhart

Cardiff, United Kingdom
London, United Kingdom
April 1980

Contents

CHAPTER I

Chromans and Tocopherols—Introduction

G. P. ELLIS

Department of Chemistry, University of Wales Institute of Science and Technology, Cardiff, U.K.

I. M. LOCKHART

BOC Limited, Special Gases, London SW19 3UF, U.K.

I. INTRODUCTION: NOMENCLATURE

The chemistry of several types of 1-benzopyrans is discussed in Volume 31 of this series.[1] These compounds are 2*H*- and 4*H*-chromenes, 3- and 4-chromanols, 3- and 4-chromanones, and chromones. The present volume seeks to complete the treatment of 1-benzopyrans in which the pyran ring does not have a double bond at the 2,3- or 3,4-positions. The compounds are therefore derivatives of one of the following systems: dihydro-1-benzopyran (chroman, **1**), dihydronaphtho[1,2-*b*]pyran (7,8-benzochroman, **2**), dihydronaphtho[2,1-*b*]pyran (5,6-benzochroman, **3**), or dihydronaphtho[2,3-*b*]pyran (6,7-benzochroman, **4**). Dihydro-1-benzopyrans may be derived from either $2H$-1-benzopyran (**5**) or $4H$-1-benzopyran (**6**) and are therefore named either 3,4- or 2,3-dihydro-1-benzopyran; the former is more common. The dihydronaphthopyrans are similarly related to the corresponding naphthopyrans. In this book, dihydro-1-benzopyrans are referred to by their simpler name, chromans.

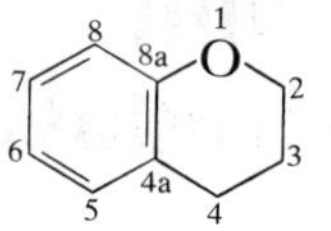

(**1**)

Ring index 1727 (2H-)
1728 (4H-)

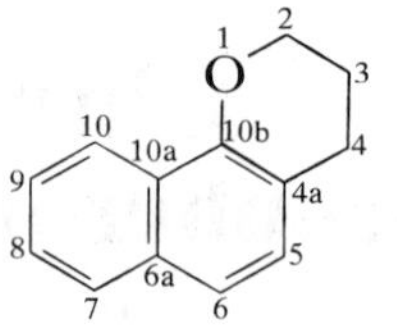

(**2**)

Ring index 3573 (2H-)
3574 (4H-)

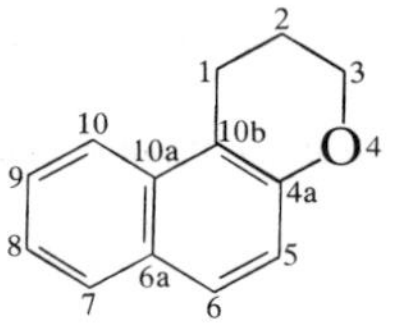

(**3**)

Ring index 3577 (1H-)
3578 (3H-)

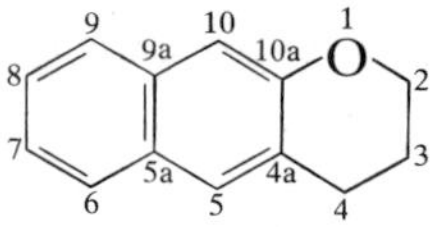

(**4**)

Ring index 3565 (2H-)
3566 (4H-)

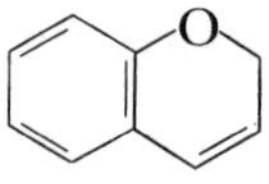

(**5**)

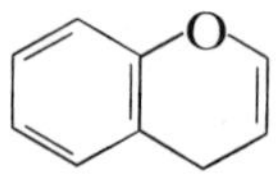

(**6**)

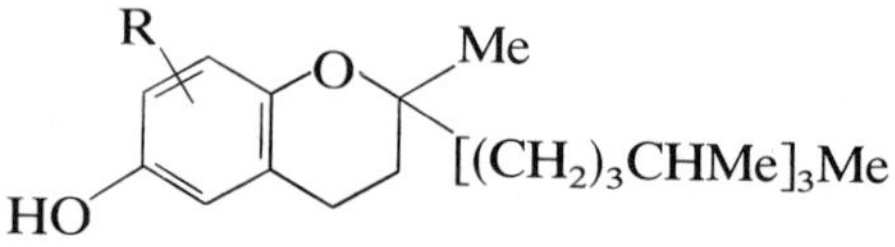

R, HO, Me, $[(CH_2)_3CHMe]_3Me$

(**7**)

The tocopherols form a group of 6-chromanols of particular importance in natural product chemistry. These compounds have the general formula (**7**), where R may be one, two, or three methyl substituents. Their chemistry is dealt with in Chapter III.

The limitations[2] that were of necessity imposed on the extent of coverage in Volume 31 also apply to this book. Thus, flavans, isoflavans, and chromans which carry an aromatic heterocyclic ring directly linked to C-2 or C-3 are not included. Such compounds are so numerous that a separate volume should be devoted to them. The cannabinoids have also been omitted as they have been well reviewed recently.[3–5]

Many ring systems are known in which another ring is spiroannulated to a chroman at C-2, C-3, or C-4. These compounds are discussed in Chapter X;

examples of their nomenclature are illustrated by 6,6′-dichloro-2,2′-spirobichroman (**8**) and 6-chloro-1′-ethylspiro[chroman-4,4′-imidazolidine]-2′,5′-dione (**9**). The naming of these compounds is similar to that currently adopted by *Chemical Abstracts*, except that the chroman part of each name is now replaced by 1-benzopyran, with 2,3-dihydro written as a prefix. As a result of this difference, the name of at least one ring system in *Chemical Abstracts* differs appreciably from that used here; for example, 2′,3′-dihydro-8′-methoxy-3-methylspiro[benzothiazole-2(3*H*),2′-chroman] (**10**) is currently indexed as 2,3-dihydro-8-methoxy-3′-methylspiro[1-benzopyran-2,2′(3′*H*)-benzothiazole] in *Chemical Abstracts*.

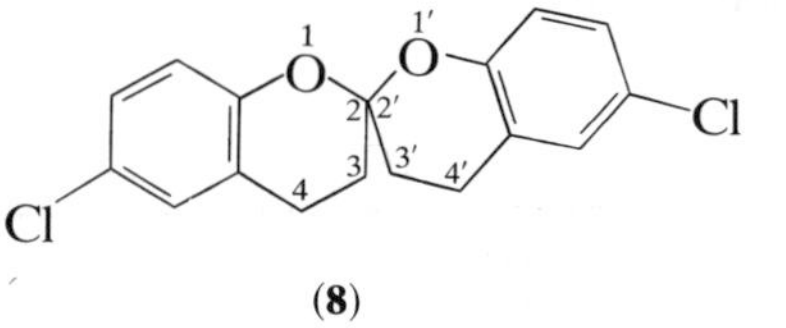

(**8**)

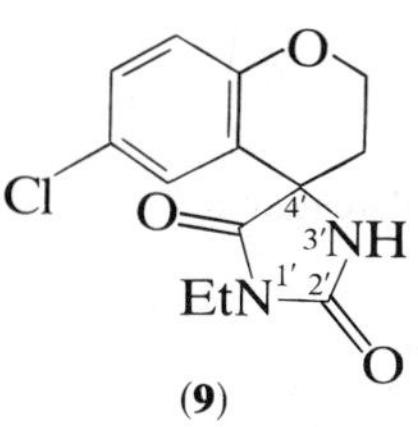

(**9**)

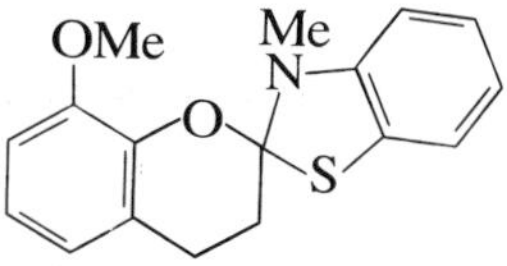

(**10**)

Over the years the nomenclature of chromans has changed considerably. It is not surprising to find trivial names used in the early literature, but some of these have persisted until very recently. For example, 1,2,4a,5,6,6a,7,8,9,10,10a,10b-dodecahydro-3-ethyl-9-methylene-3,4a,7,7,10a-pentamethylnaphtho[2,1-*b*]pyran (**11**) was called 2-methylene-8,13-epoxylabdane,[6] and 3-hydroxy-3,4,4a,5,6,8a-hexahydro-4,4a,7-trimethyl-2*H*-1-benzopyran-2-acetaldehyde (**12**) was referred to as 4-hydroxy-5,6,9-trimethyl-2-oxabicyclo[4.4.0]dec-9-en-3-ylacetaldehyde.[7] Yet another method of naming reduced chromans is illustrated by the nonsystematic name 6-carbomethoxy-*trans*-1-oxadecalin[8] for methyl *trans*-3,4,4a,5,6,7,8,8a-octahydro-2*H*-1-benzopyran-6-carboxylate (**13**).

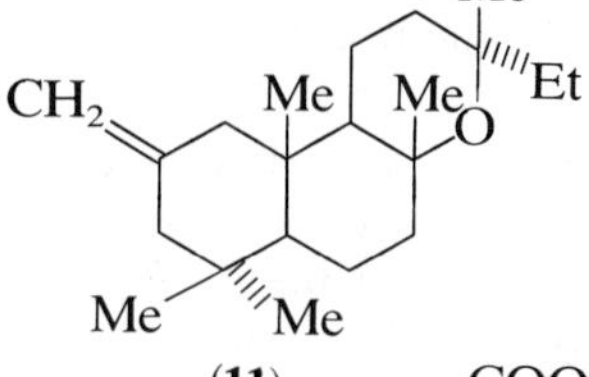

(**11**)

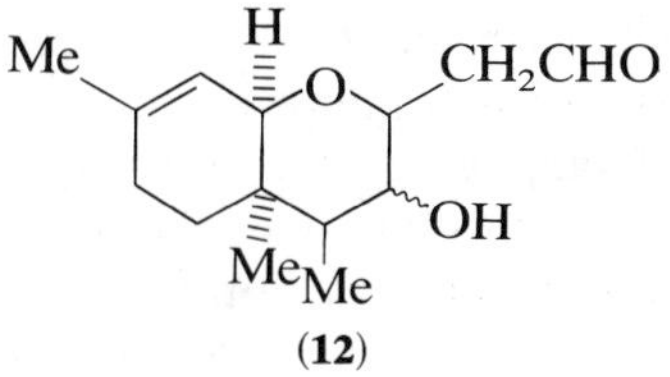

(**12**)

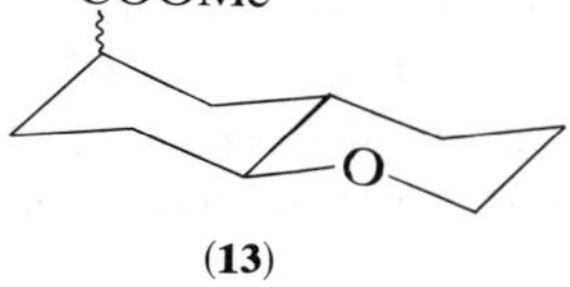

(**13**)

II. ARRANGEMENT OF TEXT

With the exception of Chapter X, which discusses all spirochromans, each chapter covers chromans containing a particular functional group. When more than one functional group is present, the compound is listed in the chapter dealing with the "principal group," which is that appearing in the latest chapter. For example, 2-hexadecyl-6-hydroxy-2,7,8-trimethylchroman-5-carboxaldehyde is listed in Table 1 of Chapter VIII (chroman carboxaldehydes) and 6-bromo-2,2-dimethyl-7-methoxychroman is listed in Table 7 of Chapter V (halogenochromans).

TABLE 1. TABLES OF COMPOUNDS: PARENTS AND DERIVATIVES

Parent compound	Derivatives
Alcohols and phenols	Acyl and sulfonyl derivatives
Aldehydes, ketones, and quinones	Acetals, ketals, oximes, semicarbazones, hydrazones, etc.
Amines	Acyl and sulfonyl derivatives, salts, and quaternary compounds
Carboxylic acids	Esters, lactones, salts, acyl halides, anhydrides, amides, anilides, toluidides, and hydrazides
Sulfonic acids	Esters, salts, amides, and halides

III. TABLES OF COMPOUNDS

Compounds that have been described in the literature are listed in each chapter according to their "principal functional groups" (see previous paragraph). For convenience the compounds are divided into subclasses, each of which has its own table of compounds. The main entries are arranged in each table according to their molecular formulas. This arrangement enables the reader to determine quickly whether or not a particular compound is known. Compounds that are regarded as derivatives of another (parent) compound are listed under the molecular formula of the parent (even if the parent compound has not been described). Table 1 shows the types of compounds that are regarded as derivatives for this purpose. For example, 4-acetamido-7-methoxy-2-methylchroman (**14**) is listed under $C_{11}H_{15}NO_2$, 7-methoxy-2-methyl-4-chromanamine (**15**), which has not yet been described (see Table 13, Chapter VI). Melting points and boiling points are given in °C.

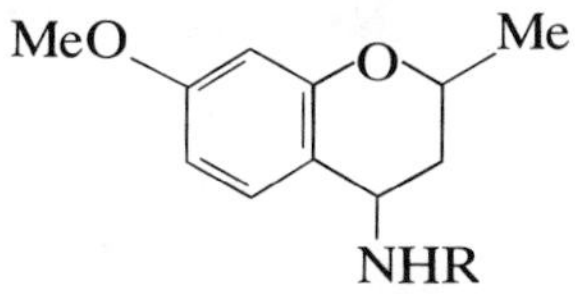

(**14**) R = Ac
(**15**) R = H

IV. THE LITERATURE OF CHROMANS AND TOCOPHEROLS

Chromans were briefly reviewed[9] up to 1948 and more recently in a multivolume work.[10] Naturally occurring chromans[11] were reviewed in 1963.

In this book, the literature has been covered through Volume 86 (1977) of *Chemical Abstracts*, but some more recent references are included.

V. ABBREVIATIONS AND CONVENTIONS

Chemical shifts in nuclear magnetic resonance spectra are expressed in terms of δ parts per million from tetramethylsilane.[12,13] The following abbreviations are used:

Ac, acetyl
Bu, butyl
Bu^t, *t*-butyl
Bz, (substituent is) attached to the benzene ring
d, (after mp) with decomposition
DDQ, 2,3-dichloro-5,6-dicyanobenzoquinone
DMSO, dimethyl sulfoxide
DMF, *N,N*-dimethylformamide
2,4-DNP, 2,4-dinitrophenylhydrazone
Et, ethyl
HMT, hexamethylenetetramine
ir, infrared
Me, methyl
m/e, mass-to-charge ratio
NBS, *N*-bromosuccinimide
nmr, nuclear magnetic resonance
Ph, phenyl
pmr, proton magnetic resonance
Pr, propyl
Pr^i, isopropyl
THF, tetrahydrofuran
Ts, toluene-4-sulfonyl
uv, ultraviolet

VI. REFERENCES

1. G. P. Ellis, Ed., *Chromenes, Chromanones, and Chromones*, John Wiley, New York, 1977.
2. G. P. Ellis, Ed., *Chromenes, Chromanones, and Chromones*, John Wiley, New York, 1977, p. 2.
3. R. Mechoulam, Ed., *Marijuana*, Academic Press, New York, 1973.
4. R. Mechoulam, N. K. McCallum, and S. Burstein, *Chem. Rev.*, **76,** 75 (1976).
5. H. G. Pars, R. K. Razdan, and J. F. Howes, *Adv. Drug Res.*, **11,** 97 (1977).
6. E. W. Colvin, S. Malchenko, R. A. Raphael, and J. S. Roberts, *J.C.S. Perkin I*, 1989, (1973).
7. M. J. Francis, P. K. Grant, K. Show, and R. T. Weavers, *Tetrahedron*, **32,** 95 (1976).
8. J. A. Kirsch and G. Schwartzkopf, *J. Org. Chem.*, **39,** 2040 (1974).
9. S. Wawzonek, in *Heterocyclic Compounds*, Vol. 2, R. C. Elderfield, Ed., Interscience, New York, 1951, Chap. 11.
10. R. Livingstone, in *Rodd's Chemistry of Carbon Compounds*, 2nd ed., Volume IVE, Elsevier, Amsterdam, 1977, p. 222.
11. F. M. Dean, *Naturally Occurring Oxygen Ring Compounds*, Butterworth, London, 1963, Chap. 7.
12. F. H. A. Rummens, *Org. Magn. Spectrosc.*, **2,** 209 (1970).
13. E. F. Mooney, *Ann. Rep. NMR Spectrosc.*, **3,** xi (1970).

CHAPTER II

Chroman, Alkyl- and Arylchromans

R. LIVINGSTONE

Department of Chemical Sciences, The Polytechnic, Huddersfield, U.K.

I. INTRODUCTION AND NOMENCLATURE

Although chroman was first prepared in 1905, little interest was shown in the compound until studies on the tocopherols (vitamin E) began to indicate that they were derivatives of chroman. The chemistry of the tocopherols is dealt with in Chapter III. Many chroman derivatives were prepared in attempts to obtain compounds that possessed vitamin E activity; for example, it was shown that 2-methyl- and 2,2-dimethylchroman were devoid of vitamin E activity. Nearly all the naturally occurring chromans have long isoprenoid chains, attached particularly at C-2, and most are members of the tocol series.

Since the trivial name chroman is still acceptable under the IUPAC Rules (see Chapter I), names derived from chroman (**1**) will be used throughout the chapter.

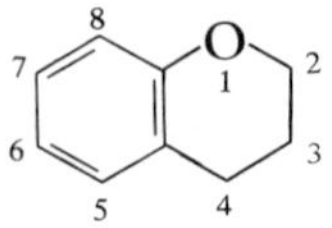

(**1**)

For the analogous compounds with a naphthalene ring fused to the heterocyclic ring, the following names will be used: 3,4-dihydro-2*H*-naphtho[1,2-*b*]pyran (**2**), 2,3-dihydro-1*H*-naphtho[2,1-*b*]pyran (**3**), 3,4-dihydro-2*H*-naphtho[2,3-*b*]pyran (**4**), and 2,3-dihydronaphtho[1,8-*bc*]pyran (**5**). Compounds of the latter type, where more than one ring is fused to the pyran ring system, are outside the scope of this volume.

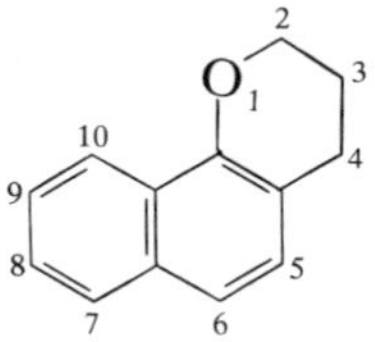

(**2**) 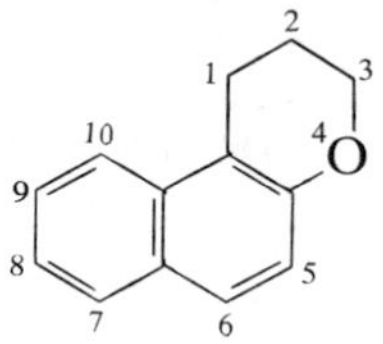 (**3**)

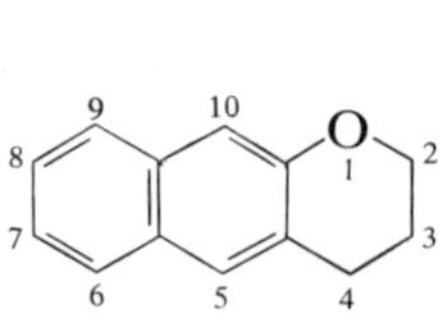

(**4**) 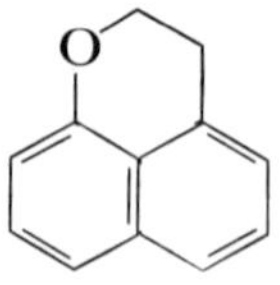 (**5**)

II. METHODS OF PREPARATION OF CHROMANS

A number of general methods of preparation can be applied to chroman and its alkyl- and aryl-substituted derivatives. These include the cyclization of derivatives of 2-propylphenol or phenyl propyl ether, treatment of 1,3-diaryloxypropane with aluminum chloride, and the reduction of coumarins and chromens. Chromans may also be obtained by dehydration and cyclization of the products formed following the Grignard reaction between dihydrocoumarins and alkyl- and arylmagnesium halides, and by the reaction between phenols and appropriate unsaturated hydrocarbons.

1. Preparation of Chroman

Chroman (**1**) was first prepared in 1905 by warming an aqueous sodium hydroxide solution of 2-(3-chloropropyl)phenol (**6**).[1] A high yield of chroman, purified by distillation from sodium, was obtained by Normant and Maitte[2] by cyclization of phenyl 3-chloropropyl ether (**7**) in the presence of stannic chloride. It had previously been found that phenyl 3-hydroxypropyl ether could similarly be cyclized by heating with zinc chloride[3] or by the action of phosphorus pentoxide without solvent or heat.[4]

OH, $(CH_2)_3X$ $\xrightarrow{OH^-}$ (**1**) $\xleftarrow{SnCl_4}$ $O(CH_2)_3Cl$

(**6**) X = Cl
(**10**) X = Br

(**7**)

PBr_3

OH, $(CH_2)_3OH$ $\xleftarrow[CuC_2O_4]{H_2}$ O, O

(**9**) (**8**)

Hydrogenation of coumarin (**8**) over copper chromite at 250° gives 2-(3-hydroxypropyl)phenol (**9**); subsequent reaction with phosphorus tribromide in benzene gives 2-(3-bromopropyl)phenol (**10**), which on treatment with an aqueous solution of sodium hydroxide affords chroman in yields of 60%[4] to 87%.[5] Chroman is not obtained by the Clemmensen reduction of dihydrocoumarin, but good yields are recorded from chroman-4-one.[4]

Maitte[6] regarded the action of stannic chloride on phenyl 3-halogenopropyl ether as the most suitable for the preparation of chroman; a yield of 85% was recorded. Deady, Topsom, and Vaughan[7] reported that the method gave variable yields of chroman which required gas chromatography for purification. They found however that the readily prepared 1,3-diphenoxypropane (**11**) reacts smoothly with aluminum chloride in boiling benzene to give chroman and phenol; chroman of high purity was readily isolated in high yield (72%), along with a good recovery of the accompanying phenol. They also showed that increasing the amount of aluminum chloride from 1.3 to 1.7 mole had little effect on the production of chroman. The mechanism of this reaction is discussed later (see Section II, 2) when it is used for the preparation of methylchromans, which contain the substituent on the benzene ring. It was also found that sulfuric acid did not promote formation of chroman and that hydrobromic acid was brought into reaction only at 250° (sealed tube) and then gave only poor yields.[7]

$$PhO(CH_2)_3OPh \xrightarrow[C_6H_6]{AlCl_3} (\mathbf{1}) + PhOH$$

(**11**)

Cyclization of 2-allylphenol (**12**, R = H) and related compounds to 2,3-dihydrobenzo[*b*]furans and chromans may be regarded as a special case of addition to an olefinic double bond. The formation of a 2,3-dihydrobenzo[*b*]furan ring would be anticipated under peroxide-free conditions; and since the allylphenol (**12**, R = H) may act similarly to dihydroquinone, treatment with hydrogen bromide in acetic acid under peroxide-free conditions or not would afford 2,3-dihydro-2-methylbenzo[*b*]furan (**13**). Even the presence of benzoyl peroxide fails to favor the formation of chroman. However when the acetyl derivative (**12**, R = Ac) of 2-allylphenol reacts with hydrogen bromide under peroxide-free conditions, 2,3-dihydro-2-methylbenzo[*b*]furan is obtained; but in the presence of benzoyl peroxide, chroman (**1**) is formed.[8] Thus the effect of the OH group of the phenol (**12**, R = H) on the course of the reaction is eliminated by its conversion to the acetate.

(**13**) $\xleftarrow{HBr, AcOH}$ (**12**) $\xrightarrow[Bz_2O_2]{HBr, AcOH}$ (**1**)

(**13**) (**12**)

The Grignard reagent formed from 6-chlorochroman on hydrolysis affords chroman in a 74% yield.[9]

Chroman is a colorless oil which is soluble in the common organic solvents and gives a red color with sulfuric acid.

2. Preparation of Alkylchromans

The monoalkyl chromans can be divided into two groups—one with the alkyl substituent attached to the benzene ring and the second with it attached to the heterocyclic ring. The former can be obtained from appropriate derivatives of benzene by methods similar to those used for the preparation of chroman, for example, by the method of Deady et al., where 1,3-di(4-tolyloxy)propane gives 6-methylchroman (92%).[7] Compounds in the second group are prepared by cyclization of the appropriate derivatives of benzene or phenol or by reduction of the corresponding 2-, 3-, or 4-substituted chromen. 4-Alkyl- or 4-arylchromans may be obtained by reaction between the necessary alkyl- or arylmagnesium halide and chroman-4-one, followed by dehydration of the resulting alcohol to give a chromen, which is hydrogenated in the presence of a catalyst.

Pure 2-methylchroman (**18**) was probably first obtained by Baker and Walker.[10] Although it had been described previously in the literature, it is very doubtful if it had been obtained in anything approaching a state of purity. Catalytic hydrogenation of 2-hydroxybenzylideneacetone (**14**) in methanol or ethanol in the presence of palladium chloride gave a mixture of 2-hydroxybenzylacetone (**15**) and 2-methoxy-2-methyl- or 2-ethoxy-2-methylchroman (**16**, R = Me or Et), respectively. The ethoxy derivative (**16**, R = Et) on boiling with acetic anhydride yielded 2-methylchrom-2-ene (**17**), which on reduction with hydrogen in acetic acid using a platinum–silica gel catalyst afforded 2-methylchroman (**18**).

OH COMe CH$_2$ CH$_2$ (**15**) — ROH, HCl / HCl, H$_2$O ⇌ — O Me OR (**16**) R = Me or Et — Ac_2O → O Me (**17**)

H$_2$, PdCl$_6$ ↑ (from **14**) ; H$_2$, AcOH, Pt ↓ (to **18**)

OH COMe CH=CH (**14**) ; O Me (**18**)

The ready formation of the six-membered cyclic acetals (**16**) was probably the first example of its kind in the aromatic series. Their formation was attributed to the presence of hydrogen chloride resulting from the reduction of the palladium chloride. This was evident from the results obtained using metallic platinum or palladized strontium carbonate catalyst. 2-Ethoxy-2-methylchroman (**16**, R = Et) was readily prepared by keeping a solution of 2-hydroxybenzylacetone (**15**) in ethanol containing a trace of hydrogen chloride at room temperature for several hours. The ethoxychroman (**16**, R = Et) was easily reconverted into 2-hydroxybenzylacetone (**15**) by heating with dilute hydrochloric acid, but it is stable to hot alkaline solutions. An ethereal solution of 2-hydroxybenzylacetone (**15**) when kept for two weeks over anhydrous sodium acetate gave 2-methylchrom-2-en (**17**).

The acetate (**19**, R = Ac) of 2-crotylphenol and hydrogen bromide in acetic acid in the presence of hydroquinone or benzoyl peroxide yields 2-methylchroman (**18**).[8]

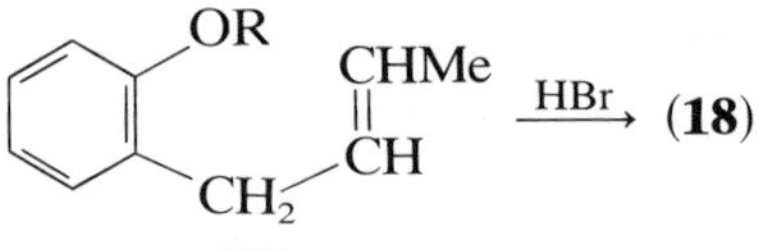

OR CHMe || CH CH$_2$ $\xrightarrow{\text{HBr}}$ (**18**)

(**19**)

The above method suggested that it might be possible to obtain 2-methylchroman from phenol (**20**, R = H) and butadiene. Passage of butadiene into phenol and ethylsulfonic acid at 32° gave a mixture of products from which 2-(but-2-en-1-yl)phenol (**19**, R = H) (5%) and 4-(but-2-en-1-yl)phenol (**21**) (21.7%) were isolated together with lower boiling point material apparently containing 2,3-dihydro-2-ethylbenzo[*b*]furan (**22**) and 2-methylchroman (**18**).[11] The use of different catalysts and reaction temperatures has been shown to direct the orientation; for example, with a complex of phosphoric acid and boron trifluoride or with alkanesulfonic acids, *p*-substitution predominates at 15–25°.[12]

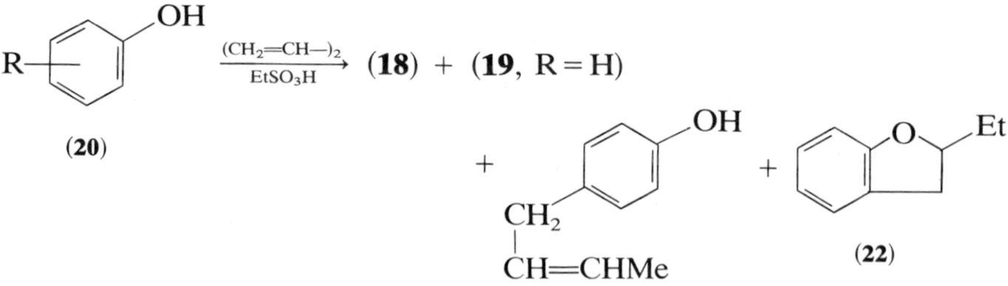

R OH (**20**) $\xrightarrow[\text{EtSO}_3\text{H}]{(\text{CH}_2{=}\text{CH}{-})_2}$ (**18**) + (**19**, R = H) + OH CH$_2$ CH=CHMe (**21**) + O Et (**22**)

Aluminum phenoxide has been shown to be a selective catalyst for the *o*-alkylation of phenol. For example, with propylene and isobutylene good yields of 2,6-dialkylated phenols are obtained, and with isoprene at temperatures above 100° in the presence of excess phenol a 36% yield of 2,2-dimethylchroman is obtained, along with another product. Under similar conditions, with more complicated olefins somewhat poorer yields of mono- and dialkylated products are obtained. Phenol reacts with butadiene to give a mixture of products containing 2-methylchroman (5%), 2-(but-2-en-1-yl)phenol (8%), and 4-(but-2-en-1-yl)phenol (57%).[13]

On bubbling butadiene into a mixture of phenol and 100% phosphoric acid at 115°, the alkali-insoluble portion (29%) of the product on distillation afforded 2-methylchroman.[14] Alkenylation of phenol by butadiene over cation exchange resin KU-2 at 60–80° gave butenylphenols (50%) and neutral products (10–40%). The latter contained 2-methylchroman (50%) and 2,3-dihydro-2-ethylbenzo[*b*]furan.[15]

2-Methylchrom-3-ene (**24**) is prepared by the Wittig reaction in 88% yield by treating 3-(2-formylphenoxy)propyltriphenylphosphonium bromide (**23**) with an excess of sodium methoxide in boiling methanol; hydrogenation over 10% palladium on charcoal at room temperature in anhydrous ether gives a quantitative yield of 2-methylchroman (**18**).[16] 2-Methylchroman-4-

one may be converted to 2-methylchroman by means of the Clemmensen reduction.[17]

$$\text{(23)}\ \xrightarrow{NaOMe}\ \text{(24)}\ \xrightarrow[Pd\text{–}C]{H_2}\ \textbf{(18)}$$

(23) $O(CH_2)_3\overset{+}{P}Ph_3\overset{-}{Br}$, CHO (24) O, Me

It is reported that phenol on treatment with the neutral sulfate of butan-1,3-diol in benzene containing aluminum trichloride affords 2-methylchroman.[18] 2-Methylchroman is also obtained by the cyclization of 2-butenylphenol,[19] formed by the dehydrogenation of 2-butylphenol on passing it over a Cu–Cr oxide catalyst at 500–600°.

2,3-Dihydro-2-ethylbenzo[*b*]furan is isomerized with triphenylmethyl perchlorate in acetic or formic acid to yield up to 27% of 2-methylchroman, but 80% of dehydrogenation products are obtained on using a mixture of triphenylmethyl perchlorate and stannic chloride.[20] 2,3-Dihydro-2-ethylbenzo[*b*]furan isomerizes at 300–400° in the presence of activated carbon to give 2-methylchroman and 2-ethylbenzo[*b*]furan,[21] and 2,3-dihydro-2,3-dimethylbenzo[*b*]furan in the presence of aluminosilicate at 433° yields 2-methylchroman (73%) and 2,3-dimethylbenzo[*b*]furan (17%).[22]

2-Ethylchroman (**26**) was first prepared in excellent yield (73%) by reacting 2-ethylchromone (**25**) with hydrogen at 165–175° over copper–chromium oxide for several hours.[23] The hydrogenation of 2-vinylchroman (**27**) using Raney nickel at room temperature and under pressure (30 atm) affords 2-ethylchroman (**26**). The 2-vinylchroman was obtained in 8–10% yield by heating[24] 2-[(dimethylamino)methyl]phenol (**28**) with butadiene in anhydrous toluene, in the presence of a small amount of dihydroquinone, in a sealed tube for several hours at 185°.

$$\text{(25)}\ \xrightarrow[Cu\text{–}Cr\ oxide]{H_2}\ \text{(26)/(27)}$$

(25) O, Et, O

(**26**) R = Et
(**27**) R = CH=CH$_2$

$$\text{(28)}\ \xrightarrow[PhMe]{(CH_2=CH)_2}\ \textbf{(27)}\ \xrightarrow[RaNi]{H_2}\ \textbf{(26)}$$

(28) OH, CH_2NMe_2

$$\text{(29)}\ \xrightarrow{H_2\ |\ Pd\text{–}C}\ \textbf{(26)}$$

(29) O, Et, R

2-Ethylchroman (**26**) is also obtained in quantitative yields by the hydrogenation of 2-ethylchrom-3-ene (**29**) in methanol over 10% palladium–carbon.[25]

3-Methylchroman (**31**) was first obtained by the cyclization of 2-(2-methylallyl)phenyl acetate (**30**) using hydrogen bromide–acetic acid in the presence of dihydroquinone or benzoyl peroxide.[8] During acetylation of 2-(2-methylallyl)phenol with ketene in the presence of a trace of sulfuric acid, a concurrent exothermic isomerization occurs to 2,3-dihydro-2,2-dimethylbenzo[*b*]furan.

OAc CH_2 C CH_2 Me —HBr, AcOH→ O R

(**30**)

(**31**) R = Me
(**32**) R = $PhCH_2$

3-Benzylchroman (**32**) has been obtained by the reduction of 3-benzylidenechroman-4-one.[26]

A number of 4-alkylchromans (**34**) have been prepared by reacting 4-bromochroman (**33**) with the appropriate Grignard reagent. 4-Bromochroman is formed when chroman (**1**) is treated with *N*-bromosuccinimide.[6]

(**1**) —NBS→ O Br —RMgX→ O R

(**33**)

(**34**)
R = Et, Pr, Bu, Ph
X = halogen

4-Alkyl-substituted chroman-4-ols (**35**) obtained by the reaction between chroman-4-one and alkylmagnesium halide afford on dehydration using either *p*-toluenesulfonic acid or anhydrous copper sulfate a mixture of 4-alkylchrom-3-ene (**36**) and 4-alkylidenechroman (**37**). The proportions of the mixtures have been determined from nmr spectral data.[27] Catalytic hydrogenation of the chrom-3-ene (**36**) affords the 4-alkylchroman.

O RCH_2 OH ⟶ O CH_2R + O CHR

(**35**)
R = alkyl or H

(**36**)

(**37**)

Dehydration of 4-alkyl-4-chromanols with perchloric acid affords a mixture of the 4-alkylchroman and the benzopyrylium salt (see Section II, 4).

The isomeric 5-, 6-, 7-, and 8-methylchromans can be prepared by the aluminum chloride-catalyzed decomposition of the appropriate 1,3-diaryloxypropanes[28] in boiling benzene. This is an extension of the method used for obtaining chroman from 1,3-diphenoxypropane.[7] 1,3-Di-*p*- (**38**) and *o*-tolyloxypropane decompose in the presence of aluminum chloride to give 6-methylchroman (**39**) and *p*-cresol, and 8-methylchroman (**48**) and *o*-cresol, respectively. Decomposition of 1,3-di-*m*-tolyloxypropane gives a mixture of 5-methylchroman (**40**) and 7-methylchroman (**41**); and although the mixture has been separated by preparative gas chromatography, the second compound eluted, namely, 5-methylchroman, was contaminated by some of the 7-isomer.

In analogous reactions, 1,4-diphenoxybutane gave no homochroman, and 1,2-diphenoxyethane was almost inert. The formation of the chroman involves substitution at an aromatic carbon atom, and this presumably implies a lateral attack on the aromatic ring resulting in ring closure synchronous with the splitting off of the phenol. The relative absence of by-products, the mild conditions employed, the use of a nonpolar solvent, and the primary alkyl groups involved also suggest that a carbonium ion intermediate is unlikely.

Me, O, CH_2, CH_2, CH_2, O, 4-MeC_6H_4 (**38**) $\xrightarrow{AlCl_3,\ C_6H_6}$ Me, O, CH_2, CH_2, CH_2, $^{+}O—AlCl_3^{-}$, 4-MeC_6H_4

↓

Me, O, +, H + 4-MeC_6H_4—O—$AlCl_3^{-}$ ⟶ Me, O (**39**)

Kinetic measurements show that, for a fixed concentration of diphenoxypropane, the rate of decomposition is at a maximum in the presence of a slight excess of aluminum chloride. The rate falls when the catalyst concentration is either decreased or increased.

The 5- and 7-methylchromans (**40**) and (**41**) have been obtained as a 1:1.46 mixture by boiling 3-methylphenyl 3-hydroxypropyl ether with phosphorus pentoxide in benzene. Fractional distillation of several combined runs gives pure 7-methylchroman. Subsequent fractions on redistillation using phenanthrene as a 'chaser' gave pure 5-methylchroman.[29]

P_2O_5, C_6H_6

(**40**) (**41**)

6-Methylchroman (**39**) and 7-methylchroman (**41**) have been prepared in yields of 85–88% from 3-(2-hydroxy-5-methylphenyl)propan-1-ol (**42**) and 3-(2-hydroxy-4-methylphenyl)propan-1-ol (**43**), respectively, by the method described previously for the preparation of chroman.[5]

(**42**) R = 4-Me

(**43**) R = 5-Me

1. PBr_3, C_6H_6 2. OH^-

HBr, AcOH, Bz_2O_2

$SnCl_4$

(**44**) (**45**)

6-Methylchroman (**39**) is obtained when the acetate of 2-allyl-4-methylphenol (**44**) reacts with hydrogen bromide in acetic acid in the presence of benzoyl peroxide,[8] an extension of a method used for the preparation of chroman. In the absence of peroxide, 2,3-dihydro-2,5-dimethylbenzo[*b*]furan is formed. Similarly 2-allyl-6-methylphenol affords 8-methylchroman (**48**). Reaction between 4-methylphenyl 3-chloropropyl ether (**45**) and stannic chloride also affords 6-methylchroman (**39**).[6]

It has been found that the alkylation of phenols with propan-1,3-diol can result in formation of some chromans. The reaction between 4-methylphenol and propan-1,3-diol at 350° in the presence of 20% by weight of zinc chloride–alumina gives a mixture of products containing 2-allyl-4-methylphenol, 3-(3-methyl-6-hydroxyphenyl)propan-1-ol, 1,3-di(3-methyl-6-hydroxyphenyl)propane, and 6-methylchroman. At 250° the reaction is very slow.[30] The halogenoalkylation of methylphenols by 1,3-dibromopropane proceeds only in the presence of metallic copper and not in the presence of phosphoric acid or zinc chloride–alumina. Heating an equimolar mixture of 4-methylphenol (**20**, R = 4-Me) and 1,3-dibromopropane (**46**) in the presence of 10% copper at 170–175° yields up to 40% 6-methylchroman (**39**).[31]

$$\underset{R=4\text{-Me}}{(\mathbf{20})} + \underset{(\mathbf{46})}{Br(CH_2)_3Br} \xrightarrow[170\text{–}175^\circ]{Cu} (\mathbf{39})$$

8-Methylchroman (**48**) has been prepared in good yield by the cyclization of 3-(2-methylphenoxy)propan-1-ol (**47**) with phosphorus pentoxide.[32] It may also be obtained by the Clemmensen reduction of 8-methylchroman-4-one (**49**).[17]

Me $O(CH_2)_3OH$ (**47**) $\xrightarrow[C_6H_6]{P_2O_5}$ Me O (**48**) $\xleftarrow[H^+]{Zn\text{–}Hg}$ Me O O (**49**)

The fact that chroman (**1**) condenses readily with various acid chlorides in carbon disulfide in the presence of aluminum chloride to yield the 6-acyl derivative affords a method for obtaining 6-alkyl-substituted chromans via the Clemmensen reduction. For instance, acetyl chloride gives 6-acetylchroman (**50**) and hence 6-ethylchroman (**51**). Similarly 6-propionylchroman yields 6-propylchroman.[4] Yields of ketone are generally about 75%.[33]

O (**1**) $\xrightarrow[CS_2]{AcCl}$ Ac O (**50**) $\xrightarrow[HCl]{Zn/Hg,}$ Et O (**51**)

3. Preparation of 4-Phenylchroman

4-Phenylchroman (**34**, R = Ph) has been obtained by reacting 4-bromochroman (**33**) with phenylmagnesium bromide.[6] It is also obtained in a 25% yield along with a large amount of tar when 3-chloro-1-phenoxy-1-phenylpropane (**52**) is heated in the presence of stannic chloride. The formation of 4-phenylchroman (**34**, R = Ph) rather than the expected 2-phenyl compound can be explained by the cleavage of the starting material by attack of stannic chloride on the ether oxygen to give a benzyl cation, which adds electrophilically to the carbon ortho to the coupled oxygen, followed by cycloelimination of halogen by complex oxygen.[34] Treatment of 1-phenoxy-1-phenylpropan-3-ol with polyphosphoric acid at 90° also yields 4-phenylchroman.[34]

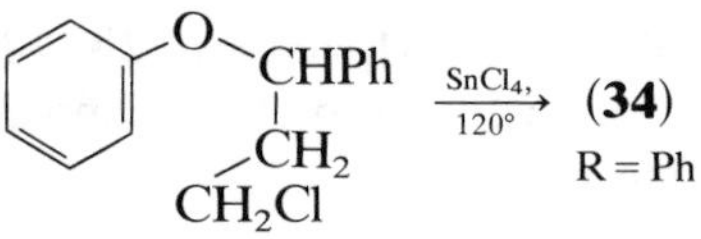

$$\text{(52)} \xrightarrow[120°]{SnCl_4,} \text{(34)}\ R = Ph$$

(52)

4. Preparation of Dialkylchromans

The dialkylchromans can be divided into a number of groups: (a) those with the two substituents on the same carbon of the heterocyclic ring; (b) those with the two substituents on different carbons of the heterocyclic ring; (c) those with one substituent on a carbon of the heterocyclic ring and one on a carbon of the benzene ring; and (d) those with both substituents on the benzene ring. Examples of the first and second groups of dialkyl compounds can be prepared by the cyclization of the appropriate derivatives of benzene and phenol or by reduction of the corresponding disubstituted chromens. The third group of dialkylchromans are generally obtained by methods similar to those used for the preparation of alkylchromans possessing the alkyl group in the heterocyclic ring, but in this instance the starting benzene or phenol derivatives contain one alkyl substituent in the appropriate position. In the majority of cases, the last group of compounds can be obtained from the appropriate derivatives of benzene by methods similar to those employed for the preparation of chroman. The most important of the dialkylchromans are those containing a *gem*-dimethyl group in the 2-position.

2,2-Dimethylchroman (**54**) was probably first prepared in 1921 by Claisen, who reported that monohydric phenols and butadienes underwent an interesting condensation and found that phenol and isoprene (**53**) on boiling in the presence of an acidic catalyst, such as zinc chloride or formic acid, gave, besides alkali-soluble products, 2,2-dimethylchroman (**54**). Its structure was proved by an unambiguous synthesis from ethyl coumarate (**55**), which was reduced to the dihydro derivative (**56**) by hydrogen using a palladium chloride catalyst and then converted to 2-hydroxy-γ-dimethyl-dihydrocinnamyl alcohol (**57**) by reaction with methylmagnesium iodide. Cyclization of the alcohol (**57**) by boiling with 25% sulfuric acid afforded 2,2-dimethylchroman (**54**).[35,36] A yield of 85% of 2,2-dimethylchroman is obtained by boiling the alcohol in acetic acid containing 20% sulfuric acid.[37]

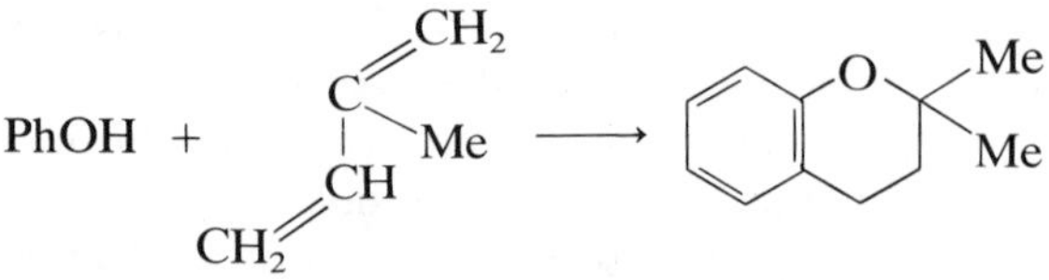

$$PhOH + \text{(53)} \longrightarrow \text{(54)}$$

(53) (54)

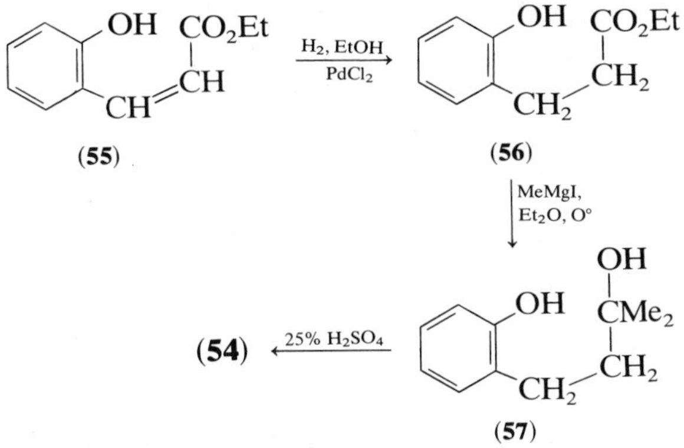

(55) $\xrightarrow[PdCl_2]{H_2, EtOH}$ (56) $\xrightarrow{MeMgI, Et_2O, 0°}$ (57) $\xrightarrow{25\%\ H_2SO_4}$ (54)

Clemo and Ghatge[38] repeated the condensation between phenol and isoprene by heating the reactants together in a sealed tube in the presence of a trace of iodine at 150°, but yields of 2,2-dimethylchroman were very low. The structure of 2,2-dimethylchroman (**54**) was verified by its dehydrogenation to 2,2-dimethylchrom-3-ene and subsequent ozonolysis to salicylaldehyde. When the condensation is catalyzed by 71% phosphoric acid at 20°, six products are obtained. The phenolic fraction yields the (3-methylcrotyl)phenols (**58**) and (**59**) and the crystalline tertiary alcohols (**60**) and (**61**). The ethereal fraction consists of 2,2-dimethylchroman (**54**) and a solid $C_{16}H_{24}O_2$ alcohol, m.p. 51–52°. A much higher yield (44%) of 2,2-dimethylchroman is obtained[40] when phenol, aluminum powder, and toluene are heated under pressure to 180° and isoprene with a very small amount of phenothiazine is pumped in over 4 hr at 120°.

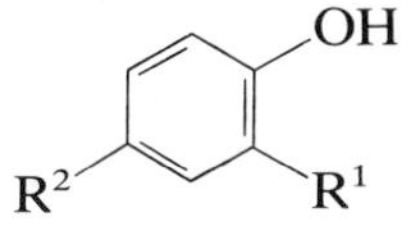

(**58**) $R^1 = CH_2CH{=}CMe_2$, $R^2 = H$
(**59**) $R^1 = H$, $R^2 = CH_2CH{=}CMe_2$
(**60**) $R^1 = (CH_2)_2CMe_2OH$, $R^2 = H$
(**61**) $R^1 = H$, $R^2 = (CH_2)_2CMe_2OH$
(**62**) $R^1 = R^2 = CH_2CH{=}CMe_2$

If phenol is treated with isoprene in the presence of aluminum phenoxide at low temperatures (20–80°) and low phenol–isoprene ratios, compounds (**54**), (**59**), (**62**), and (**63**) may be isolated from the reaction mixture. The amounts of the products vary considerably with reaction conditions, but the reaction cannot be directed to a particular product.[13] Unlike the conditions used by Bader and Bean[39] with phosphoric acid as the catalyst, none of the products had the double bond hydrated. When the aluminum phenoxide-catalyzed reaction is carried out at higher temperatures (>100°) in the presence of an excess of phenol, 2,2-dimethylchroman (**54**) is obtained in

36% yield, accompanied by the bisphenol (**64**) in 38% yield.[13] Heating phenol with KU-2 resin at 50° and then treating with isoprene at 40–50° yields up to 75% monopentenylphenols and 2,2-dimethylchroman.[41]

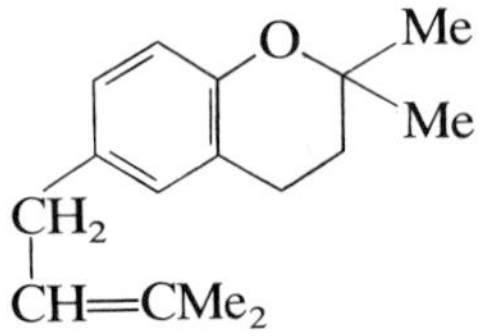

(**63**) 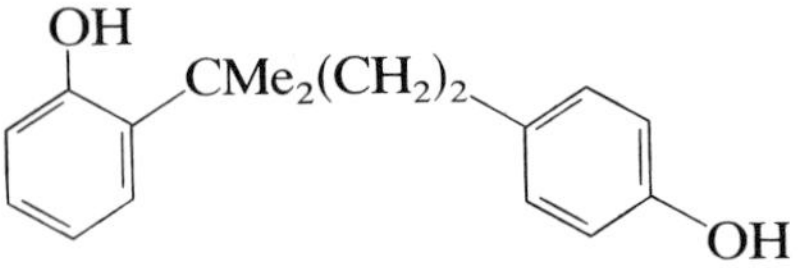 (**64**)

When phenol was treated with 1-chloro-2-methylbutene using phosphoric acid as catalyst, a mixture of products was obtained which were separated by gas chromatography and identified from ir and nmr spectral data. They included 2,2-dimethylchroman, 2,2-dimethyl-8-(3-methylbut-3-en-1-yl)chroman, 2,2-dimethyl-8-(3-methylbut-2-en-1-yl)chroman, 2,2-dimethyl-6-(3-methylbut-3-en-1-yl) chroman, 2,2-dimethyl-6-(3-methylbut-2-en-1-yl)chroman, 4-(3-methylbut-3-en-1-yl)phenol, and 4-(3-methylbut-2-en-1-yl)phenol.[42] Similar results were obtained by catalyzing the reaction between phenol and isoprene with nickelocene and phenylmagnesium bromide.[43]

During the studies of the ring closure of 2-allylphenols it was found that acetylation of 2-(3,3-dimethylallyl)phenol (**58**) with ketene (**65**) in the presence of sulfuric acid gave a mixture of the acetate (**66**) and 2,2-dimethylchroman (**54**) (68%). The latter was also obtained on ring closure of 2-(3,3-dimethylallyl)phenol, regardless of whether it was carried out in the presence of benzoyl peroxide or of dihydroquinone.[8]

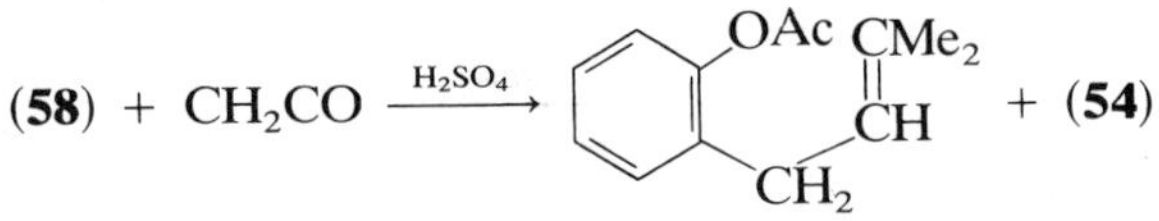

(**58**) + CH_2CO $\xrightarrow{H_2SO_4}$ (**66**) + (**54**)

(**65**) (**66**)

The reaction between various allyl diphenyl phosphates and a number of phenols usually gives dihydrobenzo[*b*]furans or chromans. The method has been used for the synthesis of phenolic natural products containing isoprenoid residues, for example, the reaction of 2,3,5-trimethylquinol with phytyldiphenyl phosphate gives α-tocopherol. The phosphate esters are prepared by stirring the appropriate alcohol, in pyridine, with diphenyl phosphorochloridate (**68**) for several hours and extracting the ester with ether. 3,3-Dimethylallyl diphenyl phosphate (**69**) obtained from 3,3-dimethylallyl alcohol (**67**) when treated with an excess of phenol at 120° gives 2,2-dimethylchroman (**54**) (46%) and diphenyl hydrogen phosphate (**70**) as the only isolable products.[44]

$$Me_2C{=}CH{-}CH_2OH + (PhO)_2P(=O)Cl \longrightarrow Me_2C{=}CH{-}CH_2O{-}P(=O)(OPh)_2$$

(**67**) (**68**) (**69**)

$$\xrightarrow[120^\circ]{PhOH} (\mathbf{54}) + (PhO)_2P(=O)OH$$

(**70**)

The photolysis of 2-(3-methylbut-2-en-1-yl)phenol gives 2,2-dimethylchroman in a 40–50% yield.[45]

Hydrogenation of 2,2-dimethylchrom-3-ene (**71**) in ethanol using platinic oxide as catalyst gives 2,2-dimethylchroman (**54**) in 92% yield.[46] It is similarly obtained from 3-bromo-2,2-dimethylchrom-3-ene (**72**).[47] There is

$$(\mathbf{71}) \xrightarrow[EtOH]{H_2, Pt} (\mathbf{54}) \xleftarrow[EtOH]{H_2, Pt} (\mathbf{72})$$

(**71**) (**72**)

no report of 2,2-dimethylchroman having been obtained by the dehydration and cyclization of 2-hydroxy-γ-dimethyldihydrocinnamyl alcohol (**57**) formed by the reaction between dihydrocoumarin (**73**) and methylmagnesium iodide. However the method was used for the preparation of 2,2-diethylchroman (**75**) from dihydrocoumarin (**73**) and ethylmagnesium

$$(\mathbf{73}) \xrightarrow[Et_2O]{EtMgBr,} (\mathbf{74}) \xrightarrow[20\%\,H_2SO_4]{AcOH,} (\mathbf{75})$$

(**73**) (**74**) (**75**)

bromide. The intermediate, 1-(2-hydroxyphenyl)-3-ethylpentan-3-ol (**74**), was dehydrated and cyclized by boiling in acetic acid and 20% sulfuric acid.[35] Other 2,2-dialkylchromans may be similarly prepared. 2,2-Di-*n*-butylchroman, for instance, was obtained in good yield by boiling the crude carbinol, 1-(2-hydroxyphenyl)-3-*n*-butylheptan-3-ol, prepared by the Grignard reaction between dihydrocoumarin and *n*-butylmagnesium bromide, with 25% sulfuric acid.[48] Attempted cyclization of the carbinol by boiling in acetic acid containing 50% sulfuric acid was unsuccessful, but it has been effected with 10% phosphoric acid.[6]

The reaction of phenol with geraniol (**76**) in the presence of 85% phosphoric acid yields a mixture containing 2-methyl-2-prenylchroman (**77**), 2- and 4-geranylphenol, and a hexahydroxanthene derivative.[49] Similar products along with phenol are obtained by the thermal reaction of triphenyl phosphite with geraniol at 200° under nitrogen in a flow system.[50]

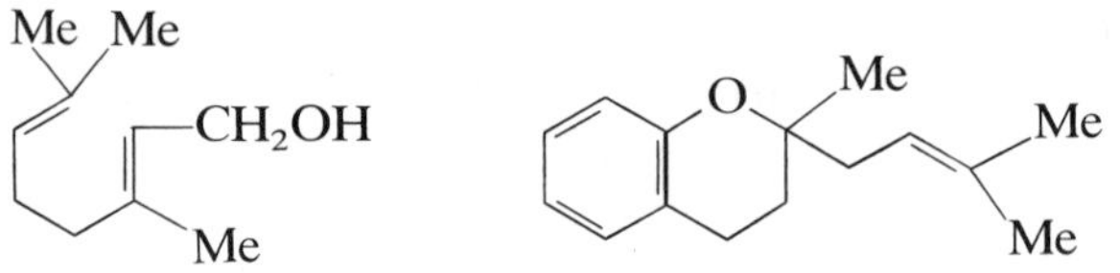

Me Me
CH_2OH
Me
(**76**)

O Me
Me
Me
(**77**)

4,4-Dimethylchroman (**80**) has been obtained by first ring opening the related dihydrocoumarin (**78**) by reduction using the Bouveault–Blanc method to yield the phenolic alcohol (**79**), which on dehydration by heating with 20% phosphoric acid gave the required chroman (**80**).[51] In a later preparation reduction of the dihydrocoumarin was carried out in THF using lithium tetrahydridoaluminate and cyclization of the diol was effected in the presence of hot 85% phosphoric acid.[52]

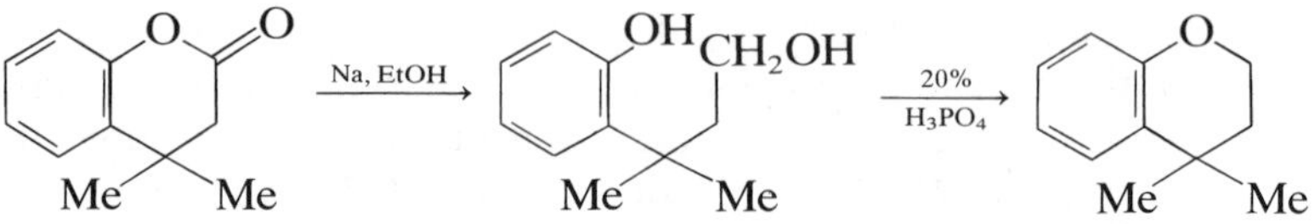

O O
Me Me
(**78**)

Na, EtOH →

OH CH_2OH
Me Me
(**79**)

20% H_3PO_4 →

O
Me Me
(**80**)

There are few dialkyl derivatives of chroman containing two alkyl groups on different carbons of the heterocyclic ring. The slow addition of 4-chloropent-2-ene (**81**) to phenol and a small amount of aluminum chloride at 48.5° yields mainly pent-3-en-2-yl phenols and a mixture (10.3%) of 2,4-dimethylchroman (**82**) and 2,3-dihydro-2-ethyl-3-methylbenzo[*b*]furan (**83**).[53] In 1925 Claisen et al. reported that the reaction between 4-bromopent-2-ene and phenol in acetone containing potassium carbonate gave a mixture of phenyl 1,3-dimethylallyl ether and 2-(1,3-dimethylallyl)phenol. The former was converted into the latter in a 70–80% yield by heating at 225° or by boiling with diethylaniline. Treatment of the allylphenol with hydrogen bromide–acetic acid resulted in the formation of 50% of either 2,4-dimethylchroman or 2,3-dihydro-2-ethyl-3-methylbenzo[*b*]furan, but no attempt was made to identify the products.[54]

The same two heterocycles, (**82**) and (**83**), are obtained in greater yields by the cyclization of *trans*-(1,3-dimethylallyl)phenol (**84**) either in the presence of hydrogen bromide or by irradiation in benzene at 20° with uv light.[55]

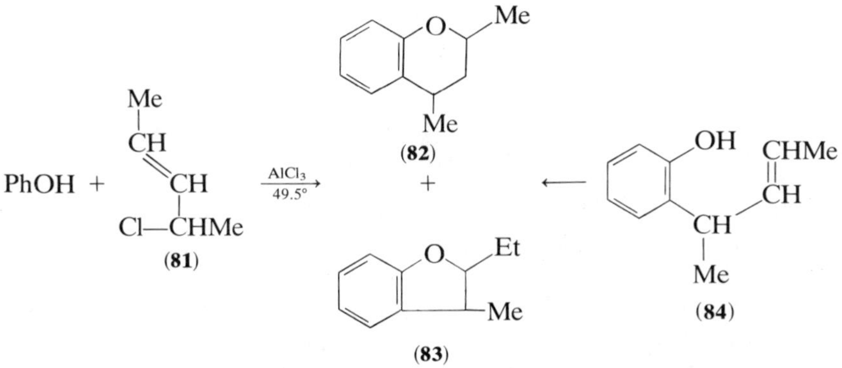

The synthesis of chromans and benzopyrylium salts has been effected by acid-catalyzed disproportionation of chrom-3-enes and dehydration of chroman-4-ols.[56] A 50:50 mixture of *cis*- and *trans*-3,4-dimethylchroman, (**88**) and (**89**), and 3,4-dimethylbenzopyrylium perchlorate (**87**) is obtained by hydride transfer when 3,4-dimethylchroman-4-ol (**86**) is treated with perchloric acid. The alcohol (**86**) is prepared by the reaction between 3-methylchroman-4-one (**85**) and methylmagnesium iodide. 3,4-Dimethylchrom-3-ene (**90**) is prepared by dehydrating the alcohol (**86**) on boiling in benzene in the presence of anhydrous copper sulfate, or by the reduction of 3,4-dimethylbenzopyrylium perchlorate (**87**) using sodium tetrahydridoborate. Catalytic hydrogenation of the chromen (**90**) in ethyl acetate over 10% palladium–carbon gives *cis*-3,4-dimethylchroman (**88**).[56] 4-Methylchroman has been similarly prepared.

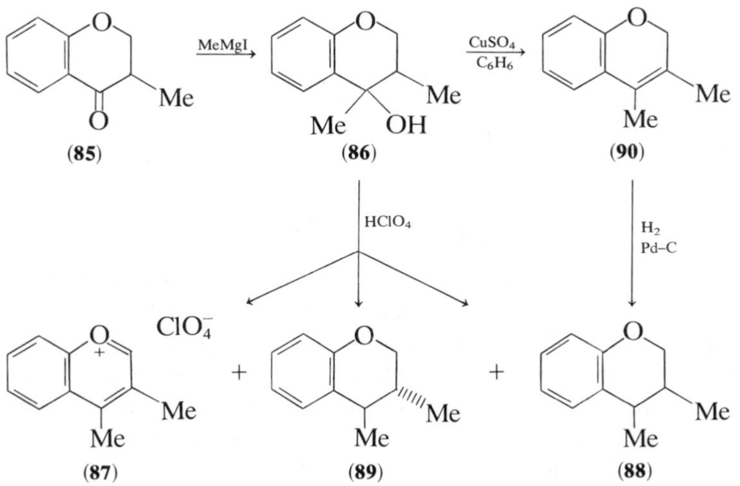

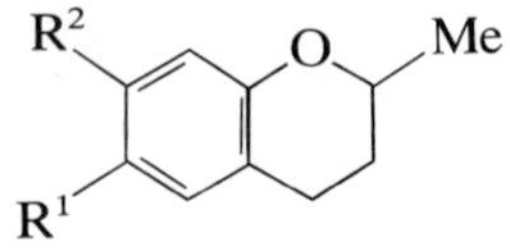

(**91**) $R^1 = Me$, $R^2 = H$
(**92**) $R^1 = H$, $R^2 = Me$

Nearly all the preparations of dialkylchromans containing a substituent attached to the heterocyclic ring and to the aromatic ring refer to 2,6- (**91**) and 2,7-dimethylchroman (**92**) and are related to those used for obtaining 2-methylchroman (**18**). The reaction between 4- or 3-methylphenol and but-1-en-4-ol in phosphoric acid[57] and butan-1,3-diol with heating in an autoclave[58] yields 2,6- and 2,7-dimethylchroman, respectively, along with alkenylated phenols. The latter method also furnishes some 4,6- or 4,7-dimethylchroman. In the case of 4-methylphenol and butan-1,3-diol, it has been shown that the best yields of 2,6-dimethylchroman (**91**) are obtained by using an excess of 4-methylphenol.[59]

The reaction between 4-methylphenol and butadiene in the presence of ethyl hydrogen sulphate at 20° yields substituted phenols and cyclic compounds, including a low yield of 2,6-dimethylchroman (**91**). This method does not afford any 2,7-dimethylchroman (**92**) from 3-methylphenol.[60] Small amounts of 2,6- and 2,7-dimethylchroman are obtained from the appropriate methylphenol and butadiene in the presence of zinc chloride on alumina, or KU-1 resin.[61,62] The alkenylation of 4- and 3-methylphenol with either crotyl alcohol or methyl vinyl carbinol in the presence of alumina or zinc chloride–alumina gives some 2,6- and 2,7-dimethylchroman, respectively, besides substituted phenols and other cyclic products.[63] 2,6-Dimethylchroman (**91**) is also obtained by the Clemmensen reduction of 2,6-dimethylchroman-4-one (**93**),[17] as is 8-ethyl-4-methylchroman (**95**) from 8-acetyl-4-methylchroman (**94**).[64]

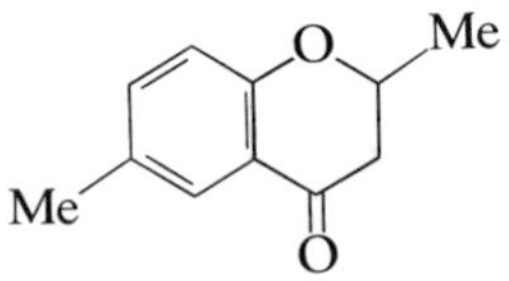

(**93**)

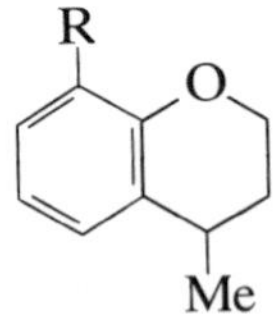

(**94**) R = Ac
(**95**) R = Et

4,6-Dimethylchroman (**98**) is formed along with the related benzopyrylium salt (**99**) by the acid-catalyzed disproportionation of the corresponding chrom-3-ene or by dehydration of 4,6-dimethylchroman-4-ol (**97**) with perchloric acid.[56] The alkenylation of 4-methylphenol using butan-1,3-diol in the presence of zinc chloride-alumina furnishes some 4,6-dimethylchroman (**98**). Similarly 3-methylphenol gives some 4,7-dimethylchroman (**101**),[58] also obtained in high yield from 3-(2-hydroxy-4-methylphenyl)-3-methylpropan-1-ol (**100**) (Section II, 1).[5]

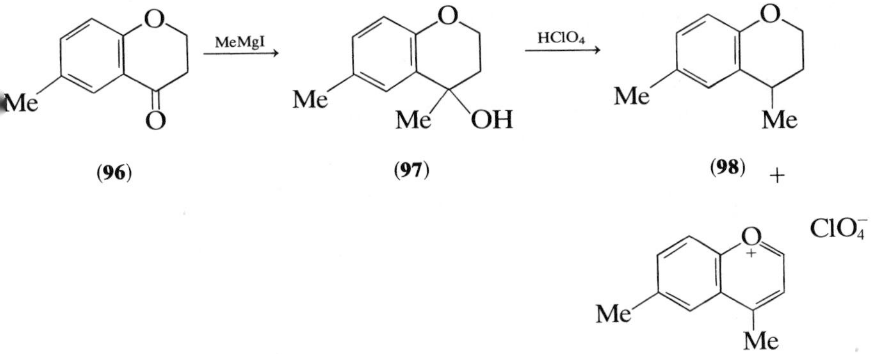

(96) → MeMgI → (97) → $HClO_4$ → (98) + (99) ClO_4^-

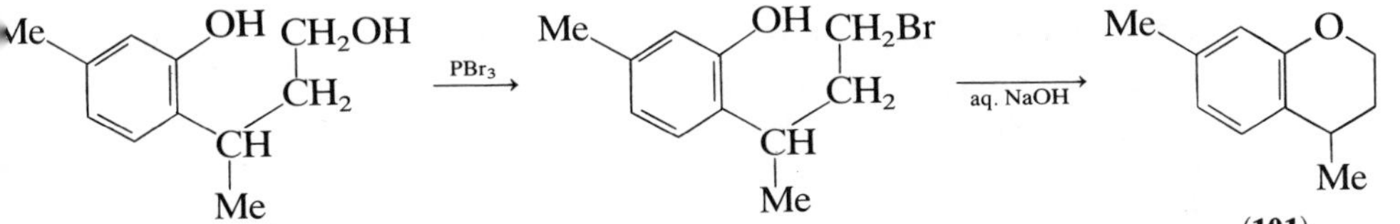

(100) → PBr_3 → → aq. NaOH → (101)

Dialkylchromans with both substituents in the benzene ring can be obtained by adopting methods used for the synthesis of chroman, starting with the appropriate dialkylbenzene derivative. 5,7-Dimethylchroman (**104**) 5,8-dimethylchroman (**105**) are prepared in 74% and 86% yield, respectively by the cyclization of 3,5-dimethylphenyl 3-hydroxypropyl ether (**102**) and 2,5-dimethylphenyl 3-hydroxypropyl ether (**103**) in boiling benzene in the presence of phosphorus pentoxide.[29]

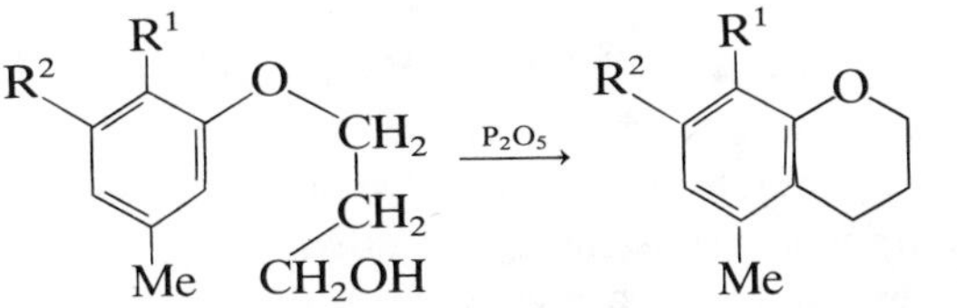

P_2O_5

(**102**) $R^1 = H$, $R^2 = Me$
(**103**) $R^1 = Me$, $R^2 = H$

(**104**) $R^1 = H$, $R^2 = Me$
(**105**) $R^1 = Me$, $R^2 = H$

5. Preparation of Diarylchromans

The only diarylchromans reported contain both aryl substituents in the heterocyclic ring and therefore fall into two classes—those with the two substituents on the same carbon atom and those with the two substituents on different carbon atoms. The 2,2- and 2,4-diaryl chromans are outside the scope of this review.

4,4-Diphenylchroman (**108**) is obtained in 24% yield when a well-stirred mixture of 3,3,3-triphenylpropan-1-ol (**106**) and lead tetraacetate in anhydrous benzene is heated in a dry atmosphere at 70° for several hours.[65] It is believed that the cyclizing intermediate is the alkoxy radical (**107**).

CH_2OH, CH_2, C, Ph, Ph $\xrightarrow{Pb(OAc)_4}$ [$CH_2O\cdot$, CH, C, Ph, Ph] $\longrightarrow$ O, Ph, Ph

(**106**) (**107**) (**108**)

cis-3,4-Diphenylchroman (**110**) and related compounds may be prepared by the hydrogenation of the appropriate 3,4-diphenyl-2*H*-chromen (**109**) in ethyl acetate in the presence of palladium or platinum catalysts.[66]

O, Ph, Ph $\xrightarrow[\text{Pt or Pd}]{H_2,\ EtAc}$ O, Ph, Ph

(**109**) (**110**)

6. Preparation of Trialkylchromans

The majority of known trialkylchromans are those containing three methyl substituents, most of which have only one substituent attached to the benzene ring, and generally this is in the 6- or 7-position. There are a few examples of compounds containing a methyl group in the 8-position but none has been reported with a methyl group in the 5-position only of the benzene ring. In the main, the other methyl groups are attached to the same carbon atom (generally C-2 but sometimes C-4), in the heterocyclic ring. Some of the methods used for obtaining trimethylchromans are derived from those used for the preparation of chroman and dimethylchromans. A series of trimethylchromans has been prepared containing two substituents on the benzene ring and one at the 2-position by the Clemmensen reduction of the related chroman-4-one.

The first trimethylchroman was reported in 1924 by Claisen,[67] who obtained 2,2,6-trimethylchroman (**111**) by reacting 4-methylphenol with isoprene in the presence of acid condensing reagents. A modification of this

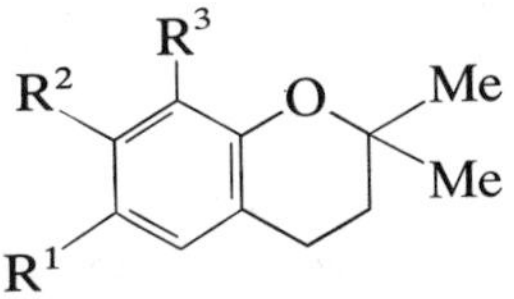

(**111**) $R^1 = Me$, $R^2 = R^3 = H$
(**112**) $R^1 = R^2 = H$, $R^3 = Me$
(**113**) $R^1 = CH_2CH{=}CMe_2$, $R^2 = R^3 = H$
(**114**) $R^1 = R^2 = H$, $R^3 = CH_2CH{=}CMe_2$
(**115**) $R^1 = R^3 = H$, $R^2 = Me$
(**116**) $R^1 = CH_2CH{=}CMe$, $R^2 = Me$, $R^3 = H$

method involves heating the reactants together in the presence of a metallic phenolate.[40]

When 4-methylphenol is allowed to react with 4-chloro-2-methylbut-2-ene in acetone containing potassium carbonate, O-alkylation takes place; but if the reaction is carried out in the absence of solvent or any other additives, only C-alkylation occurs to give 2,2,6-trimethylchroman (**111**) and 2,2,6-trimethyl-8-(3-methylbut-2-en-1-yl)chroman. Similarly, 2-methylphenol and 4-chloro-2-methylbut-2-ene afford 2,2,8-trimethylchroman (**112**) and 2,2,8-trimethyl-6-(3-methylbut-2-en-1-yl)-chroman.[68] The reaction between phenol and isoprene catalyzed by nickelocene and phenylmagnesium bromide yields a mixture containing 2,2-dimethyl-6-(3-methylbut-2-en-1-yl)chroman (**113**) and 2,2-dimethyl-8-(3-methylbut-2-en-1-yl)chroman (**114**) (see Section II, 4).

The alkenylation of 3-methylphenol with isoprene on the ion exchange resin KU-2 by mixing the reactants and resin at 50° and then stirring the mixture at 70° gives a mixture containing some alkenylated 3-methylphenols, 2,2,7-trimethylchroman (**115**), and 6-(3-methylbut-2-en-1-yl)-2,2,7-trimethylchroman (**116**). The main products are the monoalkenylated phenols.[69]

2,4,4-Trimethylchroman (**121**) may be obtained indirectly from 1,1,3-trimethylindane (**117**). The indane (**117**) is readily converted by oxygen at 90°, in the presence of alkali, into 1,3,3-trimethylindan-1-yl hydroperoxide (**118**), which on treatment with a little Fuller's earth or sulfuric acid in boiling acetone gives 2,4,4-trimethylchroman-2-ol (**119**) (40–50%) and its dehydration product 2,4,4-trimethylchrom-2-ene (**120**) (25–40%). The chromanol (**119**) on boiling with anhydrous oxalic acid in benzene affords 96% of 2,2,4-trimethylchrom-2-ene (**120**). The chromen (**120**) is only hydrogenated to 2,4,4-trimethylchroman (**121**) under comparatively drastic conditions, that is, under pressure at 100–110° over Raney nickel.[70] The amount of trimethylchromen formed during the rearrangement can not be accounted for by dehydration of the chromanol (**119**) under the reaction conditions, and the two products probably arise simultaneously from a common cationic intermediate.

(117) —O_2, 2.5% Na_2CO_3, 90°→ (118) —H^+→ (119) + (120) —H_2, Raney Ni→ (121)

(118) —H^+→ [intermediate] → [chromanylium cation] —OH^-→ (119) ⇌ (120) ←$-H^+$—

Heating a 1:1 molar mixture of 4-methylphenol and pent-1-en-5-ol in the presence of phosphoric acid at 135–155° gives 35–40% 2,4,6-trimethylchroman (**122**) and 1.1–4.8% 4-methyl-2-(1-methylbut-3-en-1-yl)phenol.[71]

(**20**) R = 4-Me, (**20**) R = 3-Me + $CH_2{=}CH{-}CH_2{-}CH_2{-}CH_2OH$ —H_3PO_4→

(**122**) R^1 = Me, R^2 = H
(**123**) R^1 = H, R^2 = Me

When 4-methylphenol and pent-1,3-diene (piperylene) are heated together at 50° in the presence of ethyl hydrogen sulfate, a mixture is obtained containing some 2,4,6-trimethylchroman (**122**), dihydrodimethylethylbenzo[*b*]furan, and high-boiling products.[40,72] Similar mixtures are obtained when the reaction is carried out with zinc chloride–alumina or KU-1 ion exchange resin as catalyst. Some 2,4,6-trimethylchroman (**122**) can be obtained from the mixture of products resulting from the reaction between pent-1,4-diene and 4-methylphenol in the presence of ethyl hydrogen sulfate at 100° or zinc chloride-alumina at 150°.[73]

When 2-chloro- or 2-bromopent-3-ene (piperylene hydrochloride or hydrobromide) is added swiftly to 4-methylphenol in acetone containing potassium carbonate and the mixture is heated in a water bath, the main product is the 2-alkenyl derivative (**124**) together with small amounts of the ether (**125**), 2,4,6-trimethylchroman (**122**), and 2,3-dihydro-3,5-dimethyl-2-ethylbenzo[*b*]furan (**126**). Slow addition of the halogeno compound results only in the formation of the alkenyl derivative (**124**) and the ether (**125**).[74] This reaction of piperylene hydrohalide was first carried out by Claisen and his co-workers[54] in 1925 using 2-bromopent-3-ene, but they did not identify the products.

(**20**) + MeCH=CHCH(Me)Br(Cl) (Me_2CO, K_2CO_3) → (**124**) + (**125**) + (**122**) + (**126**)

R = 4-Me

(**124**) (**125**) (**126**)

The reaction between 3-methylphenol and pent-1,3-diene at 150° in an autoclave in the presence of zinc chloride–alumina affords 62% phenolic products plus a nonphenolic fraction probably containing 2,4,7-trimethylchroman (**123**) and 2,3-dihydro-3,6-dimethyl-2-ethylbenzo[*b*]furan.[75] Heating a 1:1 molar mixture of pent-1-en-5-ol and 3-methylphenol in the presence of phosphoric acid at 100–135° affords 13–39% 2,4,7-trimethylchroman (**123**) and 4.7–4.9% 3-methyl-6-(1-methylbut-3-en-1-yl)phenol.[71] Higher temperatures favoured formation of the chroman.

The reaction between 3-methylphenol and pent-1,3-diene in the presence of ethyl hydrogen sulfate yields some 2,4,7-trimethylchroman (**123**) and 2,3-dihydro-3,6-dimethyl-2-ethylbenzo[*b*]furan[60] in addition to phenolic products.

The alkenylation of 3-methylphenol with 2-bromopent-3-ene under nitrogen without solvent or catalyst yields only cyclic products, mainly substituted chromans and dihydrobenzo[*b*]furans. Under the same conditions 3-methylphenol and 2-chloropent-3-ene afford 97% mono- and disubstituted phenols and ~3% cyclic products.[76]

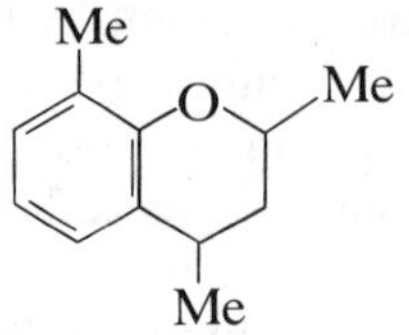

(127)

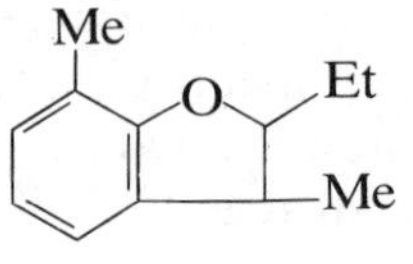

(128)

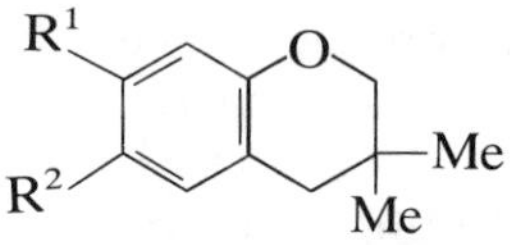

(129) $R^1 = H$, $R^2 = Me$
(130) $R^1 = Me$, $R^2 = H$

The reaction between 2-methylphenol and pent-1,3-diene in the presence of zinc chloride–alumina at 150° gives a trace of 2,4,8-trimethylchroman (**127**), but a 10% yield is obtained using KU-1 resin as catalyst at 100°. The alkenylation of 2-methylphenol with 2-chloro- or 2-bromo-pent-3-ene gives a mixture of substituted phenols and cyclic products containing some 2,4,8-trimethylchroman (**127**). Cyclization of the mixture of substituted phenols yields 43% of a mixture of 2,4,8-trimethylchroman (**127**) and 2,3-dihydro-3,7-dimethyl-2-ethylbenzo[*b*]furan (**128**).[77]

The reaction between 4-methylphenol and 2,2-dimethylpropan-1,3-diol at 350° in the presence of zinc chloride–alumina gives a mixture of substituted phenols and 3,3,6-trimethylchroman (**129**); the reaction is very slow at 250°. Similarly the reaction using 3-methylphenol yields 3,3,7-trimethylchroman (**130**). These preparations give 21–35% of the trimethylchroman, and the total yield of products is 70–90%.[30]

4,4,6-Trimethylchroman (**133**) is obtained from 3,4-dihydro-4,4,6-trimethylcoumarin (**131**). The dihydrocoumarin (**131**) is first converted into 3-(5-methyl-2-hydroxyphenyl)-3-methylbutanol (**132**) by the Bouveault–Blanc reduction. The hydroxyphenol (**132**) on heating with 20% phosphoric acid dehydrates to give 4,4,6-trimethylchroman (**133**).[51]

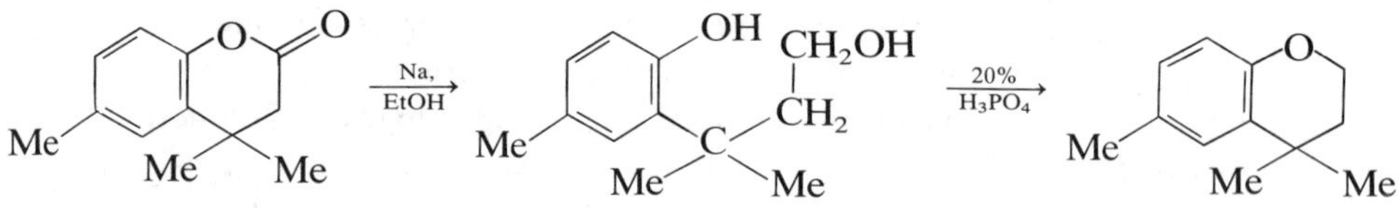

Na, EtOH; 20% H_3PO_4

(131) (132) (133)

PBr_3; $-MeBr$

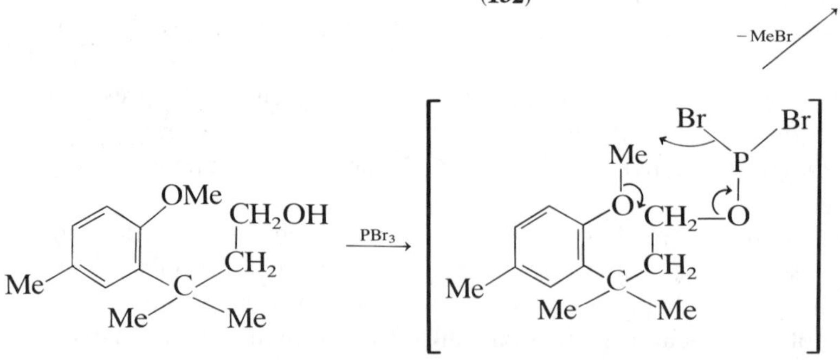

(134) (135)

4,4,6-Trimethylchroman (**133**) is also obtained by heating 2-(1,1-dimethyl-3-hydroxypropyl)-4-methylanisole (**134**) with phosphorus tribromide in benzene. The facile formation of 4,4,6-trimethylchroman (**133**) is attributed to the likely 6-centered positioning of the phosphoryl bromide (**135**) arising from the anisylpropanol (**134**), resulting in assisted intramolecular elimination of methyl bromide.[78]

The catalytic reduction of 2,8-dimethyl-5-isopropylchromone (**137**) in the presence of platinum oxide has been used to obtain 2,8-dimethyl-5-isopropylchroman (**138**), one of the few chromans found with an alkyl substituent in the 5-position. The chromone (**137**) was prepared by reacting 2-hydroxy-4-isopropyl-1-methylbenzene (carvacrol) (**136**) with ethyl acetoacetate in the presence of phosphorus pentoxide.[79]

Me OH CHMe$_2$ (**136**) $\xrightarrow{AcCH_2CO_2Et,\ P_2O_5}$ Me O Me O CHMe$_2$ (**137**) $\xrightarrow{H_2,\ PtO_2}$ Me O Me CHMe$_2$ (**138**)

7. Preparation of Trisubstituted Chromans

2,2-Dimethyl-4-phenylchroman (**142**) was unexpectedly obtained during investigations of methods of preparing a series of 2,2-dimethyl-4-phenylchroman derivatives. It was hoped that the addition of the geminal methyl to the simpler 4-phenylchroman nucleus would impart more physiological activity to the system, by analogy with certain potent naturally occurring chromans, for example, the cannabinoids. This hope was not realized. However it was found that 3-(2-halogeno-2-methylpropyl)-3-phenyl-2,3-dihydrobenzo[*b*]furan-2-one (**139**) on treatment with sodium methoxide in methanol at room temperature afforded methyl 2,2-dimethyl-4-phenylchroman-4-carboxylate (**140**). The ester was then converted into the corresponding carboxamide (**141**) with the intention of reducing it to the related amine, as had been achieved with analogous compounds containing a substituent (e.g., chloro or methoxy) in the 6-position. In this case it was found that lithium tetrahydridoaluminate reduction of the amide (**141**) persistently resulted in exclusive carbonyl cleavage to give 53% 2,2-dimethyl-4-phenylchroman (**142**). It was the only product isolated when the amide (**141**) was boiled with bis(methoxyethoxy)aluminohydride in benzene and was obtained in 70% yield.[80]

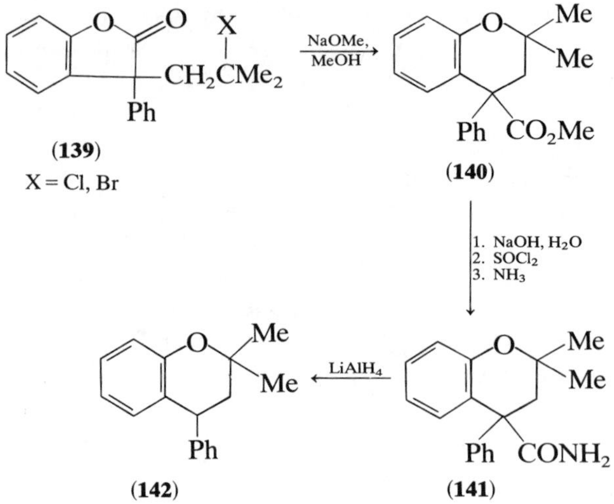

O
O
X
CH2CMe2
Ph
(139)
X = Cl, Br
NaOMe, MeOH
O
Me
Me
Ph CO2Me
(140)
1. NaOH, H2O
2. SOCl2
3. NH3
O
Me
Me
Ph CONH2
(141)
LiAlH4
O
Me
Me
Ph
(142)

8. Preparation of Tetraalkylchromans

Like the trialkyl chromans, most tetraalkylchromans contain methyl substituents, with only one of the four methyls attached to the benzene ring. In most cases the compounds possess a *gem*-dimethyl group, usually at the 2-position, but sometimes at C-4. There are examples of the preparation of tetramethylchromans containing two and three methyl groups attached to the benzene ring. In general the methods of preparing tetraalkylchromans are modifications of those used for obtaining trialkylchromans.

One of the first reported tetramethylchromans was 2,2,3,6-tetramethylchroman (**143**), obtained by the reaction between 4-methylphenol and 2,3-dimethylbutadiene in the presence of acid condensing reagents.[67]

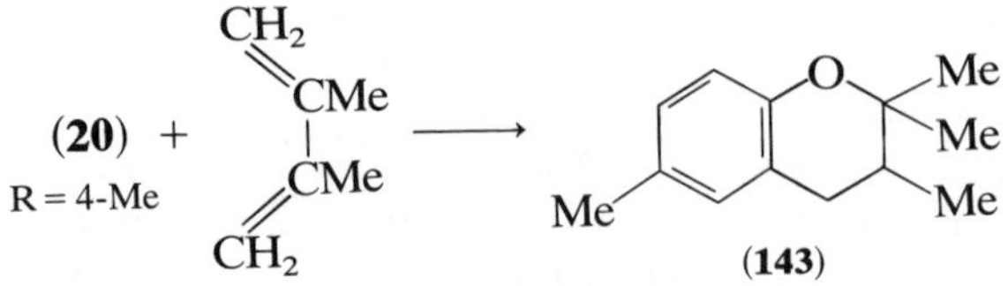

(20) +
R = 4-Me
CH2
CMe
CMe
CH2
O
Me
Me
Me
Me

(**143**)

2,2,4,4-Tetramethylchroman (**146**), in which all the methyl substituents are attached to the heterocyclic ring, is obtained by dehydration and cyclization (using 20% phosphoric acid) of 3-(2-hydroxyphenyl)-1,1,3,3-tetramethylpropan-1-ol (**145**), prepared by reaction of 3,4-dihydro-4,4-dimethylcoumarin (**144**) and methylmagnesium iodide.[51]

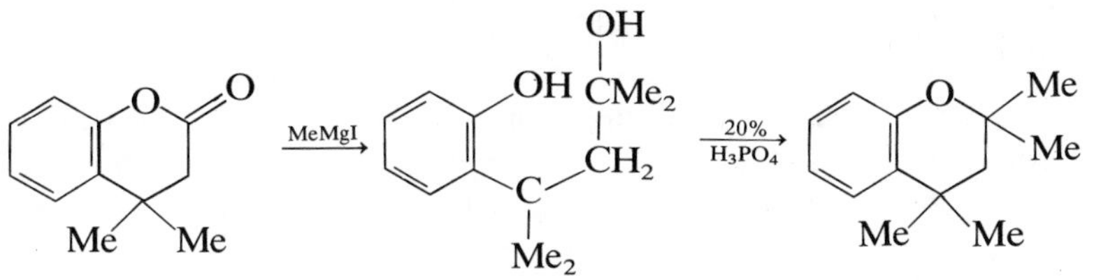

(144) (145) (146)

4-(2,3-Dimethylbut-2-en-1-yloxy)-1-methylbenzene (**147**) on heating at 220–230° in decalin rearranges to give a mixture of products containing 2,2,3,6-tetramethylchroman (**148**), 2-(2,3-dimethylbut-2-en-1-yl)-4-methylphenol (**149**), 4-methylphenol, and 2,6-bis(2,3-dimethylbut-2-en-1-yl)-4-methylphenol (**150**).[81]

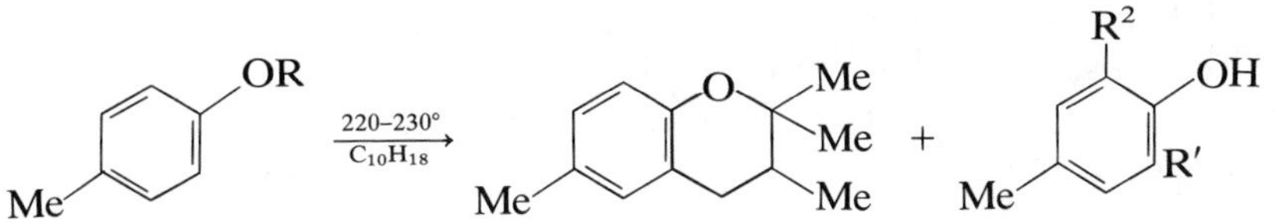

(**147**) R = —$CH_2CMe{=}CMe_2$

(**148**) 16%

(**149**) $R^1 = CH_2CMe{=}CMe_2$, $R^2 = H$ (51%)

(**150**) $R^1 = R^2 = CH_2CMe{=}CMe_2$

The alkenylation of 4-methylphenol using 1-chloro-2,3-dimethylbut-2-ene and 3-chloro-2,3-dimethylbut-1-ene at 60–70° in the presence of base affords a mixture of products containing, besides the ether (**147**) and derivative (**149**), chromans (**148**) and (**151**) and the dihydrobenzo[*b*]furans (**152**) and (**153**).[82]

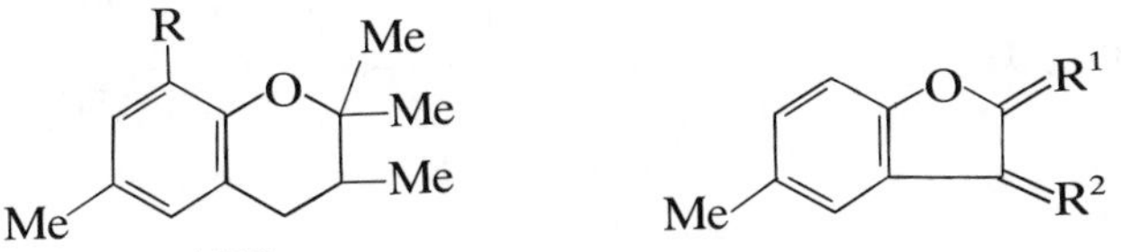

(**151**) R = —$CH_2CMe{=}CMe_2$

(**152**) $R^1 = R^2 = Me_2$

(**153**) $R^1 = Me, CHMe_2$, $R^2 = H_2$

3,5-Dimethylphenol (**154**) condenses with isoprene in the presence of zinc chloride to give 2,2,5,7-tetramethylchroman (**155**). Mercuric chloride is also an efficient reagent.[83]

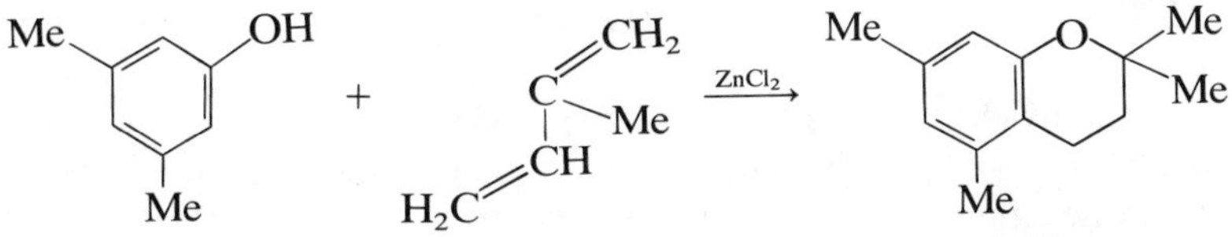

(**154**) (**155**)

The condensation of 3-methylphenol with acetone yields 2′-hydroxy-2,4,4,4′,7-pentamethylflavan (**156**), which on hydrogenolysis over Raney nickel gives 20% 3-methylphenol, 26% thymol (6-isopropyl-3-methylphenol), and 27% 2,4,4,7-tetramethylchroman (**157**).[84] The structure of flavan (**156**) has been confirmed by ir and nmr spectra.[85]

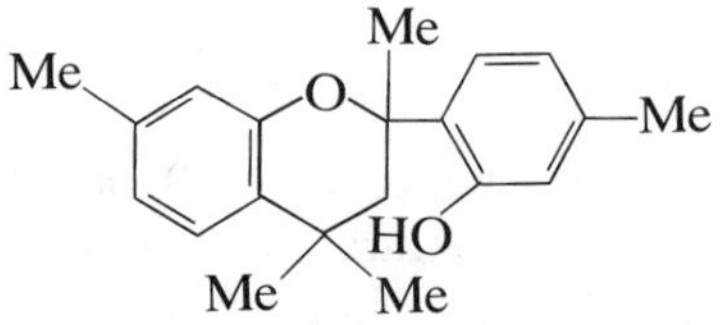

(**156**) 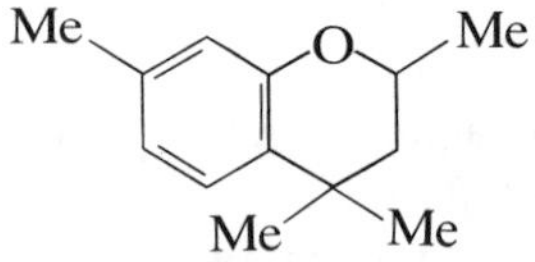 (**157**)

The condensation of 2-hydroxy-4-isopropyl-1-methylbenzene (carvacrol) (**136**) and methyl acetomethylacetate in the presence of phosphorus pentoxide gives 5-isopropyl-2,3,8-trimethylchromone (**158**), which on catalytic reduction in the presence of platinum oxide affords 5-isopropyl-2,3,8-trimethylchroman (**159**).[79]

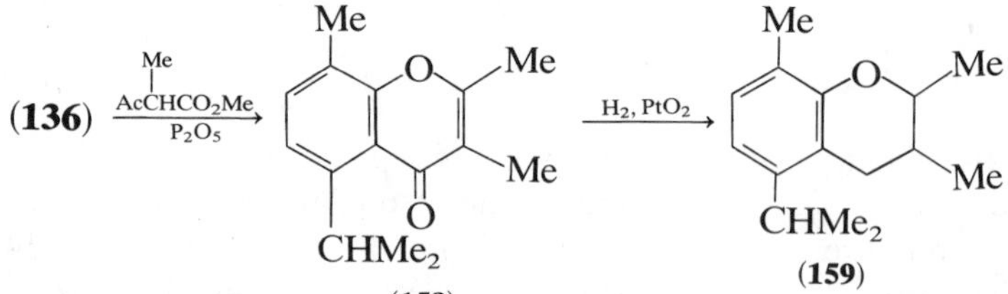

(**136**) $\xrightarrow[P_2O_5]{AcCH(Me)CO_2Me}$ (**158**) $\xrightarrow{H_2, PtO_2}$ (**159**)

The reduction of 3,4-dihydro-4,4,7,8-tetramethylcoumarin (**160**) in tetrahydrofuran with lithium tetrahydridoaluminate yields 3,3-dimethyl-3-(2-hydroxy-3,4-dimethylphenyl)propan-1-ol (**161**), which on heating with 85% phosphoric acid at 135–145° yields 4,4,7,8-tetramethylchroman (**162**) in an 87% yield.[52]

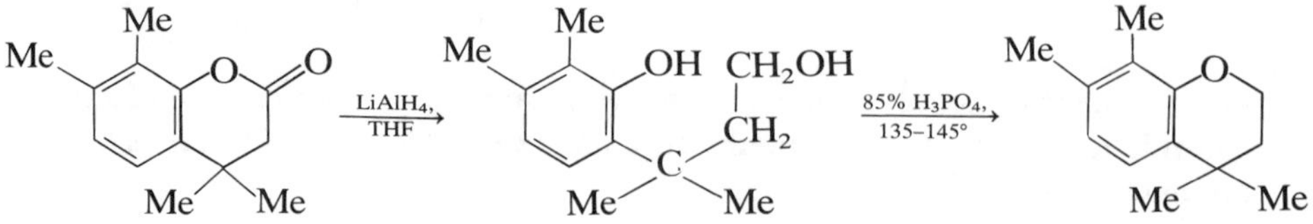

(**160**) $\xrightarrow[THF]{LiAlH_4,}$ (**161**) $\xrightarrow[135\text{–}145°]{85\%\ H_3PO_4,}$ (**162**)

A tetramethylchroman with only one methyl substituent in the heterocyclic ring is obtained by the Clemmensen reduction of the related chromanone. For example, the reduction of 2,5,7,8-tetramethylchroman-4-one (**163**) using zinc and hydrochloric acid gives 2,5,7,8-tetramethylchroman (**164**).[17]

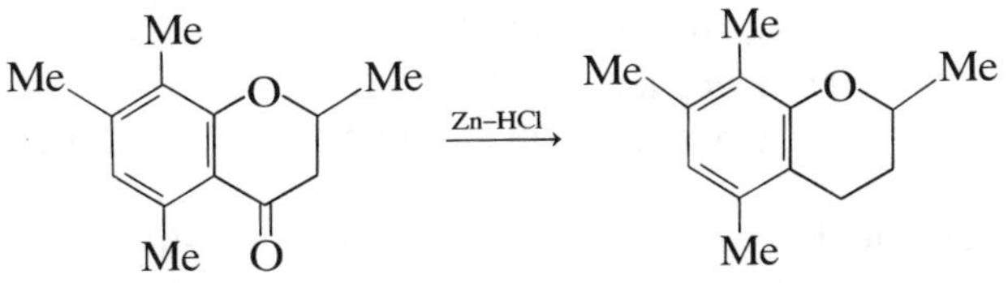

(163) $\xrightarrow{\text{Zn–HCl}}$ (164)

9. Preparation of Pentaalkylchromans

Methods for the preparation of pentamethylchromans from polymethylated phenols and of related derivatives containing a 6-hydroxy group from appropriate hydroquinones and their monoethers were investigated originally with a view to finding a possible synthesis of α-tocopherol. Toward this end certain modifications were made to the method of Claisen, of preparing chromans by condensing phenols with dienes in the presence of acidic condensing reagents.

The first reported pentamethylchroman was 2,2,3,5,7-pentamethylchroman (**165**) obtained by reacting 3,5-dimethylphenol (*m*-xylenol) (**154**) with 2,3-dimethylbutadiene in the presence of acidic condensing reagents.[67]

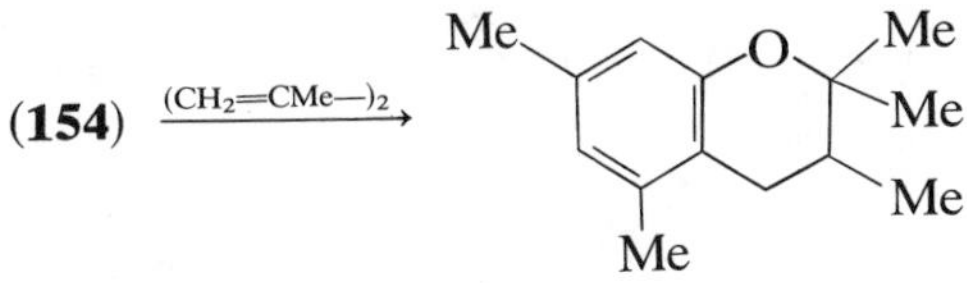

(**154**) $\xrightarrow{(CH_2{=}CMe{-})_2}$ (**165**)

A pentamethylchroman with four methyl groups attached to carbon atoms of the heterocyclic ring is obtained from 3,4-dihydro-4,4,7-trimethylcoumarin (**166**). The dihydrocoumarin (**166**) in benzene reacts with an ethereal solution of methylmagnesium iodide in an atmosphere of nitrogen to yield 2,4-dimethyl-4-(2-hydroxy-4-methylphenyl)pentan-2-ol (**167**), which when treated with hydrogen chloride in methanol undergoes cyclodehydration to give 2,2,4,4,7-pentamethylchroman (**168**).[86]

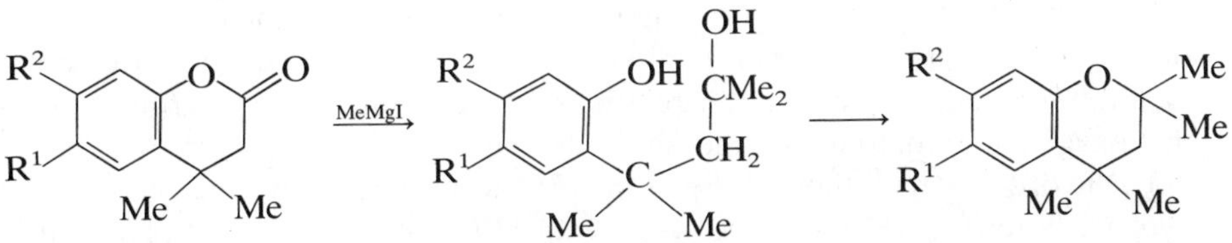

$\xrightarrow{\text{MeMgI}}$ $\longrightarrow$

(**166**) $R^1 = H$; $R^2 = Me$
(**169**) $R^1 = Me$, $R^2 = H$

(**167**) $R^1 = H$, $R^2 = Me$
(**170**) $R^1 = Me$, $R^2 = H$

(**168**) $R^1 = H$, $R^2 = Me$
(**171**) $R^1 = Me$, $R^2 = H$

2,2,4,4,6-Pentamethylchroman (**171**) is obtained in a similar way from 3,4-dihydro-4,4,6-trimethylcoumarin (**169**). In this case the cyclodehydration of the hydroxyalkylphenol (**170**) is effected by heating with 20% phosphoric acid.[51]

2,2,5,7,8-Pentamethylchroman (**173**) is related to α-tocopherol and was obtained in 11.5% yield on distilling the product formed when a mixture of 2,3,5-trimethylphenol (**172**) and isoprene in acetic acid was saturated with hydrogen chloride.[83]

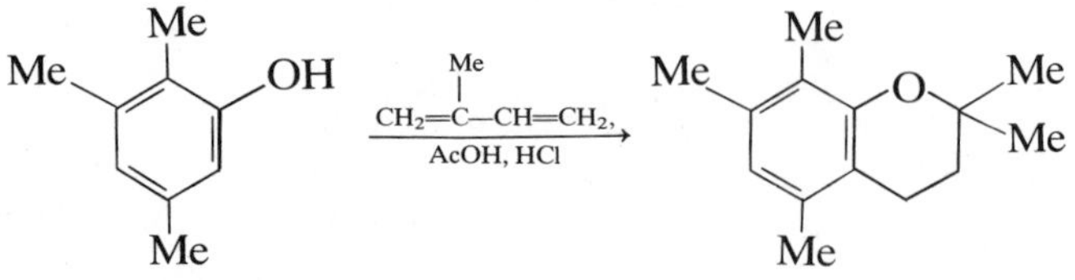

(**172**) $\xrightarrow[\text{AcOH, HCl}]{CH_2{=}C(Me){-}CH{=}CH_2,}$ (**173**)

4,4,5,7,8-Pentamethylchroman (**175**) is prepared in 82% yield by adding a solution of 3,4-dihydro-4,4,5,7,8-pentamethylcoumarin (**174**) in anhydrous tetrahydrofuran containing boron trifluoride etherate to a solution of sodium tetrahydridoborate in anhydrous diglyme at icebath temperature.[52]

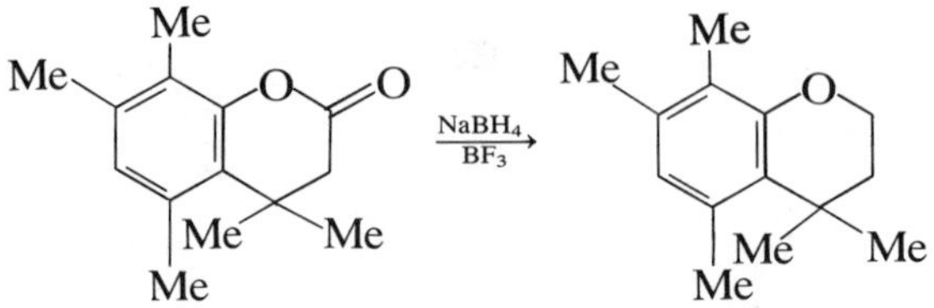

(**174**) $\xrightarrow[BF_3]{NaBH_4}$ (**175**)

Mono-, di-, tri-, tetra-, and pentasubstituted derivatives of chroman are listed in Table 5. Most of the substituents are methyl groups, but there are some where they are phenyl or other alkyl groups.

10. Preparation of Dihydronaphthopyrans

Dihydronaphthopyrans are in general prepared by the catalytic reduction of the parent naphthopyran or by the Clemmensen reduction of the naphthopyranones in which the relationship between the keto group and the oxygen heteroatom are the same as those found in chroman-4-one. 2,3-Dihydro-1*H*-naphtho[2, 1-*b*]pyran (**3**) is obtained by the catalytic hydrogenation of 3*H*-naphtho[2,1-*b*]pyran (**176**) in ethanol over platinum oxide. Other alkyl- and aryl-substituted[87] and diaryl-substituted 2,3-dihydro-1*H*-naphtho[2,1-*b*]pyrans[88–90] have been prepared by this method.

Originally 2,3-dihydro-1*H*-naphtho[2,1-*b*]pyran (**3**) was obtained by the cyclization of 2-naphthyl 3-hydroxypropyl ether.[3]

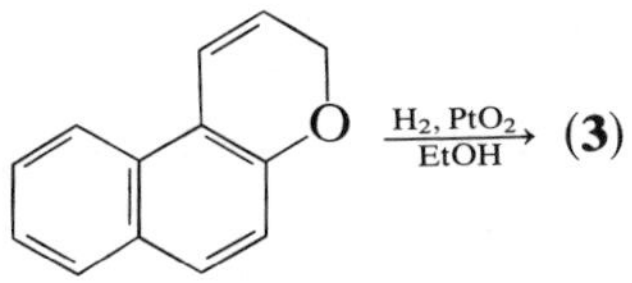

$\xrightarrow[\text{EtOH}]{H_2,\ PtO_2}$ (**3**)

(**176**)

The hydrolysis of the acid phthalate of 3-(2-hydroxy-1-naphthyl)propan-1-ol (**177**) in boiling aqueous alkali solution besides yielding the naphthol alcohol (**178**) also affords 21.7% 2,3-dihydro-1*H*-naphtho[2,1-*b*]pyran (**3**). This method of cyclization by intramolecular displacement of a carboxylate ion is successfully carried out using the acid phthalate of phenol ethanol and naphthol ethanol, when benzofurans and naphthofurans are formed without concurrent hydrolysis. When 2,3-dihydro-1*H*-naphtho[2,1-*b*]pyran (**3**) is formed, hydrolysis to the naphthol alcohol is the predominant reaction. This is the only dihydronaphthopyran to have been obtained by this method.[91]

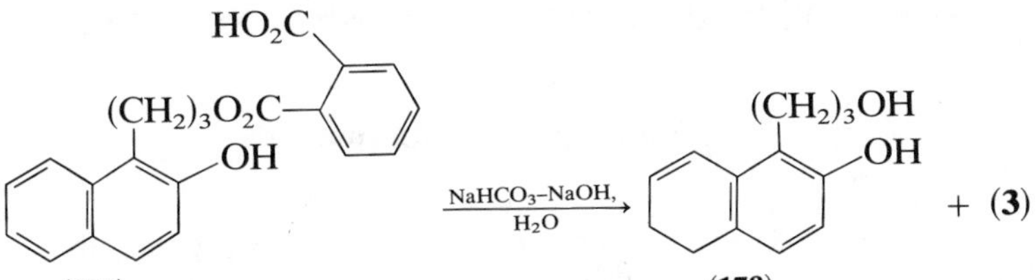

HO_2C $(CH_2)_3O_2C$ OH $\xrightarrow[H_2O]{NaHCO_3\text{–}NaOH,}$ $(CH_2)_3OH$ OH + (**3**)

(**177**) (**178**)

Both 2,3-dihydro-1*H*-naphtho[2,1-*b*]pyran (**3**) and 3,4-dihydro-2*H*-naphtho-[1,2-*b*]pyran (**2**) have been obtained in 82.5% yield by the method of Deady, Topsom, and Vaughan,[28] which involves boiling the appropriate naphthyloxypropane in benzene in the presence of aluminum chloride.

3,4-Dihydro-2*H*-naphtho[1,2-*b*]pyran (**2**) and 2,3-dihydro-1*H*-naphtho[2,1-*b*]pyran (**3**) are also obtained by the Clemmensen reduction of the related dihydropyran-4-one (**179**) and dihydropyran-1-one (**180**). The pyranones (**179**) and (**180**) when subjected to the Wolff-Kishner-Minlon reduction do not yield the expected dihydronaphthopyrans (**2**) and (**3**) but afford 1- and 2-naphthol, respectively.[92]

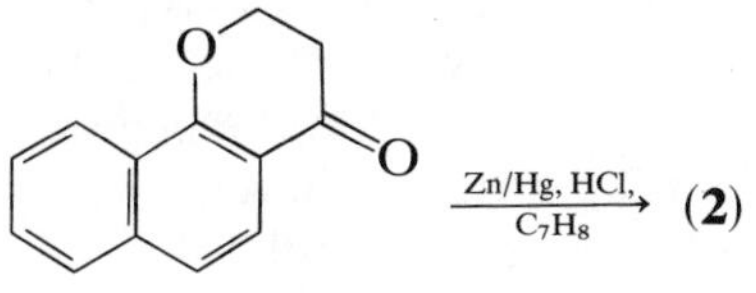

$\xrightarrow[C_7H_8]{\text{Zn/Hg, HCl,}}$ (**2**) $\xrightarrow[C_7H_8]{\text{Zn/Hg, HCl}}$ (**3**)

(**179**) 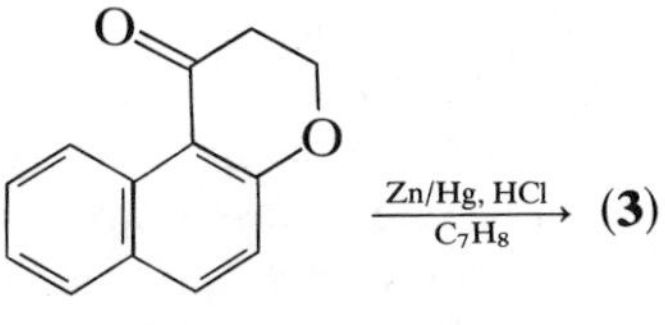 (**180**)

A 3-methylhexahydro-1*H*-naphtho[2,1-*b*]pyran, (**181**) or (**182**), obtained by McQuillin and Robinson,[93] who did not determine its structure, gives 3-methyl-2,3-dihydro-1*H*-naphtho[2,1-*b*]pyran (**183**) on dehydrogenation with palladized charcoal.

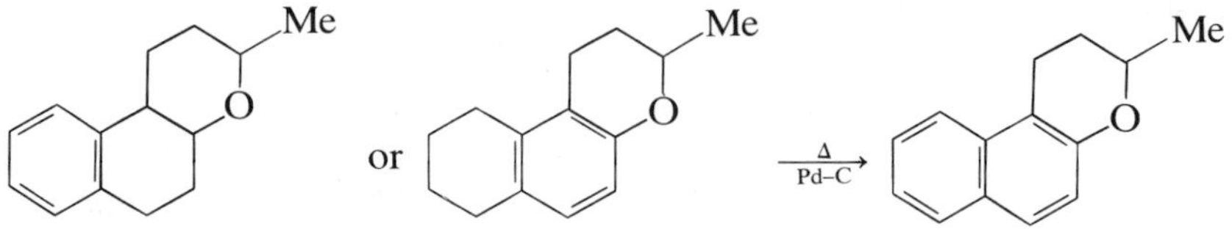

(**181**) or (**182**) $\xrightarrow[\text{Pd–C}]{\Delta}$ (**183**)

Although a number of derivatives of naphtho[2,3-*b*]pyran[94] have been prepared along with 2*H*-naphtho[2,3-*b*]pyran-2-one,[95] attempts to convert them to 3,4-dihydro-2*H*-naphtho[2,3-*b*]pyrans have not been described. Some dihydronaphthopyrans are listed in Table 6.

III. PHYSICAL CONSTANTS AND SPECTRA

1. Conformation of Chromans

The half-chair conformation for cyclohexene has been generally adopted for dihydropyran derivatives. The infrared data of a number of flavanones and isoflavanones indicate conjugation of the carbonyl group in (**184**) with the fused benzene ring. Hence the carbon atoms 3, 4, α, and β and the oxygen atoms 1 and γ occupy the same plane, and C-2 is the only cyclic atom out of that plane.

A rigid-wire model of a strainless dihydropyran ring (**185**) with the dimensions shown in the first column of Table 1 has also a "sofa" conformation in which all but one of the heterocyclic ring atoms are planar. The dimensional constants are within the accepted range, though there is doubt regarding the bond angles and bond length to be associated with atoms α and β. Alteration of these angles to 120°, combined with an increase of the angle at position 1 to 111°, does not change the conformation of the ring.

The exocyclic bonds at positions 2 and 3 in structure (**185**) are closely axial and equatorial, and the atoms in the heterocyclic ring, α excepted, are approximately at the five angles of a cyclohexane ring. The exocyclic bonds at position 4 are of the quasi type, being equally inclined to the main plane of the molecule. These quasi bonds are not fully staggered in relation to the noncyclic bonds at position 3. It can therefore be deduced that the conformation (**185**) holds for chromans.[96]

TABLE 1. DIMENSIONS DERIVED FROM A RIGID-WIRE MODEL OF DIHYDROPYRAN[96]

	Dimensions from model	Accepted values for angles and length
Angle at position 1	108°	111° ± 5°[6]
Angle at position 2	109.5°	109.5°
Angle at position 3	109.5°	109.5°
Angle at position 4	109.5°	109.5°
Angle at position α	123°	121° ± 5°[6]
Angle at position β	123°	121° ± 5°[6]
Bond β–1	1.44 Å	1.44 Å
Bond 1–2	1.44 Å	1.44 Å
Bond 2–3	1.54 Å	1.54 Å
Bond 3–4	1.54 Å	1.54 Å
Bond α–4	1.54 Å	1.54 Å
Bond α=β	1.34 Å	1.34 Å

(184) (185)

2. Infrared Spectra

The infrared spectrum of 2,2-dimethylchroman (**54**) shows peaks at 700–750 cm^{-1} due to ortho substitution, at 1380 cm^{-1} due to the *gem*-dimethyl group, and a profusion of bands near 1100 cm^{-1} associated with a cyclic ether.[13] A more detailed account of the infrared data for 2,2-dimethylchroman has been reported by Miller and Wood,[44] ν_{max} (liquid film) 2985, 2941, 2857 (alkyl), 1387, 1370 (*gem*-dimethyl), 1258 (chroman), 1220, 1157, 1124, 949, and 756 (ortho-disubstituted benzene) cm^{-1}.

3. Ultraviolet Spectra

The uv spectrum of chroman shows λ_{max} of 270, 276, and 284 nm (log ε 3.18, 3.33, and 3.40) when measured in *n*-hexane. Aromatic ethers exhibit at least two peaks in the 260–290 nm region, and in general the shorter wavelength peak is broader and more intense than the other. In the case of chroman and 8-methylchroman, the peak at the longer wavelength is the more intense.[97] The uv data of some chromans are listed in Table 2.

The uv spectra of a number of heterocyclic analogs of phenyl alkyl ethers and of choline phenyl ether have been compared, and the intensity of the

TABLE 2. ULTRAVIOLET DATA OF SOME CHROMANS

Chroman	Solvent	λ_{max} (nm)	log ε	Ref.
4-Me	Ethanol	274	3.35	56
		283	3.32	
8-Me	n-Hexane	275	3.2753	96
		278	3.2601	
		283	3.3375	
2,2-DiMe	n-Hexane	276	3.34	98
		284	3.33	
Mixture *cis*- and *trans*-3,4-DiMe	Ethanol	278	3.42	56
		285	3.38	
cis-3,4-DiMe	Ethanol	275	3.39	56
		283	3.37	
4,6-DiMe	Ethanol	281	3.49	56
		289	3.27	

Ph–O$\pi \rightarrow \pi^*$ transition has been used to determine the angle between the plane of the benzene ring and the plane containing the oxygen bonds. The calculated angles increase in the order 2,3-dihydrobenzo[*b*]furan <2,3-dihydro-7-methylbenzo[*b*]furan<chroman<homochroman. The value of the angle obtained for 2,3-dihydro-7-methylbenzo[*b*]furan is anomalously high and is explained on the basis of steric inhibition to sp^2 hybridization of oxygen. During a $\pi \rightarrow \pi^*$ transition, when the hybridization of the oxygen atom essentially changes from sp^3 to sp^2, the newly created lone pair will be in the plane of the benzene ring and will interact maximally with the 7-methyl group. In the case of 8-methylchroman the same arguments apply, but in this case the lone pair of the sp^2 hybrid state will be twisted some 34° out of plane and the interaction with the methyl group will be less.[97]

Hart and Wagner[99] have discussed the uv data of several oxygen heterocycles and their implication regarding the conformations of the respective compounds. The steady decrease in λ_{max} as the heterocyclic ring is enlarged, as shown in Table 3, can be explained in terms of increased puckering of the oxygen ring, forcing C-2 out of the plane of the benzene ring, thereby decreasing interaction between unshared pairs on the oxygen atom and the aromatic ring. This is consistent with Philbin and Wheeler's formulation (**185**) of the chroman ring,[96] where C-2 is out of the plane of the benzene ring but is easily "flipped" (through a conformation in which it is coplanar with the aromatic ring) to the opposite half-chair conformer. There is some indication that this "flip" may be hindered if a large group, for example, isopropyl, is attached at C-2. The low-wavelength band (274 nm) is the most intense band in the spectrum of 2-isopropylchroman (the isopropyl group would have to be axial in one of the half-chair conformations), whereas when the two groups attached to C-2 are identical, for instance, in 2,2-dimethylchroman, the 284 nm band is most intense.

TABLE 3. ULTRAVIOLET DATA OF SOME OXYGEN HETEROCYCLES[99]

Compound	Solvent	λ_{max} (nm)[a]
2,3-Dihydro-2-methylbenzo[*b*]furan	Hexane	281, 289
2,2-Dimethylchroman	Hexane	278, 284
2-Isopropylchroman	Cyclohexane	274, (278), 284
5,5-Dimethylhomochroman	Cyclohexane	(259), 266, 272

[a] Figures in parentheses are inflections

4. Nuclear Magnetic Resonance Spectra

The ^{1}H-nmr spectra of a number of chromans have been reported, but most of these have only been given with reference to chromans prepared as part of some other study. ^{1}H-Nuclear magnetic resonance data are listed for some chromans in Table 4.

5. Mass Spectra

The fragmentation undergone by chroman (**1**) on electron bombardment passes through what is probably the same ion (**186**) as that observed in the

Path A

m/e 115 ← *m** 99.4 — *m/e* 133 → *m* 106.5 → (**186**) *m/e* 119 → *m** 69.6 → *m/e* 91

*m** 132 — *m** 105.5

(**1**) *m/e* 134

*m** 83.8

C_6H_6 *m/e* 78

m/e 106

(**187**) *m/e* 107 —OH → *m** 58.3 → $C_6H_7^+$ *m/e* 79

*m** 75

m/e 77 $C_6H_5^+$

Path B

Scheme 1. Fragmentation of chroman in the mass spectrometer.

TABLE 4. [1]H-NUCLEAR MAGNETIC RESONANCE DATA OF SOME CHROMANS (**1**)[a]

Chroman	Position 2	3	4	5	6	7	8	Ref.
5-Methyl	*t* 3.82 (2H)	*m* 1.70 (2H)	Distorted *t* 2.28 (2H)	*s* 2.0 (3H, Me)		*m* 6.65 (3H) (positions 6–8)		29
7-Methyl	*t* 3.77 (2H)	*m* 1.62 (2H)	*t* 2.38 (2H)	*q* 6.52 (2H) (positions 5–6)		*s* 2.03 (3H, Me)	*s* 6.47 (1H)	29
2,2-Dimethyl	*s* 1.30 (6H, 2Me)	*t* 1.7 (2H)	*t* 2.7 (2H)					44
5,8-Dimethyl	*t* 4.10 (2H)	*m* 2.0 (2H)	*t* 2.58 (2H)	*s* 2.15 (3H, Me)	*q* 6.70 (2H) (positions 6–7)		*s* 2.15 (3H, Me)	29
					Phenolic OH			
2,2-Dimethyl-6-hydroxy	*s* 1.25 (6H, 2Me)	*t* 1.7 (2H)	*t* 2.6 (2H)		5.8			44
6-Hydroxy-2,2,5,8-tetramethyl	*s* 1.25 (6H, 2Me)	*t* 1.75 (2H)	*t* 2.55 (2H)		4.5			44
6-Hydroxy-2,2,5,7,8-pentamethyl	*s* 1.25 (6H, 2Me)	*t* 1.70 (2H)	*t* 1.40 (2H)		4.05			44
					Aromatic protons			
4,4-Diphenyl	*t* 4.02	*t* 2.76 (positions 3–4)		*m* 6.5–7.4(14H, C_6H_4, and 2Ph) (positions 5–8)				65
5,6,7,8-Tetrahydro	*t* 3.95 (2H)			*m* 1.90 (8H) (positions 4–8)	*m* 1.65 (4H) (positions 6–7)			29

[a] Chemical shift of protons (expressed in δppm); solvent $CDCl_3$.

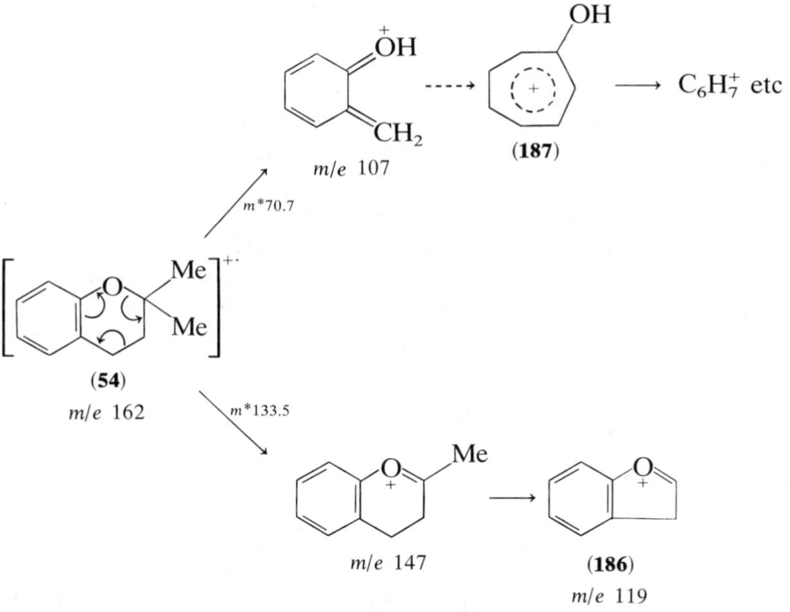

OH
$\overset{+}{O}H$
$C_6H_7^+$ etc
CH_2
(**187**)
m/e 107
*m**70.7
Me
Me
(**54**)
m/e 162
*m**133.5
Me
m/e 147
(**186**)
m/e 119

Scheme 2. Fragmentation of 2,2-dimethylchroman in the mass spectrometer.

spectrum of dihydrobenzofuran (path A) as well as by loss of C_2 (path B) that occurs both with and without hydrogen transfer to give fragments at *m/e* 106 and 107, as shown in Scheme 1. The latter fragment is presumed to be the same as the one of this mass observed in the spectrum of the methoxytoluenes and 2-hydroxy-5-methylacetophenone, and a tropylium structure (**187**) is preferred for it.[100]

The mass spectrum of 2,2-dimethylchroman (**54**) is interesting in that it is the aromatic analog of a dihydropyran and, like a dihydropyran, undergoes the path B fission but only with hydrogen transfer to give the *m/e* 107 fragment, as shown in Scheme 2.

IV. REACTIONS

Chroman is not a very reactive compound; and since the heterocyclic ring is saturated, all the known chemical reactions are associated with the benzene ring. These in the main are reduction and substitution reactions and may be compared with the results of similar reactions carried out on alkyl phenyl ether. Unlike chrom-3-en, chroman is stable to acids and oxidizing agents, but it gives a red color with sulfuric acid. It is soluble in common organic solvents.

1. Reduction

Treatment of chroman (**1**) and 2-methylchroman (**18**) with excess of sodium in liquid ammonia does not result in cleavage of the heterocyclic ring but in the formation of dihydrochromans. Although the exact structure of the dihydrochromans has not been determined, they are probably 5,8-dihydrochromans (**188**) and (**189**) formed by the Birch reduction, which ordinarily takes place in the presence of an "acid" such as methanol. This condition is fulfilled when methanol is added to decompose the excess sodium. If the fast-acting acid ammonium chloride is used to decompose the sodium, only about a fifth as much reduction of the aromatic ring takes place.[101]

1. Na, liq. NH_3
2. MeOH

(**1**) R = H
(**18**) R = Me

(**188**) R = H
(**189**) R = Me

The reduction of chroman (**1**) with lithium (6.5 equivalents) in dry ethylamine–dimethylamine (1:1 by volume) gives 85% 5,6,7,8-tetrahydrochroman (**190**). Some tetrahydrochromans are listed in Table 7. The tetrahydrochroman (**190**) with monoperphthalic acid in moist ether yields the *trans*-glycol (**191**), cleaved with lead tetraacetate to afford 6-ketononanolide (**192**), also obtained directly by treating the tetrahydrochroman (**190**) with excess 3-chloroperbenzoic acid.[29]

Li
$EtNH_2$—Me_2NH

(**1**) (**190**)

MCPBA, CH_2Cl_2

Moist Et_2O
Monoperphthalic acid

$Pb(OAc)_4$
C_6H_6

(**192**) (**191**)

A mechanism for the direct conversion suggests the intermediacy of an epoxy ether (**193**), converted by the peracid into a hydroxy perester (**194**) and hence to the ketolactone (**192**). Treatment of the glycol (**191**) with 3-chloroperbenzoic acid also gives 6-oxononanolide (**192**).[102]

(190) —MCPBA→ [(193)] —MCPBA→ (194) → (192) ←MCPBA— (191)

The ozonolysis of tetrahydrochroman, followed by treatment with zinc and acetic acid, yields the ketolactone (**192**) and *trans*-4a,8a-dihydroxyhexahydrochroman (**191**). When the ozonolysis is carried out in methanol, the principal product is 4a-hydroxy-8a-methoxyhexahydrochroman (**195**). These results suggest that the ozonolysis may occur via two pathways. The normal one may involve the formation of the ozonide (**196**), but more probably the dimeric ketoperoxide (**197**) and its isomers. The alternative one may involve the epoxy ether (**193**), because of the stabilization of the positive charge in the intermediate (**198**) by the ring oxygen. The epoxy ether (**193**) with water or methanol will be readily converted to *trans*-4a,8a-dihydroxy- or 4a-hydroxy-8a-methoxyhexahydrochroman (**191**) and (**195**), respectively.[103]

(190) —O_3→ ... (198) —$-O_2$→ (193); (193) —H_2O→ (191); (193) —MeOH→ (195); (197) → (192) ← (196)

2,2-Dimethyltetrahydrochroman (**199**) when subjected to the Koch–Haaf reaction, that is, olefin carboxylation with carbon monoxide generated in situ from sulfuric acid-formic acid, gives a mixture of two lactones, (**200**) and (**201**).[104] One of the lactones consumes alkali more rapidly than its isomer under identical conditions. Its greater ease of hydrolysis suggests that its structure is (**201**) (*cis*-decalin) and that the other lactone possesses structure (**200**) (*trans*-decalin) in which the carbonyl group is relatively more shielded to hydroxy ion attack.

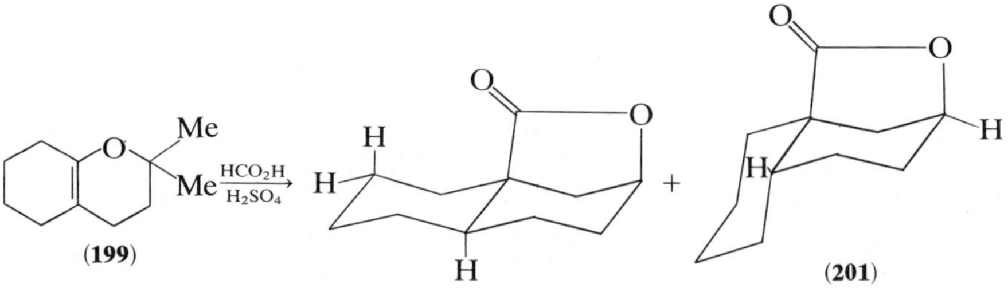

(**199**) $\xrightarrow[H_2SO_4]{HCO_2H}$ (**200**) + (**201**)

The hydrogenation of 6-methylchroman (**39**) and 7-methylchroman (**41**) in ethanol over nickel at 230° gives 72% 6- and 7-methylhexahydrochroman, (**202**) and (**203**), respectively.[105] The hydrogenation of tetrahydrochroman (**190**) over platinum affords hexahydrochroman (**204**).[106] Some hexahydrochromans are listed in Table 8.

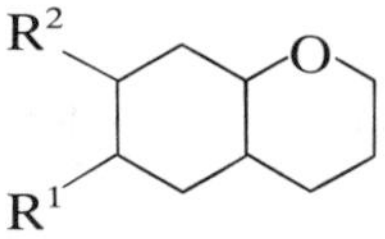

(**202**) $R^1 = Me$, $R^2 = H$
(**203**) $R^1 = H$, $R^2 = Me$
(**204**) $R^1 = R^2 = H$

2. Halogenation, Nitration, and Acylation

Chroman (**1**) reacts with *N*-bromosuccinimide in carbon tetrachloride in the presence of benzoyl peroxide to give a mixture of 4- and 6-bromochroman, (**33**) and (**205**).[2] 2,2-Dimethylchroman (**54**) has been reacted with *N*-bromosuccinimide under similar conditions. The product was not isolated, but treatment with sodium ethoxide in benzene gave the chrom-3-en (**71**), thus indicating the probable formation of 4-bromo-2,2-dimethylchroman (**206**).[38]

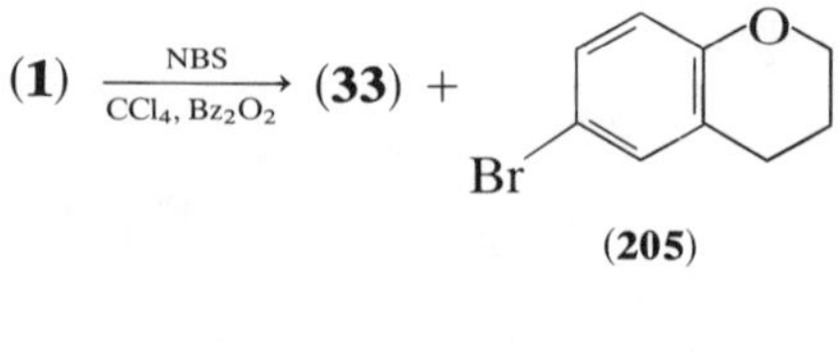

(**1**) $\xrightarrow[CCl_4,\ Bz_2O_2]{NBS}$ (**33**) + (**205**)

(54) $\xrightarrow[CCl_4, Bz_2O_2]{NBS,}$ (206) $\xrightarrow[C_6H_6]{EtONa}$ (71)

(206)

The bromination of chroman (**1**) in carbon tetrachloride by the addition of bromine to the solution gives 87% 6-bromochroman (**205**). With chlorine a mixture of 6-chlorochroman, a dichlorochroman, and a trichlorochroman is obtained. 6-Chlorochroman in 80% yield, free from other chlorinated products, is formed by boiling chroman (**1**) with sulfuryl chloride and perbenzoic acid in carbon tetrachloride.[6]

The nitration of chroman (**1**) with 60% nitric acid at 15–20° yields 45% 6-nitrochroman. If 99% nitric acid is used at 25–30°, 55% 6,8-dinitrochroman is obtained.[107] Nitro derivatives of chromans are in general readily prepared by dropwise addition of concentrated nitric acid to the appropriate chroman. 6-Methylchroman does not give a solid derivative under these conditions.[28]

6-Acylchromans are formed in about a 65% yield by treatment of a chroman with various acid chlorides in carbon disulfide at −10° in the presence of aluminum chloride.[4] The acetylation of 8-methylchroman (**48**) using acetyl chloride under the above conditions but at −5 to 0° results in the formation of 6-acetyl-8-methylchroman (**207**).[107]

(48) $\xrightarrow[AlCl_3, -5° \text{ to } 0°]{CH_3COCl}$ (207)

(207)

Heating 2,2-dimethylchroman (**54**) in a sealed tube in a water bath for several hours with acetic anhydride in the presence of zinc chloride probably gives the 8-acetyl derivative (**208**), m.p. 75°, and an acid (**209**), m.p. 181–182°, which suggests the introduction of a —CH_2—CO_2H group into the chroman.[54] The structure of the ketone (**208**) is supported by the condensation of 2- and 4-hydroxyacetophenone (**210**) with isoprene. In the former case only a trace of cyclic product is obtained, but in the latter 6-acetyl-2,2-dimethylchroman (**211**), m.p. 90°, results, which differs from the ketone described above. Further the Willgerodt reaction on 6-acetyl-2,2-dimethylchroman (**211**) gives an acid (**212**), m.p. 165°, different from that obtained in the zinc chloride–acetic anhydride reaction.[38]

(54) $\xrightarrow[ZnCl_2]{Ac_2O}$ (208) + (209)

(208) (209)

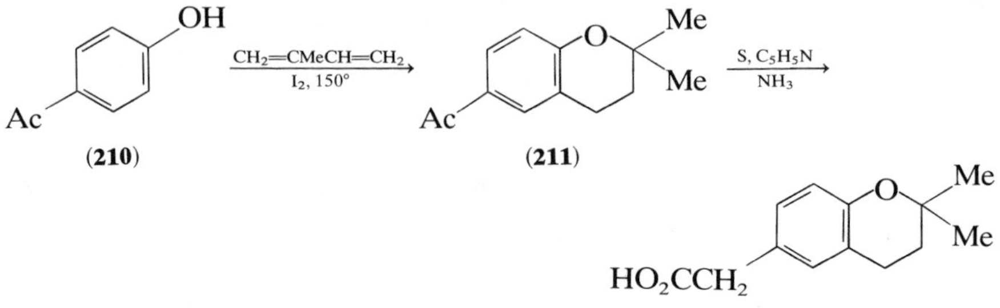

$CH_2{=}CMeCH{=}CH_2$, I_2, 150°; S, C_5H_5N, NH_3

(210) (211) (212)

3. Metallation

The metallation of 2,2-dimethylchroman (**54**), on boiling in ether with butyllithium under nitrogen, followed by oxidation at −10° in a current of oxygen, yields on hydrolysis with dilute hydrochloric acid 2,2-dimethyl-8-hydroxychroman (**213**).[37] If following metallation the mixture is poured onto solid carbon dioxide and the mixture allowed to come to room temperature, extraction with sodium hydroxide solution and subsequent acidification afford 2,2-dimethylchroman-8-carboxylic acid (**214**). The position which the lithium enters has been established by converting the phenol (**213**) into 2,2-dimethyl-8-methoxychroman and synthesizing the same compound from 8-methoxycoumarin.

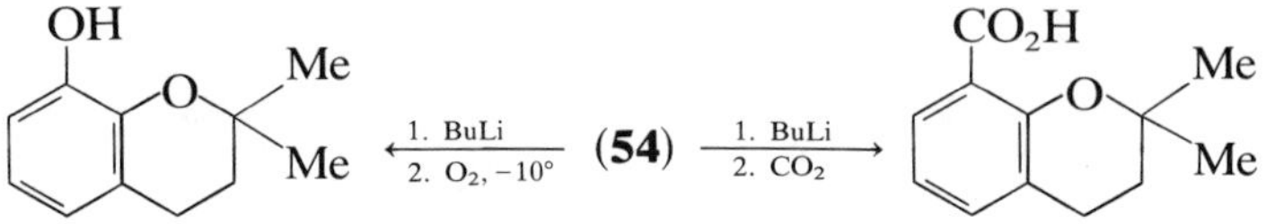

(213) ← 1. BuLi, 2. O_2, −10° — (**54**) — 1. BuLi, 2. CO_2 → (214)

4. Isomerization

Cyclization of 3-hydroxypropyl 2-*t*-butylphenyl ether (**215**) with phosphorus pentoxide following the procedure of Rindfusz[3] gives a mixture containing 8-*t*-butylchroman (**217**) only as a minor component, 6-*t*-butylchroman (**216**) as the major component, chroman (**1**), and *t*-butylbenzene (**218**).[108]

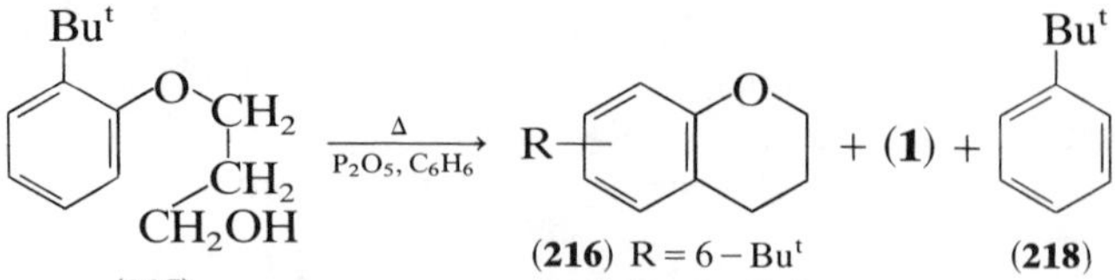

(215) —Δ, P_2O_5, C_6H_6→ (216) R = 6−Bu^t; (217) R = 8−Bu^t + (**1**) + (218)

Separation of the products by direct distillation without removal of the phosphoric acid gives the rearranged 6- to 8-isomer in a ratio of 11.5:1, whereas if the acid is removed before distillation the ratio is only 2.2:1.

Cyclization of the ether (**215**) in carbon disulfide still affords a mixture of 6- and 8-*t*-butylchroman and chroman. Thus the presence of benzene as a *t*-butyl acceptor is not necessary. The rearrangement of some of the 8-*t*-butylchroman (**217**) exemplifies a limitation on the chroman synthesis of Rindusz.[3]

5. Dehydrogenation

Chromans are dehydrogenated within 30 min when treated with triphenylmethyl perchlorate in either acetic or formic acid. Acetonitrile may also be used as the medium for the reaction. It is also found that a mixture of triphenylmethyl chloride and stannic chloride yields up to 80% of dehydrogenation products, but that chloranil requires more drastic conditions such as boiling xylene and then only affords poor yields of up to 25%. Chroman (**1**) on dehydrogenation, besides giving chromen (**219**), also yields 2,3-dihydro-2-methylbenzo[*b*]furan (**13**) and 2-methylbenzo[*b*]furan (**220**) through ring contraction. With 2,3-dihydro-2-ethylbenzo[*b*]furan (**22**) some ring expansion occurs on dehydrogenation to form 27% 2-methylchroman (**18**) besides the expected product, 2-ethylbenzofuran (**221**).[109]

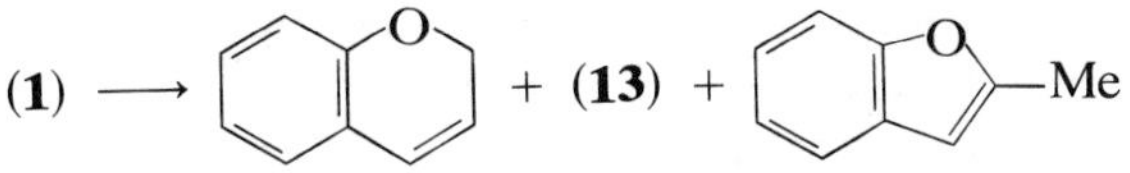

(**1**) ⟶ (**219**) + (**13**) + (**220**)

O, —Me

(**22**) ⟶ (**221**) + (**18**)

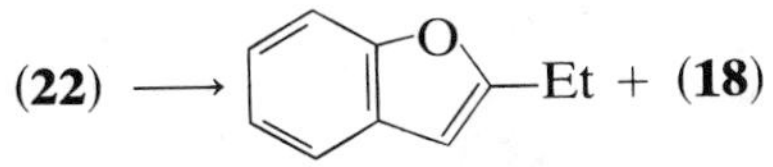

O, —Et

6. Miscellaneous

Reacting chroman with ethyl diazoacetate at 150° for several hours results in the ring expansion of the benzene ring and the formation of 16% of a dihydropyranocycloheptatrienecarboxylic acid as the main isolable crystalline product.[110] The acid rapidly decolorizes a solution of bromine in chloroform as well as potassium permanganate in acetone, and it gives a yellow color with concentrated sulfuric acid. Its uv absorption spectrum in ethanol shows peaks at 227 and 325 nm (log ε 4.07 and 4.03, respectively). The exact structure of the dihydropyranocycloheptatrienecarboxylic acid has not been determined.

The pyrolysis of chroman (**1**) at 250° on alumina gives a 10% yield of only 2,3-dihydro-2-methylbenzo[*b*]furan (**13**). At 350° the conversion is 20–25%, but besides 2,3-dihydro-2-methylbenzo[*b*]furan a number of 2-alkylphenols (**20**, R = 2-Me, 2-Et, or 2-Pr) are also formed.[111] Addition of

1.1% palladium to the alumina catalyst causes substantial reduction of the phenolic fraction, but this also increases with increase of temperature.[112]

Chroman (**1**) isomerizes at 300–400° in the presence of activated carbon to give 2,3-dihydro-2-methylbenzo[*b*]furan (**13**) and 2-methylbenzo[*b*]furan (**220**).[113]

V. USES OF ALKYL- AND ARYLCHROMANS

Chromans are condensed with aldehydes or ketones in the presence of strong mineral acids to yield thermoplastic resins, which can be molded per se or used as components of paints and varnishes. 2-Methylchroman is condensed with 36% formalin in the presence of 69% sulfuric acid at 120° to give a hard resin.[14]

A number of chroman derivatives, for example, 7,8-dimethoxy-2,5-dimethyl-6-hydroxy-2-(4-methyl-3-pentyl)chroman and its 6-propionate, and 7,8-dimethoxy-6-hydroxy-2,2,5-trimethylchroman and its 6-acetate and 6-butyrate, are useful antioxidants for fats and oils.[114] Polypropylene can be stabilized against heat and light degradation by the addition of a small amount of at least one of a group of polyalkyl-2-(2-hydroxyphenyl)- or polyalkyl-4-(2-hydroxyphenyl)-2,4,4-trimethylchromans.[115] Complex derivatives of chroman have been used in conjunction with a fatty amine in the anticorrosive treatment of sheet metal.[116]

A number of derivatives of 3,4-diphenylchromen and chroman have been prepared and their estrogenic and antifertility activity studied.[117] Of those tested for pregnancy-inhibiting activity in rats, a number showed activity, and most possessed weak estrogenic activity. The most active antifertility agents were the compounds (**222a**) and (**222b**), and the hydrochloride of the latter's 2,2-dimethyl analog (**222c**).[118]

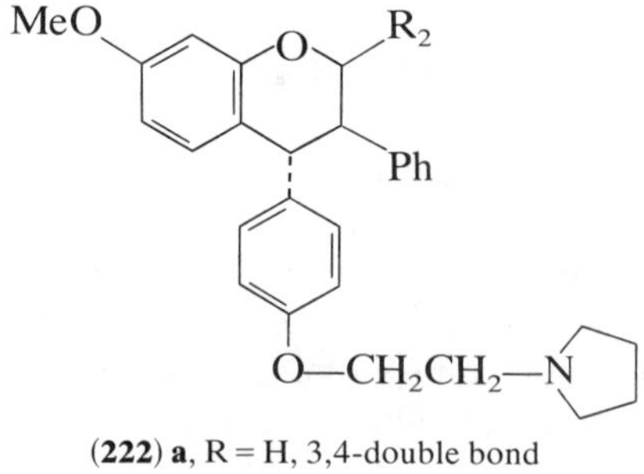

MeO, O, R_2, Ph, O—CH_2CH_2—N

(**222**) **a**, R = H, 3,4-double bond
b, R = H
c, R = Me

The tentative use of 3-methylchroman in angina pectoris has shown that 0.3–0.4 g per day improves angina in 49% of the cases.[119] A number of substituted chromans have been shown to be devoid of vitamin E activity.[120]

VI. TABLES OF COMPOUNDS

TABLE 5. CHROMAN AND ITS DERIVATIVES

Mol. formula	Substituents 2	3	4	5	6	7	8	mp or bp (mm)	Ref.
$C_9H_{10}O$								4.8	
								214 (742)	1–3
								96–98 (17)	4–7
	Picrate							78	6
$C_{10}H_{12}O$	Me							223–236 (760). 95–96 (13)	10
								225 (760)	14
								72 (3.2)	16
$C_{10}H_{12}O$		Me						102–104 (15)	8
$C_{10}H_{12}O$			Me					80–90 (5)	56
$C_{10}H_{12}O$				Me				71–72 (1)[a]	28
								56–58 (75)	29
$C_{10}H_{12}O$					Me			103–105 (16)	6, 7, 28
$C_{10}H_{12}O$						Me		67–68 (1)[a]	28
								125–126.2 (26)	29
$C_{10}H_{12}O$							Me	105–108 (15)	8
								64–67 (2)	28
								100 (10)	32
$C_{11}H_{12}O$	Vinyl							118–120 (18)	24
								113–116 (15)	
$C_{11}H_{14}O$	Me_2							225.2–225.4 (769)	35
								98.5 (11.5)	
								82–92 (2)	38
								224–226 (760)	39
								93 (10)	40
								109 (20)	13
								99 (12)	43
								67.5–68 (2)	46
								84–85 (3.5)	47

TABLE 5 (*Continued*)

Mol. formula	Substituents 2	3	4	5	6	7	8	mp or bp (mm)	Ref.
$C_{11}H_{14}O$	Me				Me			99–101 (6)	57
								113–115 (8)	60
								115–117 (9)	61
								115–125 (11)	17
$C_{11}H_{14}O$	Me					Me		118–120 (14)	57
								115–117 (13)	62
$C_{11}H_{14}O$ cis		Me	Me					106 (5)	56
$C_{11}H_{14}O$			Me_2					93 (10)	51
								44–45 (0.3)	52
$C_{11}H_{14}O$			Me		Me			105–115 (5)	56
$C_{11}H_{14}O$				Me		Me		85–86 (1)	29
$C_{11}H_{14}O$				Me			Me	114–116 (10)	29
$C_{11}H_{14}O$	Et							130–131 (26)	23
								116 (16)	24
								36–42 (0.1)	25
$C_{11}H_{14}O$			Et					113 (15)	6
$C_{11}H_{14}O$					Et			127 (17)	4
									33
$C_{12}H_{16}O$			Pr					128 (13)	6
$C_{12}H_{16}O$					Pr			136–138 (17)	4
									33
$C_{12}H_{16}O$			Me				Et	90–92 (1–2)	64
$C_{12}H_{16}O$	Me_2				Me			244–244.5 (760)	67
$C_{12}H_{16}O$	Me_2					Me		101–103 (7)	69
$C_{12}H_{16}O$	Me		Me_2					113.5 (16)	70
$C_{12}H_{16}O$	Me		Me		Me			104–106 (8)	71
								91–93 (3)	60, 72
								101–104 (7)	73
$C_{12}H_{16}O$	Me		Me			Me		125–126 (8)	60
								112–125 (9)	71

$C_{12}H_{16}O$	Me		Me				Me	76–80 (1)	72
$C_{12}H_{16}O$		Me_2			Me			82–85 (3)	30
$C_{12}H_{16}O$		Me_2				Me		87–90 (7)	30
$C_{12}H_{16}O$			Me_2		Me			112 (10)	51, 77
$C_{13}H_{18}O$			Bu					141 (15)	6
$C_{13}H_{18}O$	Et_2							128.5–128.9 (12)	36
$C_{13}H_{18}O$	Me_2	Me			Me			257–258 (760)	67
$C_{13}H_{18}O$	Me_2		Me_2					100–102 (8)	51
$C_{13}H_{18}O$	Me		Me_2			Me		94–96 (2)	85
$C_{13}H_{18}O$	Me			Me		Me	Me	46–48	17
$C_{13}H_{18}O$			Me_2			Me	Me	72–74 (0.4)	52
$C_{14}H_{20}O$	Me			Pr^i			Me	110–112 (3)	79
$C_{14}H_{20}O$	Me_2	Me		Me		Me		263–265 (760)	66
$C_{14}H_{20}O$	Me_2		Me_2		Me			108 (6)	51
$C_{14}H_{20}O$	Me_2		Me_2			Me		65–67 (0.3)	86
$C_{14}H_{20}O$	Me_2			Me		Me	Me	85 (0.1)	83
$C_{14}H_{20}O$			Me_2	Me		Me	Me	82–83 (0.3)	52
$C_{15}H_{14}O$			Ph					123 (0.6)	6
$C_{15}H_{22}O$	Pr_2							153–154 (15)	36
$C_{15}H_{22}O$	Me	Me		Pr^i			Me	102–110 (1)	79
$C_{16}H_{16}O$		CH_2Ph						52	26
$C_{17}H_{26}O$	Bu_2							165–168 (8)	48
$C_{17}H_{18}O$	Me_2		Ph					119–120	80
$C_{21}H_{18}O$			Ph_2					174–175.5	65

[a] Mixture of 5- and 7-isomers.

TABLE 6. DIHYDRONAPHTHOPYRANS

[1,2-*b*] [2,1-*b*]

Mol. formula	Isomer	Substituents 1	2	3	4	mp or bp (mm)	Ref.
$C_{13}H_{12}O$	[1,2-*b*]					134 (1)	28, 92
$C_{13}H_{12}O$	[2,1-*b*]					132 (1)	29, 91
						40	
						170 (20)	92
	picrate					127	92
$C_{14}H_{14}O$	[2,1-*b*]			Me		90–91	93
$C_{15}H_{16}O$	[2,1-*b*]			DiMe		78	47

TABLE 7. 5,6,7,8-TETRAHYDROCHROMANS

Mol. formula	Substituents 2	5	6	7	8	bp (mm)	Yield (%)	Ref.
$C_{9}H_{14}O$						68–72 (10)	84.5	29
$C_{10}H_{16}O$	Me					49–53 (6)	73[a]	29
$C_{10}H_{16}O$		Me				99–100 (24)	41	29
$C_{10}H_{16}O$			Me			79–80 (12)	81	29
$C_{10}H_{16}O$				Me		104–106 (25)	47	29
$C_{10}H_{16}O$					Me	78–82 (12)	80	29
$C_{11}H_{18}O$	DiMe					—	—	104
$C_{11}H_{18}O$		Me		Me		*b*	9	29
$C_{11}H_{18}O$		Me			Me	70–75 (0.2)	56	29

[a] Further purification difficult.
[b] Mostly starting material.

TABLE 8. HEXAHYDROCHROMANS

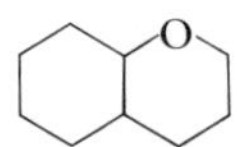

Mol. formula	Substituents 6	7	bp (mm)	Ref.
$C_{9}H_{16}O$			80 (16)	106
$C_{10}H_{18}O$	Me		100–101 (25)	105
$C_{10}H_{18}O$		Me	95–97 (25)	105

VII. REFERENCES

1. J. von Braun and A. Steindorff, *Ber.*, **38,** 850 (1905).
2. H. Normant and P. Maitte, *C.R. Acad. Sci., Paris,* **234,** 1787 (1952).
3. R. E. Rindfusz, *J. Am. Chem. Soc.*, **41,** 665 (1919).
4. G. Chatelus, *C.R. Acad. Sci., Paris,* **224,** 201 (1946).
5. P. L. De Benneville and R. Connor, *J. Am. Chem. Soc.*, **62,** 283 (1940).
6. P. Maitte, *Ann. Chim.* (Paris), **9,** 431 (1954).
7. L. W. Deady, R. D. Topsom, and J. Vaughan, *J. Chem. Soc.*, 2094 (1963).
8. C. H. Hurd and W. A. Hoffman, *J. Org. Chem.*, **5,** 212 (1940).
9. H. Normant, *C.R. Acad. Sci., Paris,* **239,** 1510 (1954).
10. W. Baker and J. Walker, *J. Chem. Soc.*, 646 (1935).
11. B. A. Arbuzov and L. A. Shapshinskaya, *Dokl. Akad. Nauk SSSR,* **110,** 991 (1956).
12. A. R. Bader, *J. Am. Chem. Soc.*, **79,** 6164 (1957).
13. K. C. Dewhirst and F. F. Rust, *J. Org. Chem.*, **28,** 798 (1963).
14. H. S. Bloch and H. E. Mammen, U.S. Patent 2,657,193 (1953).
15. V. I. Isagulyants and V. P. Evstaf'ev, *Zh. Obshch. Khim.*, **33,** 1042 (1963).
16. E. E. Schweizer, C. J. Berninger, D. M. Crouse, R. A. Davis, and R. S. Logothetis, *J. Org. Chem.*, **34,** 207 (1969).
17. O. Dann, G. Volz, and O. Huber, *Justus Liebigs Ann. Chem.*, **587,** 16 (1954).
18. J. Lichtenberger and R. Kircher, *C.R. Acad. Sci., Paris,* **229,** 1345 (1949).
19. A. A. Balandin, G. M. Marukyan, and R. G. Seimovich, *Dokl. Akad. Nauk SSSR,* **141,** 616 (1961).
20. E. A. Karakhanov, E. A. Dem'yanova, and E. A. Viktorova, *Dokl. Akad. Nauk SSSR,* **204,** 879 (1972).
21. E. A. Karakhanov, M. V. Vagabov, S. K. Dzhamalov, and E. A. Viktorova, *Khim. Geterotsikl. Soedin.*, 321 (1975); *Chem. Heterocycl. Comp.*, **11,** 278 (1975).
22. E. A. Karakhanov, S. V. Lysenko, and E. A. Viktorova, *Vestn. Mosk. Univ. Khim.*, **15,** 500 (1974).
23. R. Mozingo and H. Adkins, *J. Am. Chem. Soc.*, **60,** 669 (1938).
24. J. Brugidou and H. Christol, *C.R. Acad. Sci., Paris,* **256,** 3149 (1963); *Bull. Soc. Chim. Fr.*, 1974 (1966).
25. E. E. Schweizer, T. Minami, and S. E. Anderson, *J. Org. Chem.*, **39,** 3038 (1974).
26. P. Pfeiffer, K. Grimm, and H. Schmidt, *Justus Liebigs Ann. Chem.*, **564,** 208 (1949).
27. F. Baranton, G. Fontaine, and P. Maitte, *Bull. Soc. Chim. Fr.*, 4203 (1968).
28. L. W. Deady, R. D. Topsom, and J. Vaughan, *J. Chem. Soc.*, 5718 (1965).
29. I. J. Borowitz, G. Gonis, R. Kelsy, R. Rapp, and G. J. Williams, *J. Org. Chem.*, **31,** 3032 (1966).
30. E. A. Viktorova, N. I. Shuikin, and S. E. Popova, *Izv. Akad. Nauk SSSR, Ser. Khim.*, 1277 (1963).
31. E. A. Viktorova, N. I. Shuikin, and E. A. Karakhanov, *Izv. Akad. Nauk SSSR, Ser. Khim.*, 888 (1967).
32. E. R. Clark and S. G. Williams, *J. Chem. Soc.* (*B*), 859 (1967).
33. G. Chatelus, *Ann. Chim.*, **4,** 505 (1949).
34. M. Sliwa, H. Sliwa, and P. Maitte, *C.R. Acad. Sci. Paris, Ser. C,* **268,** 263 (1969); *Bull. Soc. Chim. Fr.*, 1540 (1972).
35. L. Claisen, *Ber.*, **54B,** 200 (1921).
36. L. I. Smith, H. E. Ungnade, and W. W. Prichard, *J. Org. Chem.*, **4,** 358 (1939).
37. M. Hallet and R. Huls, *Bull. Soc. Chim. Belg.*, **61,** 33 (1952).
38. G. R. Clemo and N. D. Ghatge, *J. Chem. Soc.*, 4347 (1955).
39. A. R. Bader and W. C. Bean, *J. Am. Chem. Soc.*, **80,** 3073 (1958).
40. Farbenfabriken Bayer A.-G., Brit. Patent 906,483, (1962).
41. V. I. Isagulp'ants and V. P. Evstaf'ev, *Zh. Org. Khim.*, **1,** 102 (1965).
42. J. Tanaka, T. Katagiri, and S. Yamada, *Yuki Gosei Kagaku Kyokai Shi,* **27,** 841 (1969).

43. K. Suga, S. Watanabe, H. Kikuchi, and K. Hijikata, *J. Appl. Chem.*, **20,** 175 (1970).
44. J. A. Miller and H. C. S. Wood, *J. Chem. Soc.* (*C*), 1837 (1968).
45. W. H. Horspool and P. L. Pauson, *J. Chem. Soc.* (*D*), *Chem. Commun.*, 195 (1967).
46. R. L. Shriner and A. G. Sharp, *J. Org. Chem.*, **4,** 575 (1939).
47. R. Livingstone, D. Miller, and S. Morris, *J. Chem. Soc.*, 3094 (1960). R. Livingstone, *J. Chem. Soc.*, 76 (1962).
48. L. I. Smith and P. M. Ruoff, *J. Am. Chem. Soc.*, **62,** 145 (1940).
49. S. Yamada, T. Katagiri, and J. Tanaka, *Yuki Gosei Kagaku Kyokai Shi.*, **29,** 81 (1971).
50. Y. Shigemasa, H. Kuwamoto, C. Sakasawa, and T. Matsuura, *Nippon Kagaku Kaishi*, 2423 (1973).
51. J. Colonge, E. Le Sech, and R. Marey, *Bull. Soc. Chim. Fr.*, 776 (1957).
52. R. T. Borchardt and L. A. Cohen, *J. Am. Chem. Soc.*, **94,** 9166 (1972).
53. E. A. Vdovtsova, L. P. Yanichkin, and I. P. Tsukervanik, *Zh. Org. Khim.*, **2,** 1279 (1966).
54. L. Claisen, F. Kremer, F. Roth, and E. Tietze, *Justus Liebigs Ann. Chem.*, **442,** 210 (1925).
55. Gy. Frater and H. Schmid, *Helv. Chim. Acta*, **50,** 255 (1967).
56. B. D. Tilak and Z. Muljiani, *Tetrahedron*, **24,** 949 (1968).
57. N. I. Shuikin, E. A. Viktorova, I. E. Pokrovskaya, and T. G. Malysheva, *Izv. Adak. Nauk SSSR, Otdel. Khim. Nauk*, 1847 (1961).
58. N. I. Shuikin, E. A. Viktorova, and I. E. Pokrovskaya, *Izv. Akad. Nauk SSSR, Otdel. Khim. Nauk*, 2192 (1961).
59. N. I. Shuikin, E. A. Viktorova, I. E. Pokrovskaya, and E. A. Karakhanov, *Izv. Akad. Nauk SSSR, Otdel. Khim. Nauk*, 122 (1962).
60. E. A. Viktorova, N. I. Shuikin, and E. A. Karakhanov, *Izv. Akad. Nauk SSSR, Ser. Khim.*, 1281 (1963).
61. E. A. Viktorova, N. I. Shuikin, and E. A. Karakhanov, *Izv. Akad. Nauk SSSR, Ser. Khim.*, 915 (1966).
62. E. A. Viktorova, N. I. Shuikin, and E. A. Karakhanov, *Izv. Akad. Nauk SSSR, Otdel. Khim. Nauk*, 1080 (1962).
63. E. A. Viktorova and N. N. Tsitsugina, *Vestn. Mosk. Univ. Khim.*, **11,** 450 (1970).
64. R. M. Lagidze and B. S. Potskhverashvili, *Soobsnch. Akad. Nauk Gruz. SSR*, **19,** 685 (1957).
65. W. H. Starnes, Jr., *J. Org. Chem.*, **33,** 2767 (1968).
66. E. Merck, Brit. Patent 1,103,094 (1968).
67. L. Claisen, Ger. Patent 374,124 (1924). *Chem. Abstr.*, **18,** P2175 (1924).
68. L. P. Kleeva, L. I. Bunina–Krivorukova, Kh. V. Bal'yan, and E. K. Aleksandrova, *Zh. Org. Khim.*, **5,** 1041 (1969).
69. V. I. Isagulyants and V. P. Evstaf'ev, *Dokl. Akad. Nauk Arm. SSR*, **38,** 235 (1964).
70. W. Webster and D. P. Young, *J. Chem. Soc.*, 4785 (1956).
71. N. I. Shuikin, E. A. Viktorova, I. E. Pokrovskaya, and T. G. Malysheva, *Izv. Akad. Nauk SSSR, Otdel. Khim. Nauk*, 1847 (1961).
72. E. A. Viktorova, N. I. Shuikin, and B. G. Bubnova, *Izv. Akad. Nauk SSSR, Otdel. Khim. Nauk*, 1657 (1961).
73. E. A. Viktorova, E. A. Karakhanov, A. N. Shuikin, and N. I. Shuikin, *Izv. Akad. Nauk SSSR, Ser. Khim.*, 523 (1966).
74. L. I. Bunina–Krivorukova, L. P. Kleeva, and Kh. V. Bal'yan, *Probl. Poluch. Poluprod. Prom. Org. Sin., Akad. Nauk SSSR, Otd. Obshch. Tekh. Khim.*, 126 (1967).
75. E. A. Viktorova, N. I. Shuikin, and G. V. Popova, *Vestnik Moskov. Univ., Ser. II, Khim.*, **15,** 62 (1960).
76. L. P. Kleeva, L. I. Bunina–Krivorukova, and Kh. V. Bal'yan, *Zh. Org. Khim.*, **5,** 1048 (1969).
77. L. P. Kleeva, L. I. Bunina–Krivorukova, and Kh. V. Bal'yan, *Zh. Org. Khim.*, **3,** 2018 (1967).
78. M. E. N. Namibudiry and G. S. K. Rao, *Chem. Ind.* (London), 518 (1975).
79. S. Mitsui, M. Sudzuki, and H. Yoshinaga, *J. Chem. Soc. Japan.* **64,** 1337 (1943).

80. H. E. Zaugg, J. E. Leonard, R. W. DeNet, and D. L. Arendsen, *J. Heterocycl. Chem.*, **11,** 797 (1974).
81. L. I. Bunina–Krivorukova, A. P. Rossinskii, and Kh. V. Bal'yan, *Zh. Org. Khim.*, **10,** 2461 (1974).
82. A. P. Rossinskii, L. I. Bunina–Krivorukova, and Kh. V. Bal'yan, *Zh. Org. Khim.*, **11,** 2364 (1975).
83. L. I. Smith, H. E. Ungnade, H. H. Hoehn, and S. Wawzonek, *J. Org. Chem.*, **4,** 311 (1939).
84. N. E. Kologrivova, T. B. Gerasimovich, Zh. A. Peregudova, and L. A. Kheifits, *Tr. Vses. Nauchn.-Issled. Inst. Sintetich. i Natural'n. Dushistvkh Veshchestv*, 3 (1961); *Chem. Abstr.*, **57,** 5877 (1962).
85. Yu. S. Dol'skaya, G. E. Svadkovskaya, and L. A. Kheifits, *Tr. Vses. Nauchn.-Issled. Inst. Sintetich. i Natural'n. Dushistykh Veshchestv*, 3 (1961); *Chem. Abstr.*, **61,** 8265b (1964).
86. W. Baker, R. F. Curtis, and J. F. W. McOmie, *J. Chem. Soc.*, 76 (1951).
87. I. Iwai and J. Ide, *Chem. Pharm. Bull.* (Tokyo), **10,** 926 (1962).
88. J. Cottam and R. Livingstone, *J. Chem. Soc.*, 5228 (1964).
89. R. Livingstone, D. Miller, and S. Morris, *J. Chem. Soc.*, 5148 (1960).
90. W. D. Cotterill, R. Livingstone, and M. V. Walshaw, *J. Chem. Soc.* (*C*), 1758 (1970).
91. G. O. Guss, *J. Am. Chem. Soc.*, **73,** 608 (1951).
92. P. Cagniant and C. Charaux, *Bull. Soc. Chim. Fr.*, 3249 (1966).
93. F. J. McQuillin and Sir R. Robinson, *J. Chem. Soc.*, 586 (1941).
94. J. Cottam and R. Livingstone, *J. Chem. Soc.*, 6646 (1965).
95. T. Boehm and E. Profft, *Arch. Pharm.*, **269,** 25 (1931).
96. E. M. Philbin and T. S. Wheeler, *Proc. Chem. Soc.*, 167 (1958).
97. E. R. Clark and S. G. Williams, *J. Chem. Soc.* (*B*), 859 (1967).
98. R. Livingstone, D. Miller, and S. Morris, *J. Chem. Soc.*, 602 (1960).
99. H. Hart and C. R. Wagner, *Proc. Chem. Soc.*, 284 (1958).
100. B. Willhalm, A. F. Thomas, and F. Gautschi, *Tetrahedron*, **20,** 1185 (1964).
101. C. D. Hurd and G. L. Oliver, *J. Am. Chem. Soc.*, **81,** 2795 (1959).
102. I. J. Borowitz, G. J. Williams, L. Gross, and R. Rapp, *J. Org. Chem.*, **33,** 2013 (1968).
103. I. J. Borowitz and R. D. Rapp, *J. Org. Chem.*, **34,** 1370 (1969).
104. M. E. N. Nambudiry and G. S. K. Rao, *Tetrahedron Lett.*, 4707 (1972).
105. P. L. De Benneville and R. Connor, *J. Am. Chem. Soc.*, **62,** 3067 (1940).
106. J. Colonge, J. Dreux, and M. Thiers, *Bull. Soc. Chim. Fr.*, 1459 (1959).
107. G. Brancaccio, G. Lettieri, and R. Viterbo, *J. Heterocycl. Chem.*, **10,** 623 (1973).
108. I. J. Borowitz and G. J. Williams, *J. Org. Chem.*, **31,** 603 (1966).
109. E. A. Karakhanov, E. A. Dem'yanova, and E. A. Viktorova, *Dokl. Akad. Nauk SSR*, **204,** 879 (1972).
110. A. W. Johnson, A. Langemann, and J. Murray, *J. Chem. Soc.*, 2136 (1953).
111. E. A. Karakhanov, N. N. Khvorostukhina, and E. A. Viktorova, *Vestn. Mosk. Univ., Khim.*, **11,** 635 (1970).
112. E. A. Karakhanov, N. N. Khvorostukhina, and E. A. Viktorova, *Vestn. Mosk. Univ., Khim.*, **12,** 502 (1971).
113. E. A. Karakhanov, M. V. Vagabov, S. K. Dzhamalov, and E. A. Viktorova, *Khim. Geterotsikl. Soedin.*, 321 (1975).
114. K. Folkes and D. E. Wolf, U.S. Patent 3,026,330 (1962).
115. D. M. Dickson, Jr., U.S. Patent 3,006,885 (1959).
116. M. Gaudoz and R. Gaudoz, Fr. Patent 1,260,649 (1962).
117. R. W. J. Carney, W. L. Bencze, J. Wojtkunski, A. A. Renzi, L. Dorfman, and G. De Stevens, *J. Med. Chem.*, **9,** 516 (1966).
118. S. Ray, P. K. Grover, V. P. Kamboj, B. S. Setty, A. B. Kar, and N. Anand, *J. Med. Chem.*, **19,** 276 (1976).
119. P. Soulie, P. Chiche, J. Carlotti, and J. Baillet, *Presse Méd.*, **62,** 847 (1954).
120. P. D. Boyer, M. Rabinovitz, and E. Liebe, *J. Biol. Chem.*, **192,** 95 (1951).

CHAPTER III

Chromanols and Tocopherols

R. M. PARKHURST and W. A. SKINNER

SRI International, Menlo Park, California

ACKNOWLEDGMENT

The authors gratefully acknowledge the assistance of librarians Elizabeth Gill and Paul Hanson, secretaries Shirley Leisses, Karen Baker, and Sandra Champion, and William W. Lee for reading the manuscript and making valuable comments.

I. INTRODUCTION

The nomenclature of the chromans, including the changes adopted by *Chemical Abstracts* in 1972, have been discussed in detail in the introduction to the preceding volume in this series, *Chromenes, Chromanones, and Chromones.*[1] The practice of the preceding volumes of using unambiguous and convenient names will be continued, that is, 3,4-dihydro-2*H*-1-benzopyran-6-ol will be referred to as chroman-6-ol.

The nomenclature of the tocopherols is perhaps more complicated, partly due to their extended literature that covers many fields of interest outside chemistry and partly due to their inherent stereochemical complexity. Recommendations on the nomenclature of the tocopherols and related compounds have been made by the IUPAC-IUB Commission on Biochemical Nomenclature.[2] The Commission, taking into account the three optically active centers of α-tocopherol, summarized their recommendations in the form of a table, Table 1. Some of the problems relating to stereochemistry have been discussed by Hoffman-Ostenhoff.[3] The IUPAC-IUB Commission has also made recommendations regarding the quinones with isoprenoid side chains (oxidation products of the tocopherols).[4]

A symposium on "Vitamin E and Metabolism" in honor of Professor H. M. Evans, a collection of invited papers presented in 1962, covers nearly 300 pages in *Vitamins and Hormones*[5] and still remains a classic review of the subject. Outstanding reviews have also appeared more recently in 1971[6]

TABLE 1. LIST OF THE TRIVIAL NAMES FOR SOME TOCOPHEROLS

Description of product	Configuration	Trivial names: Recommended	Trivial names: Others[b]
1. The compound having the configuration shown in the next column, exemplified by the only isomer of α-tocopherol as yet found in nature	2*R*,4′*R*,8′*R*	*RRR*-α-tocopherol	[*d*]-α-tocopherol
2. The isomer epimeric only at C-2 with *RRR*-α-tocopherol	2*S*,4′*R*,8′*R*	2-epi-α-tocopherol	
3. Semisynthetic α-tocopherol, such as can be produced from natural phytol	2*R*,4′*R*,8′*R* and 2*S*,4′*R*,8′*R* (mixture not necessarily in equal proportions)	2-ambo-α-tocopherol	
4. Totally synthetic α-tocopherol, such as can be produced from synthetic phytol or isophytol	a mixture (not necessarily equimolar) of all four possible racemates (*i.e.* of all the four pairs of enantiomers)	all-rac-α-tocopherol	[dl]-α-tocopherol
5. Semisynthetic α-tocopherol, obtained by hydrogenation of (2*R*)-5,7,8-trimethyltocotrienol	a mixture (not necessarily equimolar) of the four isomers 2*R*,4′*R*,8′*R*; 2*R*,4′*S*,8′*R*; 2*R*,4′*R*,8′*S*; 2*R*,4′*S*,8′*S*	4′-ambo-8′-ambo-α-tocopherol	

[a] Configuration of the natural tocopherols: tocol: $R^1 = R^2 = R^3 = H$; δ-tocopherol: $R^1 = R^2 = H, R^3 = Me$; γ-tocopherol: $R^1 = H, R^2 = R^3 = Me$; β-tocopherol: $R^1 = R^3 = Me, R^2 = H$; α-tocopherol: $R^1 = R^2 = R^3 = Me$.

[b] Not to be used in indexing without cross reference to the recommended names.

and in 1972[7] that cover a wide variety of vitamin E-related materials. The chemistry of vitamin E has been reviewed by Hjarde, Leerbeck, and Leth,[8] Isler,[9] Ames,[10] Herting,[11] Smith,[12] Rosenberg,[13] Mattill,[14] Campbell,[15] Wagner and Folkers,[16,17] Dean,[18] and Dyke.[19] Other reviews are mentioned under more specific headings in this chapter.

II. PHYSICAL AND SPECTRAL PROPERTIES

The preponderance of physical and spectral properties available for the chromanols refer specifically to the tocopherols. Comparatively little work has been published on 5-, 7-, and 8-hydroxychromans and on polyhydroxychromans. Moreover much of the available nontocopherol data were collected only because of the analogy to the more biologically important 6-hydroxychroman relative.

It can safely be concluded that research into the chemistry of vitamin E could not have progressed without the aid of model compounds. The lower molecular weight chromanols are generally colorless, crystalline materials easily soluble in a variety of organic solvents, including alcohol. Crystallization from alcohol–water and light petroleum generally aids easy purification. Most of the lower molecular weight members melt below 100°C and increasing the alkyl side chains lowers the melting point, especially for branched-chain compounds. The tocopherols are viscous oils that generally resist easy crystallization, except with their highly polar derivatives such as allophanates, and so on.

The electrical as well as dark conductivity of several tocopherols[20] has been found to be analogous to that of an organic semiconductor (10^{-14} to $10^{-11}\ \Omega^{-1}\ cm^{-1}$).

1. X-Ray Crystallography

There has been no report of an X-ray single-crystal structural determination of a chromanol that comes within the scope of this review. Dianin's compound (**1**) has been examined[21] and also the polycyclic compound (**2**).[22] Dianin's compound is a reasonable model for the chromanol ring structure.

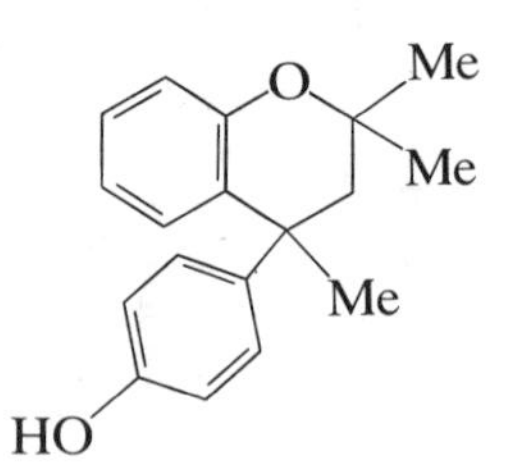

(**1**)

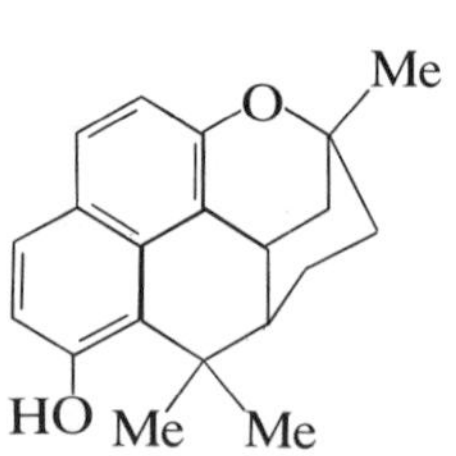

(**2**)

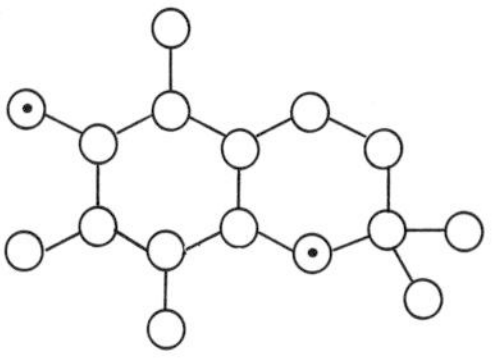

Fig. 1

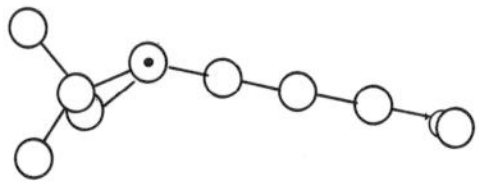

Fig. 2

The X-ray data for Dianin's compound compare reasonably well with those of a strain-free model produced with the aid of the PROPHET computer system.[23] Although the length of the C–O bond (1.5 Å) used by the Wipke program[24] seems slightly too long, no attempt was made to adjust this error. The PROPHET System was then allowed to generate[25] both α-tocopherol and its model, 2,2,5,7,8-pentamethylchroman-6-ol (Fig. 1), as viewed from above the plane of the aromatic ring and (Fig. 2) as viewed from C-8 to C-5 along the plane of the aromatic ring. The pucker of the heterocyclic ring is shown and the mass of the phytyl side chain at C-2 is calculated to have an unnoticeable depuckering effect. Figure 3 shows the bond distance and bond angle for α-tocopherol and its model, 2,2,5,7,8-pentamethylchroman-6-ol, as calculated for a strain-free system by PROPHET.

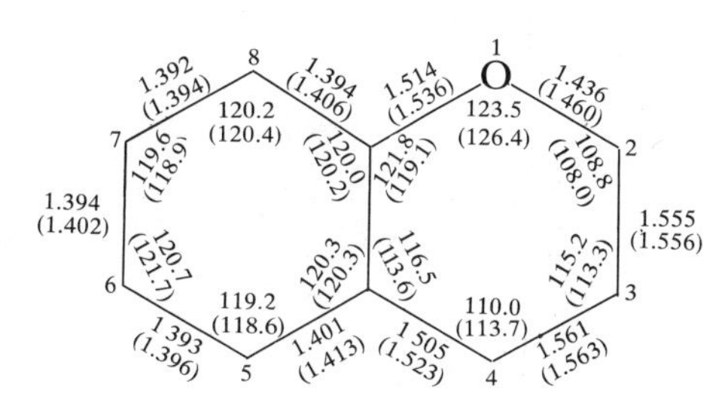

Fig. 3. Bond distance and bond angle for α-tocopherol and (in parentheses) 2,2,5,7,8-pentamethylchroman-6-ol.

The *p*-phenylazobenzoates of (2*R*,4′*R*,8′*R*)- and (2*S*,4′*R*,8′*R*)-α-tocopherol have been shown to have different X-ray powder patterns. This was an aid in relating the absolute configuration of natural α-tocopherol to that of the already determined configuration of (7*R*,11*R*)-phytol by a procedure described in detail by Schudel, Mayer, Rüegg, and Isler,[7,26]

2. Ultraviolet Spectra

Since the heterocyclic ring in the chromanols adds no further π conjugation to the system, interpretation of uv spectra can be reasonably aided by information on the alkoxyphenols. The substitution of the alkoxy group into the phenolic system produces an additional bathochromic shift of the

TABLE 2. ULTRAVIOLET ABSORPTION OF CHROMANS

Compound (Ref.)		λ_{max} (ε_{max})		Solvent
Ortho				
8-Methoxychroman[30]	210 (15,396)	277 (1848)		EtOH
2-Methoxyphenol[31]	218 (6,050)	276 (2560)		MeOH
1,2-Dimethoxybenzene[31]	225 (11,700)	275 (3920)		MeOH
1,2-Dihydroxy-4,5-dimethylbenzene[2,32]		287 (3162)		EtOH
1,2-Dihydroxybenzene[31]		278 (—)		MeOH
1,2-Dihydroxybenzene[27]	278 (2,500)	283 (2000)		Cyclohexane
Meta				
2,2-dimethyl-5-chromanol[32]	219 (15,800)	274 (2500)	281 (2820)	EtOH
2,2,7-Trimethyl-5-chromanol[33]	213 (22,000)		283 (2400)	EtOH
2,2,5-Trimethyl-7-chromanol[33]	214 (—)		283 (—)	EtOH
3-Methoxyphenol[31]		275 (—)	280 (—)	MeOH
3-Ethoxyphenol[31]		275 (—)	282 (—)	Dioxane
1,3-Dihydroxy-2,5-dimethylbenzene[31]		272 (759)	281 (692)	MeOH
1,5-Dihydroxy-2,4-dimethylbenzene[32]			286 (3388)	EtOH
1,5-Dihydroxy-3,4-dimethylbenzene[32]			282 (2239)	EtOH

Para			
5,7,8-Trimethyltocol[31]		292 (3100)	MeOH
5,8-Dimethyltocol[31]		296 (3715)	MeOH
8-Methyltocol[31]		298 (3467)	MeOH
7,8-Dimethyltocol[34]		298 (3802)	MeOH
2,2,5,7,8-Pentamethyl-6-chromanol[34]	291 (3,640)	297 (3800)	Hexane
2,2,7,8-Tetramethyl-6-chromanol[34]	294 (4,900)	300 (4800)	Hexane
2,2,5,8-Tetramethyl-6-chromanol[34]	293 (3,900)	297 (3900)	Hexane
2,2,5,7-Tetramethyl-6-chromanol[34]	288 (3,730)	294 (3640)	Hexane
2,2,7-Trimethyl-6-chromanol[34]	294 (4,160)	300 (4070)	Hexane
2,2,5-Trimethyl-6-chromanol[34]	291 (3,800)	297.5 (3730)	Hexane
2,2,8-Trimethyl-6-chromanol[34]	292 (4,160)	298 (3980)	Hexane
2,2-Dimethyl-6-chromanol[34]	295 (2,050)	306.5 (1520)	Hexane
4-Methoxyphenol[31]	225 (—)	292 (—)	MeOH
1,4-Diethoxybenzene[31]	226 (16,200)	290 (2840)	MeOH
Hydroquinone[31]	225 (5,180)	294 (2810)	MeOH
1,4-Dimethoxybenzene[31]	225 (9,670)	289 (3080)	MeOH
4-Ethoxyphenol[31]	228 (—)	294 (—)	MeOH

absorption bands of the phenol, but the further addition of alkyl groups after that has only a minor effect. The nonbonding electrons of the alkoxy or heterocyclic oxygen are available for interaction with the π electrons of the aromatic nucleus, as are those of the hydroxyl group.[27] A bathochromic shift can usually be observed as the oxygen substitution changes from ortho to meta to para (Table 2).

By far the most useful information can be obtained by conversion of the phenol to the phenolate ion in strongly basic solution (pH 13). The additional pair of nonbonding electrons of the phenolate ion that are now free to interact with the π electrons results in a relatively large bathochromic shift of λ_{max} and increase in ε_{max}.[28] If the spectrum is taken at several pH values, the presence of two species, phenol and phenolate, may also be confirmed by the appearance of an isosbestic point.[29] The pK_a of the phenol may also be characteristic for the compound involved.

3. Infrared Spectra

The most conspicuous bands in the infrared observed for the chromanols result from O–H stretching vibrations. The stretching of C–O and bending of O–H can also be observed.[28,35,36] When observed at high dilution (0.005 M) in nonpolar solvents (CCl_4) where the internal hydrogen bonding is not possible, the hydroxyl O–H stretching is found in the range of 3700–3500 cm^{-1} (2.72–2.86 μ). The convenient practice of recording ir spectra as liquid films and KBr pellets has generally precluded observation of the nonbonded O–H band for the chromanols, although one might expect this band to be lacking in chroman diols with adjacent hydroxyls and chroman-8-ols where the proton of the hydroxyl may be somewhat bound to the hetero oxygen. The frequency of vibrations is related to the distance between the proton of one hydroxyl and the oxygen of the adjacent hydroxyl, as discussed in detail by Nakanishi.[37]

The O–H band for hydrogen-bonded hydroxyls (either intermolecular or intramolecular) shifts to lower frequencies (3360–3350 cm^{-1}, 2.90–2.99 μ). These are usually the O–H bands reported for the chromanols in KBr pellets or films (Table 3). This band is also usually broader than the nonbonded one. Rosenkrantz[38] has examined a number of tocopherols and has found bands near 3.0, 6.3, and 8.0 μ in addition to unassigned bands near 8.6 and 10.9 μ that are characteristic of the tocopherol structure. He assigns the band near 6.3 μ, 1587 cm^{-1} to absorption of a conjugated C=C system and the band near 8.0 μ, 1250 cm^{-1} to the phenolic C–O absorption. The 8.0 μ band was shifted by esterification and methylation. Green et al.[39] showed that this band was also shifted by substitution on the aromatic ring depending on both the number and locations of the methyl groups in the tocopherol series. They also examined the bands in the 1110–1250 cm^{-1} (9.01–8.0 μ) region for a number of tocopherols and related compounds.

TABLE 3. THE INFRARED SPECTRA OF 2,2-DIMETHYLCHROMANOLS AND RELATED ALKOXYPHENOLS

Compound (Ref.)	Absorbance (cm^{-1})
2,2-Dimethyl-5-hydroxychroman[42]	3300–3150 (OH), 1610, 1590, 1505 (Ar)
5-Hydroxy-2,2,7-trimethylchroman[33b]	3333 (OH), 1379, 1368 (Me gem)
m-Methoxyphenol[43b]	3350 (OH), 2915, 2810, 1590, 1490, 1460, 1440, 1323, 1300, 1290
6-Hydroxy-2,2,5-trimethylchroman[34,45]	3415 (OH), 3450, 2980, 1485, 1350, 1260, 1200, 805
6-Hydroxy-2,2,7-trimethylchroman[34a,45]	3370 (OH), 3380, 2980, 2925, 1620 (Ar), 1515, 1420, 1190, 885, 875
6-hydroxy-2,2,8-trimethylchroman[34a,45]	3420 (OH), 3270, 2960, 1600, 1460, 1365, 1220, 1200, 1155, 1120, 920
6-Hydroxy-2,2,5,7-tetramethylchroman[34a]	3320 (OH), 1590 (Ar)
6-Hydroxy-2,2,5,8-tetramethylchroman[34a]	3300 (OH), 1590 (Ar)
6-Hydroxy-2,2,7,8-tetramethylchroman[34a]	3430 (OH), 1615 (Ar)
6-Hydroxy-2,2,5,7,8-pentamethylchroman[34a]	3330 (OH)
p-Methoxyphenol[43]	3340 (OH), 2960, 2915, 1520, 1480, 1460, 1400, 1385, 1310, 1265
5,7-Dihydroxy-2,2-dimethylchroman[44]	3420, 3260 (OH), 2915, 1635, 1610, 1520 (Ar), 1470, 1392, 1380, 1360 (Me gem), 1290
7-Hydroxy-2,2,5-trimethylchroman[33b]	3390 (OH), 1383, 1368 (Me gem)
o-Methoxyphenol[43]	3530 (OH), 1600, 1500, 1490, 1470, 1450, 1360, 1305, 1263
o-Ethoxyphenol[43]	3533 (OH), 1600, 1500, 1490, 1470, 1450, 1400, 1380, 1360, 1305, 1263
5-Methyltocol[45b]	3400, 2920
7-Methyltocol[45b]	3400, 2900
8-Methyltocol[45b]	3310, 2920

[a] KBr disk [b] liquid film

Holman and Edmondson[40] have examined the near-infrared spectra of α-tocopherol (1.0–2.8 μ) and report bands at 1.42, 1.74, 2.07 and 2.76 μ. The 2.07 μ absorption is assigned to the phenolic OH and is narrow and quite intense. The general subject is reviewed by Wheeler.[41]

4. Proton Magnetic Resonance Spectra

A discussion of the proton nmr of the chromanols should start with their one common denominator—the phenolic proton. The phenolic proton peak may be a sharp singlet, or, more usually, a broad singlet that lies between 4 and 8 ppm, depending on concentration, solvent, and temperature. Its chemical shift is in effect an equilibrium position between bonded and nonbonded protons in very rapid exchange, with bonded protons being less shielded and consequently farthest downfield. At higher temperature the equilibrium postion is shifted toward nonbonded and consequently upfield. Less polar solvents decrease hydrogen bonding and cause an upfield shift.

Dilution of the sample with a low-polarity solvent decreases hydrogen bonding and causes a shift toward higher fields providing the bonding is intermolecular. Since the increasing use of Fourier transform (FT) nmr on smaller samples in more dilute solutions, the dilution effect can be very marked with consequent shift of the hydroxylic proton so far upfield that it becomes buried in the alkyl region. The removal of this peak via deuterium exchange may be most helpful in distinguishing it.

The aromatic protons are deshielded by the diamagnetic anisotropy of the π electrons and fall around 5–8 ppm. Alkyl, alkoxyl, and hydroxyl groups all have an increasing shielding effect on the ring protons. The effects are felt mostly at the proton ortho to the substituent, with somewhat less effect at the para position and they are approximately additive. The aromatic protons of a large number of alkyl and alkoxyl phenols have been assigned,[46] and a few of these are compared with a few chromanols[34] in Table 4. The nmr of the monomethyl tocols seems not to have been reported under comparable conditions, but it would be predicted they would be less shielded by about 0.1 ppm—the aromatic proton lying between the 7-hydroxyl and the oxygen of the chroman ring in chroman-7-ol should be most shielded. The aromatic proton of phloroglucinol is near 6 ppm in acetone-d_6. The methylene next to the benzene ring is most affected by the deshielding of the circulating π electrons and can be found around 2.62 ppm in α-tocopherol, compared with the less shielded tetralin methylenes at 2.76 and 2.73 ppm for 6-methoxytetrahydroquinoline. The methylene at the 3-position of the

TABLE 4. CHEMICAL SHIFT OF AROMATIC PROTONS (IN CCl_4 IN ppm FROM TMS)[34,46]

Compound	Proton	δ
Benzene	—	7.25
o-Xylene	—	6.95
o-Ethoxyphenol	—	6.72
3,4-Xylenol	—	6.82
p-Dimethoxybenzene	—	6.68
2,5-Dimethoxytoluene	—	6.52
2,6-Dimethylphenol	3,5 H	6.90
	4 H	6.72
2,3,6-Trimethylphenol	4 H	6.50
	5 H	6.72
2,4,6-Trimethylphenol	3 H	6.56
	5 H	
2,3,5-Trimethylphenol	6 H	6.25
	4 H	6.40
7,8-Dimethyltocol	5 H	6.1
5,8-Dimethyltocol	7 H	6.3
5,7-Dimethyltocol	8 H	6.3

chroman ring is found at 1.77 ppm, while those of tetralin are at 1.79 ppm and those of tetrahydroquinoline are at 1.93 ppm. The 2-methylene is deshielded by the adjacent oxygen of the chroman ring at 4.08 ppm, like the methylene of *p*-ethoxyphenol at 3.93 ppm. The 3- and 4-methylenes of the chroman ring show as triplets ($J = 6$–7 Hz)[34] in all the tocopherol model compounds, probably due to equilibration in the pucker of the ring system, resulting in a A_2X_2 system. The phytyl side chain of (2*R*,4′*R*,8′*R*)-α-tocopherol either fails to lock the chroman ring into a rigid semichair form or, if it does, it forms an AA′XX′ system that appears as a *deceptively simple spectrum*;[47,48] in this case again two triplets ($J = 7$ Hz).[49]

The methyls attached to the aromatic ring of α-tocopherol are not separated and all fall at 2.12 ± 0.03 ppm, although, in tocopherol acetate they are well separated in CCl_4 at 60 MHz, 1.90, 1.95, and 2.06 ppm.

The nmr spectra of methylated hydroquinones and chromans have been completely investigated with the intent of developing a method for determination of the various tocopherols. The method developed uses the difference in the chemical shift of the ring methyls in going from CCl_4 to pyridine. The methyls adjacent to the hydroxyl of the tocopherol–pyridine complex are most deshielded; the method may lack sensitivity but could be applied to a wide variety of compounds.[50] The use of tris(dipivalomethanato)europium as a shift reagent for the identification of the various tocopherols has been reported by Tsukida and Ito.[51]

5. Carbon-13 Magnetic Resonance Spectra

In recent years natural-abundance carbon-13 nmr spectroscopy (cmr) has advanced to the point of a routine laboratory tool. This has been made possible by the advance in equipment design such that samples of 5–10 mg can be handled without difficulty in a modern FT spectrometer.

The nmr spectra of carbon have many advantages over the proton spectra. The field of interest is no longer 10 ppm but over 200 ppm, and it is often possible to count carbon peaks individually for compounds up to 500 in molecular weight. The chemical shift is more characteristic of the skeletal structure and less dependent on solvent polarity. Experimentally, it has been found that the effects of substituents on ^{13}C chemical shifts are largely additive, and thus the use of a table allows the prediction of many cmr shifts on a semiempirical basis.[52]

The first glance at the proton nmr of α-tocopherol does not reveal its aromatic character—this disadvantage does not exist with cmr since all aromatic carbons are easily seen, even though they are not attached to a hydrogen.

If a proton is attached, a cmr spectrum can be run either ^1H decoupled so each carbon will be a single peak, or "off-resonance ^1H decoupled," so that

TABLE 5. ^{13}C CHEMICAL SHIFTS AND $Eu(fod)_3$-INDUCED SHIFTS IN 2,2-DIMETHYLCHROMANOLS[a]

Carbon no.	TriMe (α)		5,8-DiMe (β)		7,8-DiMe (γ)		5,7-DiMe		None	
	δ[b]	▲Eu[c]	δ[b]	▲Eu[c]	δ[b]	▲Eu[c]	δ[b]	▲Eu[c]	δ[b]	▲Eu[c]
8*a*	145.5	0.9	145.9	1.0	145.7	1.2	147.2	1.0	147.5	1.3
6	144.4	5.2	145.6	5.3	146.1	6.9	145.0	4.8	148.4	7.6
8	122.4	1.4	123.8	1.3	125.6	1.3	116.1[d]	1.3	117.6[d]	1.4
7	121.1	2.6	115.5[d]	2.7	121.8	2.0	122.4	2.5	115.5[d]	2.9
5	118.6	2.7	119.4[d]	1.7	112.3[d]	3.7	122.0	2.5	114.5[d]	2.3
4*a*	116.9	1.3	120.0	1.1	117.9	1.5	117.8	1.2	121.7	1.4
2	72.3	0.4	72.4	0.4	73.3	0.4	72.6	0.4	74.0	0.4
3	33.1	0.3	33.0	0.4	33.0	0.3	33.0	0.3	32.8	0.4
4	21.1[a]	0.5	21.1[e]	0.5	22.6	0.3	20.8[e]	0.4	22.6	0.5
2*a*, 2*b*	26.7	0.2	26.6	0.2	26.9	0.2	26.5	0.2	26.7	0.3
7*a*	12.1[e]	2.2	—	—	11.9[e]	2.4	16.0	2.0	—	—
8*b*	11.8[e]	0.4	15.8	0.3	11.9[e]	0.6	—	—	—	—
5*a*	11.3[e]	2.2	11.0[e]	1.9	—	—	11.4[e]	2.0	—	—

[a] Each compound (1 mole) was dissolved in $CDCl_3$ (1 cm^3), and $Eu(fod)_3$ (200 mg) was added to the solution after determination of δ values.
[b] Chemical shifts in ppm for TMS.
[c] ▲Eu [the differential shift value induced by $Eu(fod)_3$] = $\delta_{Eu} - \delta$.
[d] Chemical shift values for protonated aromatic carbons.
[e] A steric effect was observed.

C, CH, CH_2, and CH_3 will now appear as singlets, doublets, and so on. There are many other special techniques such as measurement of relaxation times and the use of shift reagents that make cmr and pmr valuable complementary tools. The use of stable isotope labeling, including ^{13}C and ^{2}H, has revolutionized the study of reaction mechanisms, biosynthesis, and metabolic pathways, since labor-intensive degradation is no longer required for positioning the label.[53]

The ^{13}C nmr of the tocopherols and their model compounds, the 2,2-dimethylchromanols, has been extensively investigated by Matsuo and Urano[54] using the $Eu(fod)_3$-induced shifts, since europium ions coordinate selectively to a phenolic oxygen. The use of deuterium labeling also aided in assigning carbon shifts. These assignments are given in Tables 5 and 6. It should be noted that the C-4 methylene signals in the 5-methylated chromanols were about 1.5 ppm upfield from those of 5-protonated compounds. This shift difference is also found in the tocopherols and is attributed to steric compression. Steric compression shifts of 4 ppm are also observed for the methyl groups flanked either by CH_2 or CH_3, rather than a proton. The methyl at C-2 of the chroman ring in the tocopherols receives a steric compression shift 2.5 ppm upfield relative to the model compounds due to the effects of the isoprenyl chain.

TABLE 6. ^{13}C CHEMICAL SHIFTS AND $Eu(fod)_3$-INDUCED SHIFTS IN TOCOPHEROLS[a]

Carbon no.	α		β		γ		δ	
	δ[b]	▲Eu[c]	δ[b]	▲Eu[c]	δ[b]	▲Eu[e]	δ[b]	▲Eu[c]
8*a*	145.4	0.9	145.7	1.1	145.5	1.1	145.9	1.1
6	144.4	4.7	145.5	6.1	146.0	5.8	147.5	7.0
8	122.3	1.2	123.8	1.4	125.5	1.1	127.1	1.2
7	121.0	2.5	115.4[d]	3.2	121.5	1.9	115.9[d]	2.4
5	118.5	2.5	119.2	2.0	112.0	3.3	112.7[d]	2.6
4*a*	117.0	1.2	120.1	1.2	118.0	1.2	121.1	1.1
2	74.3	0.4	74.4	0.4	75.3	0.4	75.5	0.3
1′	39.8	0.3	39.8	0.3	40.0	0.3	40.0	0.2
11′	39.4	0.1	39.4	0.1	39.4	0.1	39.4	0.0
3′, 5′, 7′, 9′	37.5	0.1	37.4	0.1	37.5	0.1	37.5	0.0
4′, 8′	32.7	0.1	32.7	0.1	32.8	0.0	32.7	0.1
3	31.6	0.3	31.5	0.4	31.4	0.3	31.4	0.3
12′	26.0	0.1	28.0	0.1	27.9	0.1	28.0	0.0
10′	24.8	0.1	24.8	0.1	24.8	0.1	24.8	0.0
6′	24.5	0.1	24.5	0.1	24.4	0.1	24.5	0.0
2*a*	23.6	0.2	23.8	0.3	24.0	0.2	24.0	0.2
12′*a*, 13′	22.6	0.1	22.7	0.0	22.6	0.0	22.7	0.0
2′	21.0	0.2	21.0	0.2	21.0	0.2	21.0	0.2
4	20.8[e]	0.4	20.8[e]	0.5	22.3	0.4	22.7	0.3
4′*a*, 8′*a*	19.7	0.1	19.7	0.1	19.7	0.1	19.7	0.0
7*a*	12.1[e]	2.1	—	—	11.9[e]	2.0	—	—
8*b*	11.8[e]	0.4	15.8	0.3	11.9[e]	0.5	16.0	0.4
5*a*	11.2[e]	2.0	11.0[e]	2.2	—	—	—	—

[a] For key, see footnotes (a) through (e) in Table 5.

6. Mass Spectra

An extensive study of aromatic ethers in which the oxygen forms part of a ring was conducted by Williams, Thomas, and Gautschi.[30] They have used the appearance of metastable peaks to explain the complex breakdown of both the chromans and chromanols by the pathways shown in Schemes 1 and 2.

Nilsson et al.,[55] have determined the spectra of the tocol model compound 2,2-dimethylchroman-6-ol and all seven mono-, di-, and trimethyl aromatic ring-substituted derivatives. They found the breakdown pathway for these model compounds to be the same as those shown in Scheme 2 for 2-substituted chromanols, although Williams et al.[30] report that 2,2-dimethylchroman always fragments with hydrogen transfer to form the M-55 fragment. Both groups report the loss of methyl at the 2-position, and Nilsson et al. point out that this is equivalent to the loss of the side chain that

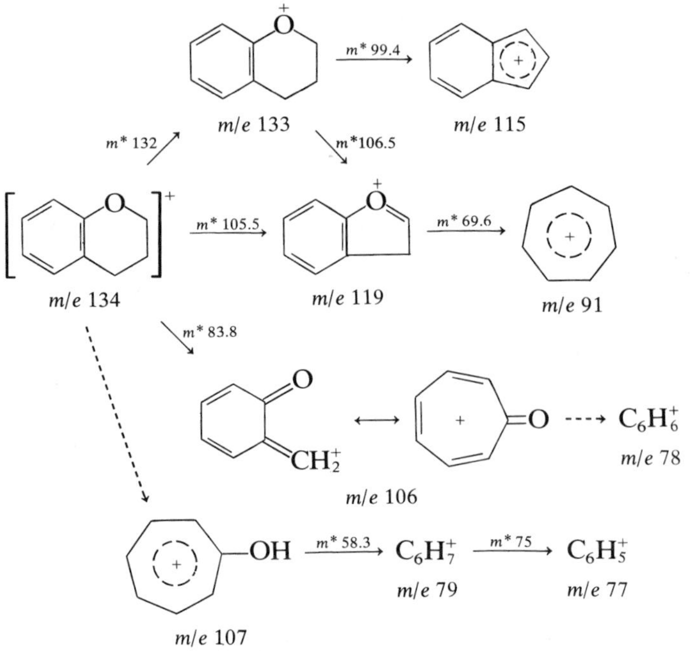

m* 99.4

m/e 133 m/e 115

m* 132 m*106.5

m* 105.5 m* 69.6

m/e 134 m/e 119 m/e 91

m* 83.8

$CH_2^{+\cdot}$ O $C_6H_6^{+\cdot}$

m/e 78

m/e 106

OH m* 58.3 $C_6H_7^+$ m* 75 $C_6H_5^+$

m/e 79 m/e 77

m/e 107

Scheme 1

occurs in the fragmentation of γ-tocotrienol and further loss of a two-carbon fragment gives the fragment *m/e* 135, Scheme 2.

Trudell, Woodgate, and Djerassi[56] have reported failing to find the metastable peak at 105.5, Scheme 1, and have shown via ^{13}C and ^{2}H labeling that the formation of ion *m/e* 119 is produced by at least three independent competing pathways resulting in the loss of carbons 2, 3, and 4 of the chroman in the ratio of 2:1:1. Pathways were proposed in each case. Scheppele and his colleagues[57] suggest that the difference in the mode of fragmentation may be critically dependent on the C-6 hydroxyl. Verhé et al.[58] report nearly the same peak intensities for the *m/e* 123 [100] and 122 [32] for 2,2-dimethylchroman-5-ol as Nilsson's team reported for the 6-hydroxy compound, 123 [100] and 122 [42].

The mass spectra of α-, β-, and γ-tocopherols are discussed by Scheppele et al.,[57] who point out that ion intensities, but not reaction pathways, are strongly dependent on temperature and confirm the previously discussed pathways relating to the model compounds. They found the loss of ketene from the molecular ion to be the dominant process for fragmentation of the tocopherol acetates. Rao and Perkins[59] have used a one-step method to estimate and identify the TMS derivatives of tocopherols and tocotrienols using gas chromatography and mass spectrometry, although thin-layer chromatography is used as a pretreatment of the unsaponifiable material when β- and γ-tocopherols are present together in crude vegetable oils. The

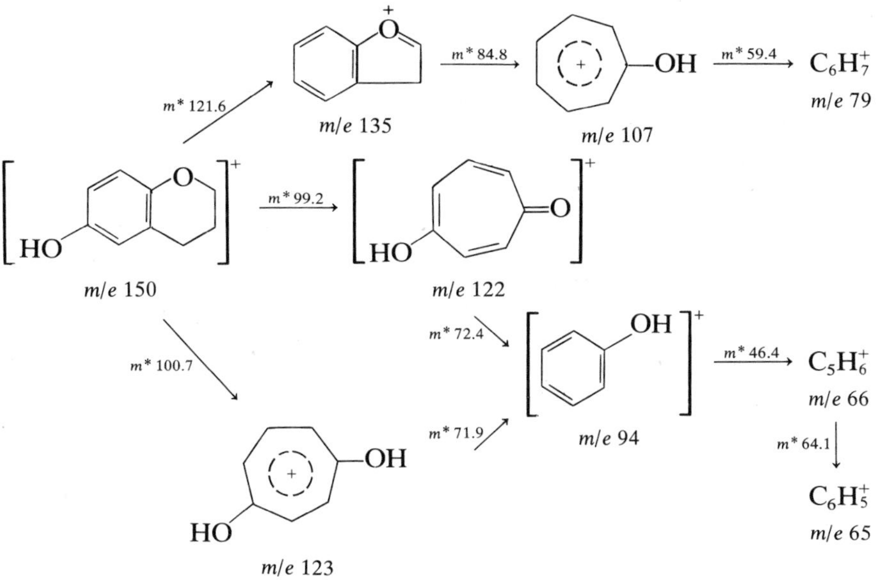

Scheme 2

mass spectra of the TMS derivative of the synthetic 'orthotocopherol,' 2,5,6,7-tetramethyl-2-[4′,8′,12′-trimethyl(tridecyl)]chroman-8-ol, were taken by Vance and Bentley[60] and closely resembled those of α-tocopherol and six other structural isomers.

III. ANALYTICAL METHODOLOGY

1. Oxidative Methods

The qualitative and quantitative determination of the various tocopherols has been the subject of extensive study and review[61–65] since the early discovery of its reducing properties.[67] Oxidizing agents such as gold chloride,[66] ceric sulfate,[68] lead tetraacetate,[68] nitric acid,[69] silver nitrate,[70] potassium ferricyanide,[71,72] and ferric ions[61,73–75] have been employed in analytical schemes. Nitric acid and silver nitrate, under proper conditions, give the orthoquinone (**3**), a red-colored compound that can be determined by colorimetric methods. Emmerie and Engel[73] used α,α'-dipyridyl, which gives a red color in the presence of an organic compound that produces a colored complex with the ferrous ions produced by a compound's oxidation with ferric ions. 1,10-Phenanthroline, 2,2′,2″-terpyridine, *o*-nitrosoresorcinol monomethyl ether, tripyridyltriazine, and diphenyl(batho)-phenanthroline have also been used;[61] the latter two materials gave increased sensitivity.[74,75] Oxidative methods generally lack specificity and require careful separation and purification before the colorimetric step is applied. Carefully controlled electrolytic oxidation would seem to have advantages over chemical oxidative methods. The polarographic oxidation of tocopherol at the dropping-mercury anode was first introduced by Smith, Spillane, and Kolthoff.[76] Knobloch, Machá and Mňouček[77] obtained half-wave potentials in alcoholic acetate buffer at pH 7 for α-, β-, γ-, and δ-tocopherol. McBride and Evans[78] were able to make an analysis of individual tocopherols (except that β and γ were superimposed) by the use of a glassy carbon voltametric method.

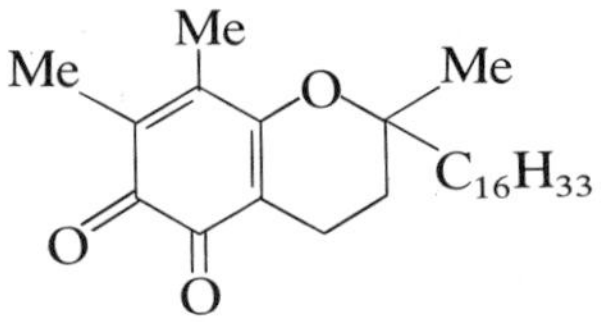

Me Me O Me $C_{16}H_{33}$ O O

(3)

Parker[79] studied the reversible two-electron oxidation of 2,2,5,7,8-pentamethylchroman-6-ol in acetonitrile at a stationary platinum electrode. Bates[80] compared the oxidation potentials of the tocopherols with their antioxidant properties. Examination of the polarographic oxidation products of tocopherol indicates that it is oxidized directly to the tocopherylquinone.[82] Atuma and Lindquist[81] have developed an analytical method for the determination of the individual tocopherols in foods, oils, and pharmaceuticals using a carbon paste electrode; β- and γ-tocopherol are not distinguished from each other.

2. Coupling Methods

Quaife[83] found that the unmethylated 5-position in γ-tocopherol could be coupled with a diazotized aromatic amine over a wide pH range to give a red dye that could be measured colorimetrically, while β-tocopherol failed to couple at all. Weisler, Robeson, and Baxter[84] used the pH color dependence of the coupling products of γ- and δ-tocopherols to assay these materials independently. The coupling method also lacks specificity and suffers from several other disadvantages.[64,85] Tocopherols that are not substituted in the positions ortho to the hydroxyl group form nitroso derivatives, and Quaife[86] used the low absorption of these derivatives at 405–415 mμ for their assay. Chromatographic separation of the nitroso derivatives has been accomplished and their assay[87] made more sensitive by the subsequent application of the Emmerie–Engel colorimetric method.

3. Ultraviolet Methods

The ultraviolet absorption maxima of the various tocopherols lie in a narrow range, making assay of individuals in the group impractical without prior separation. This method is also confounded by low specificity and low sensitivity. The tocopherylquinones absorb much more intensely, but the absorption maxima are found at shorter wavelengths thus adding to the background absorption. Duggan[88] showed that tocopherol gives an intense ultraviolet fluorescence, with a maximum around 330 mμ, when excited at the optimum of 295 mμ. The fluorescence of the phenazine derivatives formed by the condensation of the orthoquinones with *o*-phenylenediamine has also been used for an assay procedure.[61]

4. Chromatography

The chromatographic analysis of tocopherol mixtures on paper impregnated with liquid paraffin or zinc carbonate was in its day a widely used method. The two-dimensional paper-chromatographic method introduced by Green et al.[89] became the basis for the standard official procedure for the separation of the individual tocopherols.[65] The use of paper impregnated with olive oil served as a basis for the study of R_m (phenol)[90] and complex structure (tocopherols, among other natural products). Paper chromatography today is mostly of historical interest.

Column chromatography closely parallels thin-layer techniques, although columns predate the latter. Diatomaceous earth impregnated with liquid paraffin,[91] aluminum oxide, zinc carbonate, celite,[92] and magnesium phosphate[93,94] columns have seen early use. Thick layers of silica gel[93] on glass plates[94] were used to separate the individual tocopherols in peanut oil.[75] Dilley and Crane[95] used silica gel thin-layer plates for the analysis of individual tocopherols in plant material. The separated tocopherols were removed from the plate, and the tocopherylquinone formed by gold chloride oxidation of the sample was determined spectrophotometrically, first as the quinone and then after reduction with sodium borohydride.

Whittle and Pennock[96] have used two-dimensional thin-layer chromatography. The column chromatography of tocopherols has been reviewed and the use of an alumina column has been described in detail.[97] A bibliography of liquid column chromatography of the tocopherol group[98] and the paper and thin-layer chromatography has also been covered separately.[99] A column of microparticle silica 30 cm × 4 mm (i.d.) and a mobile phase of chloroform : *n*-hexane : THF (70 : 30 : 1) was used in a high-pressure system to separate a synthetic mixture of vitamins A, D, and E.[100] Vatassery and Hagen[101] used columns of 180 × 0.2 cm packed with 37–50 μm silica-coated glass beads and 0.6% methanol in hexane at 200 psi to determine α-tocopherol in rat brain. The procedure requires samples of 50–100 mg tissue, and saponification is not required. The tocopherol is measured by its fluorescence according to Duggan.[88]

5. Gas Chromatography

Nicolaides[102] first described the gas chromatography of tocopherol in 1960. Two years later Wilson, Kodicek, and Booth[103] reported an extensive study of the tocopherols and tocotrienols, using silicone polymer (SE-30), fluoroalkyl silicone polymer (QF-1), and poly(ethylene glyco adipate), that indicated the valuable role gas chromatography would play in the future of tocopherol analysis, not only with regard to increased resolution but also high sensitivity (less than 2 μg tocopherol).

Gas chromatography seems well adapted for separation of tocopherols substituted with various numbers of methyl groups, but difficulties arise in the separation of isomers, especially β- and γ-tocopherol, Kofler et al.[104] report the partial separation on Apiezon-N stationary phase, and Nair et al.[105] separated these isomeric dimethyltocols with a SE-52/XE-60 biphase column after first converting them to the para-quinones. The complete separation of the isomeric monomethyltocols as their TMS ethers has been conducted on the biphase column.[106] The TMS ethers of tocopherols[107] and some tocotrienols were separated on SE-30 and Apiezon columns.[108] Gas chromatography of natural and synthetic δ- and ε-tocopherol on an Apiezon-N column was used to help establish the all-trans configuration of the side chains.[109] A gas-chromatographic method[110] for the analysis of the vitamin E content of pharmaceutical products, that outperforms the chemical method was adopted by the Association of Official Analytical Chemists.[63]

The gas chromatography of the tocopherols has been the subject of several excellent reviews.[61,63,107,111–113] Nair and Luna[114] used a gas chromatograph directly coupled to a mass spectrometer for the determination of α-tocopherol in heart muscle. They also collected samples of the effluent for ir spectroscopy in KBr pellets. A method has been worked out by Nelson and Milun[115] for the analysis of soya sludges and residues. Although the method does not separate β- and γ-tocopherol, effluent compounds are trapped from the gas chromatograph (gc) for spectral analysis by both ir and mass spectrometry (ms).

One of the most serious problems involved with any analytical procedure for vitamin E is losses due to oxidation, that is, dimers and trimers. Attempts to minimize these losses have involved the addition of antioxidants or working rapidly in the cold, precautions that are not conducive to complete extraction from tissue samples. Lunan, Wenn, and Thomas[116] have overcome these problems by the addition of a known amount of deuterium-labeled tocopherol to the sample early in the analytical process, thus providing each sample with its internal standard. The analysis is then based on the difference in mass peaks for labeled and nonlabeled tocopherol after gc–ms of the TMS ethers. Lunan[117] has indicated that a comparison of this method with other standard methods has shown the variability in the standard methods with the type of tissue being analyzed.

IV. SYNTHESIS

The reaction of Grignard reagents on coumarins often gives 2,2-disubstituted 2*H*-1-benzopyrans,[118] and 4-substituted 2*H*-benzopyrans may be obtained from 4-chromanones with Grignard reagents followed by dehydration.[119] Hydrogenation of the nonaromatic double bond of the 2*H*-1-benzopyrans[120] to yield chromans and chromanols has been described at length. Hydrogenation and Clemmensen reduction have been used under a variety of conditions to give chromans.[121] Since these methods have already been eloquently reviewed, they will not be further discussed here.

The synthesis section is separated into five main headings: Grignard reactions, reduction methods, acid condensations, base condensations, and other methods. No special significance should be assumed from the hierarchy of this section. Each method has its special applications and advantages, and only a few examples of each are given.

1. Grignard Reactions

A. *Coumarins*

One of the general methods of making chromans, the treatment of coumarins with Grignard reagents,[122] works well for the synthesis of chromanols usually from the corresponding methoxy starting material. Many coumarins have been synthesized.[124] Fukami, et al.[123] produced 2,2-dimethyl-5-methoxychrom-3-ene (**6**) and the corresponding 7-methoxy compound by treating the respective methoxy coumarins (**4**) with methylmagnesium iodide in boiling benzene for 4 hr, but the reaction failed in boiling ether. The intermediate (**5**) was formed after $\frac{1}{2}$ hr and could be closed to (**6**) on refluxing with acetic acid.

Although it is known that Grignard reagents may add 1,4 to α,β-unsaturated ketones, the cis double bond[125] and the small size of the methyl group[126] encourage formation of the gem dimethyl compound also indicated by the 1380–1360 cm^{-1} absorption bond in the ir spectra. Hydrogenation with Pd–C in ethanol at room temperature gave (**7**). All attempts to demethylate (**6**) gave resinous products, and standing in 20% sulfuric acid at room temperature gave what they presumed was a dimer. Huls[127] prepared (**7**) by hydrogenation followed by Grignard reagent. Demethylation of (**7**) was accomplished in 42% yield with HI.

When dihydroumbelliferone acetate was treated with excess methylmagnesium iodide, a 70% yield of 2,2-dimethylchroman-7-ol was obtained.[128] John et al.[129] obtained about the same yield of 2,2,5,7,8-pentamethylchroman-6-ol by treating 6-hydroxy-5,7,8-trimethyldihydrocoumarin in a benzene–anisole mixture with excess methylmagnesium iodide in ether at 40°C over a 3-hr period, showing that blockage of a hydroxy group is not necessarily required.

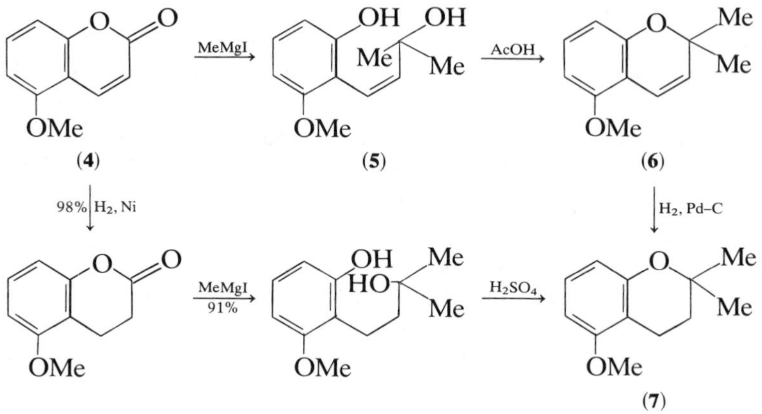

(**4**) → MeMgI → (**5**) → AcOH → (**6**)

(**4**) → 98% H_2, Ni → → MeMgI 91% → → H_2SO_4 → (**7**) ← H_2, Pd–C ← (**6**)

B. *Phenylethyl ketones*

John and Schmeil[130] treated the product obtained from zinc dust–dilute acetic acid reduction of 1-γ-oxobutyl-3,4,6-trimethylbenzoquinone with methylmagnesium iodide in boiling ether for 3 hr to give 2,2,5,7,8-pentamethylchroman-6-ol. The use of methyl Grignard reagent on (**8**) gave the alcohol (**9**), which was demethylated and ring closed in 10% HBr acetic acid by refluxing overnight to give (**10**) in 55% yield.[34]

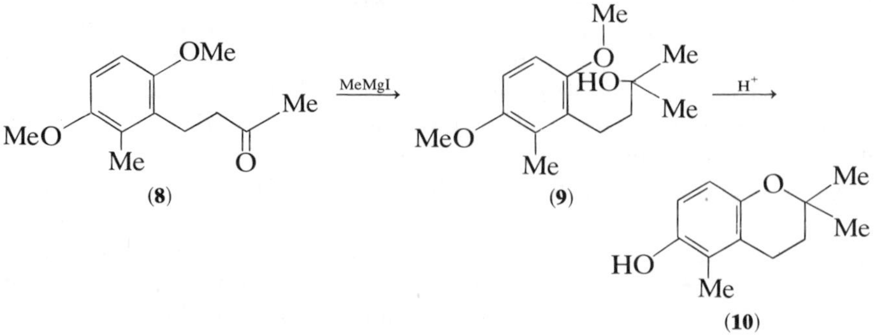

(**8**) → MeMgI → (**9**) → H^+ → (**10**)

C. *Ring Formation*

McHale, Mamalis, Green, and Marcinkiewicz[131] formed 7-methyltocol by the use of Grignard reagents starting from 5-bromotoluquinol dimethyl ether (**11**). Nilsson et al.[34] used a slightly modified procedure to give 2,2,7-trimethylchroman-6-ol in 37% yield on going from step (**13**) to (**15**).

Me, OMe, MeO, Br (**11**) —1. Mg; 2. CH_2CH_2 (O)→ Me, OMe, OH, MeO (**12**)

PBr_3 ↓

Me, OMe, Br, MeO (**13**)

↓ 1. Mg 2. 6,10,14-Me_3-pentadecan-2-one

Me, OMe, Me, HO, $C_{16}H_{33}$, MeO (**14**) (not purified) —HBr, AcOH→ Me, O, Me, $C_{16}H_{33}$, HO (**15**)

2. Reduction of the Heterocyclic Ring

The reactions of lithium aluminum hydride closely resemble those of Grignard reagents and differ principally in the more aggressive behavior of the hydride, resulting in a lessening of side reactions and steric influences.[132,133] Lithium aluminum hydride, borohydride, zinc amalgam HCl, and catalytic hydrogenation have been used to produce 3- and 4-chromanols, and this has been discussed at length.[134] Catalytic hydrogenation of 7-hydroxy-4-chromanone with Raney nickel at 40–45°C for 24 hr in ethanol gave directly a 77% yield of chroman-7-ol.[135] However Bridge et al,[136] obtained a mixture of 2,2-dimethylchroman-7-ol and unchanged starting material in their reduction of 2,2-dimethyl-7-hydroxy-4-chromanone using palladium on charcoal in acetic acid, which also was used successfully to obtain 2,3-dimethyl-7,8-diacetoxychroman, 2-methylchroman-7-ol, 2,3-dimethylchroman-6-ol, and 2-methylchroman-7,8-diol from their respective 4-chromanones.[137] Platinum[129] and copper chromite[138] have also been used as catalysts. The hydrogenation of 4-chromanones directly to the chromans must be considered as special cases rather than a general procedure.[139]

The dehydration of the 3- and 4-chromanols to chromans has been reviewed.[140]

Karrer and Banerjea[141] obtained *o*-hydroxycinnamyl alcohol by treatment of coumarin with $LiAlH_4$ at room temperature for 20 min, while treatment for 24 hr gave a 50% yield of 3-(*o*-hydroxyphenyl)propanol. Methods have already been discussed for ring closure to the chromans.

Hydrogenation of the chromenes has already been mentioned, and these procedures are further discussed by Schweizer and Meeder-Nycz.[142] Special attention has been given to the reduction of coumarin.[15] The Clemmensen reduction of 4-chromanones with zinc amalgam and hydrochloric acid directly to the chromans has been widely used, but this method is said to be unpredictable.[139]

3. Acid Condensation Methods

Direct alkylation of phenols involving alkyl halides, olefins, alcohols, esters, and other reactants capable of forming carbonium ions is the most obvious and widely used method for the formation of the chroman ring system. While the carbonium ion may be the attacking group in some cases, this may not always be the case, and electron-donating groups such as hydroxyl will increase the nucleophilicity of the aromatic reactant and increase the S_N2 character and speed of the reaction.

Phenols can be alkylated even in alkaline solution, and the reaction is first order in both reactants, involving inversion of configuration at the center of attachment (S_N2). The aryl ether is not an intermediate in the reaction.[143,144] Hart, Spliethoff, and Eleuterio[145] found that the para-alkylated products from the reactions of phenol and 2,6-xylenol with optically active α-phenylethyl chloride had undergone inversion at the asymmetric carbon (S_N2), while the ortho alkylation products from phenol, *p*-cresol, and *p*-chlorophenol all had retained configuration at this asymmetric center. They suggested this as evidence for an S_N1-type mechanism in which the hydroxyl group solvates the chloride ion as it is displaced (Scheme 3).

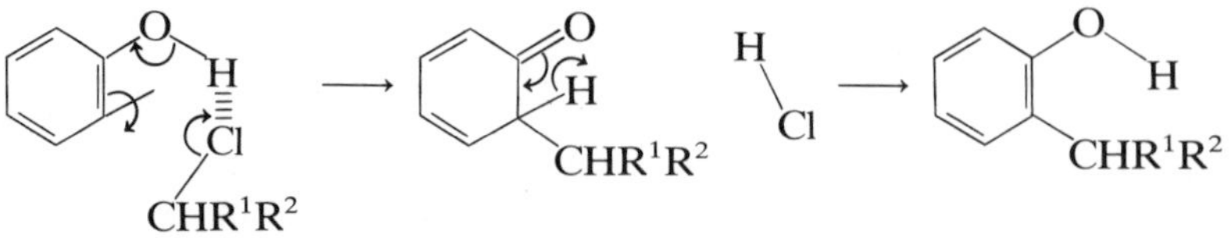

O H Cl CHR^1R^2 → O H CHR^1R^2 H Cl → O H CHR^1R^2

Scheme 3

It is then not surprising to find direct alkylation of phenols carrying additional hydroxy groups as the most used method of obtaining chromanols. The difficulty has been to stop the reaction before one obtains polyalkylated products and to obtain products alkylated in the desired position. After direct nuclear alkylation of the phenol, ring closure generally follows in accordance with Markovnikoff's rule. The condensation of trimethylhydroquinone with crotyl bromide gives mostly the chroman (**16**) but also some coumaran (**17**). Smith[12] has reviewed the various conditions that determine chroman versus coumaran formation.

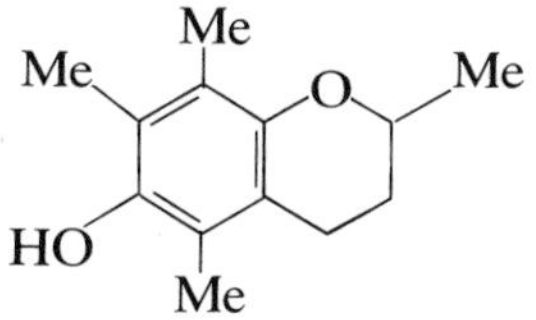

(**16**)

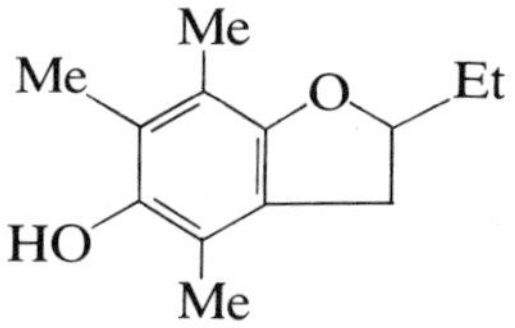

(**17**)

Hydroquinone and its methylated derivatives have figured heavily in many early syntheses of 6-chromanols, tocopherols, and various model compounds. The condensation of hydroquinone with 1,3-diols,[146] allyl alcohols,[5,34] halides, or dienes[147] using zinc halide in glacial acetic acid were in their time notoriously successful. Many new syntheses were developed; however, unfortunately much of this work was intended to circumvent a patent position rather than supply technical advantage. Yet today the condensation of hydroquinone with allyl alcohol or its equivalent is one of the most intensely studied reactions. Almost everything has been used for its catalysis: $SiO_2:Al_2O_3$ (87:13), As_2O_3, TiO_2, P_2O_5, $ZnCl_2$, Fe_2O_3–Al_2O_3–SiO_2, $NiSO_4$–Al_2O_3, zeolite (H form), zeolite Na-4, bentonite,[148] BF_3,[146] $AlCl_3$,[149] $SnCl_4$, $MeSO_3H$, *p*-toluenesulfonic acid, alkali metal acid sulfates, and $ZnCl_2$–$KHSO_4$.[150] The synthesis of α-tocopherol has been reviewed.[151]

Wehrli et al.[147] have reported in 1971 that a 1:1 complex of trimethylhydroquinone with BF_3 or $AlCl_3$ reacts very smoothly and almost instantly with isophytol (**18**) at low temperature (−20° to −60°) to give pure (±)-α-tocopherol. The choice of solvent is critical; chlorinated hydrocarbons, nitroparaffins, nitrotoluenes, nitrobenzenes, or mixtures of these solvents are preferred, while the reaction fails in ether, THF, and more polar solvents due to complex formation with the solvent instead of the hydroquinone. The mechanism that they propose for the reaction is shown in Scheme 4. They[147] claim yields of 82% based on isophytol and purities above 98%. The purity of the product is claimed to be its main advantage.

Nelan[152] has proposed a continuous process for making (±)-α-tocopherol acetate by allowing a solution of trimethylhydroquinone, phytol, or isophytol and $BF_3:Et_2O$ in acetic acid to flow over a packed column heated to 110–120°C at a rate as to distil most of the solvent. The product collected

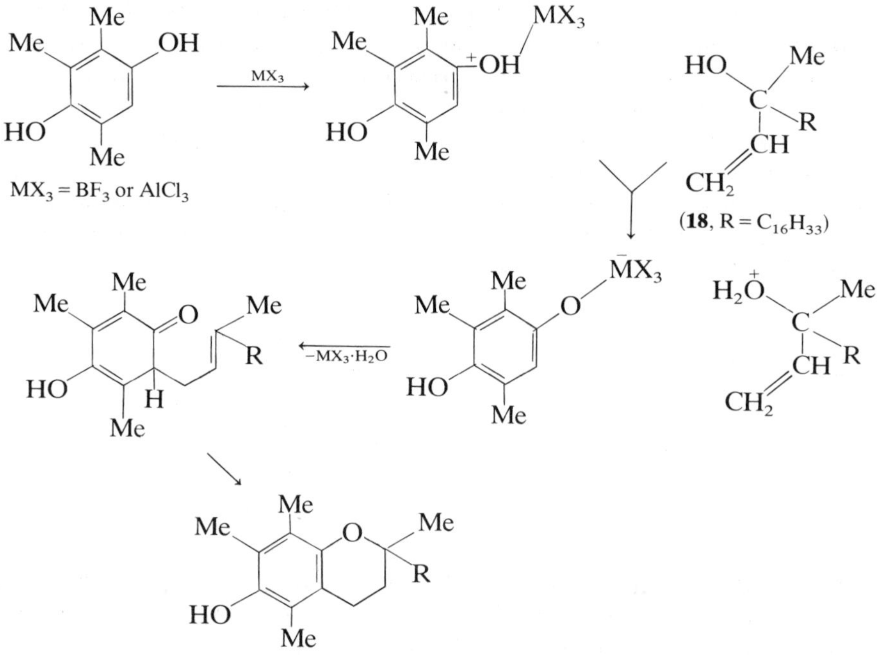

$MX_3 = BF_3$ or $AlCl_3$

(**18**, $R = C_{16}H_{33}$)

Scheme 4

at the bottom was treated with acetic anhydride in pyridine and gave an 87% yield of product of 98% purity.

2,2,5,7,8-Pentamethylchroman-6-ol, the model compound for α-tocopherol, was first synthesized by Smith et al.[153] by condensation of isoprene with trimethylhydroquinone in acetic acid and $ZnCl_2$. Its easy synthesis and purification by crystallization makes it an ideal model for the study of the chemistry of α-tocopherol. The model compounds for the other tocopherols with unmethylated ring positions may be more difficult to obtain, since there is the possibility of the formation of several isomers as well as dichromans.

Molyneux and Jurd[154] condensed 2-methylbut-3-ene-2-ol in 5% aqueous citric acid with phloroglucinol and found that the products isolated by distillation at 155–170°/0.2 mm were an equal mixture of 5,7-dihydroxychromans (**19**) and the dipyran (**22**), while the alternately possible dipyran (**24**) was not formed. Resorcinol under similar conditions of both reaction and isolation by high-temperature distillation gave only the chroman-7-ol and none of the alternately possible 5-hydroxy compound.

Miller and Wood[33] have condensed 3,3-dimethylallyl diphenyl phosphate, made from 3,3-dimethylallyl alcohol and diphenyl phosphorochloridate with orcinol and obtained two chromans (**20**) and (**21**) and an unidentified dipyran, (**23**) or (**25**). Phloroglucinol under similar conditions gave the hydrobenzodipyran (**22**) and the benzotripyran.

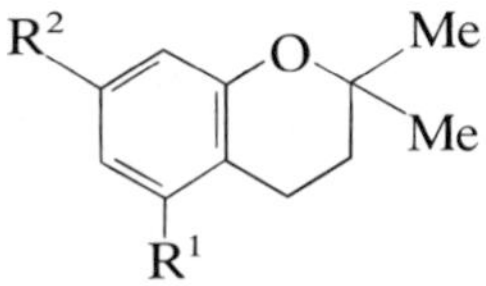

(**19**) $R^1 = R^2 = OH$
(**20**) $R^1 = OH, R^2 = Me$
(**21**) $R^1 = Me, R^2 = OH$

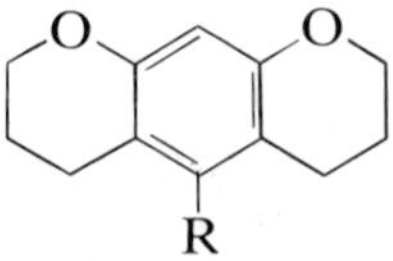

(**22**) R = OH
(**23**) R = Me

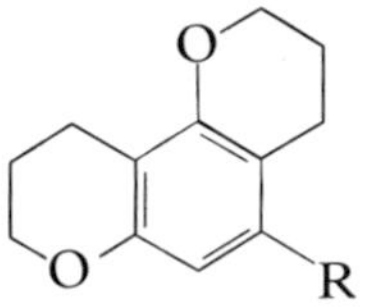

(**24**) R = OH
(**25**) R = Me

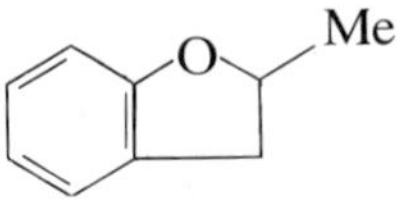

(**26**)

The phosphate esters of geraniol or farnesol, natural phytyl alcohols, were used in similar experiments to produce a number of compounds including (2*RS*,4′*R*,8′*R*)-α-tocopherol in 90% yield. The disadvantages of this reaction are clearly shown, however, by the reaction of allyl diphenyl phosphate with phenol to give allyl phenyl ether (34%) and *O*-allylphenol (25%). The latter product on ring closure does not give a chroman but 2-methylbenzofuran (**26**), the expected product if a carbonium intermediate was involved.

4. Base-Catalyzed Condensations

It has already been pointed out that many base-catalyzed ring cyclizations may be useful for the formation of the pyran ring system. These reactions are related to well-known 'name' reactions and are not further discussed except for a few specific examples that have been directly linked to the chromanols listed in the tables for this chapter.

One of the problems with the acid-catalyzed condensations of monomethylhydroquinone with phytol was the complex mixture of 5-, 7-, and 8-methyltocols produced in the approximate ratio of 1:2:1.[34,155]

A British team[156] overcame this difficulty by using a rather involved route to 5-methyltocol (**34**) starting with 2,3-dimethylhydroquinone dimethyl ether (**27**). Nilsson, Sievertsson, and Selander[34] using a similar approach condensed (**28**) with acetoacetate ester under basic conditions to obtain 41% of (**35**) after acid decarboxylation. Treatment with methyl Grignard reagent and then HBr in acetic acid gave the 2,2,5-trimethylchroman-6-ol (**36**) in 55% yield.

5. Other Methods

The difficulties in the synthesis of the monomethyltocols has led to yet another synthesis involving the formation of η-allylnickel complexes[45] with bromo-p-diacetoxytoluenes to give the corresponding alkyl-substituted p-diacetoxytoluenes, which were converted in high yield to 6-chromanols by hydrolysis and cyclization with stannous chloride and hydrochloric acid. The alkylation step gave yields from 52 to 93%, while cyclization yields ranged from 78 to 97%. The nickel complex was made by adding the 1-bromo-3-unsaturated compound to tetracarbonylnickel in benzene at 50°C over several hours. The solution was concentrated under reduced pressure and

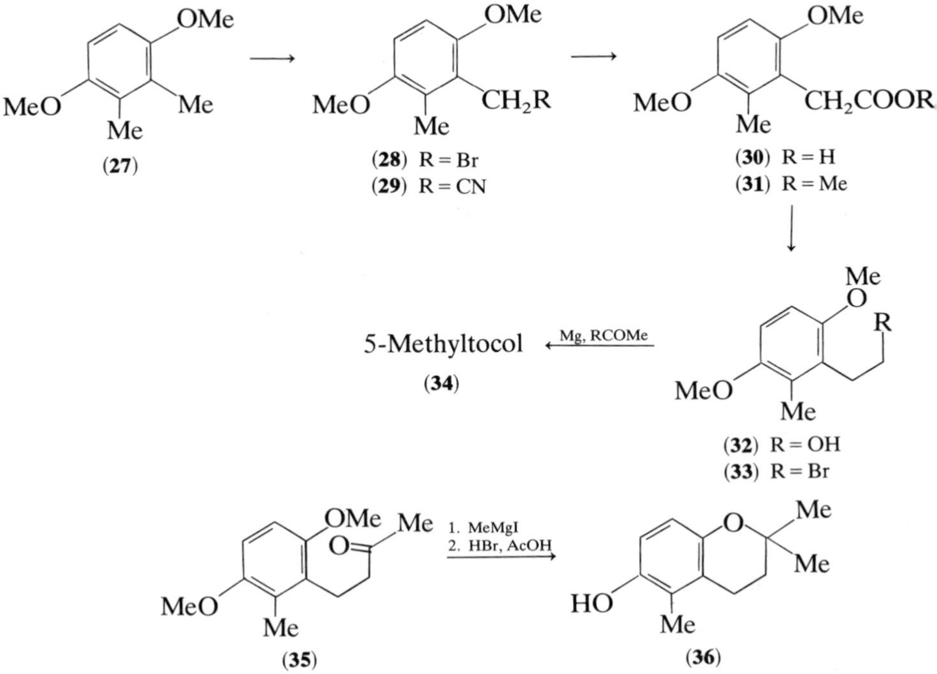

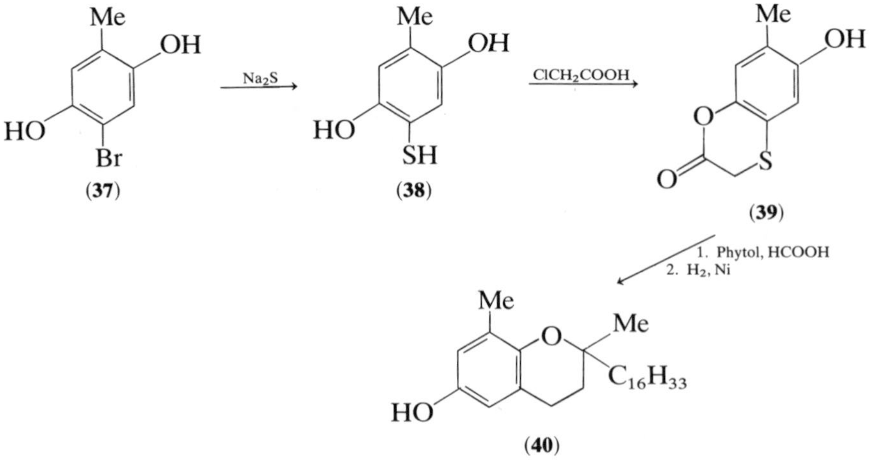

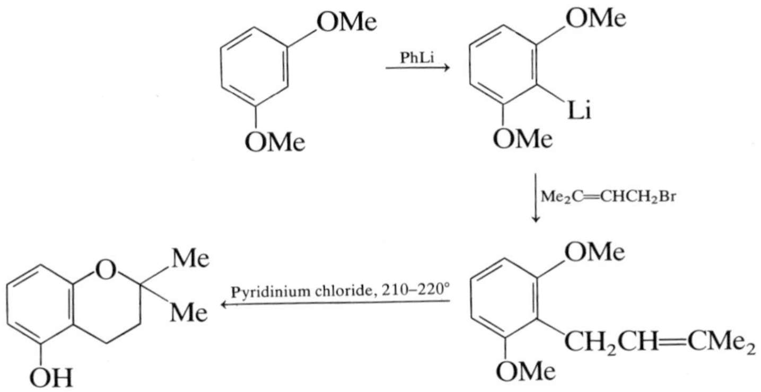

Scheme 5

dissolved in hexamethylphosphoramide (HMPA) and the bromodiacetoxytoluene added. The formation of the diacetoxyalkyltoluene took 5 hr at 50°C. The hydrolysis and cyclization are similar to those discussed previously using zinc chloride.

Green and co-workers[157] blocked the alternate routes for chroman ring formation using a unique scheme. Unfortunately the yield was only 22% from (**39**) to (**40**), but the material was free of isomeric products. Hallet and Huls[158] obtained a low yield of 2,2-dimethyl-8-chromanol by forming the lithium salt of 2,2-dimethylchroman by treating the compound with butyllithium in refluxing ether and finally oxidizing the lithium salt at −10°C with oxygen. Boltze and Dell[159] have prepared 2,2-dimethyl-5-chromanol by direct metallation of resorcinol dimethyl ether, Scheme 5.

Smith, Hoehn, and Ungnade[160] made two different approaches toward introducing the hydroxyl group into a chroman directly. They started with 2,2,5,7,8-pentamethylchroman forming the 6-bromo derivative in carbon tetrachloride. The remaining part of the process–formation of the Grignard reagent and its oxidation to the 6-chromanol with molecular oxygen—was not nearly so successful (8% yield).

The other route, Scheme 6, via the amino derivative through the diazonium salt to the 6-chromanol failed first to couple with diazotized sulfanilic acid; and finally, even though the 6-nitrochroman was obtained in good yield, it failed to be reduced with tin and hydrochloric acid or hydrogenation. Hromatka[161] formed the chroman ring on the aminophenol (the amino group was protected by formylation) using $ZnCl_2$–HCO_2H and the appropriate unsaturated alcohol as previously described. The formyl groups were removed and the amine was diazotized and converted to the hydroxy compound. Also mentioned is the nitrophenol, but no reduction procedure is mentioned or yields given. Baker et al.[162] obtained a quantitative yield of (**43**) by reductive acetylation of the quinone (**42**) with zinc dust in acetic anhydride–acetic acid mixtures.

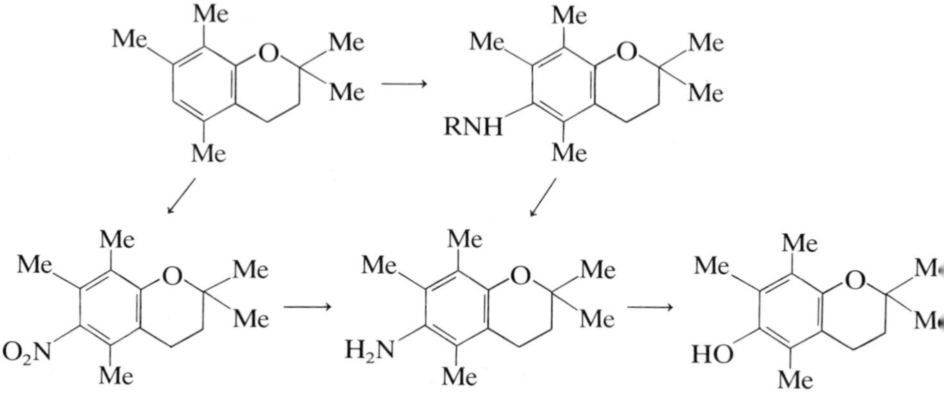

Scheme 6

An interesting new source for the phytyl side chain of tocopherol has been developed starting from shale oil.[162a] Pristane (2,6,10,14-tetramethylpentadecane), isolated via a thiourea adduct, is allowed to undergo microbial oxidation to pristanol.[162b] Easy chemical steps take pristanol to pristyl iodide, pristene-1, phytone, and finally phytol.

Cohen and Saucy[162c] have developed a synthesis of optically active vitamin E from a Wittig reaction of the side-chain triphenylphosphonium halide with the optically active chroman-2-aldehyde. The optically active aldehyde is produced from active 2-methyl-5-oxotetrahydro-2-furoic acid in a scheme involving over a dozen steps.

The reaction of chromanols with formaldehyde[163,164] and other aldehydes[165] or cyanide[166] has been used extensively to upgrade the α-tocopherol content of natural tocol mixtures[167] and also to produce the 5-labeled vitamin.[168] The reaction can take place under acid conditions (i.e., chloromethylation[169] or hydroxymethylation with HCHO and boric acid[170,171]) or basic conditions (i.e., hydroxymethylation).[165]

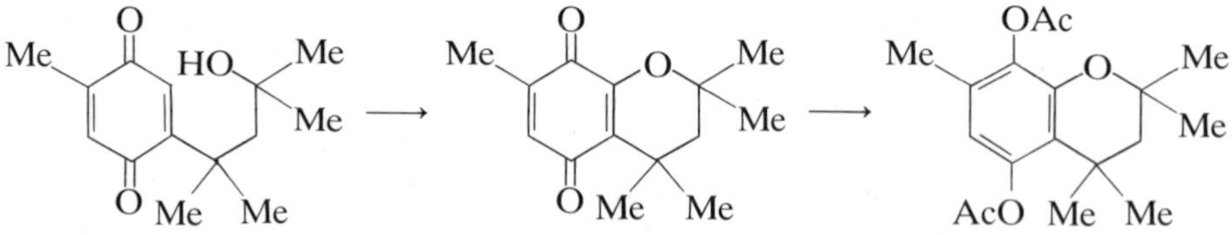

(41) **(42)** **(43)**

V. CHEMICAL REACTIONS

Reactions of the tocopherols and other hydroxychromans can be classified as reactions involving the phenolic hydroxyl group, reactions on the aromatic ring, reactions involving the aromatic methyl groups, and reactions of the oxygen ring system. A review[172] of reaction products of tocopherols covers the literature through 1970.

1. Oxidation Reactions Involving the Phenolic Hydroxy Group and 5-Methyl Group of 6-Hydroxy Derivatives

Because of their role as antioxidants, the oxidation of chroman-6-ols has been extensively studied. Oxidation can occur with various chemical oxidizing agents or even with air. These oxidations can result in a variety of oxidation products, depending upon the oxidizing agents, solvents, and conditions of the oxidation.

Most of these oxidation reactions have been covered in the aforementioned review[172] but bear repeating here. Much of the early research with the natural tocopherols involved their oxidation.

Oxidation of α-tocopherol with $FeCl_3$ resulted in formation of α-tocopherylquinone,[173] which could also be produced by oxidation of α-tocopherol with silver nitrate.[174] In addition, silver nitrate oxidation of α-tocopherol or the model chroman (2,2,5,7,8-pentamethylchroman-6-ol) yielded a red-colored product. This product was also formed in alcoholic solution.[175] This latter reaction formed the basis of the Furter–Meyer colorimetric method for tocopherol analysis,[69] while the ferric chloride oxidation of tocopherols was used in the Emmerie–Engel analytical method.[176]

The structure of this red oxidation product was shown by Smith, Irwin, and Ungnade[177] to be an orthoquinone (α-tocored). John and Emte[178] isolated a hydroxy-*p*-quinone upon oxidation of α-tocopherol with silver nitrate in boiling ethanol. A structure was proposed for this product which was purple colored in base and yellow in acid. Frampton et al.[179,180] later showed that this product could be formed by oxidation of α-tocopherol in methanol at 50°C with $FeCl_3$ and proved its structure unequivocally (α-tocopurple) (Fig. 4). Tocored could also be produced by $FeCl_3$ oxidation under these conditions. These oxidations under relatively mild conditions, resulting in the loss of aromatic methyl groups, are interesting and quite unusual. The extreme reactivity of the 5-methyl group of α-tocopherol and its model chromans becomes obvious as one studies their oxidation.

Boyer[181] isolated a product formed by oxidation of α-tocopherol with $FeCl_3$, which was converted by acid to α-tocopherylquinone. The correct structure for this product, "α-tocopheroxide," was shown by Martius and Eilingsfeld[182] not to be an epoxide but an ethoxide. Oxidation of α-tocopherol with tetrachloro-*o*-quinone or *N*-bromosuccinimide resulted in the formation of 9-hydroxy-α-tocophorone.[183] In the presence of alcohols, compounds related to the Martius compound, namely, alkoxides, were formed.

Free radical-initiated oxidation of α-tocopherol and its model compounds

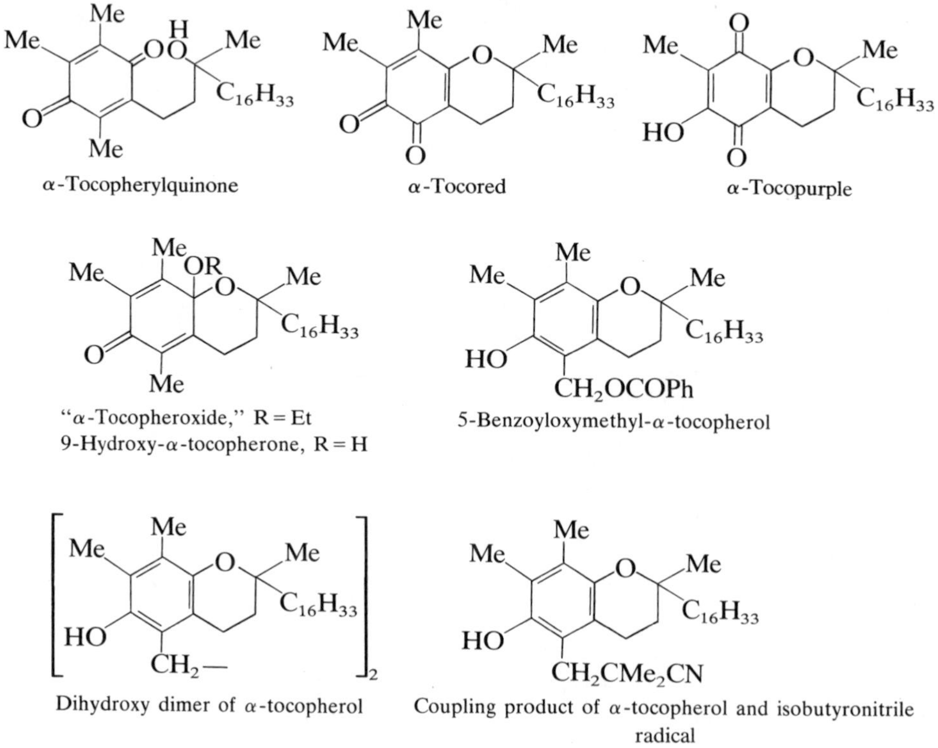

α-Tocopherylquinone

α-Tocored

α-Tocopurple

"α-Tocopheroxide," R = Et
9-Hydroxy-α-tocopherone, R = H

5-Benzoyloxymethyl-α-tocopherol

Dihydroxy dimer of α-tocopherol

Coupling product of α-tocopherol and isobutyronitrile radical

Fig. 4. Compounds related to tocopherols.

has led to interesting information about the mechanism of these oxidations and to new oxidation products. Inglett and Mattill[184] studied the oxidation of α- and γ-tocopherol and 2,2,5,7,8-pentamethylchroman-6-ol with benzoyl peroxide at 30°C. This yielded α-tocopherylquinone and the 5-benzoyloxymethyl derivative. γ-Tocopherol yielded α-tocored under these conditions. Goodhue and Risley[185] studied the oxidation of α-tocopherol in hydrocarbon solvents using benzoyl peroxide and obtained the same 5-benzoyloxymethyl derivative. In the presence of alcohols, benzoyl peroxide oxidation led to the 8*a*-alkoxy-α-tocopherones.[186] Oxidation of 2,2,5,7,8-pentamethylchroman-6-ol with azobisisobutyronitrile, another free-radical initiator, yielded a dihydroxy dimeric product and a coupling product from the initiator and the phenoxy radical.[187]

Oxidation of α-tocopherol or 2,2,5,7,8-pentamethylchroman-6-ol with alkaline ferricyanide yielded a spiro dimer.[188] When *p*-benzoquinone was used, three products resulted: the spiro dimer, α-tocopherylethane (dihydroxy dimer), and the trimer.[189] The tocopherylethane and the trimer are also formed when the spiro dimer and monomer are refluxed in toluene.[190] It has been suggested that a quinone methide[191] is formed from α-tocopherol upon oxidation by the spiro dimer and that this quinone methide undergoes a Diels–Alder reaction with another molecule of the dimer to form trimer. Nilsson et al.[192] found that the quinone methide is an intermediate when α-tocopherol is oxidized with alkaline ferricyanide. These interconversion products were investigated further using ^{14}C-labeled tocopherol.[193]

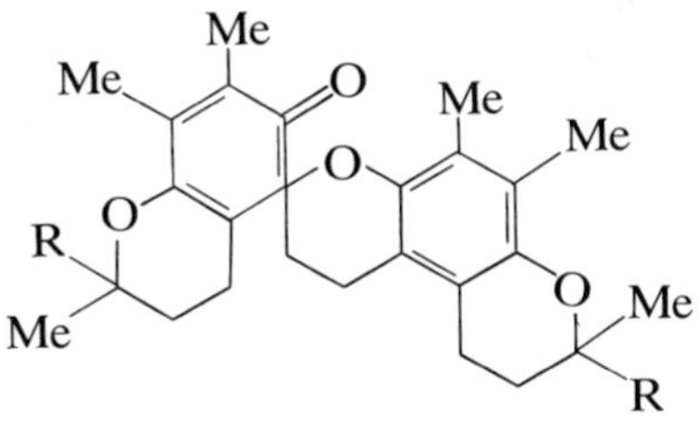

α-Tocopheryl spiro dimer, $R = C_{16}H_{33}$

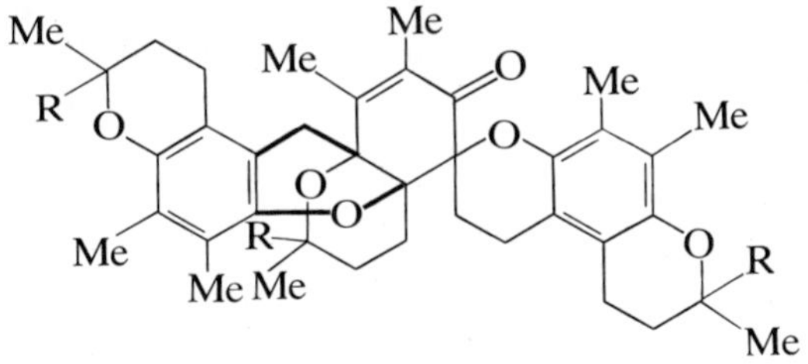

(**43a**) α-Tocopheryl trimer, $R = C_{16}H_{33}$
(**43b**) R = Me
(**43c**) R = H

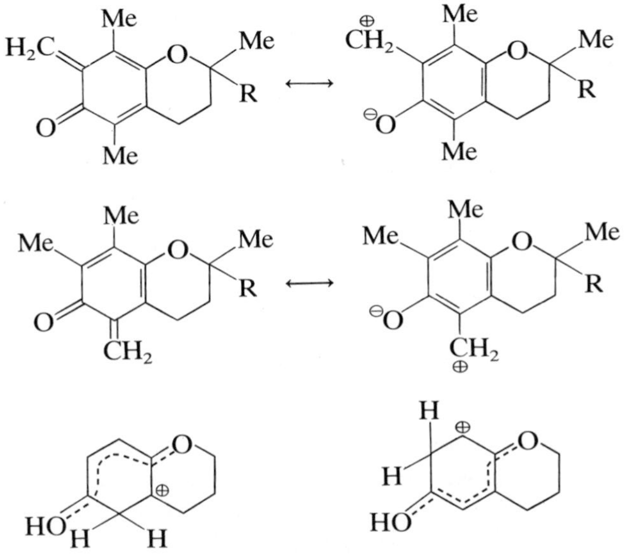

Scheme 7

The regiospecificity of the chroman-6-ol nucleus toward oxidation or electrophilic substitution has been discussed by Behan et al.[194] In the case of α-tocopherol, oxidation of the 5-methyl group occurs readily, but not that of the other methyl groups. In α-tocopherol oxidative coupling occurs at the 5-position. In other tocopherols the 5-position is involved in attack even in preference to other positions where no methyl substitution exists. This preference of the 5-position for oxidation occurs with a variety of oxidizing agents, for example, alkaline ferricyanide, benzoyl peroxide, silver salts, iron salts, quinones, and uv[195] or γ-irradiation.[196] Behan, Dean and Johnstone[194] discuss the activation of the 5-methyl group in terms of the quinone methide intermediate transition states (Scheme 7) and their aromatic stabilization. Oxidative couplings still occur in the absence of a 5-methyl group because of the depicted transition states.

Oxidation of α-tocopherol in air at 80°C in the presence of linoleic acid on silica gel adsorbent formed an additional compound.[197] Epoxy tocopheryl ethers are formed from α-tocopherol and linoleyl hydroperoxide.[198] Tocopheryl ethers are also formed from autoxidation of linoleate in the presence of α-tocopherol, when primarily four products, (**44**), (**45**), (**46**), and (**47**), are formed.[199]

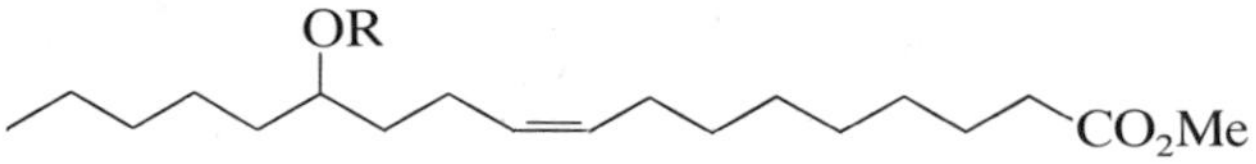

OR CO_2Me (**44**)

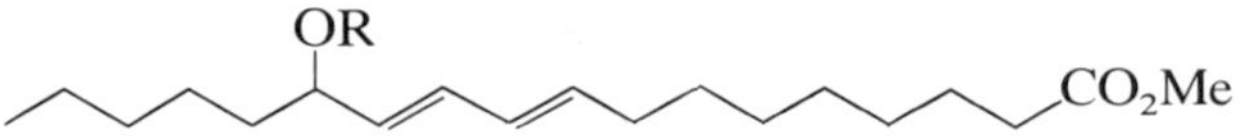

OR CO_2Me (**45**)

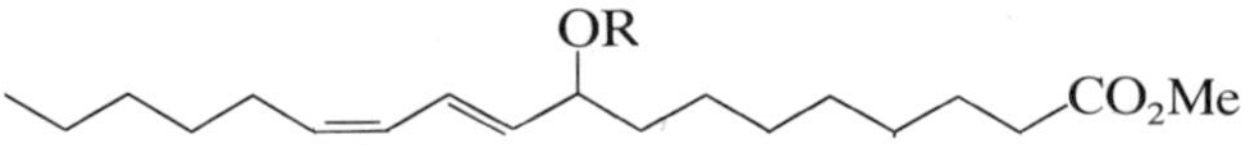

OR CO_2Me (**46**)

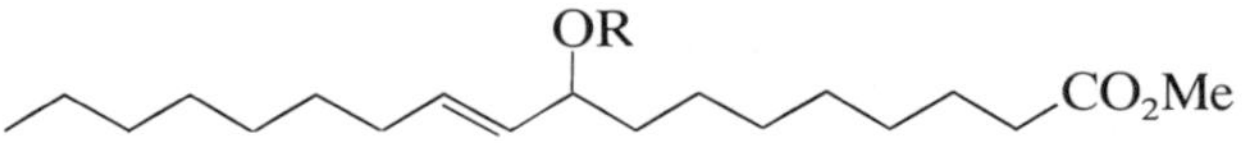

OR CO_2Me (**47**)

R = α-tocopheryl

The reaction of α-tocopherol with various alkyl radicals (ethyl, *n*-propyl, isopropyl, *n*-butyl, and *sec*-butyl) has been studied recently.[200] Two types of products were formed, alkyl ethers (**48**) and cyclohexadienones (**49**). The structures were identified by the use of mass-spectral and nmr analyses.

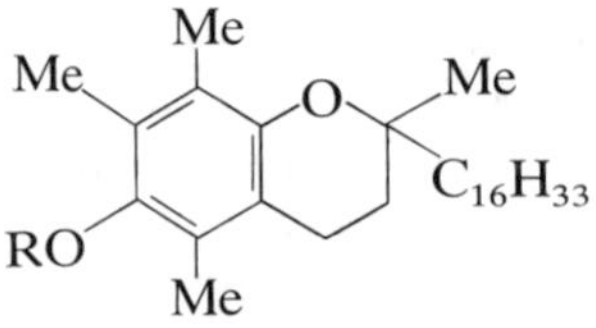

Me Me Me O RO Me $C_{16}H_{33}$ (**48**)

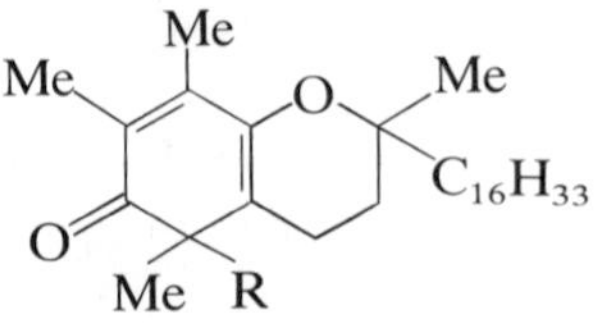

Me Me Me O O Me R $C_{16}H_{33}$ (**49**)

The oxidation of α-, γ-, and δ-tocopherols by trimethylamine oxide has been studied by Ishikawa and Yuki.[201,202] In addition to the compounds shown in Scheme 8, some other aldehydes were produced (Chapter VIII, Section II, 3A and 3D).

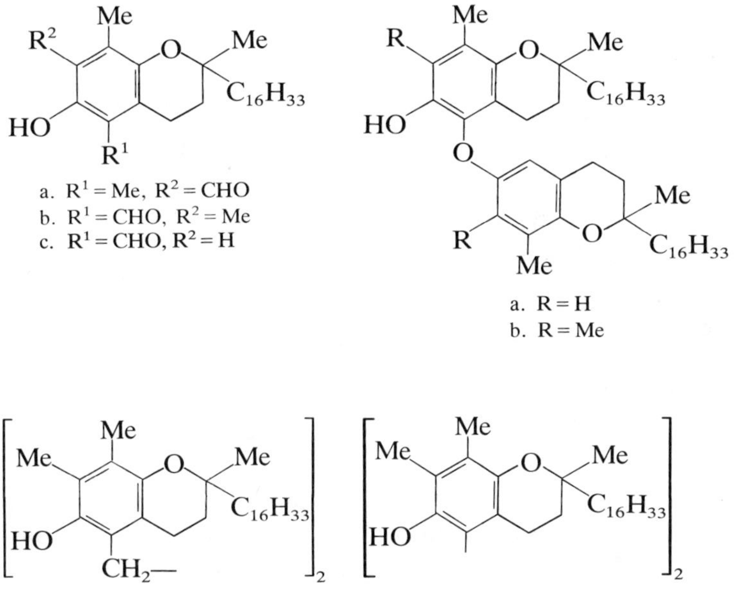

a. $R^1 = Me$, $R^2 = CHO$
b. $R^1 = CHO$, $R^2 = Me$
c. $R^1 = CHO$, $R^2 = H$

a. R = H
b. R = Me

Scheme 8

In *Vitamins and Hormones* Boguth[203] discusses the activity of vitamin E related to the hydroxy function of the molecule. Oxidation of α-tocopherol in nonpolar solvents and in polar solvents is discussed separately. Oxidation in nonpolar solvents results in some of the products previously discussed, for example, spirodienone ether (α-tocopheryl spiro dimer), 1,2-bis-γ-tocopherylethane (dihydroxy dimer), and trimer. Boguth discusses his own studies using the stable free radical 1,1-diphenyl-2-picrylhydrazyl (DPPH·) as initiator of oxidation. Oxidation of α-tocopherol in polar solvents forms $8a$-alkoxy-α-tocopherone and α-tocopherylquinone.

Using proflavin as the sensitizer and visible light, α-tocopherol was photo-oxidized in methanol to $4a$,5-epoxy-$8a$-methoxy-α-tocopherone, $8a$-methoxy α-tocopherone, γ-tocopherylquinone-2,3-oxide, and α-tocopherylquinone.[204] Singlet oxygen oxidation of α-tocopherol[205] and other tocopherols[206] was further investigated by Grams using methylene blue as the photosensitizer or generating singlet oxygen chemically from $Ca(OCl)_2$–H_2O_2.

Novel dimers and trimers were isolated[189,207–209] when chroman-6-ols were oxidized with alkaline ferricyanide or p-benzoquinone. Oxidation of these chromanols gave two different types of products, depending on the substitution pattern of the chromanol. When alkaline ferricyanide was used, chromanols with an unsubstituted 5-position formed spiroketal trimers (**50**) by carbon–carbon and carbon–oxygen coupling. When the 5-position was occupied by a methyl group, the spiro dimers (**51**) were produced by

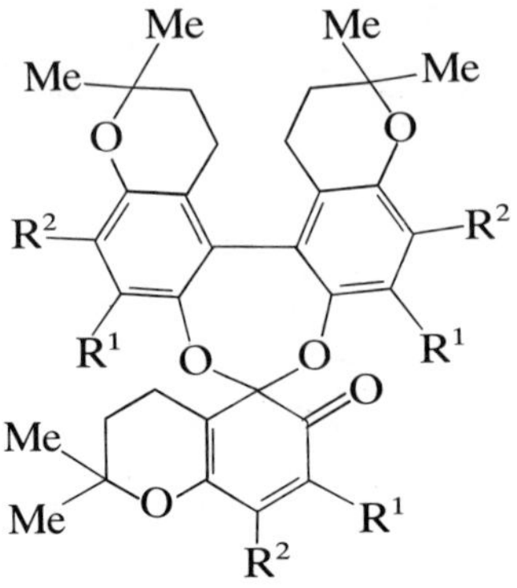

(**50**) a. $R^1 = R^2 = Me$
b. $R^1 = H, R^2 = Me$
c. $R^1 = R^2 = H$

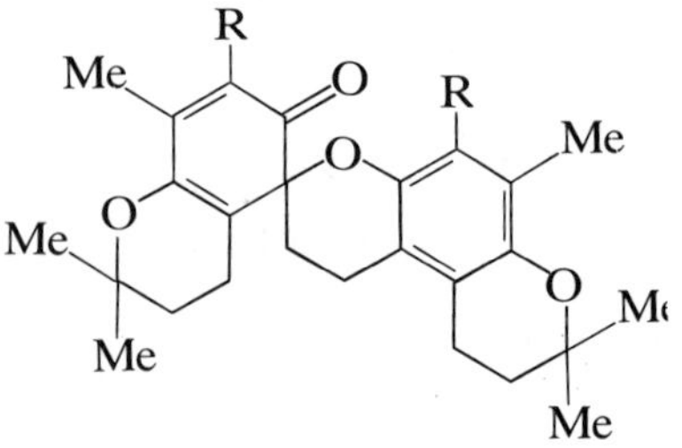

(**51**) R = H or Me

reactions involving both carbon–oxygen and benzylic coupling. Benzoquinone oxidation of 6-chromanols with an unsubstituted 5-position gave dimers (**52**) and (**53**) by coupling exclusively at the 5-position. When the chromanol has a 5-methyl group, benzylic coupling occurs via the methyl group yielding two dimers and a trimer (**43b 43c, 54–56**).

A urinary metabolite of α-tocopherol was synthesized by oxidation[210] of 2,5,7,8-tetramethyl-2-(β-carboxyethyl)chroman-6-ol. This oxidation yielded the lactone of 2-(3-hydroxy-3-methyl-5-carboxypentyl)-3,5,6-trimethylbenzoquinone.

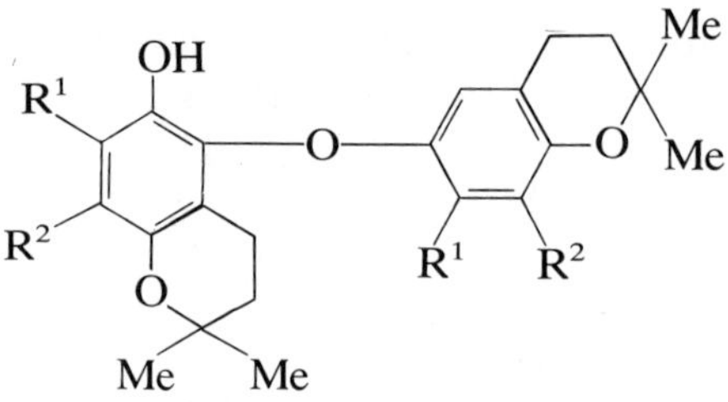

(**52**) $R^1 = R^2 = Me$
$R^1 = H, R^2 = Me$
$R^1 = R^2 = H$

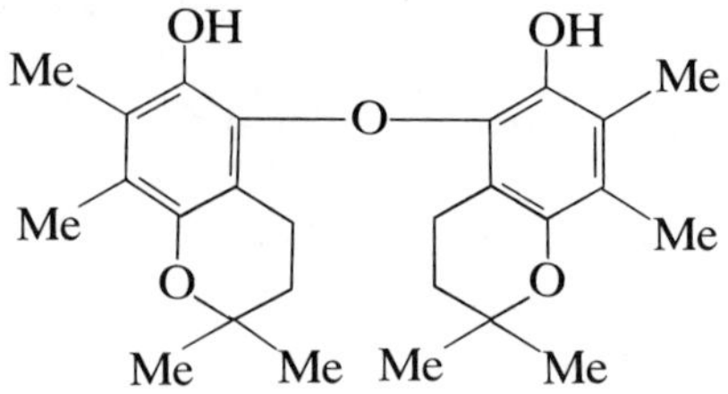

(**53**)

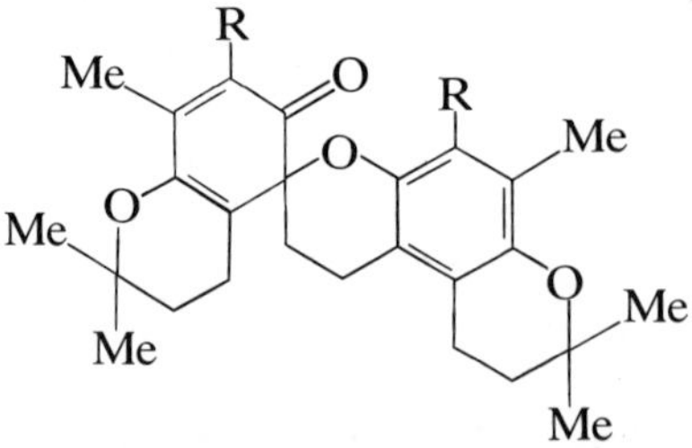

(**54**) R = H or Me

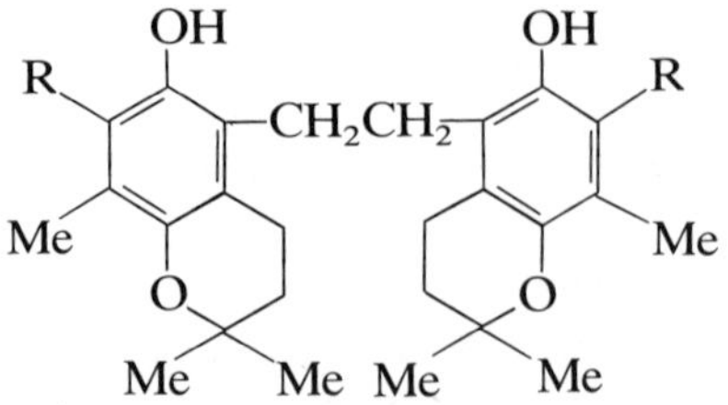

(**55**) R = H
(**56**) R = Me

2. Esterification of the Phenolic Hydroxyl Group

Numerous esters of α-tocopherol have been prepared. Phosphate esters were prepared from α-tocopheryl quinone by reaction of the quinone with dibenzyl phosphite. By this method α-tocopherylquinol monodibenzyl phosphate and α-tocopheryl monohydrogen phosphate were prepared.[211] Reaction of α-tocopherol with $POCl_3$ followed by diethanolamine[211] resulted in α-tocopheryl phosphate diethanolamine salt.[212] Refluxing α-tocopherol with nicotinic acid in the presence of $SOCl_2$ yielded α-tocopheryl nicotinate.[213] A number of phosphate derivatives of α-tocopherol and its model were prepared.[172] These included 5-hydroxymethyl-2,2,7,8-tetramethylchroman-6-ol.

α-Tocopherol has been reacted with isocyanato esters of several amino acids, dimethyl α-isocyanatoglutarate, and dimethyl α-isocyanatosuccinate and the esters saponified to give water-soluble compounds.[214] Reaction of α-tocopherol with $POCl_3$ in the presence of triethylamine yielded α-tocopheryl phosphorodichloridate, which was too unstable to isolate but which would react with ethyleneimine to yield the *N,N,N′,N′*-bis(ethylene)-phosphorodiamidate.[215] Weichet et al.[216] synthesized 2,5,7,8-tetramethyl-2-(β-carboxyethyl)-6-acetoxychroman from trimethylhydroquinone, γ-vinyl-γ-hydroxyvalerolactone, acetic acid, anhydrous zinc chloride, BF_3 etherate, and acetic anhydride.

3. Halogenation of Hydroxychromans

Bromination of α-tocopherol yielded the 5-bromomethyl derivative, which was converted to the spirodienone dimer upon treatment with 1 *N* KOH.[186] Treatment of α-tocopherylquinone with acetyl chloride in benzene yielded the 5-chloromethyl-6-acetoxy derivative.[217] The latter product was converted to a variety of 5-methyl-substituted derivatives. Also 5-acetoxymethyltocopheryl acetate was converted to 5-bromomethyl-α-tocopheryl acetate. Similarly prepared were 5-bromomethyl-γ-tocopheryl acetate and 5-bromomethyl-δ-tocopheryl acetate.[218]

Chlorination of 2,2,5,7,8-pentamethylchroman-6-ol and its acetate gave the 5-chloromethylchromanol and its acetate.[219] The chromanol was more reactive to chlorinating agents ($SOCl_2$, SO_2Cl_2, Cl_2, PCl_5) than was the acetate.

A study of the directive effects in the bromination of bicyclic phenols, including chromans, showed that bromination of chroman-6-ol yielded the 5-bromo derivative.[220] It was found that the heterocyclic ring in chromanol had a stronger directing effect than the alicyclic ring in tetralol. Bromination of 2,2,5,8-tetramethylchroman-6-ol in CCl_4 at 0°C afforded 7-bromo-2,2,5,8-tetramethylchroman-6-ol in 68% yield,[221] and of 2,2,7,8-tetramethylchroman-6-ol gave 5-bromo-2,2,7,8-tetramethylchroman-6-ol in

88% yield. Similar bromination of 2,2,8-trimethylchroman-6-ol gave 79% 5-bromo-2,2,8-trimethylchroman-6-ol and 4% 5,7-dibromo-2,2,8-trimethylchroman but none of the 7-bromo derivative.

4. Reactions on Available Positions of the Aromatic Ring of Hydroxychromans

Chloromethylation of β-, γ-, and δ-tocopherols has been achieved in good yields.[222–224] These derivatives have been reduced to α-tocopherol as a means of conversion of other tocopherols to the more biologically active α-tocopherol. Hydroxymethylation of these same compounds was also described.[225] Direct hydroxymethylation of β- and γ-tocopherols and their model compounds by the use of boric acid as catalyst gave 7-hydroxymethyl-β-tocopherol, 5-hydroxymethyl-γ-tocopherol, 7-hydroxymethyl-2,2,5,8-tetramethylchroman-6-ol, and 5-hydroxymethyl-2,2,7,8-tetramethylchroman-6-ol in 83, 86, 75, and 78% yields, respectively.

Formylation of the 5-position of β-, γ-, and δ-tocopherols has also been achieved,[226,227] as has their nitrosation.[228] Marcinkiewicz[229] studied the products obtained by reaction of isomeric tocopherols with sodium nitrite and concluded that they were nitrotocopherols, not nitroso compounds, as claimed by Quaife.[228] Aminomethylation is also described by Weisler.[230]

5. Complexes of Hydroxychromans

Robeson and Nelan[231] have prepared solid piperazine complexes from chroman-6-ols.[232] The molar ratio of chromanol to piperazine is 2:1. Water splits the complex into its components.

6. Reactions Involving the Chroman Heterocyclic Ring

Oxidation of α-tocopherol acetate with dichlorodicyanohydroquinone yielded the 3,4-dehydro acetate. The same compound could be obtained via bromination, followed by dehydrohalogenation.[233] Cleavage of the heterocyclic ring of 2,2-dimethylchromans is difficult. Molyneux[234] found that treatment of 6-acetyl-7,8-dimethoxy-2,2-dimethylchroman with BCl_3 yielded the triol (**57**). Reaction of 6-acetyl-2,2-dimethylchroman-7,8-diol gave the same product.

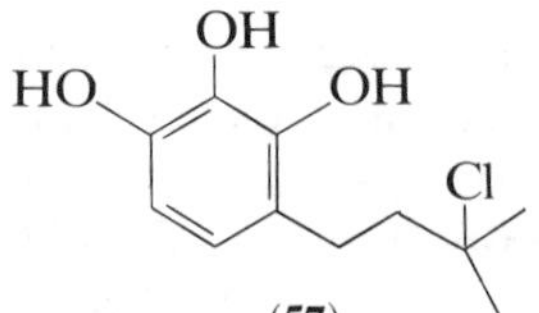

OH
HO OH
Cl

(**57**)

VI. NATURAL PRODUCTS—BIOSYNTHESIS

The tocopherols occur naturally, being found in a variety of plant species. They perform the role of natural antioxidants and protect the lipids in plants from excessive oxidation. Because of the interest in biosynthetically derived syntheses of natural products, a brief review of the biosynthetic pathways to the tocopherols should be instructive.

Whistance and Threlfall[235] have studied the biosynthesis of chromanols and phytoquinones in maize shoots and ivy. They found that the nucleus and one nuclear methyl group of α-, γ-, and δ-tocopherol or α-tocopherylquinone could be formed from the aromatic ring and the β-carbon atom, respectively, of either phenylalanine or tyrosine. Previous experiments by this group[236,237] had shown the incorporation of (2-^{14}C)mevalonic acid, (methyl-^{14}C, $^{3}H_3$)methionine, (G-^{14}C)shikimic acid, and *p*-hydroxy(U-^{14}C)-benzoic acid, L-(U-^{14}C)-tyrosine, and L-(U-^{14}C)phenylalanine into these compounds (Scheme 9). The methyl groups in the 5- and 7-positions of α-tocopherol probably arise from methionine.

Pennock, Hemming, and Kerr[238] had earlier proposed that the final stages of α-tocopherol biosynthesis involved methylation of δ-tocotrienol to β-or γ-tocotrienol and a further methylation to α-tocotrienol, followed by a reduction to α-tocopherol. Wellburn[239] found that tritium from labeled NADPH was incorporated into tocopherols when δ-tocotrienol was incubated with *Ficus elasticus* leaves or latex.

At this time there seem to be two biosynthetic pathways, both starting with homogentisic acid. Condensation with geranylgeraniol pyrophosphate gives 6-geranylgeranyltoluquinol and δ-tocotrienol, while the other route

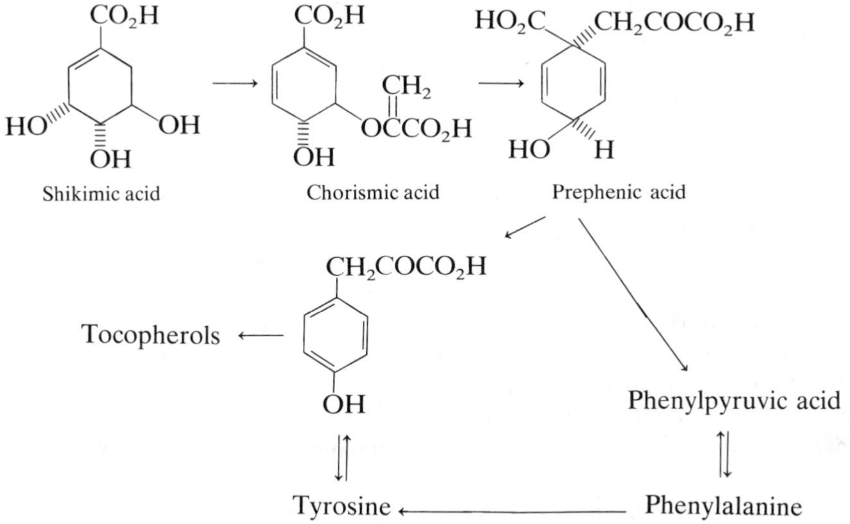

Shikimic acid → Chorismic acid → Prephenic acid

Prephenic acid → p-HOC$_6$H$_4$CH$_2$COCO$_2$H → Tocopherols

Prephenic acid → Phenylpyruvic acid ⇌ Phenylalanine → Tyrosine ⇌ p-HOC$_6$H$_4$CH$_2$COCO$_2$H

Scheme 9

involves condensation with phytyl pyrophosphate to give 6-phytyltoluquinol and β-tocopherol. The first pathway is referred to as the tocotrienol path, and the second the tocopherol path. Reduction of the side chain in the tocotrienols allows conversion to the tocopherols. The tocotrienol pathway is not present in leaf cells and has a more limited distribution. Within the cell there seem to be two sites of tocopherol biosynthesis, inside and outside the chloroplast.[241]

All these results have been summarized by Threlfall.[240,241] However up to the present no concerted attempts to mimic this biosynthetic scheme to produce the tocopherols synthetically have been conducted.

VII. BIOLOGICAL SIGNIFICANCE

The biological significance of the naturally occurring tocopherols, especially α-tocopherol (vitamin E), has been debated since their discovery. Some of the significant findings are briefly discussed here.

Bieri and Farrell[242] have recently brought into focus the many nutritional and biological aspects of vitamin E. The effect of vitamin E deficiency on cellular membranes has been reviewed.[243]

Ivanov, Merzlyak, and Tarusov[244] have reviewed the role of vitamin E in a number of important problems, including aging, radiation injury, malignant growth, and liver damage.

Vitamin E deficiency in laboratory animals results in a variety of symptoms: hemolysis of erythrocytes,[245,246] swelling, and respiratory decline of mitochondria,[247] labilization of lysosomes[248] and lowering of drug hydroxylation in microsomes.[249–251]

The following lesions have been ascribed to deficiency of vitamin E: gestation resorption, testicular degeneration, muscular dystrophy, encephalo malacia, exudative diathesis (chick), browning of the uterus, necrotic liver degeneration, kidney autolysis, red cell hemolysis, and gastric ulcers. While the specificity of some, if not all, of these changes have been questioned, some have been standardized as a bioassay.[252]. Blood hemolysis, gestation resorption, liver storage, and, most recently, the body-wall outgrowth in the rotifer (*Asplanchna sieboldi*) have been used[253] to estimate tocopherol 10^6 times more dilute than the detection limits of previously reported bioassays. Goodhue[254] has developed a bioassay that uses a soil organism tentatively identified as a *Mycobacterium*. The growth rate is α-tocopherol dependent as well as stereospecific when other nutrients are excluded from the growth media. A thorough discussion of the antioxidant theory of the biological action of α-tocopherol has been presented by Green,[255,256] who concluded that this theory does not satisfactorily explain the biological role of this vitamin.

Despite the confusion regarding the role of vitamin E and even its value in some cases, it is interesting to note that no lesion has been reported due to hypervitaminosis in the case of vitamin E. Mice are able to withstand doses of 50 g/kg daily for periods of two months.[251–261]

The specific role of vitamin E in heme synthesis discovered by Dinning[262] and clarified further by Murty, Coasi, Brooks, and Nair is of great importance.[263] A recent finding is that α-tocopherol protects mice against the cardiotoxic effects of the anticancer drug adriamycin.[264]

Schroeder[265] has found several analogs of α-tocopherol that lack the terpenoid side chain to be potent, noncompetitive inhibitors of cyclic AMP and cyclic GMP phosphodiesterases from beef liver. The effect of α-tocopherol disodium phosphate on cyclic nucleotide phosphodiesterases from a soluble supernatant fraction of rat liver has been studied.[266]

VIII. TABLES OF COMPOUNDS

TABLE 7. 5-HYDROXYCHROMANS

Mol. formula	Substituents						m.p.	Ref.
		2	2′	3	4	7		
$C_{11}H_{14}O_2$		Me	Me				122	32, 42, 58, 127, 267
	Acetate						63	32
	Benzoate						78	32
	3,5-Dinitrobenzoate						133	267
$C_{12}H_{16}O_2$	Me	Me	Me			Me	Oil	33
$C_{14}H_{20}O_2$						C_5H_{11}		268
$C_{16}H_{22}O_2$		Me	Me			[a]		270
$C_{16}H_{24}O_2$		Me	Me			C_5H_{11}		269
$C_{17}H_{24}O_2$		Me	[b]			Me		271
$C_{17}H_{26}O_2$		Me	Me			C_5H_{11}		269
$C_{21}H_{32}O_2$		Me	[b]			C_5H_{11}		272, 273
	3,5-Dinitrocarbanilate						97	272, 273
$C_{21}H_{34}O_2$		Me	Me	Bu	Me	C_5H_{11}		269
$C_{21}H_{34}O_2$		Me	[c]			C_5H_{11}		272
	3,5-Dinitrocarbanilate						128	272
$C_{22}H_{32}O_2$		Me	[d]			Me		274, 275
$C_{22}H_{32}O_2$		Me	[e]			Me		275
	Acetate							275
$C_{22}H_{32}O_2$		Me	[f]			Me		275
	Acetate							275
$C_{22}H_{32}O_2$		Me	[g]			Me		33, 276
$C_{25}H_{40}O_2$		Me	Me		[h]	[i]		277
$C_{26}H_{42}O_2$		Me	Me		[j]	[i]		277
	Acetate							277

[a] $CH_2{=}CMe_2$
[b] $CH_2CH_2CH{=}CMe_2$
[c] $(CH_2)_3CHMe_2$
[d]

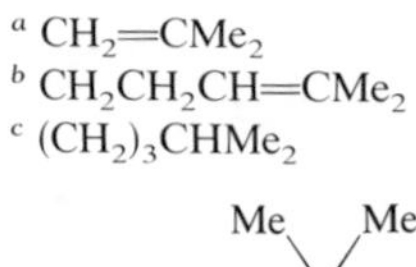

[e] CH_2CH_2– (2,6,6-trimethylcyclohex-1-enyl; Me, Me, Me)

[f] CH_2CH_2– (2,6,6-trimethylcyclohex-2-enyl; Me, Me, Me)

[g] $[CH_2CH_2CH{=}CMe]_2Me$
[h] Cyclopentyl
[i] $CHMeCHMe(CH_2)_4Me$
[j] Cyclohexyl

TABLE 8. 6-HYDROXYCHROMANS AND SOME TOCOPHEROLS

Mol. formula	Substituents 2	2′	3	4	4′	5	7	8	m.p. or b.p. (mm)	Ref.
$C_9H_{10}O_2$										307–309
Acetate										307–309
$C_{10}H_{12}O_2$	Me									137, 310
4-Nitrobenzoate									120	137
$C_{10}H_{12}O_2$						Me				311
$C_{11}H_{14}O_2$	Me	Me							75	34, 54, 310
$C_{11}H_{14}O_2$	Me		Me						150	137
4-Nitrobenzoate									105	137
$C_{11}H_{14}O_2$						Me		Me	127	208
$C_{11}H_{14}O_2$							Et			312
$C_{11}H_{14}O_2$							Me	Me	143	208
$C_{12}H_{15}DO_2$	Me	Me					D	Me		209
$C_{12}H_{16}O_2$	Me	Me	Me							45
$C_{12}H_{16}O_2$	Me	Me		Me					108	313
$NH_2CONHCO$									182	313
$C_{12}H_{16}O_2$	Me	Me				Me			61.3	45
$C_{12}H_{16}O_2$	Me	Me					Me		88	45
$C_{12}H_{16}O_2$	Me	Me						Me	85.5	45, 315
$C_{12}H_{16}O_2$	Me					Me		Me	128	314
$C_{12}H_{16}O_2$						Me	Me	Me		316
$C_{13}H_{18}O_2$	Me	Me				Et				317
$C_{13}H_{18}O_2$	Me	Me				Me	Me			310
$C_{13}H_{18}O_2$	Me	Me				Me		Me		310
$C_{13}H_{18}O_2$	Me	Me					Me	Me	85.5	193, 318
Acetate									71.5	318
$C_{13}H_{18}O_2$	Me					Me	Me	Me	143	67
	$2\text{-}^{14}CH_3$									193
Acetate									102	318
Allophanate										174

TABLE 8. (*Contd.*)

Mol. formula		Substituents 2	2′	3	4	4′	5	7	8	m.p. or b.p. (mm)	Ref.
$C_{14}H_{20}O_2$		Me	Me				Me	Me	Me	95	320
		5-$^{14}CH_3$									193
	$ClCH_2CO$									114	321
	Acetate									93.5	321
	$NH_2CONHCO$										129
	Ethylcarbonate									50	321
	$COOHCH_2CH_2CO$									138	321
	Nicotinate									139	322
	4-Bromobenzoate									159	129
	3,5-Dinitrophenylcarbamate									208	323
	3,4-Diacetoxycinnamate										324
	Phosphate									175	172
$C_{14}H_{20}O_2$		Me		Me			Me	Me	Me	108	129
$C_{14}H_{20}O_2$		Et					Me	Me	Me	116	325
$C_{14}H_{20}O_2$					Me	Me	Me	Me	Me	122 (0.3)	326
$C_{15}H_{18}O_2$		C≡CH	Me				Me	Me	Me		329
	Acetate									124	329, 330
$C_{15}H_{20}O_2$		CH═CH$_2$	Me				Me	Me	Me		329
$C_{15}H_{22}O_2$		Et	Me				Me	Me	Me	62.5	327, 328
	3,5-Dinitrophenylcarbamate									201.5	327
	R(−)										26
	Acetate									99	329
	S(−)										329
$C_{15}H_{22}O_2$	Me	Me	Me		Me			Pr^i			331
$C_{15}H_{22}O_2$	Me	Me	Me					Bu^t			332
$C_{16}H_{22}O_2$		[a]	Me								276
	4-Nitrophenylphosphate									97	276
$C_{16}H_{22}O_2$	[b]	[b]	Me				Me	Me	Me		333
	3,5-Dinitrobenzoate										333

$C_{16}H_{24}O_2$	Me	Pr		Me	Me	Me	59	328
$C_{16}H_{24}O_2$	Me	Pr[i]		Me	Me	Me	80	334
$C_{16}H_{24}O_2$	Me	Me	Me		Bu[t]			317, 331
$C_{16}H_{24}O_2$	Me	Me		Me	Bu[t]		144	332
$C_{17}H_{18}O_2$	Me	Me			Ph			317
$C_{17}H_{24}O_2$	Me	*b*		Me		Pr[i]		333
$C_{17}H_{24}O_2$	Me	Me				*c*		334
$C_{17}H_{24}O_2$	Me	*d*		Me	Me	Me	73	334
$C_{17}H_{26}O_2$	Me	*e*		Me	Me	Me	44.5	323
3,5-Dinitrophenylcarbamate							190	323
$C_{17}H_{26}O_2$	Me	Bu[t]		Me	Me	Me	105	334
$C_{17}H_{26}O_2$	Me	Me	Pr[i]		Pr[i]			331
$C_{18}H_{26}O_2$	Me	Me		*f*	Me	Me		335
$C_{18}H_{28}O_2$	Me	Me	Pr[i]		Bu[t]		149	331
$C_{19}H_{28}O_2$	Me	*a*		Me	Me	Me		336, 337
Acetate								338
$C_{19}H_{28}O_2$	Me	*g*		Me	Me	Me		33
3,5-Dinitrobenzoate								33
$C_{19}H_{30}O_2$		C_9H_{19}	Me				178–183 (0.2)	339
$C_{19}H_{30}O_2$	Me	*h*		Me	Me	Me		340
Acetate								341
$NH_2CONHCO$							201	340
$C_{19}H_{30}O_2$	Me	Me			C_7H_{15}	Me	61	342
$C_{20}H_{24}O_2$	Me	Me			*i*			343
$C_{20}H_{24}O_2$	Me	Me		$PhCH_2$	Me	Me		344
$C_{20}H_{30}O_2$	Me	*c*		Me		Pr[i]		333
$C_{20}H_{32}O_2$	Me	Me	Me		*k*			346, 349
$C_{20}H_{32}O_2$	Me	Me	Me		C_8H_{17}		331, 347, 348	331, 347, 348
$C_{20}H_{32}O_2$	Me	Me			C_9H_{19}			345
$C_{20}H_{32}O_2$	Me	Me				C_9H_{19}		345
$C_{21}H_{26}O_2$	Me	Me			*a*			343
$C_{21}H_{30}O_2$	Me	*l*						276
$C_{22}H_{32}O_2$	Me	*m*		Me		Me		286
4-Phenylazobenzoate								286
$C_{22}H_{32}O_2$	Me	*n*		Me	Me	Me		336

TABLE 8. (*Contd.*)

Mol. formula		Substituents 2	2′	3	4	4′	5	7	8	m.p. or b.p. (mm)	Ref.
$C_{22}H_{36}O_2$		Me	[o]				Me	Me	Me		350
$C_{22}H_{36}O_2$		Me	C_8H_{17}					Bu[t]			351
$C_{22}H_{36}O_2$		Me	Me		Pr[i]			[k]		151	331, 352
$C_{22}H_{36}O_2$		Me	Me		Pr[i]			C_8H_{17}		82.5	353
$C_{23}H_{38}O_2$		$C_{13}H_{27}$			Me					200 (0.2)	339
$C_{24}H_{36}O_2$		Me	[l]				Me	Me	Me		333
$C_{24}H_{40}O_2$		Me	[o]				Me	Me	Me		355
	Acetate										355
	Allophanate									170	356
$C_{25}H_{38}O_2$		Me	[l]				Me		Pr[i]		333
$C_{25}H_{42}O_2$		$C_{15}H_{31}$			Me					194 (0.2)	339
$C_{25}H_{42}O_2$		$C_{12}H_{25}$					Me	Me	Me	61	357
	Acetate									43	357
	Allophanate									180	129
$C_{26}H_{44}O_2$		Me	[p]							165–175 (0.001)	358, 359
(tocol)	Acetate									52	362
										180–185 (0.001)	359
	Allophanate (2RS,4′R,8′R)									150	359
	Propionate									193 (0.05)	363
	3,5-Dinitrophenylcarbamate (2*RS*,4′*R*8′*R*)									97	358
	4-Nitrophenylcarbamoyl (2*RS*,4′*R*,8′*R*)									95	359
	4-Phenylazobenzoate									38	358
$C_{26}H_{44}O_2$		Me	$C_{13}H_{27}$				Me	Me	Me	65	364
	Acetate									52	364
$C_{26}H_{44}O_2$		Me	Me				[q]	Me	Me		365
$C_{27}H_{46}O_2$		Me	[p]				Me				
	(2*RS*)										103, 155
	(2*R*)										156

	Acetate							103
	1:1 Pyridine complex							50
	4-Phenylazobenzoate						70	156
$C_{27}H_{46}O_2$		Me	p		Me			
	(2*RS*)							103, 131
	(2*R*)							
	Acetate							103, 157, 287
	1:1 Pyridine complex							50
	3,5-Dinitrophenylcarbamoyl						117	131
	4-Nitrophenylcarbamoyl						112	155
	4-Phenylazobenzoate						56	131
$C_{27}H_{46}O_2$ (δ-tocopherol)		Me	p			Me		103, 367, 368
	Acetate (2*RS*)							157
	Acetate (2*R*)							157
	4-Phenylazobenzoate (2*R*,4′*R*,8′*R*)						42	368, 369
	4-Phenylazobenzoate (2*RS*,4′*R*,8′*R*)						250 (0.001)	157
$C_{28}H_{30}O_2$		Me	r	Me	Me	Me		329
	Acetate						104.5	329
$C_{28}H_{48}O_2$		Me	p	Me	Me		−4	131
(2*RS*)	Formate						223 (0.1)	371
	Acetate						180–210 (0.1)	89, 370, 371
	Allophanate						150	359, 372–375
	3,5-Dinitrophenylcarbamate						65	131
	4-Nitrophenylcarbamate						90	359, 376
	4-Phenylazobenzoate						61	131
$C_{28}H_{48}O_2$ (β-tocopherol)		Me	p		Me		Me	
	(2*R*,4′*R*,8′*R*)						140–170 (0.001)	173, 293, 376–381
	(2*RS*,4′*R*,8′*R*)							
	Acetate (2*R*,4′*R*,8′*R*)							103
	Allophanate (2*R*,4′*R*,8′*R*)						147	382, 384, 385
	Allophanate (2*RS*,4′*R*,*S*′*R*)						155	366
	$COOHCH_2CH_2CO$ (2*RS*,4′*R*,8′*R*)						Oil	381
	S-Benzylisothiuronium salt						106	381
	4-Nitrobenzoate (2*R*,4′*R*,8′*R*)						40	387

TABLE 8. (*Contd.*)

Mol. formula	Substituents 2	2′	3	4	4′	5	7	8	m.p. or b.p. (mm)	Ref.
$C_{28}H_{48}O_2$ (β-tocopherol) (*Contd.*)										
	4-Nitrophenylcarbamoyl (2*R*,4′*R*,8′*R*)								90	366
	4-Nitrophenylcarbamoyl (2*RS*,4′*R*,8′*R*)								112	376
	4-Phenylazobenzoate (2*R*,4′*R*,8′*R*)								71	70
	9,10-Dioxo-9,10-dihydroanthracen-2-carbonyl (2*R*,4′*R*,8′*R*)								79	388
	3,5-Dinitrophenylcarbamate								153–155	323
$C_{28}H_{48}O_2$ (γ-tocopherol)	Me	p					Me	Me		
	(2*R*,4′*R*,8′*R*)								−2	70, 389
	(2*RS*,4′*R*,8′*R*)									70, 389
	Acetate									103
	Acetate (2*RS*)								150–160 (0.01)	390
	Allophanate (2*R*)								140	384, 391
	Allophanate (2*RS*)								146	392
	MeOCO									393
	$COCH_2OPO(OH)_2$									394
	COOEt									393
	$COOHCH_2CH_2CO$								Oil	381
	S-Benzylisothiuronium salt (2*R*)								105	381
	S-Benzylisothiuronium salt								106	381
	BuOCO									363
	3,5-Dinitrophenylcarbamoyl (2*RS*)								145	323
	4-Nitrophenylcarbamate (2*R*)								121	381
	4-Nitrophenylcarbamate (2*RS*)								102	376, 381
	$COCH_2OP(=O)(OCH_2Ph)_2$									394
	Palmitoyl (2*R*)								45	70

$C_{28}H_{48}O_2$		Me	$C_{15}H_{31}$			Me	Me	Me	68	357
	Disodium phosphate									146
	Acetate								63	377
$C_{28}H_{48}O_2$		Me	*s*			Me	Me	Me		
	Allophanate								172	395
$C_{28}H_{48}O_2$		Me	Me		Pr^i		$C_{14}H_{29}$			331, 396
$C_{29}H_{40}O_2$		Me	*t*			Me	Me	Me		149, 333
	3,5-Dinitrobenzoate									149, 333
$C_{29}H_{48}O_2$		Me	*u*			Me	Me	Me		329, 423
	Acetate									423
$C_{29}H_{48}O_2$	Me	Me	*p*				—$(CH_2)_3$—			423
	Acetate								213 (0.04)	424
	Allophanate								149	424
$C_{29}H_{50}O_2$		Me	*p*		Me		Me	Me		425
$C_{29}H_{50}O_2$		Me	*p*			*v*				426
$C_{29}H_{50}O_2$		*p*		Me		Me	Me	Me		473
$C_{29}H_{50}O_2$		*p*	Me					Pr^i		474
$C_{29}H_{50}O_2$		Me	n-$C_{16}H_{33}$			Me	Me	Me	68	146, 430
	Disodium phosphate									146
	Acetate								58	146, 357
	Allophanate								176	427
	$COCH_2OMe$								50	428
	3,5-Dinitrobenzoate								75	429
$C_{29}H_{50}O_2$ (α-tocopherol)		Me	*p*			Me	Me	Me		See Table 9
$C_{30}H_{46}O_2$		Me	*p*				*w*	*w*		331
$C_{30}H_{50}O_2$		Me	*p*			*x*				475
$C_{30}H_{50}O_2$		Me	*p*			*y*				475
$C_{30}H_{50}O_2$		Me	*p*			*z*				475
$C_{30}H_{50}O_2$		Me	*p*				*z*			475
$C_{30}H_{50}O_2$		Me	*p*			*v*		Me		476
$C_{30}H_{50}O_2$		Me	*p*				*v*	Me		476
$C_{30}H_{50}O_2$		Me	*p*				*aa*	*aa*		424
	Acetate								220–240 (0.04)	424
	Allophanate								162	424

TABLE 8. (*Contd.*)

Mol. formula		Substituents 2	2′	3	4	4′	5	7	8	m.p. or b.p. (mm)	Ref.
$C_{30}H_{52}O_2$		Me	$C_{17}H_{35}$				Me	Me	Me	73	357, 364
	Acetate									65	357, 364
$C_{30}H_{52}O_2$	Et	Et	[p]				Me	Me	Me		477
	Acetate									208 (0.05)	477
	Allophanate									152	477
$C_{30}H_{52}O_2$		Me	[p]				Et	Et			357
	Allophanate									107	357
$C_{30}H_{52}O_2$		Me	[p]				Et	Me	Me		323, 479
	3,5-Dinitrophenylcarbamate									48	323, 479
$C_{30}H_{52}O_2$		Me	[p]				Me	Et	Me		323
	3,5-Dinitrophenylcarbamate									69	323
$C_{30}H_{52}O_2$		Me	[p]				Me	Me	Et		323
	3,5-Dinitrophenylcarbamate									60	323
$C_{30}H_{52}O_2$		Me	[p]				Me		Pr^i		480
	Acetate										480
$C_{31}H_{48}O_2$		Me	[p]				Me	[w]	[w]		481
	4-Nitrobenzoate									85	482
$C_{31}H_{52}O_2$		Me	[p]				[z]		Me		476
$C_{31}H_{52}O_2$		Me	[p]					[z]	Me		67
$C_{31}H_{52}O_2$		Me	[p]				[v]	Me	Me		67
$C_{31}H_{52}O_2$		Me	[p]				Me	[v]	Me		67
	Allophanate									165	67

$C_{31}H_{54}O_2$	Pr		[p]	Me	Me	Me		477
	Acetate						215 (0.005)	477
	Allophanate						148	477
$C_{31}H_{54}O_2$		Me	[p]	Et	Et	Me		483
$C_{31}H_{54}O_2$		Me	[p]	Me		Pr^i		474
$C_{32}H_{54}O_2$		Me	[p]	[x]	Me	Me		67
$C_{32}H_{54}O_2$		Me	[p]	Me	[x]	Me		67
	Allophanate						142	67
$C_{32}H_{54}O_2$		Me	[p]	[z]	Me	Me		426
$C_{32}H_{54}O_2$		Me	[p]	Me	[z]	Me		426
$C_{32}H_{56}O_2$		Me	[p]	Et	Et	Et		484
	Acetate						140–160 (0.01)	484
$C_{34}H_{52}O_2$ (all trans)		Me	[bb]	Me	Me	Me		485, 491
	Acetate							485
$C_{34}H_{60}O_2$		Me	[cc]	Me	Me	Me		203, 490, 491
$C_{35}H_{52}O_2$		Me	[dd]	[ee]				475
	Acetate							475
$C_{35}H_{52}O_2$		Me	[dd]		[ee]			475
	Acetate							475
$C_{36}H_{64}O_2$	[ff]		[ff]	Me	Me	Me	28	129
	Allophanate						116	129
$C_{39}H_{60}O_2$		Me	[gg]	Me	Me	Me		491
$C_{39}H_{70}O_2$		Me	[hh]	Me	Me	Me		203, 490, 491
$C_{44}H_{68}O_2$		Me	[ii]	Me	Me	Me		491
$C_{44}H_{80}O_2$		Me	[jj]	Me	Me	Me		203, 490, 491
$C_{44}H_{80}O_2$	[kk]		[kk]	Me	Me	Me		429
	Allophanate						109	429
$C_{49}H_{76}O_2$		Me	[ll]	Me	Me	Me		491
$C_{49}H_{90}O_2$		Me	[mm]	Me	Me	Me		203, 490, 491

TABLE 8. (*Contd.*)

Mol. formula	Substituents 2	2′	3	4	4′	5	7	8	m.p. or b.p. (mm)	Ref.
$C_{53}H_{82}O_2$ (2*R*, all trans)	Me	*nn*					Me	Me		288
$C_{53}H_{82}O_2$ (2*RS*)	Me	*nn*					Me	Me		288, 487–489
$C_{54}H_{84}O_2$	Me	*nn*				Me	Me	Me		491
$C_{54}H_{100}O_2$	Me	*oo*				Me	Me	Me		203, 288, 490, 491
$C_{58}H_{90}O_2$	Me	*pp*					Me	Me		486
$C_{58}H_{100}O_2$	Me	*qq*					Me	Me		486
$C_{59}H_{92}O_2$	Me	*pp*				Me	Me	Me		486, 491
$C_{59}H_{110}O_2$	Me	*qq*				Me	Me	Me		203, 490, 491

[a] $(CH_2)_2CH{=}CMe_2$
[b] cyclopropyl
[c] $CMe_2CH_2CH{=}CH_2$
[d] $CH{=}CMe_2$
[e] CH_2CHMe_2
[f] $CH_2CH{=}CMe_2$
[g] $CHEtCMe{=}CH_2$
[h] $(CH_2)_3CHMe_2$
[i] CMe_2Ph
[j] $CHMe(CH_2)_2Ph$
[k] $CMe_2CH_2CMe_3$
[l] $(CH_2CH_2CH{=}CMe)_2Me$
[m] $[(CH_2)_2CH{=}CMe]_2H$
[n] Me Me Me
[o] $[(CH_2)_3CHMe]_2Me$
[p] $[(CH_2)_3CHMe]_3Me$(phytyl)
[q] $CHPh_2$
[r] $CH_2CH{=}CPh_2$
[s] $(CH_2)_2CHMe[(CH_2)_3CHMe]_2Me$
[t] $CH{=}CHCH{=}CMeCH{=}CH–$ Me Me Me
[u] $CH{=}CHCH_2CHMe[(CH_2)_3CHMe]_2Me$
[v] $CH_2CH{=}CH_2$
[w] Fused benzo ring
[x] $CH_2CH{=}CHMe$
[y] $CH{=}CHEt$
[z] $CHMeCH{=}CH_2$
[aa] Fused cyclohexane ring
[bb] $[(CH_2)_2CH{=}CMe]_4Me$
[cc] $[(CH_2)_3CHMe]_4Me$
[dd] $[(CH_2)_3CHMe]_3Me$
[ee] $CH_2CH{=}CHPh$
[ff] $(CH_2)_{11}Me$
[gg] $[(CH_2)_2CH{=}CMe]_5Me$
[hh] $[(CH_2)_3CHMe]_5Me$
[ii] $[(CH_2)_2CH{=}CMe]_6Me$
[jj] $[(CH_2)_3CHMe]_6Me$
[kk] $(CH_2)_{15}Me$
[ll] $[(CH_2)_2CH{=}CMe]_7Me$
[mm] $[(CH_2)_3CHMe]_7Me$
[nn] $[(CH_2)_2CH{=}CMe]_8Me$
[oo] $[(CH_2)_3CHMe]_8Me$
[pp] $[(CH_2)_2CH{=}CMe]_9Me$
[qq] $[(CH_2)_3CHMe]_9Me$

TABLE 9. α-TOCOPHEROL ($C_{29}H_{50}O_2$) AND ITS 6-*O*-DERIVATIVES (SEE TABLE 1, SECTION I FOR STRUCTURAL FORMULA)

Mol. formula	Derivative	m.p. or b.p. (mm)	Ref.
$C_{29}H_{50}O_2$	(2*R*,4′*R*,8′*R*)	2.5–3.5 200–210(0.1) (0.1)	26, 70, 146, 153, 305 325, 327–329, 350, 362, 372, 389, 397–422
	(2*S*,4′*R*,8′*R*)	Oil	70
	(2*RS*,4′*R*,8′*R*)	Oil	70
	(2*R*,4′*RS*,8′*RS*)	Oil	70
	(2*S*,4′*RS*,8′*RS*)	Oil	70
	(2*RS*,4′*RS*8′*RS*)	Oil	70
	3,4-Di-^{3}H		432
	5-$^{14}CH_3$		170, 433
	5-$^{14}CH_2{}^3H$		168, 435
	5-$^{14}CH_3$(2*RS*,4′*R*,8′*R*)		168, 434
	7-$^{14}CH_3$		436
$C_{29}H_{51}O_5P$	Phosphate (2*RS*,4′*R*,8′*R*)		437
	Phosphate sodium salt		437
	Phosphate disodium salt		437
$C_{30}H_{49}NO_2$	NCO		438
$C_{30}H_{51}NO_3$	H_2NCO		438
$C_{30}H_{51}NO_2S$	H_2NCS		438
$C_{30}H_{52}N_2O_3$	HOHNC(NH)		438
$C_{31}H_{51}ClO_3$	Chloroacetate		439
$C_{31}H_{52}O_3$	Acetate, 3,4-di-^{14}C		431
$C_{31}H_{52}O_3$	Acetate (2*R*,4′*R*,8′*R*)	28	440, 441
$C_{31}H_{52}O_3$	Acetate (2*S*,4′*R*,8′*R*)	23	442
$C_{31}H_{52}O_3$	Acetate (2*RS*,4′*R*8′*R*)	224 (0.3)	362, 369, 443
$C_{31}H_{52}O_4$	$HOCH_2CO$ (glycolate)	52	444
$C_{31}H_{52}N_2O_4$	Allophanate (2*R*,4′*R*,8′*R*)	160	422, 445
$C_{31}H_{52}N_2O_4$	Allophanate (2*RS*,4′*R*,8′*R*)	176	370
$C_{31}H_{52}N_2O_4$	Allophanate (2*RS*,4′*RS*,8′*RS*)	168	339
$C_{31}H_{53}O_7P$	$COCH_2OPO(OH)_2$		446
$C_{31}H_{55}O_6P$	$PO(OH)OCH_2CH_2OH$ (2*RS*,4′*R*,8′*R*)		447
$C_{32}H_{54}O_3$	Propionate (2*RS*,4′*R*,8′*R*)	228 (0.5)	369
$C_{32}H_{54}O_3$	Propionate, 3,4-di-^{14}C,		432
$C_{32}H_{54}O_4$	Ethyl carbonate	200 (0.1)	363
$C_{32}H_{57}O_6P$	$PO(OH)OCH_2CH_2OMe$ (2*RS*,4′*R*,8′*R*)		447
$C_{33}H_{54}O_5$	Hemisuccinate (2*R*,4′*R*,8′*R*)	78	70
$C_{33}H_{54}O_5$	Hemisuccinate (2*S*,4′*R*,8′*R*)	51	442
$C_{33}H_{54}O_5$	Hemisuccinate (2*RS*,4′*R*,8′*R*)	76	448
	Lithium salt		450
	Magnesium salt	65	449
	Calcium salt		450
	Lysine salt		451
	Ornithine salt		451
	Arginine salt		451
$C_{33}H_{54}O_4$	Acetoacetate		452
$C_{33}H_{55}NO_5$	Unknown isomeric DL-hemiaspartate	173	453
$C_{33}H_{55}NO_5$	Derivatives of (2*RS*,4′*R*,8′*R*)	213	453

TABLE 9. (*Contd.*)

Mol. formula	Derivative	m.p. or b.p. (mm)	Ref.
$C_{33}H_{56}O_3$	$CO(CH_2)_2Me$ (2*RS*,4′*R*,8′*R*)	230 (0.3)	443
$C_{33}H_{56}N_3O_2$	$-C(=NH)-N\langle CH_2CH_2NHCH_2\rangle$ (ring: —C(=NH)—N⟨ ⟩NH)		454
$C_{33}H_{57}NO_5$	$COCH_2NMe_2$	Oil	438, 439
	Hydrochloride salt	193	438, 439
	Hydrobromide salt	203.5	438, 439
	Sulfate salt	225–240(d)	438, 439
$C_{33}H_{57}N_2O_3P$	$(CH_2CH_2N)-P(=O)-(NCH_2CH_2)$ (bis-aziridinyl: $\overset{CH_2}{\underset{CH_2}{\vert}}$>N—P(O)—N<$\overset{CH_2}{\underset{CH_2}{\vert}}$)		215
$C_{34}H_{53}NO_7Na_2$	$CONHCH(CO_2Na)CH_2CO_2Na$ (2*R*,4′*R*,8′*R*)	234	455
$C_{34}H_{55}NO_7$	$CONHCH(CO_2H)CH_2CO_2H$	234	214, 455
$C_{34}H_{58}O_3$	Isovalerate (2*RS*,4′*R*,8′*R*)	116(?)	450
$C_{34}H_{58}N_2O_3$	$C(=NH)-N\langle CH_2CH_2OCH_2CH_2\rangle$ (morpholino: C(=NH)—N⟨ ⟩O)		438
$C_{34}H_{58}O_4$	*t*-Butylcarbonate	Oil	393
$C_{34}H_{60}N_2O_2$	$C(=NH)NEt_2$		438
$C_{34}H_{61}O_6P$	$PO(OH)O(CH_2)_5OH$	Oil	447
$C_{35}H_{34}NO_3$	Nicotinate (2*R*,*S*4′*R*,8′*R*)		456, 457, 503
	Nicotinate methiodide	201	457
	Nicotinate methyl-*p*-toluenesulfonate salt	142	457
$C_{35}H_{55}NO_3$	*p*-Aminobenzoate (2*R*,4′*R*,8′*R*)	163	6
$C_{35}H_{55}NO_3$	*p*-Aminobenzoate (2*RS*,4′*R*,8′*R*)	159	6
$C_{35}H_{55}NO_3$	*p*-Aminobenzoate (2*RS*,4′*RS*,8′*RS*)	159	6
$C_{35}H_{56}O_3$	$Me(CH{=}CH)_2CO$		458
$C_{35}H_{57}ClO_4$	$ClCO(CH_2)_4CO$		458
$C_{35}H_{57}NO_7$	$CONHCH(CO_2H)CH_2CH_2CO_2H$ (2*R*,4′*R*,8′*R*)	107	214, 455
	Disodium salt	292	455
$C_{35}H_{59}ClO_4$	$ClCH_2CMe_2CHOHCO$		458
$C_{35}H_{61}NO_3$	$COCH_2NEt_2$	Oil	439
	Hydrochloride salt	157	439
	Hydrobromide salt	171	439
	Tartrate salt	170	439
	Citrate salt		439
	Phosphate salt		439
	Sulfate salt		439

TABLE 9. (*Contd.*)

Mol. formula	Derivative	m.p. or b.p. (mm)	Ref.
$C_{36}H_{52}N_2O_7$	3,5-Dinitrobenzoate (2*R*,4′*R*,8′*R*)	87	401, 459
$C_{36}H_{52}N_2O_7$	3,5-Dinitrobenzate (2*RS*,4′*R*,8′*R*)	63–67	392, 401
$C_{36}H_{53}O_4$	Salicylate		460
$C_{36}H_{53}NO_5$	4-Nitrobenzoate (2*R*,4′*R*,8′*R*)	66	382, 461
$C_{36}H_{53}N_3O_7$	3,5-Dinitrophenylcarbamoyl (2*R*,4′*R*,8′*R*)	131	323
$C_{36}H_{53}N_3O_7$	3,5-Dinitrophenylcarbamoyl (2*RS*,4′*R*,8′*R*)	145–147	323
$C_{36}H_{54}O_3$	Benzoate (2*RS*,4′*R*,8′*R*)	225 (91)	369
$C_{36}H_{54}N_2O_5$	4-Nitrophenylcarbamate (2*RS*,4′*R*,8′*R*)	131	401
$C_{36}H_{54}N_2O_5$	4-Nitrophenylcarbamate (2*R*,4′*R*,8′*R*)	131	384, 401, 422
$C_{36}H_{55}NO_3$	4-Aminobenzoate (2*R*,4′*R*,8′*R*)	163	462
$C_{37}H_{53}CaO_5$	Phthalic acid half ester Ca salt (2*RS*,4′*R*,8′*R*)		449
$C_{37}H_{58}O_9$	Diacetyl tartrate (2*RS*,4′*R*,8′*R*)		463
$C_{38}H_{56}O_3$	Cinnamate (2*RS*,4′*R*,8′*R*)		463
$C_{38}H_{56}O_5$	3,4-Dihydroxycinnamate	152	464
$C_{38}H_{59}NO_3$	4-Dimethylaminobenzoate (2*RS*,4′*R*,8′*R*)	42	465
$C_{39}H_{60}ClO_4$	$COCMe_2OC_6H_4$-4-Cl		466
$C_{39}H_{63}BrO_5S$	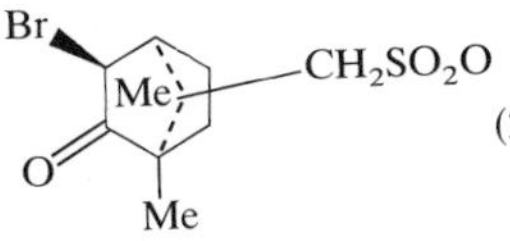 [structure: Br, Me, CH_2SO_2O, O, Me] (2R, 4′R, 8′R)	52	392
$C_{40}H_{60}O_5$	3,4-Dimethoxycinnamate		464
$C_{41}H_{60}O_7$	3-Methoxy-4-methoxycarbonylcinnamate		464
$C_{41}H_{72}O_3$	Lauroyl (2*RS*,4′*R*,8′*R*)	270 (0.15)	467
$C_{42}H_{57}N_3O_5$	$4-NO_2C_6H_4N{=}NC_6H_4$-4-CO		458
$C_{42}H_{58}N_2O_3$	4-Phenylazobenzoate (2*R*,4′*RS*,8′*RS*)	58.5–60	329
	(2*R*,4′*R*,8′*R*)	64–65	329, 468
	(2*S*,4′*R*,8′*R*)	76–79	329, 468
	(2*S*,4′*RS*,8′*RS*)	60.5	329
$C_{42}H_{60}O_7$	3,4-Diacetoxycinnamate		464
$C_{42}H_{62}O_7$	3-Methoxy-4-ethylcarbonylcinnamate		464
$C_{42}H_{64}O_5$	3,4-Diethoxycinnamate		464
$C_{42}H_{64}O_5$	$Ph(CH_2)_3OCO(CH_2)_2CO$ (2*R*,4′*R*,8′*R*)		469
$C_{43}H_{58}Cl_2O_5$	$[4-ClC_6H_4O]_2CHCO$		466
$C_{43}H_{63}O_5P$	Dibenzyl phosphate		470
$C_{44}H_{58}ClF_3O_4$	$COCH(C_6H_4$-4-Cl$)OC_6H_4$-3-CF_3		466
$C_{43}H_{63}ClO_4$	$COCMe_2OC_6H_4$-4-C_6H_4-4-Cl		466
$C_{45}H_{66}O_7P$	$COCH_2OPO(OCH_2Ph)_2$		446
$C_{45}H_{60}O_3$	Palmitate (2*R*,4′*R*,8′*R*)	43	70
$C_{45}H_{60}O_3$	Palmitate (2*RS*,4′*R*,8′*R*)	38	70
$C_{45}H_{60}O_3$	Palmitate (2*RS*,4′*RS*,8′*RS*)	38	414
$C_{46}H_{65}ClO_4$	$COCMe_2OCH_2C_6H_4$-4-C_6H_4-4-C_6H_4-4-Cl		466
$C_{47}H_{78}O_3$	Linolenoyl (2*R*,4′*R*,8′*R*)	Oil	471
$C_{47}H_{80}O_3$	Linoleoyl (2*R*,4′*R*,8′*R*)	Oil	471
$C_{47}H_{82}O_3$	Octadec-6*c*-enoyl (2*R*,4′*R*,8′*R*)	Oil	471
$C_{47}H_{82}O_3$	Oleoyl (2*R*,4′*R*,8′*R*)	Oil	471

TABLE 9. (*Contd.*)

Mol. formula	Derivative	m.p. or b.p. (mm)	Ref.
$C_{47}H_{82}O_3$	Elaidoyl (2*R*,4′*R*,8′*R*)	27.5	471
$C_{47}H_{84}O_3$	Stearoyl (2*R*,4′*R*,8′*R*)	48.5	471
$C_{47}H_{84}O_3$	Stearoyl (2*RS*,4′*RS*,8′*RS*)	40.5	471
$C_{49}H_{70}O_4$	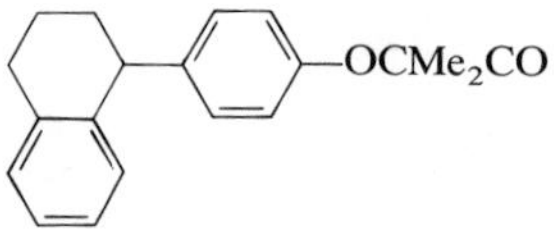 —$OCMe_2CO$		466
$C_{51}H_{90}O_3$	Erucoyl (2*R*,4′*R*,8′*R*)	−6 to −4	471
Polymer	$COCH_2CH_2(OCH_2CH_2)_nOMe$		472
Polymer	$P(OH)OO(CH_2CH_2O)_nH$ (2*RS*,4′*R*,8′*R*)		447

TABLE 10. TOCOTRIENOLS

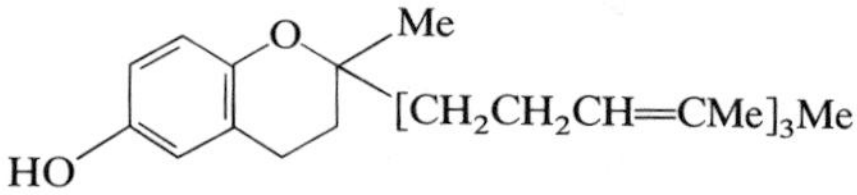

HO, O, Me, $[CH_2CH_2CH{=}CMe]_3Me$

Mol. formula	Derivative	Substituents 5	7	8	m.p or b.p. (mm)	Ref.
$C_{26}H_{38}O_2$						96, 284
$C_{27}H_{40}O_2$			Me			107
$C_{27}H_{40}O_2$				Me		107
$C_{27}H_{40}O_2$	*REE*			Me		285
$C_{28}H_{42}O_2$	2*RS*	Me		Me	140 (0.005)	109, 286
	Acetate					109
	4-Phenylazobenzoate				52	109
$C_{28}H_{42}O_2$			Me	Me		109, 238, 287–291
$C_{29}H_{44}O_2$	*REE*	Me	Me	Me	32	292
$C_{29}H_{44}O_2$[a]	2*R*	Me	Me	Me		292
	Acetate					109
	4-Phenylazobenzoate				85.5	109
$C_{29}H_{44}O_2$	2*RS*	Me	Me	Me		109
	Acetate					293, 294
	4-Phenylazobenzoate				85.5	293, 294

[a] Side chain at C-2 is $[(CH_2)_2CH{=}CMe]_2(CH_2)_3CMe{=}CH_2$.

TABLE 11. 7-HYDROXYCHROMANS

Mol. Formula	Derivative	Substituents 2	2′	4	5	6	8	m.p. or b.p. (mm)	Ref.
$C_9H_{10}O_2$								91	135, 295, 296
	Acetate							50	135, 296
$C_{10}H_{12}O_2$		Me						72	137, 297
	Crotonoyl							78	298
$C_{11}H_{14}O_2$		Me	Me					73	154, 270, 299–302
								140–145 (0.2)	
	$CON(CH_2CH_2Cl)_2$							Oil	302
	4-Nitrobenzoate							126	136
$C_{11}H_{14}O_2$						Et		70	135, 303
								105–115 (0.9)	135, 303
$C_{11}H_{14}O_2$		Me				Me		127	298
$C_{11}H_{14}O_2$		Me					Me	73	298
$C_{12}H_{16}O_2$		Me	Me	Me				108	136, 271, 304, 305
	4-Nitrobenzoate							137	136, 271, 304, 305
$C_{12}H_{16}O_2$		Me	Me		Me			139 (5)	271, 304
$C_{16}H_{22}O_2$		Me	Me			[a]			270
$C_{20}H_{28}O_2$		[b]							33
$C_{21}H_{32}O_2$		Me	[b]		[c]				306
$C_{22}H_{32}O_2$		Me	[b]		Me				33, 276

[a] $CH_2CH{=}CMe_2$ [b] $CH_2CH_2CH{=}CMeCH_2CH_2CH{=}CMe_2$ [c] C_5H_{11}

TABLE 12. 8-HYDROXYCHROMANS

Mol. formula	Substituents						m.p. or b.p. (mm)	Ref.
	2	2′	4	5	6	7		
$C_9H_{10}O_2$							97 (1.2)	278
	CONHMe						158	279
$C_{10}H_{12}O_2$	Me							279, 280
	CONHMe						97	279, 280
$C_{11}H_{14}O_2$	Me	Me					93 (2)	60
	Acetate							281
	CONHMe						148	281, 283
	$CONHCH_2OH$							281, 283
	$CONMe_2$							281, 283
	$CONHCH_2C{\equiv}CH$							281, 283
	$CONHCOCH_2Cl$							281, 283
	$CONHCH_2CH{=}CH_2$							281, 283
	CONHPr							281, 283
	4-Nitrobenzoate						101	281, 283
$C_{11}H_{14}O_2$	Me		Me					282
$C_{12}H_{16}O_2$	Me	Me		Me				282
	CONHME							283
$C_{12}H_{16}O_2$	Me	Me				Me		283
	CONHMe							283
$C_{13}H_{18}O_2$	Me	Me			Et			280
	CONHMe							280
$C_{15}H_{22}O_2$	Pr	Pr						280
	CONHMe							280
$C_{15}H_{22}O_2$	Me	Me			Bu^i			280
	CONHMe							280
$C_{29}H_{50}O_2$	Me	[a]		Me	Me	Me	Oil	60
	Acetate						Oil	60

[a] $[(CH_2)_3CHMe]_3Me$ (phytyl)

TABLE 13. CHROMANS CONTAINING HYDROXYALKYL, THIOL, SELENYL, OR MORE THAN ONE HYDROXYL GROUP

Mol. formula	Substituents										m.p. or b.p. (mm)	Ref.
	2	2′	3	3′	4	4′	5	6	7	8		
$C_{10}H_{12}O_2$								CH_2OH			159–160	492
$C_{10}H_{12}O_3$	Me								OH	OH	122–123	137
$C_{11}H_{14}O_3$	Me	Me					OH	OH			162–163	493, 494
$C_{11}H_{14}O_3$	Me	Me					OH		OH		163–164	154, 495–497
	Diacetate										84	495
	6,8-Bisphenylazo										256	496
$C_{11}H_{14}O_3$	Me	Me							OH	OH	99–100	154, 498
	Diacetate											498
	Dibenzoate										152	498
$C_{11}H_{14}O_3$	Me		Me						OH	OH	60–61	137
	Diacetate										105–106	137
	Di-4-nitrobenzoate										169–170	137
$C_{12}H_{16}O_2$	CH_2OH							Me		Me		499
$C_{12}H_{16}O_2$			CH_2OH					Me		Me		499
$C_{12}H_{16}O_3$					Me	Me	CH_2OH	OH				326
$C_{12}H_{16}O_3$	Me						OH	OH	Me	Me		175
	Diacetate										82	175
$C_{13}H_{18}O_2$	$(CH_2)_2OH$				Me				Me			500
$C_{13}H_{18}O_3$			CH_2OH				Me	OH	Me		171–172	501, 502
$C_{13}H_{18}O_3$	Me	Me					OH	OH	Me	Me		175
$C_{14}H_{20}O_2$	Me	CH_2OH			Me	Me			Me		47	162
$C_{14}H_{20}O_3$	Me	CH_2OH					Me	OH	Me	Me	111–113	504
$C_{14}H_{20}O_3$	Me	Me					CH_2OH	OH	Me	Me		505
$C_{14}H_{20}O_3$	Me	Me					$^{14}CH_2OH$	OH	Me	Me		193
	Cyclic phosphate										230–233	172
	6-Phosphate										145–151	172
	Diacetate											162
	5-Benzoate										129–132	507, 508
											125–126	508

TABLE 13. (*Contd.*)

Mol. formula		Substituents 2	2′	3	3′	4	4′	5	6	7	8	m.p. or b.p. (mm)	Ref.
$C_{14}H_{20}O_3$ (*Contd.*)													
	5-Benzoate 6-acetate											105–106	507
	Dinicotinate												503
$C_{14}H_{20}O_3$		Me	Me					Me	OH	CH_2OH	Me		505
	7-Acetate												505
$C_{14}H_{20}O_3$						Me	Me	CH_2OH	OH	Me	Me		107
$C_{14}H_{20}O_3$		Me	Me			Me	Me	OH		Me	OH		162
	Diacetate											95	162
$C_{14}H_{20}O_4$		Me	Me					CH_2OH	OH	CH_2OH	Me		193
	7-Acetate												193
$C_{15}H_{16}O_3$		Me	Me					OH	OH	[a]	[a]		506
	Diacetate											162–162.5	506
$C_{15}H_{16}O_3$		Me	Me					OH	[a]	[a]	OH		506
	Diacetate											169.8–170	506
$C_{15}H_{22}O_3$		Me	$(CH_2)_2OH$					Me	OH	Me	Me		49
$C_{16}H_{24}O_2$		Me	Me						[b]			51–52	508
$C_{16}H_{16}O_3$						[c]				OH		127	509
	Diacetate											116	509
$C_{16}H_{22}O_3$		Me	Me		[d]			OH		OH		127–128	510
$C_{16}H_{22}O_3$		Me	Me						[e]	OH	OH	160–170	154
$C_{18}H_{28}O_3$		Me	Me						OH	C_7H_{15}	OH	79	342
$C_{19}H_{28}O_3$		Me	[f]					Me	OH	Me	Me	207 (0.001)	292
$C_{19}H_{30}O_3$		Me	[g]					Me	OH	Me	Me	95–98	341
	Acetate											175 (0.08)	341
$C_{21}H_{26}O_3$		Me	Me					[h]	OH	Me	Me	170–172	533
$C_{21}H_{28}O_3$		[i]	Me					Me	OH	Me	Me		511
$C_{22}H_{24}O_3$		[j]	Me					Me	OH	Me	Me	211d	511
$C_{22}H_{28}O_3$		Me	Me					[k]	OH	Me	Me	117–118	219, 533
$C_{22}H_{28}O_3$		Me	Me					[l]	OH	Me	Me	150	219, 533

$C_{23}H_{30}O_3$		Me	Et	[l]	OH	Me	Me	137	533
$C_{23}H_{30}O_3$		Me	Me	[m]	OH	Me	Me	154–155	219, 533
$C_{23}H_{30}O_4$		Me	Me	[n]	OH	Me	Me		219, 533
$C_{23}H_{30}O_4$		Me	Me	[o]	OH	Me	Me	233–235	219, 533
$C_{24}H_{26}O_3$		Me	Me	[p]	OH	Me	Me	185–186	219, 533
$C_{24}H_{32}O_3$		Me	Pr	[k]	OH	Me	Me		219, 533
$C_{24}H_{32}O_3$		Me	Pr	[l]	OH	Me	Me		219, 533
$C_{24}H_{36}O_3$		Me	[q]	Me	OH	Me	Me		292
$C_{24}H_{40}O_3$		Me	[r]	Me	OH	Me	Me	170 (0.001)	341
	Acetate							175 (0.001)	341
$C_{25}H_{34}O_3$		Me	Pr	[m]	OH	Me	Me		219, 533
$C_{26}H_{30}O_3$		Me	Pr	[p]	OH	Me	Me		219, 533
$C_{27}H_{30}O_3$		Me	Me	[s]	OH	Me	Me	169–178	219, 533
$C_{27}H_{40}O_3$		Me	[t]		OH		Me		512
$C_{27}H_{42}O_3$		Me	[u]	Me	OH	Me	Me		333
$C_{27}H_{46}O_3$		Me	$C_{16}H_{33}$		OH	CH_2OH			513
	6-Acetate 7-dibenzylphosphate								513
	6-acetate 7-phosphate								514
	diacetate								514
$C_{27}H_{46}O_4$		Me	$C_{16}H_{33}$	OH	OH	OH	Me		503
$C_{28}H_{32}O_3$		Me	[v]	Me	OH	Me	Me	130–131	329
$C_{28}H_{44}O_3$		Me	[u]	Me	OH		Pr^i		333
$C_{28}H_{48}O_3$		Me	$C_{16}H_{33}$	CH_2OH	OH		Me		515
	5-Disodium phosphate								516
	5-Phosphate								516
	6-Acetate								516
	6-Acetate 5-phosphate								516, 517
	5,6-Diacetate								518
	5,6-Dinicotinate								503
	5,6-Dibenzoate								518
	6-Acetate 5-dibenzylphosphate								517
$C_{29}H_{44}OS$		Me	[w]	Me	SH	Me	Me		518
$C_{29}H_{44}O_3$		Me	[x]	Me	OH	Me	Me		292
	6-Acetate								292

TABLE 13. (*Contd.*)

Mol. formula		Substituents 2	2′	3	3′	4	4′	5	6	7	8	m.p. or b.p. (mm)	Ref.
$C_{29}H_{50}OS$		Me	$C_{16}H_{33}$					Me	SH	Me	Me		519
	Acetate												519
	^{35}S-Acetate												521
$C_{29}N_{50}OSe$		Me	$C_{16}H_{33}$					Me	SeH	Me	Me		522
$C_{29}H_{50}O_3$		Me	y					Me	OH	Me	Me	95–98	41
												207 (0.001)	41
	6-Acetate											198 (0.001)	41
	Diacetate											175 (0.001)	41
$C_{29}H_{50}O_3$		Me	$C_{16}H_{33}$					CH_2OH	OH	Me	Me	Gum	49
	6-Acetate												523
	6-Acetate 5-phosphate												517
	6-Acetate 5-isothiuronium HCl											130–135	217
	Diacetate												516
	6-Acetate 5-dimethylphosphate												516
	6-Nicotinate												503
	Dipropionate												520
	5-Benzoate												184
	Dinicotinate												503
	Dibenzoate												517
	6-Acetate 5-dibenzylphosphate												517
	Triphosphate												49
$C_{29}H_{50}O_3$		Me	$C_{16}H_{33}$					Me	OH	CH_2OH	Me		330, 524
	6-Acetate												330, 524
	7-Acetate												525
	6-Acetate 7-phosphate												514
	6-Acetate 7-phosphate disodium salt												514
	Nicotinate												503
	Dinicotinate												503
	7-Dibenzylphosphate 6-acetate												514

$C_{29}H_{50}O_4$		Me	$C_{16}H_{33}$	CH_2OH	OH	CH_2OH	Me	520
	6-Acetate							330
	7-Acetate							330, 520
	Triacetate							520
$C_{31}H_{48}O_3$		Me	$C_{16}H_{33}$	CH_2OH	OH	[a]	[a]	513
	6-Acetate	5-hydroxymethylphosphate						513
	Diacetate							513

[a] Fused benzo ring
[b] $(CH_2)_2CMe_2OH$
[c] $4\text{-}HOC_6H_4CH_2$
[d] $CMe{=}CMe_2$
[e] $CH_2CH{=}CMe_2$
[f] $CH_2CH_2CH(OH)CMe{=}CH_2$
[g] $(CH_2)_3CHMeCH_2OH$
[h] 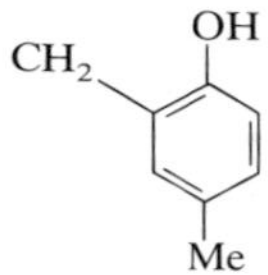
[i] 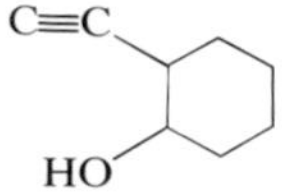
[j] $C{\equiv}CCH(OH)Ph$
[k] 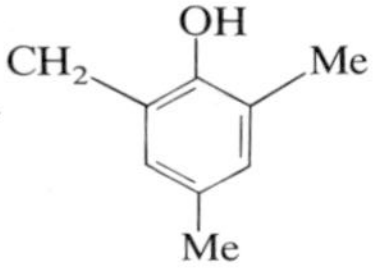
[l] 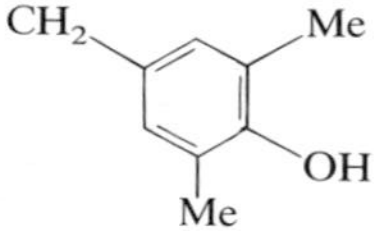
[m] 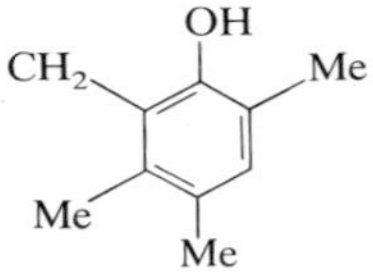
[n] 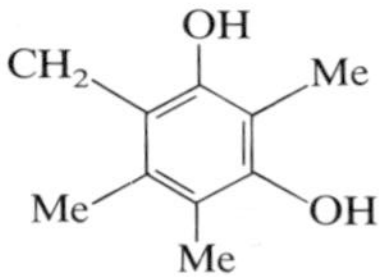
[o] 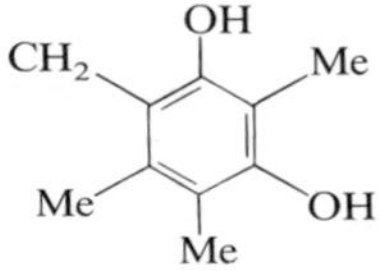
[p] 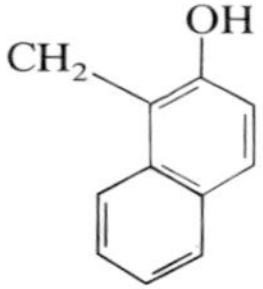
[q] $(CH_2)_2CH{=}CMe(CH_2)_2CH(OH)CMe{=}CH_2$
[r] $[(CH_2)_3CHMe]_2CH_2OH$
[s] 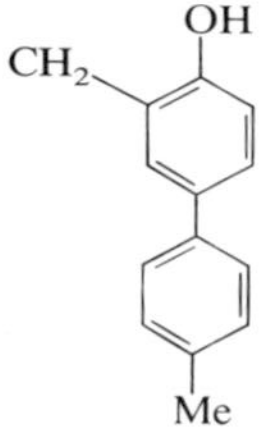
[t] $(CH_2CH_2CH{=}CMe)_2CH_2CH_2CH{-}CMe_2$ (epoxide, O bridging)
[u] $CH_2CMeCH_2CH_2CH{=}CMeCH_2CH_2CH{=}CMe_2$
[v] $CH_2CH_2CPh_2OH$
[w] $(CH_2CH_2CH{=}CMe)_3Me$
[x] $(CH_2CH_2CH{=}CMe)_2CH_2CH_2CH(OH)CMe{=}CH_2$

TABLE 14. DIMERS WITH DIRECT AROMATIC LINKAGE

Mol. formula	Linkage	Substituents 2	2′	5	6	7	8	m.p.	Ref.
$C_{22}H_{26}O_4$	5,5′				OH	Me	Me		208
$C_{24}H_{30}O_4$	5,5′	Me	Me		OH		Me		337
$C_{26}H_{34}O_4$	5,5′	Me	Me		OH	Me	Me	192–194	207, 209, 337
$C_{26}H_{34}O_4$	7,7′	Me	Me	Me	OH		Me		189
$C_{54}H_{90}O_4$	5,5′	Me	$C_{16}H_{33}$		OH		Me		189, 201, 337, 5
		Diacetate							189, 201, 337, 5
$C_{56}H_{94}O_4$	5,5′	Me	$C_{16}H_{33}$		OH	Me	Me		189

TABLE 15. BISCHROMANS CONNECTED WITH AN OXYGEN ATOM

Mol. formula	Structure	Substituents 2	2′	5′	7′	8′	7	8	m.p.	Ref.
$C_{22}H_{26}O_4$	**I**				Me	Me	Me	Me		208
$C_{22}H_{26}O_4$	**I**	Me	Me							207
$C_{24}H_{30}O_4$	**I**	Me	Me		Me		Me			209
$C_{24}H_{30}O_4$	**I**	Me	Me			Me		Me		207, 209
$C_{24}H_{30}O_4$	**II**[a]	Me	Me	Me						209
$C_{26}H_{34}O_4$	**I**	Me	Me		Me	Me	Me	Me	185–188	207, 209
$C_{26}H_{34}O_4$	**II**[a]	Me	Me	Me		Me		Me		189
$C_{54}H_{90}O_4$	**I**	Me	$C_{16}H_{33}$			Me		Me		189, 201
$C_{55}H_{92}O_4$	**I**	Me	$C_{16}H_{33}$	Me			Me	Me		189
$C_{56}H_{94}O_4$	**I**	Me	$C_{16}H_{33}$	Me	Me	Me	Me	Me		526, 536
	I[b]									526

[a] 5-Me
[b] Acetate

TABLE 16. BISCHROMANS CONNECTED WITH A METHYLENE GROUP

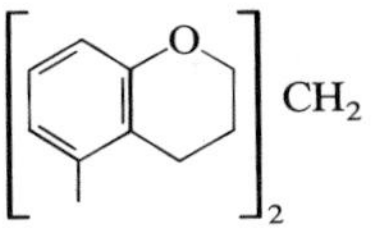

Mol. formula	Structure	Substituents 2	2′	6	7	8	m.p.	Ref.
$C_{27}H_{36}O_4$	**I**	Me	Me	OH	Me	Me	220 230–232	219, 507, 533
	I[a]							219
$C_{29}H_{40}O_4$	**I**	Me	Et	OH	Me	Me	185–187	219
$C_{31}H_{44}O_4$	**I**	Me	Pr	OH	Me	Me	173–175	533
$C_{57}H_{96}O_4$	**I**	Me	$C_{16}H_{33}$	OH	Me	Me		170
	I[a]							170
$C_{57}H_{96}O_4$	**II**[b]	Me	$C_{16}H_{33}$	OH		Me		170

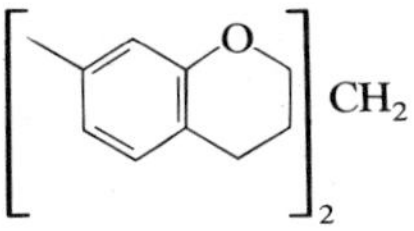

[a] Diacetate derivative
[b] 5,5′-Me_2

TABLE 17. OTHER BISCHROMANS

Mol. formula	X[a]	Substituents A	Substituents B	m.p.	Ref.
$C_{24}H_{30}O_4$	$(CH_2)_2$	2,2-Me_2-6-OH	2,2-Me_2-6-OH		209
$C_{26}H_{34}O_4$	$(CH_2)_2$	2,2,7-Me_3-6-OH	2,2,7-Me_3-6-OH	178–180	527
$C_{26}H_{34}O_4$	$(CH_2)_2$	2,2,8-Me_3-6-OH	2,2,8-Me_3-6-OH	183–185	207, 527
$C_{28}H_{38}O_4$	$(CH_2)_2$	2,2,7,8-Me_4-6-OH	2,2,7,8-Me_4-6-OH	186–188	190, 193, 201, 207, 528, 529
		Diacetate		217–218.5	188, 191
$C_{54}H_{90}O_4$	6,8′-CH_2O	2-Me-2-$C_{16}H_{33}$-8-OH	2′,6′-Me_2-2′-$C_{16}H_{33}$		532
$C_{54}H_{90}O_4$	$(CH_2)_2$	2-Me-2-$C_{16}H_{33}$-6-OH	2-Me-2-$C_{16}H_{33}$-6-OH		189
$C_{56}H_{94}O_4$	$(CH_2)_2$	2,7-Me_2-2-$C_{16}H_{33}$-6-OH	2,7-Me_2-2-$C_{16}H_{33}$-6-OH		530
$C_{56}H_{94}O_4$	$(CH_2)_2$	2,8-Me_2-2-$C_{16}H_{33}$-6-OH	2,8-Me_2-2-$C_{16}H_{33}$-6-OH		530
$C_{58}H_{98}O_4$	$(CH_2)_2$	2,7,8-Me_3-2-$C_{16}H_{33}$-6-OH	2,7,8-Me_3-2-$C_{16}H_{33}$-6-OH	Wax	190, 195
		Diacetate		74–75	531

[a] Connected at 5,5′ unless otherwise stated.

IX. REFERENCES

1. G. P. Ellis, Ed., *Chromenes, Chromanones, and Chromones* John Wiley, New York, 1977.
2. IUPAC-IUB Commission on Biochemical Nomenclature (CBN), *Eur. J. Biochem.*, **46,** 217 (1974).
3. O. Hoffmann-Ostenhoff, *Am. J. Clin. Nutr.* **27,** 1105 (1974).
4. IUPAC-IUB Commission on Biochemical Nomenclature (CBN), *Arch. Biochem. Biophys.*, **165**(1), 1 (1974).
5. R. S. Harris and I. G. Wood, Ed., *Vitamins and Hormones*, Vol. 20, Academic Press, New York, 1962.
6. H. Mayer and O. Isler, in *Methods in Enzymology–Vitamins and Coenzymes*, Vol. XVIIIC, Section XII, D. B. McCormick and L. D. Wright, Eds., Academic Press, New York, 1971, pp. 241–396.
7. P. Schudel, H. Mayer, and O. Isler, *The Vitamins*, W. H. Sebrell, Jr., and R. S. Harris, Eds., Academic Press, New York, 1972, Chap. 16, Section II, p. 168.
8. W. Hjarde, E. Leerbeck, and T. Leth, *Acta Agr. Scand. Suppl.*, **19,** 87 (1973); *Chem. Abstr.*, **80,** 23604r (1974)
9. O. Isler, *Experientia*, **33,** 1 (1977).
10. S. R. Ames, *Ann. Rev. Biochem.*, **27,** 371 (1958).
11. D. C. Herting, *Kirk-Othmer Encyclopedia of Chemical Technology*, 2nd ed., Vol. 21, Wiley–Interscience, New York, 1970, p. 574.
12. L. I. Smith, *Chem. Rev.* **27,** 287 (1940).
13. H. R. Rosenberg, *Chemistry and Physiology of the Vitamins*, Wiley–Interscience, New York, 1945.
14. H. A. Mattill, in *The Vitamins*, Vol. III, M. H. Sebrell, Jr. and R. S. Harris, Eds., Academic Press, New York, 1954, Chap. 17.
15. N. Campbell, in *Chemistry of Carbon Compounds*. Vol. IVB, E. H. Rodd, Ed., Elsevier, Amsterdam, 1959, p. 809.
16. A. F. Wagner and K. Folkers, in *Medicinal Chemistry*, A. Burger, Ed., Wiley–Interscience, New York, 1960.
17. A. F. Wagner and K. Folkers, in *Vitamins and Coenzymes*, Wiley–Interscience, New York, 1964.
18. F. M. Dean, *Naturally Occurring Oxygen Ring Compounds*, Butterworth, London, 1963.
19. S. F. Dyke, *The Chemistry of the Vitamins*, Wiley–Interscience, New York, 1965, p. 256.
20. W. Boguth, G. Herrmann, and H. Niemann, *Experientia*, **24,** 232 (1967).
21. J. L. Flippen, J. Karle, and I. L. Karle, *J. Am. Chem. Soc.*, **92,** 3749 (1970).
22. J. R. Cannon, I. A. McDonald, A. F. Sierakowski, A. H. White, and A. C. Willis, *Austr. J. Chem.*, **28,** 57 (1975).
23. W. F. Raub, *Fed. Proc.*, **33,** 2390 (1974).
24. The organization and analysis of the data base associated with this investigation were carried out using the PROPHET system, a unique national computer resource sponsored by the National Institutes of Health. Information about PROPHET, including how to apply for access, can be obtained from the Director, Chemical/Biological Information-Handling Program, Division of Research Resources, National Institutes of Health, Bethesda, Maryland 20014.
25. K. F. Kuhlmann, personal communication.
26. H. Mayer, P. Schudel, R. Rüegg, and O. Isler, *Helv. Chim. Acta*, **46,** 963 (1963). *Chem. Abstr.*, **59,** 3872 (1963).
27. R. A. Friedel and M. Orchin, *Ultraviolet Spectra of Aromatic Compounds*, Wiley, New York, 1951.
28. R. M. Silverstein and G. C. Bassler, *Spectrometric Identification of Organic Compounds*, Wiley, New York, 1963.
29. S. K. Freeman, Ed., *Interpretive Spectroscopy*, Reinhold, New York, 1965.

30. B. Williams, A. F. Thomas, and F. Gautschi, *Tetrahedron*, **20,** 1185 (1964).
31. J. G. Grasseli and W. M. Ritchey, Ed., *CRC Atlas of Spectral Data and Physical Constants for Organic Compounds*, Vols. II and IV, CRC Press, Cleveland, 1972.
32. H. Fukami and M. Nakajima, *Agr. Biol. Chem.*, **25,** 247 (1961); *Chem. Abstr.* **55,** 14448 (1961).
33. J. A. Miller and H. C. S. Wood, *J. Chem. Soc.* (*C*), 1837 (1968).
34. J. L. G. Nilsson, H. Sievertsson, and H. Selander, *Acta Chem. Scand.*, **22,** 3160 (1968).
35. J. R. Dyer, *Applications of Absorption Spectroscopy of Organic Compounds*, Prentice-Hall, Englewood Cliffs, New Jersey, 1965.
36. M. Gianturco, in *Interpretive Spectroscopy*, S. K. Freeman, Ed., Reinhold, New York, 1965, Chap. 2.
37. K. Nakanishi, *Infrared Absorption Spectroscopy*, Holden-Day, San Francisco, 1969.
38. H. Rosenkrantz, *J. Biol. Chem.*, **173,** 439 (1948).
39. J. Green, D. McHale, S. Marcinkiewicz, P. Mamalis, and P. R. Watt, *J. Chem. Soc.*, 3362 (1959).
40. R. T. Holman and P. R. Edmondson, *Anal. Chem.*, **28,** 1533 (1956).
41. O. H. Wheeler, *Chem. Rev.* **59,** 629 (1959).
42. R. Verhé, N. Schamp, and L. DeBuyck, *Synthesis*, 392 (1975).
43. *The Sadtler Standard Spectra*, Sadtler Research Laboratories, Philadelphia, Pennsylvania, 1964.
44. J. Polonsky, *Bull. Soc. Chim. Fr.* 914 (1956).
45. S. Inoue, K. Saito, K. Kato, S. Nozaki, and K. Sato, *J. Chem. Soc., Perkin I*, 2097 (1974).
46. K. A. Kun and H. G. Cassidy, *J. Org. Chem.*, **26,** 3223 (1961).
47. C. N. Banwell in D. W. Mathieson, Ed., *Nuclear Magnetic Resonance for the Organic Chemist*, Academic Press, London, 1967, Chap. 6, p. 110.
48. L. M. Jackman and S. Sternhell, *Applications of Nuclear Magnetic Resonance Spectroscopy in Organic Chemistry*, Pergamon Press, Oxford, 1969, Chaps. 2 and 3.
49. N. Cohen, W. F. Eichel, R. J. Lopresti, C. Neukom, and G. Saucy, *J. Org. Chem.*, **41,** 3511 (1976).
50. H. Finegold and H. T. Slover, *J. Org. Chem.*, **32,** 2557 (1967).
51. K. Tsukida and M. Ito, *Experientia*, **27,** 1004 (1971).
52. G. C. Levy and G. L. Nelson, *Carbon-13 Nuclear Magnetic Resonance for the Organic Chemist*, Wiley–Interscience, New York, 1972.
53. M. Tanabe, in *Biosynthesis*, Vol. 4, T. A. Geissman, Ed., The Chemical Society, London, 1973, Chap. 7.
54. M. Matsuo and S. Urano, *Tetrahedron*, **32,** 229 (1976); *Chem. Abstr.*, **84,** 165060 (1976).
55. J. L. G. Nilsson, S. Agurell, H. Selander, H. Sievertsson, and I. Skanberg, *Acta Chem. Scand.*, **23,** 1832 (1969).
56. J. R. Trudell, S. D. S. Woodgate, and C. Djerassi, *Org. Mass Spectrom.* **3,** 753 (1970).
57. S. E. Scheppele, R. K. Mitchum, C. J. Rudolph, Jr., K. J. Kinneberg, and G. V. Odele, *Lipids*, **7,** 297 (1972).
58. R. Verhé, N. Schamp, L. DeBuyck, N. De Kimpe, and M. Sadones, *Bull. Soc. Chim. Belg.*, **84,** 747 (1975).
59. M. K. G. Rao and E. G. Perkins, *J. Agr. Food Chem.*, **20,** 240 (1972).
60. J. Vance and R. Bentley, *Bio-Org. Chem.* **1,** 345 (1971).
61. M. Kofler, P. F. Sommer, H. R. Bolliger, B. Schmidli, and M. Vecchi, *Vitamins and Hormones*, **20,** 407 (1962).
62. W. Müller-Mulot, *J. Am. Oil Chem. Soc.*, **53,** 732 (1976).
63. A. J. Sheppard, A. R. Prosser, and W. D. Hubbard, *J. Am. Oil Chem. Soc.*, **49,** 619 (1972).
64. R. W. Lehman, *Methods Biochem. Anal.*, **2,** 153 (1955).
65. Vitamon E Panel, *Analyst*, **84,** 356 (1959).
66. P. Karrer and H. Fritzsche, *Helv. Chim. Acta*, **21,** 1234 (1938).

67. P. Karrer, R. Escher, H. Fritzsche, H. Keller, B. H. Ringier, and H. Salomon, *Helv. Chim. Acta*, **21,** 939 (1938).
68. M. Kofler, *Verhandl. Schweiz. Naturforsch. Ges.*, **6,** 239 (1941).
69. M. Furter and R. E. Meyer, *Helv. Chim. Acta*, **22,** 240 (1939).
70. J. G. Baxter, C. D. Roseson, J. D. Taylor, and R. W. Lehman, *J. Am. Chem. Soc.*, **65,** 918 (1943).
71. W. A. Skinner and R. M. Parkhurst, *J. Chromatogr.*, **13,** 69 (1964).
72. A. Vinet and P. Meunier, *Bull. Soc. Chim. Biol.*, **23,** 217 (1941).
73. A. Emmerie and C. Engel, *Nature*, **142,** 873, (1938).
74. C. C. Tsen, *Anal. Chem.*, **33,** 849 (1961).
75. P. A. Sturm, R. M. Parkhurst, and W. A. Skinner, *Anal. Chem.*, **38,** 1244 (1966).
76. L. I. Smith, L. J. Spillane, and I. M. Kolthoff, *J. Am. Chem. Soc.*, **64,** 447 (1942).
77. E. Knobloch, F. Machá, K. Mňouček, *Chem. Listy*, **46,** 718 (1952).
78. H. D. McBride and D. H. Evans, *Anal. Chem.*, **45,** 446 (1973).
79. V. D. Parker, *J. Am. Chem. Soc.*, **91,** 5380 (1969).
80. J. Bates, *Fette Seifen, Anstrichm.*, **56,** 484 (1954).
81. S. S. Atuma and J. Lindquist, *Analyst*, **98,** 886 (1973).
82. W. H. Harrison, J. E. Glander, E. R. Blakley, and P. D. Boyer, *Biochim. Biophys. Acta*, **21,** 150 (1956).
83. M. L. Quaife, *J. Am. Chem. Soc.*, **66,** 308 (1944).
84. L. Weisler, C. D. Robeson, and J. G. Baxter, *Anal. Chem.*, **19,** 906 (1947).
85. J. Green and S. Marcinkiewicz, *Analyst*, **84,** 297 (1959).
86. M. L. Quaife, *J. Biol. Chem.*, **175,** 605 (1948).
87. S. Marcinkiewicz and J. Green, *Analyst*, **84,** 304 (1959).
88. D. E. Duggan, *Arch. Biochem. Biophys.*, **84,** 116 (1959).
89. J. Green, S. Marcinkiewicz, and P. R. Watt, *J. Sci. Food Agr.*, **6,** 274 (1955).
90. J. Green, S. Marcinkiewicz, and D. McHale, *J. Chromatogr.*, **10,** 158 (1963).
91. P. W. R. Eggitt and F. W. Norris, *J. Sci. Food Agr.*, **6,** 689 (1955).
92. J. G. Bieri, C. J. Pollard, I. Prange, and H. Dam, *Acta Chem. Scand.*, **15,** 783 (1961).
93. F. Bro-Rasmussen and W. Hjarde, *Acta Chem. Scand.*, **11,** 34 (1957).
94. F. Bro-Rasmussen and W. Hjarde, *Acta Chem. Scand.*, **11,** 44 (1957).
95. R. A. Dilley and F. L. Crane, *Anal. Biochem.*, **5,** 531 (1963).
96. K. J. Whittle and J. F. Pennock, *Analyst*, **92,** 423 (1967).
97. D. L. Laidman and G. S. Hall, in *Methods in Enzymology*, Vol. XVIII, D. B. McCormick and L. D. Wright, Eds., Academic Press, New York, 1971.
98. Z. Deyl and J. Kopecký, Ed., *J. Chromatogr. Suppl.*, **6,** 667 (1976).
99. K. Macek, I. M. Hais, J. Kopecký, V. Schwarz, J. Gasparič, and J. Churáček, Eds., *J. Chromatogr. Suppl.*, **5,** 321 (1976).
100. K. A. Tartivita, J. P. Sciarello, and B. C. Rudy, *J. Pharm. Sci.*, **65,** 1024 (1976).
101. G. T. Vatassery and D. F. Hagen, *Anal. Biochem.*, **79,** 129 (1977).
102. N. Nicolaides, *J. Chromatogr.*, **4,** 496 (1960).
103. P. W. Wilson, E. Kodicek, and V. H. Booth, *Biochem. J.*, **84,** 524 (1962); *Chem. Abstr.*, **57,** 16984 (1962).
104. M. Kofler, P. F. Sommer, H. R. Bollinger, B. Schmidli, and M. Vecchi, *Vitamins and Hormones*, **20,** 430 (1962).
105. P. P. Nair, I. Sarlos, and J. Machiz, *Arch. Biochem. Biophys.*, **114,** 488 (1966).
106. P. P. Nair and J. Machiz, *Biochim. Biophys. Acta*, **144,** 446 (1967).
107. H. T. Slover, L. M. Shelley, and T. L. Burks, *J. Am. Oil Chem. Soc.*, **44,** 161 (1967).
108. H. T. Slover, R. J. Valis, and J. Lehmann, *J. Am. Oil Chem. Soc.*, **45**(8), 580 (1968).
109. P. Schudel, H. Meyer, J. Metzger, R. Rüegg, and O. Isler, *Helv. Chim. Acta*, **46,** 2517 (1963).
110. H. C. Pillsbury, A. J. Sheppard, and D. A. Libby, *J. Assoc. Offic. Anal. Chem.*, **50,** 809 (1967).

111. A. J. Sheppard, A. R. Prosser, and W. D. Hubbard, in *Methods in Enzymology*, Vol. XVIII, D. B. McCormick and L. D. Wright, Eds., Academic Press, New York, 1971.
112. S. R. Ames, in *The Vitamins*, 2nd ed., Vol. 5, W. H. Sebrell, Jr., and R. S. Harris, Eds., Academic Press, New York, 1972, p. 225.
113. R. H. Bunnell, in *The Vitamins*, 2nd ed., Vol. 6, P. Gyorgy and W. N. Pearson, Eds., Academic Press, New York, 1967, p. 261.
114. P. P. Nair and Z. Luna, *Arch. Biochem. Biophys.*, **127,** 413 (1968).
115. J. P. Nelson and A. J. Milun, *J. Am. Oil Chem. Soc.*, **45,** 848 (1968).
116. K. Lunan, A. Wenn, and D. W. Thomas, Private communication.
117. K. Lunan, private communication.
118. E. E. Schweizer and D. Meeder-Nycz, in *Chromenes, Chromanones, and Chromones*, G. P. Ellis, Ed., John Wiley, New York, 1977, p. 44.
119. E. E. Schweizer and D. Meeder-Nycz, in *Chromenes, Chromanones, and Chromones*, G. P. Ellis, Ed., John Wiley, New York, 1977, p. 45.
120. E. E. Schweizer and D. Meeder-Nycz, in *Chromenes, Chromanones, and Chromones*, John Wiley, New York, 1977, p. 70
121. I. M. Lockhart, in *Chromenes, Chromanones, and Chromones*, G. P. Ellis, Ed., John Wiley, New York, 1977, p. 305.
122. E. E. Schweizer and D. Meeder-Nycz, in *Chromenes, Chromanones, and Chromones*, G. P. Ellis, Ed., John Wiley, New York, 1977, Chap. II, p. 43.
123. H. Fukami, M. Nakayama, and M. Nakajima, *Agr. Biol. Chem.* (Tokyo), **25,** 243 (1961). *Chem. Abstr.*, **55,** 14447 (1961).
124. S. Sethna and R. Phadke, in *Organic Reactions*, Vol. VII, R. Adams, A. H. Blatt, A. C. Cope, F. C. McGrew, C. Niemann, and H. R. Snyder, Eds., Wiley, New York, 1953, Chap. 1.
125. R. L. Shriner and A. G. Sharp, *J. Org. Chem.*, **4,** 575 (1939).
126. A. Löwenbein, E. Pongrácz, and E. A. Spiess, *Ber.*, **57B,** 1517 (1924).
127. R. Huls, *Bull. Classe. Sci. Acad. Roy. Belg.*, **39,** 1064 (1953); *Chem. Abstr.*, **49,** 5459 (1955).
128. E. Späth and W. Močnik, *Ber.*, **70B,** 2276 (1937); *Chem. Abstr.*, **32,** 551 (1938).
129. W. John, P. Günther, and M. Schmeil, *Ber.*, **71B,** 2637 (1938).
130. W. John and M. Schmeil, *Ber.*, **72B,** 1653 (1939); *Chem. Abstr.*, **34,** 106 (1940).
131. D. McHale, P. Mamalis, J. Green, and S. Marcinkiewicz, *J. Chem. Soc.*, 1600 (1958).
132. F. A. Hochstein, *J. Am. Chem. Soc.*, **71,** 305 (1949).
133. W. G. Brown, *Organic Reactions* Vol. VI, R. Adams, H. Adkins, A. H. Blatt, A. C. Cope, F. G. McGrew, C. Niemann, and H. R. Snyder, Eds., John Wiley, New York, 1951, Chap. 10.
134. I. M. Lockhart, in *Chromenes, Chromanones, and Chromones*, G. P. Ellis, Ed., John Wiley, New York, 1977, Chap. III, 149–154.
135. P. Naylor, G. R. Ramage, and F. Schofield, *J. Chem. Soc.*, 1190 (1958).
136. W. Bridge, A. J. Crocker, I. Cubin, and A. Robertson, *J. Chem. Soc.*, 1530 (1937).
137. T. Kawai, T. Skimizu, and H. Chiba, *J. Pharm. Soc. Japan*, **75,** 274 (1955); *Chem. Abstr.*, **50,** 1787 (1956).
138. J. R. Beck, R. Kwok, R. N. Booher, A. C. Brown, L. E. Patterson, P. Pranc, B. Rockey, and A. Pohland, *J. Am. Chem. Soc.*, **90,** 4706 (1968).
139. I. M. Lockhart, in *Chromenes, Chromanones, Chromones*, G. P. Ellis, Ed., John Wiley, New York, 1977, Chap. V, pp. 305–307.
140. E. E. Schweizer and D. Meeder-Nycz, in *Chromenes, Chromanones and Chromones*, G. P. Ellis, Ed., John Wiley, New York, 1977, pp. 45–46.
141. P. Karrer and P. Banerjea, *Helv. Chim. Acta*, **32,** 1692 (1949).
142. E. E. Schweizer and D. Meeder-Nycz, in *Chromenes, Chromanones, and Chromones*, G. P. Ellis, Ed., John Wiley, New York, 1977, Chap. II, pp. 70–71.
143. H. Hart and H. S. Eleuterio, *J. Am. Chem. Soc.*, **76,** 516 (1954).

144. J. Hine, *Physical Organic Chemistry*, McGraw-Hill, New York, 1956, p. 335.
145. H. Hart, W. L. Spliethoff, and H. S. Eleuterio, *J. Am. Chem. Soc.*, **76,** 4547 (1954).
146. L. Bláha, J. Hodrová, and J. Weichet, *Coll. Czech. Chem. Commun.*, **24,** 2023 (1959); *Chem. Abstr.*, **53,** 20042 (1959).
147. P. A. Wehrli, R. I. Fryer, and W. Metlesics, *J. Org. Chem.*, **36,** 2910 (1971).
148. Y. Ichikawa, Y. Yamanaka, N. Yamamoto, T. Takeshita, T. Niki, T. Yamaji, Japan Patent 75 89,372 (1975); *Chem. Abstr.*, **84,** 74471 (1976).
149. S. Kitamura and N. Morioka, Japan Patent 74 05,344 (1974); *Chem. Abstr.*, **81,** 169671 (1974).
150. Y. Tachibana, Japan Patent 76 133,269 (1976); *Chem. Abstr.*, **86,** 190280 (1977).
151. B. Stalla Bourdillon, *Ind. Chim. Belge*, **35,** 13 (1970); *Chem. Abstr.*, **72,** 100898 (1970).
152. D. R. Nelan, U.S. Patent 3,444,213 (1969); *Chem. Abstr.*, **71,** 13238 (1969).
153. L. I. Smith, H. E. Ungnade, H. H. Hoehn, and S. Wawzonek, *J. Org. Chem.*, **4,** 311 (1939).
154. R. J. Molyneux and L. Jurd, *Tetrahedron*, **26,** 4743 (1970).
155. S. Marcinkiewicz, D. McHale, P. Mamalis and J. Green, *J. Chem. Soc.*, 3377 (1959).
156. D. McHale, P. Mamalis, J. Green and S. Marcinkiewicz, *J. Chem. Soc.*, 3358 (1959).
157. J. Green, D. McHale, P. Mamalis, and S. Marcinkiewicz, *J. Chem. Soc.*, 3375 (1959).
158. M. Hallet and R. Huls, *Bull. Soc. Chim. Belg.*, **61,** 33 (1952).
159. K. H. Boltze and H. D. Dell, *Angew, Chem. Internat. Edn.*, **5,** 415 (1966).
160. L. I. Smith, H. H. Hoehn, and H. E. Ungnade, *J. Org. Chem.*, **4,** 351 (1939).
161. O. Hromatka, U.S. Patent 2,354,317 (1944), *Chem. Abstr.*, **39,** 3537 (1945).
162. W. Baker, R. F. Curtis, J. F. W. McOmie, L. W. Olive, and V. Rogers, *J. Chem. Soc.*, 1007 (1958); W. Baker, R. F. Curtis, and J. F. W. McOmie, *J. Chem. Soc.*, 76 (1951).
162a. T. Iida, private communication, April 5 1979.
162b. K. Kakajima, A. Sato. T. Misono, R. Iida, and K. Nagayasu, *Agr. Biol. Chem.*, **38,** 1859 (1974).
162c. N. Cohen and G. Saucy, U.S. Patent 4,113,740 (1978).
163. K. C. D. Hickman and L. Weisler, U.S. Patent 2,486,540 (1949); *Chem. Abstr.*, **44,** 1234 (1950).
164. J. Green, and S. Z. Marcinkiewicz, Brit. Patent 827,391 (1960); *Chem. Abstr.*, **54,** 12502 (1960).
165. L. Weisler, U.S. Patent 2,640,058 (1953); *Chem. Abstr.*, **48,** 7643 (1954).
166. L. Weisler, U.S. Patent 2,592,629 (1952); *Chem. Abstr.*, **47,** 1193 (1953).
167. O. Isler, P. Schudel, H. Mayer, J. Würsch, and R. Rüegg, *Vitamins and Hormones*, **20,** 389 (1962).
168. E. A. Evans and R. F. Phillips, *J. Labelled Comp.*, **5,** 12 (1969).
169. F. C. Fuson and C. H. McKeever, in *Organic Reactions*, Vol. I, R. Adams, Ed., John Wiley, New York, 1942, Chap. III.
170. T. Nakamura and S. Kijima, *Chem. Pharm. Bull.*, **20,** 1681 (1972); *Chem. Abstr.*, **77,** 126366 (1972).
171. T. Nouaka and S. Kijima, Japan Patent, 76 65,761 (1976); *Chem. Abstr.*, **86,** 29962 (1977).
172. W. A. Skinner and R. M. Parkhurst, *Lipids*, **6,** 240 (1971).
173. W. John, *Z. Physiol. Chem.*, **252,** 222 (1938).
174. W. John, E. Dietzel, and W. Emte, *Z. Physiol. Chem.*, **257,** 173 (1939).
175. W. John and W. Emte, *Z. Physiol. Chem.*, **261,** 24 (1939).
176. A. Emmerie and C. Engel, *Rev. Trav. Chim. Pays-Bas*, **58,** 283 (1939).
177. L. I. Smith, W. B. Irwin, and H. E. Ungnade, *J. Am. Chem. Soc.*, **66,** 2424 (1941).
178. W. John and W. Emte, *Z. Physiol. Chem.*, **268,** 85 (1941).
179. V. L. Frampton, W. A. Skinner, and P. S. Bailey, *J. Am. Chem. Soc.*, **76,** 282 (1954).
180. V. L. Frampton, W. A. Skinner, P. Cambour, and P. S. Bailey, *J. Am. Chem. Soc.*, **82,** 4632 (1960).

181. P. D. Boyer, *J. Am. Chem. Soc.*, **73,** 733 (1951).
182. C. Martius and H. Eilingsfeld, *Justus Liebigs Ann. Chem.*, **607,** 159 (1957); *Chem. Abstr.*, **52,** 5400 (1958).
183. W. Durckheimer and L. A. Cohen, *Biochem. Biophys. Res. Commun.*, **9,** 262 (1962).
184. G. E. Inglett and H. A. Mattill, *J. Am. Chem. Soc.*, **77,** 6552 (1955).
185. C. T. Goodhue and H. A. Risley, *Biochem. Biophys. Res. Commun.*, **17,** 549 (1964).
186. C. T. Goodhue and H. A. Risley, *Biochem.*, **4,** 854 (1965).
187. W. A. Skinner, *Biochem. Biophys. Res. Commun.*, **15,** 469 (1964).
188. P. Schudel, H. Mayer, J. Metzger, R. Rüegg, and O. Isler, *Helv. Chim. Acta*, **46,** 636 (1963).
189. J. L. G. Nilsson, G. D. Daves, and K. Folkers, *Acta Chem. Scand.*, **22,** 207 (1968).
190. W. A. Skinner and P. Alaupovic, *Science*, **140,** 803 (1963).
191. W. A. Skinner and R. M. Parkhurst, *J. Org. Chem.*, **29,** 3601 (1964).
192. J. L. G. Nilsson, J. O. Branstad, and H. Sievertsson, *Acta Pharm. Suec.*, **5,** 509 (1968).
193. J. L. G. Nilsson and H. Sievertsson, *Acta Pharm. Suec.*, **5,** 517 (1968).
194. J. M. Behan, F. M. Dean, and R. A. W. Johnstone, *Tetrahedron*, **32,** 167 (1976).
195. C. Cassagne and J. Baraud, *Bull. Chim. Soc. Fr.*, 1470 (1968).
196. F. W. Knapp and A. L. Tappel, *J. Am. Oil Chem. Soc.*, **38,** 151 (1961); *Chem. Abstr.*, **55,** 11881 (1961).
197. W. L. Porter, A. S. Henick, and L. A. Levasseur, *Lipids*, **8,** 31 (1973).
198. H. W. Gardner, K. Eskins, G. W. Grams, and E. E. Inglett, *Lipids*, **7,** 324 (1972).
199. G. K. Koch, R. K. W. Han, J. J. L. Hoagenboom, M. Mutter, and H. van Tiborg, *Chem. Phys. Lipids*, **17,** 85 (1976).
200. S. Urano, S. Yamanoi, Y. Hattori, and M. Matsuo, *Lipids*, **12,** 105 (1977).
201. Y. Ishikawa and E. Yuki, *Agr. Biol. Chem.*, **39,** 851 (1975).
202. Y. Ishikawa, *Agr. Biol. Chem.*, **38,** 2545 (1974).
203. W. Boguth, *Vitamins and Hormones*, **27,** 1 (1969).
204. G. W. Grams, K. Eskins, and G. E. Inglett, *J. Am. Chem. Soc.*, **94,** 866 (1972).
205. G. W. Grams, *Tetrahedron Lett.* 4823 (1971).
206. G. W. Grams and K. Eskins, *Biochemistry*, **11,** 606 (1972).
207. J. L. G. Nilsson, H. Sievertsson, and H. Selander, *Tetrahedron Lett.*, 5023 (1968).
208. J. L. G. Nilsson and H. Sievertsson, *Acta Chem. Scand.*, **24,** 2885 (1970).
209. J. L. G. Nilsson, H. Sievertseon, and H. Selander, *Acta Chem. Scand.*, **23,** 859 (1969).
210. J. Weichet, L. Bláha, and B. Kakac, *Chem. Listy*, **52,** 722 (1958).
211. K. J. M. Andrews, Brit. Patent 893,172 (1962); *Chem. Abstr.*, **57,** 11131 (1962).
212. D. Ishisaka and K. Murakami, Japan. Patent 1737 (1962); *Chem. Abstr.*, **58,** 3400 (1963).
213. T. Matsura, T. Fujita, K. Miyao, and K. Naito, Japan. Patent 64 24,968 (1964); *Chem. Abstr.*, **62,** 10417 (1965).
214. W. J. Humphlett and C. V. Wilson, *J. Org. Chem.*, **26,** 2511 (1961).
215. G. Sosnovsky and M. Konieczny, *Z. Naturforsch. B. Anorg. Chem., Org. Chem.*, **31B,** 1379 (1976).
216. J. Weichet, L. Bláha, and B. Kakac, *Chem. Listy*, **24,** 1689 (1959).
217. W. A. Skinner, R. M. Parkhurst, J. Scholler, and K. Schwarz, *J. Med. Chem.*, **12,** 64 (1961).
218. T. Nakamura and S. Kijima, Japan. Patent, 68 08,109 (1968); *Chem. Abstr.*, **69,** 96469 (1968).
219. K. Murase, J. Matsumoto, K. Tamazawa, K. Takahashi, and M. Murakami, *Yamanouchi Seiyaku Kenkyu Hokoku*, **2,** 66 (1974); *Chem. Abstr.*, **84,** 30811 (1976).
220. J. L. G. Nilsson, H. Selander, H. Sievertsson, J. Skanberg, and K. G. Svensson, *Acta Chem. Scand.*, **25,** 94 (1971).
221. J. L. G. Nilsson, H. Sievertsson, and H. Selander, *Acta Pharm. Suec.*, **6,** 585 (1969).
222. L. Weisler, U.S. Patent 2,486,539 (1949); *Chem Abstr.*, **44,** 2037 (1950).
223. L. Weisler and A. J. Chechak, U.S. Patent 2,486,542 (1949); *Chem. Abstr.*, **44,** 2037 (1950).

224. J. G. Baxter, M. A. Stern, and L. Weisler, U.S. Patent, 2,486,541 (1949); *Chem. Abstr.*, **44,** 2038 (1950).
225. L. Weisler, U.S. Patent 2,673,858 (1954); *Chem. Abstr.*, **49,** 5533 (1955).
226. L. Weisler, U.S. Patent 2,592,628 (1952); *Chem. Abstr.*, **47,** 1192 (1953).
227. L. Weisler, U.S. Patent 2,592,630 (1952); *Chem. Abstr.*, **47,** 1192 (1953).
228. M. L. Quaife, U.S. Patent 2,499,778 (1950); *Chem. Abstr.*, **44,** 4376 (1950).
229. S. Marcinkiewicz, *Acta Pol. Phar.*, **24,** 375 (1967); *Chem. Abstr.*, 68, 68827 (1968).
230. L. Weisler, U.S. Patent, 2,519,863 (1950); *Chem. Abstr.*, **45,** 669 (1951).
231. C. D. Robeson and D. R. Nelan, U.S. Patent 3,153,040 (1964); *Chem. Abstr.*, **61,** 16079 (1964).
232. C. D. Robeson and D. R. Nelan, U.S. Patent 3,153,053 (1964); *Chem. Abstr.*, **61,** 16079 (1964).
233. F. Hoffman-LaRoche and Company, A. G., Belg. Patent 635,999 (1964); *Chem. Abstr.*, **61,** 13285 (1964).
234. R. J. Molyneux, *J. Chem. Soc. Chem. Commun.*, 318 (1974).
235. G. R. Whistance and D. R. Threlfall, *Biochem. Biophys. Res. Commun.*, **28,** 295 (1967).
236. G. R. Threlfall, G. R. Whistance, and T. W. Goodwin, *Biochem. J.*, **102,** 49P (1967); *Chem. Abstr.*, **66,** 113001 (1967).
237. G. R. Whistance, D. R. Threlfall, and T. W. Goodwin, *Biochem. Biophys. Res. Commun.*, **23,** 849 (1966).
238. J. F. Pennock, F. W. Hemming, and J. S. Kerr, *Biochem. Biophys. Res. Commun.*, **17,** 542; *Chem. Abstr.*, **62,** 4253 (1964).
239. A. R. Wellburn, *Phytochemistry*, **9,** 743 (1970).
240. D. R. Threlfall, *Vitamins and Hormones*, **29,** 153 (1971).
241. W. Janiszowsha and J. F. Pennock, *Vitamins and Hormones*, **34,** 77 (1976).
242. J. G. Bieri and P. M. Farrell, *Vitamins and Hormones*, **34,** 31 (1976).
243. I. Molenaar, J. Vos, and F. A. Hommes, *Vitamins and Hormones*, **30,** 45 (1972).
244. I. I. Ivanov, M. N. Merzlyak, and B. N. Tarusov, *Trans. Moscow Soc. Natur.*, **52,** 30 (1975); *Chem. Abstr.*, **83,** 159227 (1975).
245. H. S. Jacob and S. E. Lux, *Blood*, **32,** 549 (1968).
246. J. G. Bieri and R. K. H. Poukka, *J. Nutr.*, **100,** 557 (1970).
247. K. Schwarz, *Ann. N.Y. Acad. Sci.*, **203,** 45 (1972).
248. H. Zalkin, A. L. Tappel, K. A. Caldwell, S. Shibko, J. P. Desai, and T. A. Holiday, *J. Biol. Chem.*, **237,** 2678 (1962).
249. M. P. Carpenter, *Ann. N.Y. Acad. Sci.*, **203,** 81 (1972).
250. K. Fukuzawa and M. Uchiyama, *J. Nutr. Sci. Vitaminol.*, **19,** 433 (1973).
251. A. T. Diplock, *Am. J. Clin. Nutr.*, **27,** 995 (1974).
252. C. I. Bliss and P. György, in *The Vitamins*, Vol. VI, 2nd ed., P. György and W. N. Pearson, Eds., Academic Press, New York, 1967, p. 304.
253. M. E. Kaboy and J. J. Gilbert, *Lipids*, **12,** 875 (1977).
254. C. T. Goodhue, *Biochemistry*, **4,** 1822 (1965).
255. J. Green, *Ann. N.Y. Acad. Sci.*, **202,** 29 (1972).
256. J. Green, in *The Vitamins*, Vol. 5, W. H. Sebrell, Jr., and R. S. Harris, Eds., Academic Press, New York, 1972, Chap. 16. pp. 259–272.
257. M. K. Horwitt and K. E. Mason, in *The Vitamins*, Vol. V, W. H. Sebrell and R. S. Harris, Eds., Academic Press, New York, 1972, Chap. 16. p. 310.
258. M. K. Horwitt, C. C. Harvey, and E. M. Harmon, *Vitamins and Hormones*, **26,** 487 (1968).
259. G. D. Fitch, *Vitamins and Hormones*, **26,** 501 (1968).
260. K. E. Mason and M. K. Horwitt, *The Vitamins*, Vol. 5, W. H. Sebrell, Jr., and R. S. Harris, Eds., Academic Press, New York, 1972, pp. 272, 293.
261. S. R. Ames, *The Vitamins*, Vol. 5, W. H. Sebrell, Jr., and R. S. Harris, Eds., Academic Press, New York. 1972, p. 312.
262. J. S. Dinning, *Vitamins and Hormones*, **20,** 511 (1962).
263. H. S. Murty, P. I. Coasi, S. K. Brooks and P. P. Nair, *J. Biol. Chem.*, **245,** 5498 (1970).

264. C. E. Myers, W. McGuire, and R. Young, *Cancer Treatment Rep.* **60,** 961 (1976); *Chem. Abstr.*, **85,** 171675 (1976).
265. J. Schroeder, *Biochem. Biophys. Acta*, **343,** 173 (1974).
266. T. Sakai, T. Okano, H. Makino, and T. Tsudzuki, *J. Cyclic, Nucleot. Res.*, **2,** 163 (1976).
267. V. I. Gunar, and S. I. Zavyalov, *Isvest. Akad. Nauk, SSSR Otdel. Khim. Nauk*, 937 (1960); *Chem. Abstr.*, **54,** 24700 (1960).
268. F. J. E. M. Kuppers, C. A. L. Bercht, C. A. Salemink, Ch. J. J. R. Lousberg, J. K. Terlouw and W. Heerma, *J. Chromatogr.*, **108,** 375 (1975).
269. O. P. Malik, R. S. Kapil, and N. Anand, *Indian J. Chem.*, **14B,** 449 (1976).
270. S. Yamada, O. Futara, T. Katagiri, and J. Tanaka, *Nippon Kagaku Kaishi*, **10,** 1987 (1972); *Chem. Abstr.*, **78,** 16079 (1973).
271. G. Manners, L. Jurd, and K. Stevens, *Tetrahedron*, **28,** 2949 (1972). *Chem. Abstr.*, **77,** 62140 (1972).
272. Y. Gaoni and R. Mechoulan, *J. Am. Chem. Soc.*, **93,** 217 (1971).
273. R. Mechoulam and B. Yagen, *Tetrahedron Lett.*, **60,** 5349 (1969).
274. S. Nozoe and K. T. Suzuki, *Tetrahedron*, **27,** 6063 (1971).
275. S. Nozoe and K. Hirai, *Tetrahedron*, **27,** 6073 (1971).
276. Wellcome Foundation Ltd., *Neth Appl.* 6,508,323 (1965); *Chem. Abstr.*, **65,** 2229 (1966).
277. P. E. Bender and B. Loev, *U.S. Publ. Pat. Appl.* B 380,446 (1975); *Chem. Abstr.*, **82,** 156076 (1975).
278. P. S. Bramwell and A. O. Fitton, *J. Chem. Soc.*, 3882 (1965).
279. P. S. Bramwell, A. O. Fitton, and G. R. Ramage, Brit. Patent 1,116,562 (1968). *Chem. Abstr.*, **69,** 67355 (1968).
280. FMC Corp., Neth. Appl. 6.608,777 (1967); *Chem. Abstr.*, **67,** 100004 (1967).
281. P. E. Drummond and P. H. Schroeder (to FMC Corp.), U.S. Patent 3,557,150 (1971); *Chem. Abstr.*, **74,** 141533 (1971).
282. E. A. Vdovtsova and G. F. Fedorova, *Zh. Org. Khim.*, **6,** 1230 (1970); *Chem. Abstr.*, **73,** 66178 (1970).
283. R. Heiss, A. Seyberlich, I. Hammann, and W. Behrenz, Brit Patent 1,126,140 (1968); *Chem. Abstr.*, **69,** 106557 (1968).
284. Y. Kitahara, T. Kato. and Y. Kakinuma, *Kagaku To Seibutsu*, **14,** 472 (1976); *Chem. Abstr.*, **85,** 155100 (1976).
285. T. Kato, A. S. Kumanireng, I. Ichinose, Y. Kitahara, K. Kakinuma, M. Nishihira, and M. Kato, *Experientia*, **31,** 433 (1975).
286. D. McHale, J. Green, S. Marcinkiewicz, J. Feeney, and L. H. Sutcliffe, *J. Chem. Soc.*, 784 (1963).
287. J. Green and S. Marcinkiewicz, *Nature*, **177,** 86 (1956).
288. M. Mayer, J. Metzger, and O. Isler, *Helv. Chim. Acta*, **50,** 1376 (1967).
289. R. L. Coop and J. F. Pennock, cited as unpublished work (1964) in Ref. 238.
290. J. Green, P. Mamalis, S. Marcinkiewicz, and D. McHale, *Chem. Ind.* (London), 73 (1960).
291. P. J. Dunphy, K. J. Whittle, J. F. Pennock, and R. A. Morton, *Nature* (London), **207,** 521 (1965).
292. J. W. Scott, F. T. Bizzarro, D. R. Parrish, and G. Saucy, *Helv. Chim. Acta*, **59,** 290 (1976).
293. O. Isler, H. Mayer, J. Metzger, R. Rüegg, and P. Schudel, *Angew. Chem.*, **75,** 1030 (1963).
294. J. Bunyan, D. McHale, J. Green, and S. Marcinkiewicz, *Br. J. Nutr.* **15,** 253 (1961); *Chem. Abstr.*, **55,** 21290 (1961).
295. B. Graffe, M. C. Sacquet, and P. Maitte, *J. Heterocycl. Chem.*, **12,** 247 (1975).
296. F. Prillinger and H. Schmid, *Monatsh.*, **72,** 427 (1939); *Chem. Abstr.*, **34,** 1017 (1940).
297. A. S. Mujumdar and R. N. Usgaonkar, *Indian J. Chem.*, **10,** 672 (1972).

298. O. Dann, G. Volz, and O. Huber, *Justus Liebigs Ann. Chem.*, **587,** 16 (1954).
299. E. F. Elslager and F. H. Tendick, *J. Med. Pharm. Chem.*, **5,** 546 (1962).
300. I. I. Dragotă, L. V. Feyns, and I. Niculescu-Duvăz, *Rev. Roum. Chim*, **20,** 665 (1975); *Chem. Abstr.*, **83,** 193007 (1975).
301. J. N. Chatterjea, K. D. Banerji, and N. Prasad, *Ber.*, **96,** 2356 (1961). *Chem. Abstr.*, **59,** 12774 (1963).
302. R. Verhé and N. Schamp, *Experientia*, **29,** 784 (1973).
303. H. Cairns, D. Hunter, and J. King, U.S. Patent 3,860,617 (1975); *Chem. Abstr.*, **79,** 5261 (1973).
304. P. R. Iyer and G. D. Shah, *Indian J. Chem.*, **6,** 227 (1968).
305. J. A. Miller and H. C. S. Wood, *Chem. Commun.* 39 (1965).
306. L. Crombie and R. Ponsford, *Chem. Commun.*, 894 (1968).
307. B. Willhalm, A. F. Thomas, and F. Gautchi, *Tetrahedron*, **20,** 1185 (1964).
308. S. Ivanov and B. Konova, *Dokl. Bolg. Akad. Nauk*, **29** (3), 371 (1976); *Chem. Abstr.*, **85,** 59635 (1976).
309. D. R. Threlfall, *Asp. Terp. Chem. Biochem. Proc. Phytochem. Symp.*, 1970; *Chem. Abstr.*, **76,** 43938 (1972).
310. P. Hartmann and E. Roos (to Farbenfabriken Bayer A. G.), Ger. Patent 2,101,480 (1972); *Chem. Abstr.*, **77,** 151934 (1972).
311. B. O. Linn (to Merck and Co.), U.S. Patent 3,160,637 (1964); *Chem. Abstr.*, **62,** 5256 (1965).
312. A. O. Fitton and G. R. Ramage, *J. Chem. Soc.*, 5426 (1963).
313. F. Bergel, A. Jacob, A. R. Todd, and T. S. Work, *J. Chem. Soc.*, 1375 (1938).
314. P. Mamalis, J. Green, S. Marcinkiewicz, and D. McHale, *J. Chem. Soc.*, 3362 (1959).
315. K. Sato and S. Inoue, Japan. Patent 74 00,267 (1974); *Chem. Abstr.*, **80,** 108376 (1974).
316. H. G. Reppe, H. Pohlemann, M. Seefelder, and H. Christoph, Ger. Patent 1,149,165 (1963); *Chem. Abstr.*, **59,** 5326 (1963).
317. T. Okutsu, N. Tsuji, R. Ohi, and T. Kondo (to Fuji Photo Film Co., Ltd.), Ger. Offen. 2,210,368 (1972); *Chem. Abstr.*, **78,** 59275 (1973).
318. L. I. Smith and R. W. H. Tess, *J. Am. Chem. Soc.*, **66,** 1523 (1944).
319. L. I. Smith and J. A. King, *J. Am. Chem. Soc.*, **63,** 1887 (1941).
320. H. Selander and J. L. G. Nilsson, *Acta Pharm. Suec.*, **9,** 125 (1972).
321. L. I. Smith, *J. Am. Chem. Soc.*, **64,** 1084 (1942).
322. M. Murakami, I. Tamazawa, and Z. Tsuchikata, Japan. Patent 68 28,824 (1968); *Chem. Abstr.*, **70,** 68182 (1969).
323. L. I. Smith and J. A. Sprung, *J. Am. Chem. Soc.*, **64,** 433 (1942).
324. M. Murakami, T. Sato, A. Matsumoto, Y. Hirata, H. Iwamoto, and H. Suzuki, Japan. Patent 13,104 (1967); *Chem. Abstr.*, **68,** 29599 (1968).
325. L. I. Smith, H. E. Ungnade, J. R. Stevens, and C. C. Christman, *J. Am. Chem. Soc.*, **61,** 2615 (1939).
326. R. T. Borchardt and L. A. Cohen, *J. Am. Chem. Soc.*, **95,** 8308 (1973).
327. L. I. Smith and J. A. Sprung, *J. Am. Chem. Soc.*, **65,** 1276 (1943).
328. L. I. Smith and H. C. Miller, *J. Am. Chem. Soc.*, **64,** 440 (1942).
329. H. Mayer, P. Schudel, R. Rüegg, and O. Isler, *Helv. Chim. Acta*, **46,** 650 (1963).
330. T. Nakamura and K. Kijima, Japan. Patent 72 47,034 (1972); *Chem. Abstr.*, **78,** 111129 (1973).
331. M. H. Stern, *Eastman Org. Chem. Bull.*, **42,** 2 (1970).
332. K. Sawatari, T. Nishino, S. Oda, K. Suenobu, T. Miura, and T. Nakamori, Japan. Kokai 74 72,338 (1974); *Chem. Abstr.*, **82,** 44424 (1975).
333. A. A. Svishchuk and N. N. Vysotskaya, *Ukr. Khim. Zh.*, **41,** 506 (1975); *Chem. Abstr.*, **83,** 79031 (1975).
334. M. Murakami, A. Matsumoto and K. Murase, Japan Patent 69 27,725 (1969); *Chem. Abstr.*, **72,** 43453 (1970).

335. L. I. Smith, W. B. Irwin, and H. E. Ungnade, *Science*, **90,** 334 (1939).
336. G. V. Protopopova, N. S. Lagoshina, L. I. Reidalova, N. N. Vysotskii, and A. A. Svishchuk, *Fiziol. Akt. Veshchestva*, **7,** 142 (1975); *Chem. Abstr.*, **83,** 142932 (1975).
337. T. Nakamura and S. Kijima, *Chem. Pharm. Bull.*, **20,** 1297 (1972).
338. J. Green and D. McHale, Br. Patent 949,715 (1964); *Chem. Abstr.*, **60,** 13227 (1964).
339. A. H. Cook, I. M. Heilbron, and F. B. Lewis, *J. Chem. Soc.*, 659 (1942).
340. L. I. Smith, H. E. Ungnade, J. R. Stevens, and C. C. Christman, *J. Am. Chem. Soc.*, **61,** 2615 (1939).
341. J. Weichet, L. Bláha, and B. Kakac, *Coll. Czech. Chem. Commun.*, **31,** 4598 (1966); *Chem. Abstr.*, **66,** 18647 (1967).
342. C. Cardani and G. Gasnati, *Rend. Ist. Lomb*, **91,** 624 (1957).
343. E. Roos, R. Nast, P. Hartmann, W. Redetsky, and G. P. Langner, Ger. Offen. 2,315,349 (1974); *Chem. Abstr.*, **82,** 59538 (1975).
344. T. Nakamura and S. Kijima, *Chem. Pharm. Bull.*, **19,** 2318 (1971).
345. Y. Ooishi, M. Yamada, H. Amano, and T. Nishimura, Ger. Offen. 2,420,066 (1974); *Chem. Abstr.*, **82,** 162977 (1975).
346. T. Okutsu, N. Tsuji, R. Ohi, and T. Kondo, Japan. Appl. 71 11,076 (1971); *Chem. Abstr.*, **78,** 59275 (1973).
347. R. Oi, Japan. Kokai 76 68,569 (1976); *Chem. Abstr.*, **86,** 16540 (1977).
348. G. J. Lestina and M. H. Stern, Ger. Offen. 2,140,309 (1972); *Chem. Abstr.*, **76,** 160803 (1972).
349. J. R. Schaeffer and C. T. Goodhue, *J. Org. Chem.*, **36,** 2563 (1971).
350. K. Nakajima and S. Kitamura, Japan. Patent 67 5334 (1967); *Chem. Abstr.* **67,** 90673 (1967).
351. Y. Ichikawa, Y. Yamanaka, M. Yamamoto, T. Takeshita, T. Niki, and T. Yamaji, Ger. Offen. 2,404,621 (1975); *Chem. Abstr.*, **84,** 59792 (1976).
352. V. A. Hoyle, Jr., and W. V. McConnell, U.S. Patent 3,476,772 (1969); *Chem. Abstr.*, **72,** 31617 (1970).
353. K. Sawatari, T. Nishino, S. Oda, K. Suenobu, T. Miura, and T. Nakamori, Japan. Kokai 74 72,338 (1974); *Chem. Abstr.*, **82,** 44424 (1975).
354. Teijin Ltd., Fr. Demande 2,259,822 (1975); *Chem. Abstr.*, **85,** 46898 (1976).
355. H. Frick, N. Halder, and W. Vogler, Ger. Offen. 2,160,103 (1972); *Chem. Abstr.*, **77,** 151936 (1972).
356. P. Karrer and K. A. Jensen, *Helv. Chim. Acta*, **21,** 1622 (1938).
357. J. Weichet and J. Hodrova, *Chem. Listy*, **51,** 133 (1957); *Chem. Abstr.*, **51,** 14656 (1957).
358. P. Mamalis, D. McHale, J. Green, and S. Marcinkiewicz, *J. Chem. Soc.*, 1850 (1958).
359. H. K. Pendse and P. Karrer, *Helv. Chim. Acta*, **40,** 1837 (1957).
360. A. Jacob, F. K. Sutcliffe, and A. R. Todd, *J. Chem. Soc.*, 327 (1940).
361. J. Green, D. McHale, S. Marcinkiewicz, P. Mamalis, and P. R. Watt. *J. Chem. Soc.*, 3362 (1959).
362. J. Weichet and J. Hodrovă, *Coll. Czech. Chem. Commun.*, **22,** 595 (1957).
363. S. Kitamura, Y. Fujita, and M. Morioka, Japan. Patent 70 21,713 (1970); *Chem. Abstr.*, **73,** 87794 (1970).
364. J. Weichet, Pharmacotherapeutia Collect. Papers, Anniversary Research Inst. Pharm. Biochem., 10th, Prague, 1960, p. 49, published 1961; *Chem. Abstr.*, **56,** 10291 (1962).
365. L. I. Smith and R. W. H. Tess, *J. Am. Chem. Soc.*, **66,** 1526 (1944).
366. P. Karrer and H. Fritzsche, *Helv. Chim. Acta*, **22,** 260 (1939).
367. P. Karrer and P. C. Dutta, *Helv. Chim. Acta*, **31,** 2080 (1948).
368. M. H. Stern, C. D. Robeson, K. Weisler, and J. G. Baxter, *J. Am. Chem. Soc.*, **69,** 869 (1947).
369. F. Hoffman-LaRoche and Company A. G., Swiss Patent 214,043 (1941); *Chem. Abstr.*, **36,** 4977 (1942).
370. L. F. Fieser, M. Tishler, and N. L. Wendler, *J. Am. Chem. Soc.*, **62,** 2861 (1940).

371. O. Hromatka, Ger. Patent 736,797 (1943), Addition to Ger. Patent 730,790; *Chem. Abstr.*, **38,** 2974 (1944).
372. F. Bergel, A. M. Copping, A. Jacob, A. R. Todd, and T. S. Work, *J. Chem. Soc.*, 1382 (1938).
373. Hoffmann-LaRoche, Ger. Patent 731,972 (1938).
374. P. Karrer and O. Isler (to Hoffman-LaRoche), U.S. Patent 2,411,967 (1938); *Chem. Abstr.*, **41,** 1713 (1947).
375. *Beilstein* E III/IV **17**(2) 1425 (1974).
376. A. Jacob, M. Steiger, A. R. Todd, and T. S. Work, *J. Chem. Soc.*, 542 (1939).
377. P. Karrer, H. Salomon, and H. Fritzsche, *Helv. Chim. Acta*, **21,** 309 (1938).
378. W. John, E. Dietzel, P. Günther, and W. Emte, *Naturwissenschaften*, **26,** 366 (1938).
379. W. John, E. Dietzel, and P. Günther, *Hoppe-Seyler's Z. Physiol. Chem.*, **252,** 208 (1938).
380. F. Bergel, A. R. Todd, and T. S. Work, *J. Chem. Soc.*, 253 (1938).
381. O. H. Emerson and L. I. Smith, *J. Am. Chem. Soc.*, **62,** 1869 (1940).
382. W. John, *Hoppe-Seyler's Z. Physiol. Chem.*,
383. O. H. Emerson, *J. Am. Chem. Soc.*, **60,** 1741 (1938).
384. O. H. Emerson, G. A. Emerson, A. Mohammad, and H. M. Evans, *J. Biol. Chem.*, **122,** 99 (1937).
385. A. R. Todd, F. Bergel, and T. S. Work, *Biochem. J.*, **31,** 2257 (1937).
386. A. Issidorides, *J. Am. Chem. Soc.*, **73,** 5146 (1951).
387. W. John, *Z. Physiol. Chem.*, **252,** 201 (1938).
388. A. Itiba, *Sci., Papers Inst. Phys. Chem. Res.*, **34,** 121 (1938); *Chem. Abstr.*, **32,** 3331 (1938); *Beilstein*, E II/IV **17**/2, p. 1429.
389. C. D. Robeson, *J. Am. Chem. Soc.*, **65,** 1660 (1943).
390. P. Karrer, H. Fritzsche, and R. Escher, *Helv. Chim. Acta*, **22,** 661 (1939).
391. A. R. Todd, F. Bergel, H. Waldmann, and T. S. Work, *Biochem. J.*, **31,** 2247 (1937); *Chem. Abstr.*, **32,** 2992 (1938).
392. P. Karrer, H. Koenig, B. H. Ringier, and H. Salomon, *Helv. Chim. Acta*, **22,** 1139 (1939).
393. S. Kitamura, Y. Fujita, and M. Morioka, Japan. Patent 70 38,336 (1970); *Chem. Abstr.*, **74,** 87830 (1971).
394. T. Nakamura, S. Umeda, and T. Matsuura, Japan. Patent 70 21,711 (1970); *Chem. Abstr.*, **73,** 77055 (1970).
395. W. John and H. Herrmann, *Z. Physiol. Chem.*, **273,** 191 (1942); *Chem. Abstr.*, **37,** 3092 (1943).
396. W. C. Gray and F. A. Stahly, Fr. Patent 1,596,533 (1970); *Chem. Abstr.*, **74,** 93458 (1971).
397. C. S. McArthur and E. M. Watson, *Science*, **86,** 35 (1937).
398. E. Fernholz, *J. Am. Chem. Soc.*, **59,** 1154 (1937).
399. O. H. Emerson, *Science*, **88,** 40 (1938).
400. E. Fernholz, *J. Am. Chem. Soc.*, **60,** 700 (1938).
401. P. Karrer, H. Fritzsche, B. H. Ringier, and H. Salomon, *Helv. Chim. Acta*, **21,** 520 (1938).
402. P. Karrer, H. Fritzche, B. H. Ringier, and H. Salomon, *Helv. Chim. Acta*, **21,** 820 (1938).
403. L. I. Smith, H. E. Ungnade, and W. W. Prichard, *Science*, **88,** 37 (1938).
404. M. Matsui and S. Kitamura, *Agr. Biol. Chem.*, **29,** 978 (1965).
405. T. Ichakawa and T. Kato, Japan. Patent 11064 (1967); *Chem. Abstr.*, **67,** 10003 (1967).
406. T. Ichakawa and T. Kato, Japan. Patent 11065 (1967); *Chem. Abstr.*, **67,** 10001 (1967).
407. K. Nakagawa and S. Muraki, Japan. Patent 11993 (1963); *Chem. Abstr.*, **59,** 13953 (1963).
408. K. Nakagawa, Japan. Patent 18338 (1966); *Chem. Abstr.*, **67,** 90966 (1967); *ibid.*, **66,** 28937 (1967).
409. O. Ehrmann, Ger. Patent 1,015,446 (1957); *Chem. Abstr.*, **54,** 578 (1960).

410. L. I. Smith and H. E. Ungnade, *J. Org. Chem.*, **4,** 298 (1939).
411. F. von Werder, U.S. Patent 2,230,659 (1941); *Chem. Abstr.*, **35,** 3270 (1941).
412. H. Mayer, P. Schudel, R. Rüegg, and O. Isler, *Chimia*, **16,** 367 (1962).
413. J. D. Surmatis and J. Weber, U.S. Patent 2,723,278 (1955); *Chem. Abstr.*, **50,** 10794 (1956).
414. M. E. Maurit, G. V. Smirnova, E. A. Parfenov, I. K. Sarycheva, and N. A. Preobrazhenskii, *Dokl. Akad. Nauk SSSR*, **140,** 1330 (1961); *Chem. Abstr.*, **56,** 8672 (1962).
415. J. A. Aeschlimann, U.S. Patent 2,307,010 (1943); *Chem. Abstr.*, **37,** 3567 (1943).
416. P. Karrer and O. Isler, U.S. Patent 2,411,969 (1941); *Chem. Abstr.*, **41,** 1713 (1947).
417. P. Karrer, R. G. Legler, and G. Schwab, *Helv. Chim. Acta*, **23,** 1132 (1940).
418. W. John and H. Pini, *Hoppe-Seyler's Z. Physiol. Chem.*, **273,** 225 (1942).
419. P. Karrer and B. H. Ringier, *Helv. Chim. Acta*, **22,** 610 (1939).
420. P. Karrer and O. Isler, U.S. Patent 2,411,967 (1938); *Chem. Abstr.*, **41,** 1713 (1947).
421. F. Bergel, A. Jacob, A. R. Todd, and T. S. Work, *Nature* (London), **142,** 36 (1938).
422. H. M. Evans, O. H. Emerson, and G. A. Emerson, *J. Biol. Chem.*, **113,** 319 (1936).
423. G. Saucy, J. W. Scott, and D. R. Parrish, Ger. Offen. 2,364,165 (1974); *Chem. Abstr.*, **81,** 169670 (1974).
424. P. Karrer and A. Kugler, *Helv. Chim. Acta*, **28,** 436 (1945).
425. J. Harrill and E. D. Gifford, *J. Nutr.*, **89,** 247 (1966); *Chem. Abstr.*, **65,** 14173 (1966).
426. D. McHale, S. Marcinkiewicz and J. Green, *J. Chem. Soc.* (*C*), 1427 (1966).
427. W. John and F. H. Rathmann, *Ber.*, **74,** 890 (1941).
428. I. G. Farbenind, Ger. Patent 757928 (1939); *Chem. Abstr.*, **39,** 1256 (1945).
429. W. John, *Z. Physiol. Chem.*, **268,** 104 (1941).
430. W. John and P. Günther, *Ber.*, **74,** 879 (1941).
431. U. Gloor, J. Wuersch, U. Schwieter, and O. Wiss, *Helv. Chim. Acta*, **49,** 2303 (1966).
432. T. Nakamura, Y. Aoyama, T. Fujita, and G. Katsui, *Lipids*, **10,** 627 (1975). *Chem. Abstr.* **84,** 25708 (1976).
433. P. Johnson and W. F. R. Pover, *Life Sci.*, **1,** 115 (1962); *Chem Abstr.*, **57,** 8983 (1962).
434. A. A. Svishchuk and E. A. Tikhomirova, *Ukr. Khim. Zh.*, **29,** 1070 (1963); *Chem. Abstr.*, **60,** 11972 (1964).
435. M. Zielinski, in *The Chemistry of the Quinonoid Compounds*, Part 2, S. Patai, Ed., John Wiley, New York, 1974; Chap. 12.
436. D. McHale and J. Green, *Chem. Ind.* (London), 366 (1964).
437. U. V. Solmssen and J. Lee, U.S. Patent 2,457,932 (1949); *Chem. Abstr.*, **43,** 4702 (1949).
438. H. Scholz, Ger. Offen. 2,331,300 (1975); *Chem. Abstr.*, **83,** 10526 (1975).
439. E. Merck to Chemische Fabrik., Brit. Patent 811,895 (1959); *Chem. Abstr.*, **53,** 18965 (1959).
440. C. D. Robeson, *J. Am. Chem. Soc.*, **64,** 1487 (1942).
441. B. C. Rudy and B. Z. Senkowski, *Analytical Profiles of Drug Substances*, **3,** 111 (1974).
442. C. D. Robeson and D. R. Nelan, *J. Am. Chem. Soc.*, **84,** 3196 (1962).
443. V. Demole, *Z. Vitaminforsch.*, **8,** 338 (1939); *Chem. Abstr.*, **33,** 5510 (1939).
444. T. Nakamura, S. Umeda, and S. Kijima, Japan Patent 70 12,738 (1970); *Chem. Abstr.*, **73,** 66431 (1970).
445. P. Karrer and H. Salomon, *Helv. Chim. Acta*, **21,** 514 (1938).
446. T. Nakamura, S. Umeda, and T. Matsuura, Japan. Patent 70 21,711 (1970); *Chem. Abstr.*, **73,** 77055 (1970).
447. Y. Mushika and N. Yoneda, *Chem. Pharm. Bull.*, **19,** 687 (1971); *Chem. Abstr.*, **76,** 14747 (1972).
448. T. Nakamura and S. Kijima, U.S. Patent 3,459,774 (1969); *Chem. Abstr.*, **71,** 112815 (1969).
449. Eisai Co. Ltd., Neth. Appl. 6,607,090 (1966); *Chem. Abstr.*, **66,** 85895 (1967).
450. F. L. Grinberg, G. M. Lushchevskaya, and B. G. Savinov, *Vitaminy, Akad. Nauk Ukr. SSSR, Inst. Biokhim.*, **3,** 50 (1958); *Chem. Abstr.*, **53,** 3207 (1959).

451. S. Kudo, K. Fujii, and Y. Takemoto, Japan Patent 71 29,734 (1971); *Chem. Abstr.*, **75,** 133016 (1971).
452. S. Kijima, S. Imabayashi, and T. Nakamura, Japan Kokai 74 13,179 (1974); *Chem. Abstr.*, **80,** 108375 (1974).
453. S. Tachibana and K. Miyao, Japan. Patent 68 03,782 (1968); *Chem. Abstr.*, **70,** 38096 (1969).
454. P. A. Marr and S. J. Ciurca, Jr., Eastman Kodak Co., Fr. Demande 2,232,776 (1975); *Chem. Abstr.*, **83,** 50711 (1975).
455. W. J. Humphlett and C. V. Wilson, U.S. Patent 2,875,195 (1959); *Chem. Abstr.*, **53,** 13039 (1959).
456. S. A. Roncales, Span. Patent 3 92,676 (1973); *Chem. Abstr.*, **80,** 83332 (1974).
457. K. Miyao and M. Toda, Japan. Patent 69 11,674 (1969); *Chem. Abstr.*, **71,** 102074 (1969).
458. A. A. Svishchuk, N. N. Vysotskii, and O. D. Overchuk, *Farm. Zh. (Kiev)*, **29**(6), 36 (1974); *Chem. Abstr.*, **82,** 73228 (1975).
459. Hoffman-LaRoche, *D.R.P.* 712,744 (1939); Swiss Patent 208,086; *Chem. Abstr.*, **37,** 4408 (1943); *ibid.*, **35,** 3773 (1941).
460. T. Nakamura, S. Umeda, and S. Kijima, Japan. Patent 71 39,863 (1971); *Chem. Abstr.*, **76,** 34105 (1972).
461. W. Thiele, *D.R.P.* 7 24,268 (1942); *Chem. Abstr.*, **37,** 5831 (1943).
462. P. Schudel, H. Mayer, and O. Isler, in *The Vitamins*, Vol. V, W. H. Sebrell, Jr., and R. S. Harris, Eds., Academic Press, New York, 1972, Chap. 16.
463. F. L. Grinberg, G. M. Lushchevskaya, and B. G. Savinov, *Vitaminy Adak. Nauk Ukr. S.S.R., Inst. Biokhim*, **3,** 50 (1958); *Chem. Abstr.*, **53,** 3207 (1959).
464. Fujisawa Pharm. Co., Ltd., Neth. Appl. 6,414,867 (1965); *Chem. Abstr.*, **63,** 16310 (1965).
465. J. A. Aeschlimann, U.S. Patent 2,393,134 (1946); *Chem. Abstr.*, **40,** 3573 (1946).
466. T. P. Mulholland-Cunningham, Brit. Patent 1,394,260 (1975); *Chem. Abstr.*, **83,** 97026 (1975).
467. F. Hoffmann-LaRoche A.G., Ger. Patent 711,243 (1941); *Chem. Abstr.*, **37,** 4207 (1943).
468. P. Schudel, H. Mayer, J. Metsger, R.. Rüegg, and O. Isler, *Helv. Chim. Acta*, **46,** 333 (1963).
469. G. W. Ahrens, U.S. Patent 3,803,179 (1974); *Chem. Abstr.*, **81,** 140868 (1974).
470. K. J. M. Andrews, Brit. Patent 893,172 (1962); *Chem. Abstr.*, **57,** 11131 (1962).
471. P. F. G. Praill, *J. Chem. Soc.*, 3100 (1959).
472. T. Kakagawa and Y. Mori, *Yakugaku Zasshi*, **79,** 591 (1959); *Chem. Abstr.*, **54,** 638 (1960).
473. F. Hoffman-LaRoche A.G., *Fr. M* 3517 (1965); *Chem. Abstr.*, **64,** 7979 (1966).
474. A. A. Svishchuk and E. D. Basalkeich, *Fisiol. Aktiv Veshchestva Repub. Mezhvedom. Sb.*, **No. 3,** 256 (1971); *Chem. Abstr.*, **77,** 88217 (1972).
475. K. H. Baggaley, S. G. Brooks, J. Green, and B. T. Redman, *J. Chem. Soc. Chem. Commun.*, 6 (1970).
476. J. Green, S. Marcinkiewicz, and D. McHale, *J. Chem. Soc. (C)*, 1422 (1966).
477. P. Karrer and M. Ståhelin, *Helv. Chim. Acta*, **28,** 438 (1945).
478. P. Karrer and R. Schäpfer, *Helv. Chim. Acta*, **24,** 298 (1941).
479. L. I. Smith and W. B. Renfrow, *J. Am. Chem. Soc.*, **64,** 445 (1942).
480. A. A. Svishchuk and E. D. Basalkevich, *Ukr. Khim. Zh. (Russ. Ed.)*, **41,** 70 (1975); *Chem. Abstr.*, **83,** 9690 (1975).
481. P. Mamot, R. Azerad, P. Cohen, M. Vilkas, and E. Lederer, *IUPAC Symposium, Kyoto, Japan, Apr.* 1964, B-15-3.
482. M. Tishler, L. F. Fieser, and N. L. Wendler, *J. Am. Chem. Soc.*, **62,** 1982 (1940).
483. M. Gogolewski, *Rocz. Wyzsz. Szk. Roln. Poznaniu*, **47,** 75 (1970); *Chem. Abstr.*, **76,** 111766 (1972).

484. M. Gogolewski, *Rocz. Wyzsz. Szk. Roln. Poznaniu*, **35,** 69 (1967); *Chem. Abstr.*, **72,** 43340 (1970).
485. T. Ichakawa and T. Kato, *Bull. Chem. Soc. Japan*, **41,** 1224 (1968).
486. R. L. Rowland, *J. Am. Chem. Soc.*, **80,** 6130 (1958).
487. C. Etman-Gervais, *C.R. Acad. Sci. (Paris) Ser. D*, **282,** 1171 (1976); *Chem. Abstr.*, **85,** 17072 (1976).
488. R. W. Battle, J. K. Gaunt, and D. L. Laidman, *Biochem. Soc. Trans.*, **4,** 484 (1976); *Chem. Abstr.*, **85,** 90239 (1976).
489. M. Gogolewski, *Rocz. Akad. Roln. Poznaniu, Pr. Habilitacyjue*, **44,** 55 (1973); *Chem. Abstr.*, **81,** 117146 (1974).
490. W. Boguth, R. Repges, and R. Zell, *Int. Z. Vitamin Forsch.*, **40,** 323 (1970); *Chem. Abstr.*, **73,** 119899 (1970).
491. W. Metlesics and P. Wehrli, Ger. Offen. 1,909,164 (1969); *Chem. Abstr.*, **72,** 12569 (1970).
492. G. Baddeley, N. H. P. Smith, and M. A. Vickars, *J. Chem. Soc.*, 2455 (1956).
493. F. N. Lahey and R. V. Stick, *Austr. J. Chem.*, **26,** 2311 (1973); *Chem. Abstr.*, **79,** 126678 (1973).
494. H. B. Bhat and K. Venkataraman, *Tetrahedron*, **19,** 77 (1963).
495. R. Mittledorf and W. Riedl, *Chem. Ber.*, **93,** 309 (1960).
496. A. McGookin, A. Robertson, and E. Tittensor, *J. Chem. Soc.*, 1579 (1939).
497. M. L. Wolfrom and B. S. Wildi, *J. Am. Chem. Soc.*, **73,** 235 (1951).
498. L. Jurd, K. Stevens, and G. Manners, *Tetrahedron Lett.*, 2275 (1971).
499. K. Hultzsch, *Ber.*, **74B,** 898 (1941); *J. Prakt. Chem.*, **158,** 275 (1941); *Chem. Abstr.*, **35,** 6941 (1941); *Beilstein*, E III—IV, **17,** p. 1382.
500. D. J. Goldsmith and C. T. Helms, Jr., *Syn. Commun.*, **3,** 231 (1973).
501. N. Nakabayashi, G. Wegner, and H. G. Cassidy, *J. Org. Chem.*, **33,** 2539 (1968).
502. G. Wagner, N. Nakabayashi, and H. G. Cassidy, *J. Org. Chem.*, **32,** 3155 (1967).
503. T. Nakamura and S. Kijima, Ger. Offen. 2,331,958 (1974); *Chem. Abstr.*, **80,** 82664 (1974).
504. J. W. Scott, H. Harley, W. M. Cort, D. R. Parrish, and G. Saucy, *J. Am. Oil Chem. Soc.*, **51,** 200 (1974).
505. T. Nakamura and S. Kijima, Japan. Patent 73 92,372 (1973); *Chem. Abstr.*, **80,** 133252 (1974).
506. S. C. Hooker, *J. Am. Chem. Soc.*, **58,** 1190 (1936).
507. W. A. Skinner and R. M. Parkhurst, *J. Org. Chem.*, **31,** 1248 (1966).
508. A. R. Bader and W. C. Bean, *J. Am. Chem. Soc.*, **80,** 3073 (1958).
509. D. M. Lynch and J. W. Cole, Fr. Patent 1,522,359 (1968); *Chem. Abstr.*, **71,** 70392 (1969).
510. E. Collins and P. V. R. Shannon, *J. Chem. Soc., Perkin Trans. I*, 419 (1973).
511. A. A. Svishchuk, F. L. Grinberg, E. D. Basalkevich, and N. K. Makhnovskii, *Fiziol. Artiv. Veschchestva, Akad. Nauk Ukr. SSR Repub. Mezhvedom. S. B.*, 150 (1966); *Chem. Abstr.*, **67,** 64174 (1967).
512. T. Kato, A. S. Kumanireng, I. Ichinose, Y. Kitahara, Y. Kakinuma, and Y. Kato, *Chem. Lett.*, 335 (1975).
513. T. Nakamura, S. Umeda, and T. Matsuura, Japan. Patent 70, 12,142 (1970); *Chem. Abstr.*, **73,** 25705 (1970).
514. T. Nakamura and S. Umeda, Japan. Patent 70 12,143; *Chem. Abstr.*, **73,** 45347 (1970).
515. T. Kakamura and S. Kijima, *Chem. Pharm. Bull.*, **19,** 2318 (1971).
516. T. Nakamura and S. Kijima, *Chem. Pharm. Bull.*, **20,** 794 (1972).
517. T. Nakamura and S. Kishima, Japan. Patent 69, 27,642 (1969); *Chem. Abstr.*, **72,** 43927 (1970).
518. S. Blachser and W. Boguth, *Int. Z. Vitaminforsch.*, **39,** 163 (1969); *Chem. Abstr.*, **71,** 79530 (1969).

519. S. A. Csallany and H. H. Draper, *Int. J. Vitam. Nutr. Res.*, **41,** 368 (1971); *Chem. Abstr.*, **75,** 140632 (1971).
520. T. Nakamura and S. Kijima, Japan. Patent 69 17,906 (1969); *Chem. Abstr.*, **71,** 112813 (1969); Japan. Patent 69 17,907 (1969); *Chem. Abstr.*, **71,** 112814 (1969).
521. K. Hamamura, T. Matuura, and S. Kujima, *Radioisotopes*, **19,** 465 (1970); *Chem. Abstr.*, **75,** 20690 (1971).
522. S. A. Csallany and H. H. Draper, *Int. J. Vitam. Nutr. Res.*, **41,** 516 (1971).
523. T. Nakamura, and M. Kohei, Japan. Patent 66 5936 (1966); *Chem. Abstr.*, **65,** 5498 (1966).
524. L. Weisler, U.S. Patent 2,640,048 (1953); *Chem. Abstr.*, **48,** 7643 (1954).
525. R. E. Erickson, A. F. Wagner, and K. Folkers, *J. Am. Chem. Soc.*, **85,** 1535 (1963).
526. D. McHale and J. Green, *Chem. Ind.* (London), 982 (1963).
527. J. L. G. Nilsson, H. Sievertsson, and H. Selander, *Acta Chem. Scand.*, **23,** 268 (1969).
528. M. Fujimaki, K. Kanamaru, T. Kurata, and O. Igarashi, *Agr. Biol. Chem.*, **34,** 1781 (1970).
529. W. A. Skinner and P. Alaupovic, *J. Org. Chem.*, **28,** 2854 (1963).
530. J. L. G. Nilsson, G. D. Daves, Jr., and K. Folkers, *Acta Chem. Scand.*, **22,** 200 (1968).
531. D. R. Nelan and C. D. Robeson, *J. Am. Chem. Soc.*, **84,** 2963 (1962).
532. H. Seino, Y. Sugeta, S. Watanabe, and Y. Abe, *Yukagaku*, **22,** 145 (1973); *Chem. Abstr.*, **78,** 146390 (1973).
533. M. Murakami, K. Takahashi, A. Matsumoto, K. Murase, K. Tamazawa, S. Kawamura, and T. Shigeo, U.S. Patent 3,476,777 (1969); *Chem. Abstr.*, **72,** 31614 (1970).
534. J. Terao, S. Kawamura, and M. Murakami, *Yamanouchi Seiyaku Kenkyu Hokoku*, **2,** 74 (1974); *Chem. Abstr.*, **83,** 145893 (1975).
535. M. Murakami and K. Murase, Japan. Patent 70 11,692 (1970); *Chem. Abstr.*, **73,** 45347 (1970).
536. M. Komoda and I. Harada, *J. Am. Oil Chem. Soc.*, **46,** 18 (1969).

CHAPTER IV

Alkoxychromans

R. LIVINGSTONE

The Polytechnic, Huddersfield, U.K.

I. INTRODUCTION

The majority of alkoxychromans found in the literature are those which either contain one or more methoxy substituents on the benzene ring or an ethoxy or methoxy substituent at the 2-position. Those containing an ethoxy or methoxy substituent at either the 3- or 4-position generally possess a hydroxyl substituent at the 4-position or a halogeno substituent at the 3-position, respectively. The nomenclature used for these compounds is the same as that for the chromans and related compounds described in Chapter I. No aryloxychromans appear to have been described.

Along with a number of chromans, some methoxychromans, especially 6-methoxychromans, were prepared in attempts to find suitable routes to the synthesis of the tocopherols and also during efforts to find related compounds possessing vitamin E activity.

II. PREPARATION OF METHOXYCHROMANS

Chromans with methoxy substituents attached to the benzene ring are in general prepared by methods similar to those used for the related chromans. The necessary starting benzene derivatives possess methoxy or hydroxyl substituents at the appropriate positions, the hydroxyl groups being converted to the required methoxy substituents after the formation of the hydroxychroman. They may also be obtained by Clemmensen reduction of the related chroman-4-one. Those containing a methoxy or other alkoxy substituent attached to the heterocyclic ring are in general formed by addition of the appropriate alcohol to the double bond of a chrom-2-ene, by replacement of a halogen in the 4-position by an alkoxy group, and by ring opening of a 3,4-epoxychroman on reaction with an alcohol, under the appropriate neutral, acidic, or basic conditions.

1. Preparation of Monomethoxy- and Monoethoxychromans

Probably the first monomethoxychroman to be prepared containing no other substituent was 7-methoxychroman (**2**). It was isolated as an oil following the reduction of 7-methoxychromone (**1**), using hydrogen in the presence of palladium.[1] 7-Methoxychroman (**2**) is also obtained on reducing 7-methoxychroman-4-one (**3**) over Raney nickel in ethanol.[2]

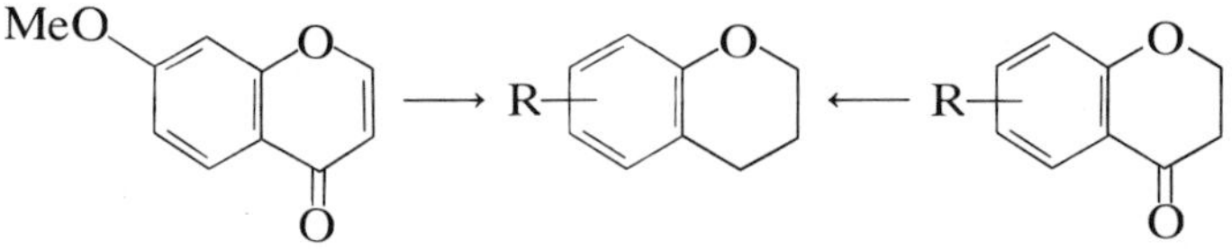

MeO

(**1**)

(**2**) R = 7-OMe
(**4**) R = 6-OMe
(**5**) R = 8-OMe
(**7**) R = 6-OH
(**13**) R = 5-OMe
(**17**) R = 6-Br

(**3**) R = 7-OMe
(**14**) R = 6-OMe
(**15**) R = 8-OMe

In 1964 Willhalm, Thomas, and Gautschi prepared 6-methoxychroman (**4**) and 8-methoxychroman (**5**) as part of a study of the mass spectra of a number of aromatic ethers containing the oxygen atom in a ring. The former compound (**4**) was obtained by heating 1-(4-methoxyphenoxy)-3-chloropropane (**6**) with stannic chloride and distilling the neutral product after its separation from 6-hydroxychroman (**7**), which was also formed during the reaction. Methylation of 6-hydroxychroman (**7**) with dimethyl sulfate and alkali gives 6-methoxychroman (**4**). 8-Methoxychroman (**5**) was prepared in a similar manner to 6-methoxychroman by the cyclization of 1-(2-methoxyphenoxy)-3-chloropropane (**8**) and was chromatographed on alumina to give a pure product.[3]

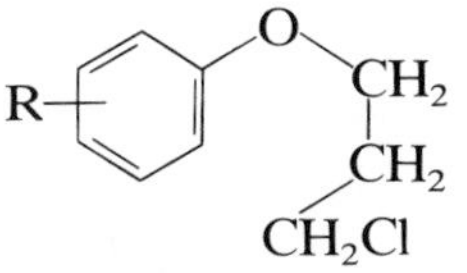

(**6**) R = 4-OMe
(**8**) R = 2-OMe

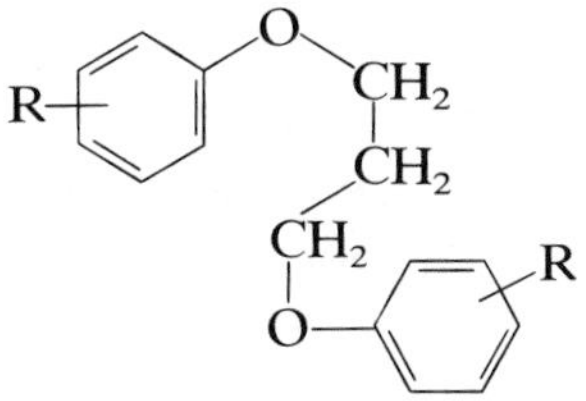

(**9**) R = H
(**10**) R = 4-OMe
(**11**) R = 2-OMe
(**12**) R = 3-OMe

The method used for the preparation of chroman, by the aluminum chloride-catalyzed decomposition of 1,3-diphenoxypropane (**9**), has been applied to furnish methoxychromans with the methoxy substituent attached to the benzene ring. It provides a simple and useful method for the preparation of 6- and 8-methoxychroman, (**4**) and (**5**), from 1,3-di(4-methoxyphenoxy)- and 1,3-di(2-methoxyphenoxy)propane, (**10**) and (**11**), respectively. In the reported preparation, 8-methoxychroman (**5**) was not isolated but was converted to the dinitro derivative. Decomposition of the *m*-substituted 1,3-di(3-methoxyphenoxy)propane (**12**) yields a mixture of 5- and 7-methoxychroman, (**13**) and (**2**). The two isomers can be separated by preparative gas chromatography, but their retention times are similar, and considerable 'trailing' of the first peak occurs. The first compound, 7-methoxychroman (**2**), to come off the column can be obtained pure, but the second is contaminated by the first. Additional passages through the chromatograph would probably afford pure 5-methoxychroman (**13**). The reported yields from this method of preparation are good, especially for 6-methoxychroman (**4**).[4]

6-Methoxychroman,[5] 7-methoxychroman,[6] and 8-methoxychroman[7] may be obtained by the Clemmensen reduction of the related methoxychroman-4-ones (**14**), (**3**), and (**15**), using amalgamated zinc and hydrochloric acid.

The preparation of 4-ethoxychroman (**19**) by treatment of a mixture of 4- and 6-bromochroman, (**16**) and (**17**) (obtained by reacting chroman and *N*-bromosuccininimide) with sodium ethoxide was first reported in 1952. It was apparently formed as a by-product during the preparation of chrom-3-ene (**18**) by the dehydrobromination of the 4-bromo isomer in the mixture[8] and was isolated by distillation. Willhalm et al.[3] reported that the sample of 4-ethoxychroman they obtained by this method was not pure. Purification was effected by chromatography in light petroleum (b.p. 30–50°) on alumina. Slow distillation at 91–93°/15 mm results in decomposition of 4-ethoxychroman (**19**) to give chromen (**18**).[9]

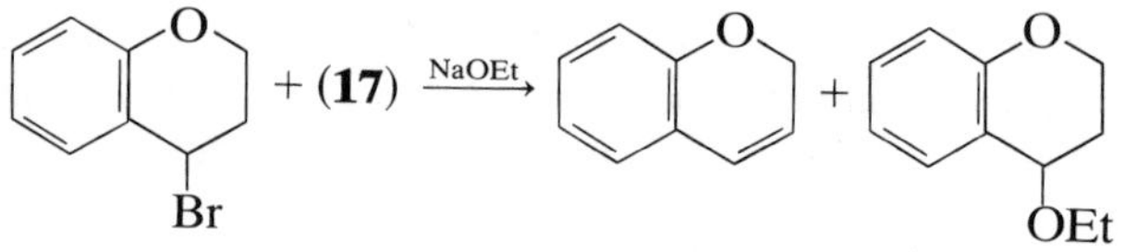

+ (**17**) $\xrightarrow{NaOEt}$ +

Br OEt

(**16**) (**18**) (**19**)

Chrom-2-ene (**20**) on boiling for several hours with twice its volume of ethanol and a drop of hydrochloric acid yields 2-ethoxychroman (**21**).[9] Mannich bases of phenol condense with olefinic compounds, such as ethyl vinyl ether, isobutene, styrene, and butadiene, to give 2-substituted chromans. The Mannich base 2-(dimethylaminomethyl)phenol (**22**) on heating at 180° for several hours in a sealed tube with ethyl vinyl ether, in the presence of a small amount of dihydroquinone, affords 2-ethoxychroman (**21**) in 30% yield.[10]

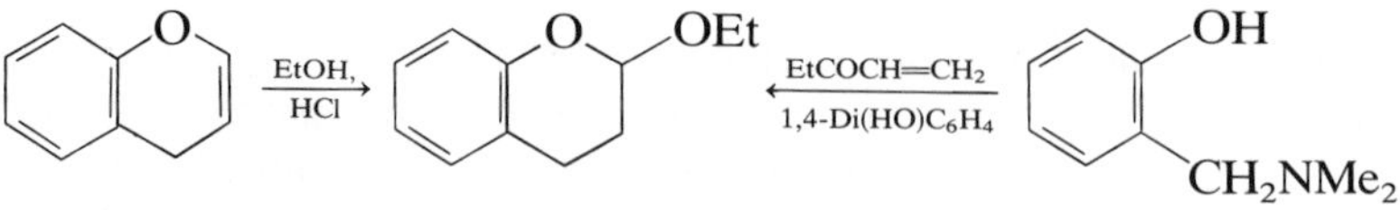

$\xrightarrow[HCl]{EtOH,}$ OEt $\xleftarrow[1,4\text{-Di(HO)}C_6H_4]{EtCOCH{=}CH_2}$ OH

CH_2NMe_2

(**20**) (**21**) (**22**)

The literature reports a number of monomethoxychromans containing a methoxy group attached to the benzene ring with either a methyl group at the 2-position or a methyl or ethyl group attached to the benzene ring. Generally the monoalkoxymonoalkylchromans containing the substituents attached to the heterocyclic ring are those with either a methoxy or ethoxy group at the 2- or 4-position and a methyl group at the 2-position.

The catalytic reduction of 2-hydroxybenzylideneacetone (**23**) in methanol or ethanol in the presence of palladium chloride gives a mixture of 2-hydroxybenzylacetone (**24**) and 2-methoxy- or 2-ethoxy-2-methylchroman, (**25**) or (**26**), respectively. The products are separated by dissolving the former compound in aqueous sodium hydroxide. This ready formation of six-membered cyclic acetals is closely analogous to the production of the methyl glycosides of the sugars and was probably the first example of its kind in the aromatic series. Its formation is attributed to the presence of hydrogen chloride arising from the reduction of the palladium chloride.[11] The 2-alkoxy compound is readily prepared on treating 2-hydroxybenzylacetone with the appropriate alcohol containing hydrogen chloride. The reverse reaction is effected by heating the 2-alkoxy compound with dilute hydrochloric acid.

2-Methoxy-2-methylchroman (**25**) is also obtained in 82% yield by reacting phenol with 2-methoxybutadiene in benzene, in the presence of a small amount of 2,5-di-*t*-butylhydroquinone and triethylamine in a sealed tube at 150° for several hours. Polymerization of the diene takes place in the absence of base. The 2-methoxy-2-methylchroman (**25**) is presumably furnished by cyclization of the rearranged product (**28**) formed from the initial

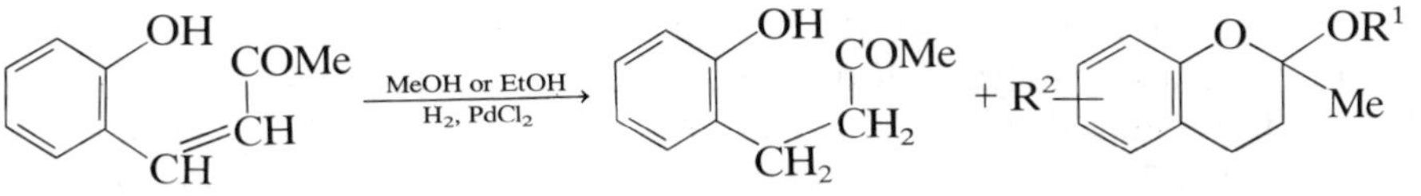

(23) → [MeOH or EtOH; H_2, $PdCl_2$] → (24) + (25)–(29)

(25) $R^1 = Me, R^2 = H$
(26) $R^1 = Et, R^2 = H$
(29) $R^1 = Me, R^2 = 8\text{-}Me$

reaction product (**27**). The reaction of 2-methylphenol with 2-methoxybutadiene takes place less readily. In this case the addition of triethylamine to the reaction mixture is not necessary or desirable as the base exerts a very pronounced inhibitory effect. In the absence of base at 160°, 2,8-dimethyl-2-methoxychroman (**29**) is obtained in 76% yield.[12]

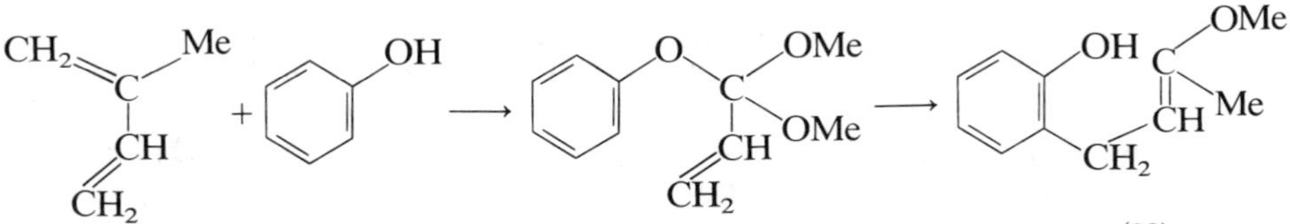

(27) → (28) → (25)

6-Methoxy-2-methylchroman (**30**) is prepared by the Clemmensen reduction of the related chroman-4-one, and 7-methoxy-2-methylchroman (**31**) is prepared by the methylation of 7-hydroxy-2-methylchroman (**32**), obtained originally by the Clemmensen reduction of 7-hydroxy-2-methylchroman-4-one. A number of 6- and 7-methoxy-2-methylchromans containing two and three methyl groups attached to the benzene ring have been obtained by methylation of the corresponding 6- and 7-hydroxychromans.[13]

Two methods have been used for the preparation of 8-methoxy-2-methylchroman (**33**). The first involved treating 3-chloro-1-(2-hydroxy-3-methoxyphenyl)butane (**34**) with sodium hydroxide solution. The second involved boiling 1-(2-hydroxy-3-methoxyphenyl)butan-3-ol (**35**) with phosphorus pentoxide in benzene, when a 70% yield was obtained.[14]

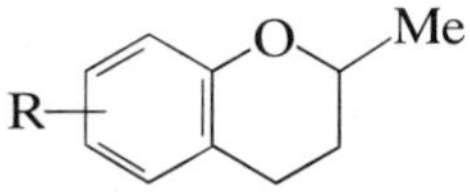

(**30**) R = 6-OMe
(**31**) R = 7-OMe
(**32**) R = 7-OH
(**33**) R = 8-OMe

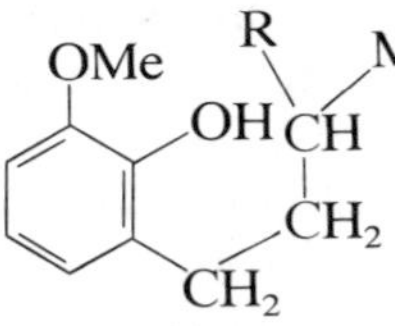

(**34**) R = Cl
(**35**) R = OH

As reported in Chapter II, chromans and benzopyrylium salts may be formed by acid-catalyzed disproportionation of chrom-3-enes and dehydration of chroman-4-ols. Similarly 6- and 7-methoxy-4-methylchroman-4-ols (**36**) and (**37**) on treatment with perchloric acid give the corresponding chromans (**38**) and (**39**) and benzopyrylium perchlorates (**40**) and (**41**). It is found that 6-methoxy-4-methylpyrylium perchlorate decomposes within a few hours of formation but that the 7-methoxy salt is much more stable, presumably because of the stabilization of the cation by the 7-methoxy group.[15]

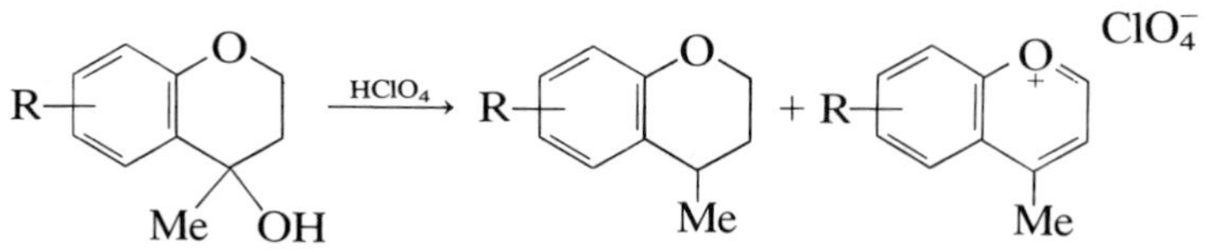

$\xrightarrow{HClO_4}$ ClO_4^-

(**36**) R = 6-OMe
(**37**) R = 7-OMe

(**38**) R = 6-OMe
(**39**) R = 7-OMe

(**40**) R = 6-OMe
(**41**) R = 7-OMe

Clemmensen reduction of 8-acetyl-6-methoxychroman-4-one (**42**) affords 8-ethyl-6-methoxychroman (**43**).[16] 5-Ethyl- and 6-ethyl-8-methoxychroman, (**44**) and (**45**) are obtained by the Clemmensen reduction of 5-ethyl- and 6-ethyl-8-methoxychroman-4-one (**46**) and (**47**), respectively.[7]

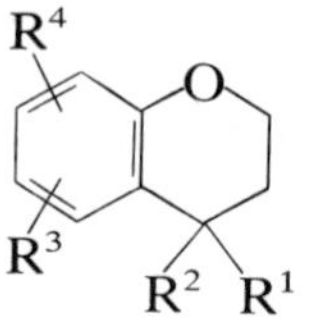

(**42**) $R^1R^2 = O$, $R^3 = $ 6-OMe, $R^4 = $ 8-Ac
(**43**) $R^1 = R^2 = H$, $R^3 = $ 6-OMe, $R^4 = $ 8-Et
(**44**) $R^1 = R^2 = H$, $R^3 = $ 5-Et, $R^4 = $ 8-OMe
(**45**) $R^1 = R^2 = H$, $R^3 = $ 6-Et, $R^4 = $ 8-OMe
(**46**) $R^1R^2 = O$, $R^3 = $ 5-Et, $R^4 = $ 8-OMe
(**47**) $R^1R^2 = O$, $R^3 = $ 6-Et, $R^4 = $ 8-OMe

It has been demonstrated that the *o*-quinone methide can be trapped by reactive dienophiles to give a variety of chromans. Most preparations of this type described in the literature start with the readily available 4-*t*-butyl-2,6-dimethylphenol (**48**), and the 2-substituted chromans formed contain a 6-*t*-butyl and an 8-methyl substituent. For example, 6-*t*-butyl-2-ethoxy-8-methylchroman (**52**) is obtained by stirring together the phenol (**48**), ethyl vinyl ether, and silver oxide. The first step in the oxidation of the phenol (**48**) is the removal of the phenolic hydrogen to yield a phenoxy radical (**49**) that dimerizes to give an *o*-quinol ether (**50**), with which it exists in equilibrium. This system disproportionates to yield the phenol (**48**) and an *o*-quinone methide (**51**), which is trapped by the ethyl vinyl ether to yield 6-*t*-butyl-2-ethoxy-8-methylchroman (**52**). If one of the substituents on the original phenol is a methoxy group, then dialkoxy derivatives of chroman

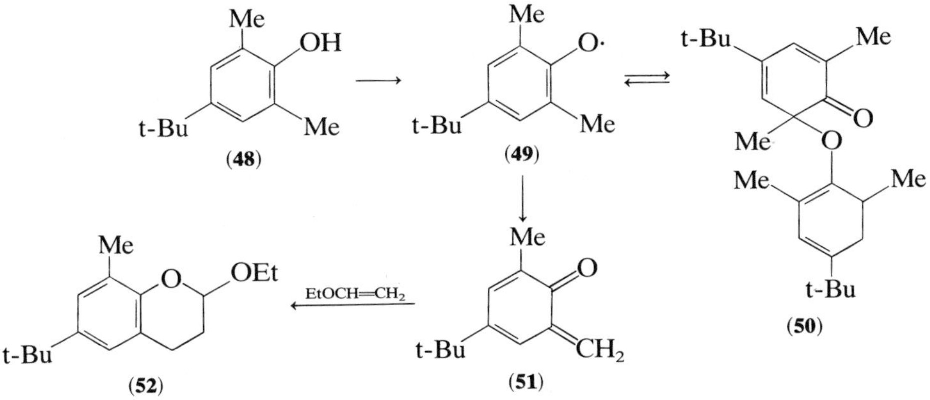

can be prepared; for instance, 4-methoxy-2,6-dimethylphenol with ethyl vinyl ether affords 2-ethoxy-6-methoxy-8-methylchroman.[17]

4-(4-Methoxyphenyl)-2-hydroxy-3-methylchroman (**53**) has been prepared by uv irradiation of *trans*-anethole and salicylaldehyde in benzene under nitrogen at room temperature. Subsequent methylation with dimethyl sulfate under alkaline conditions afforded 4-(4-methoxyphenyl)-2-methoxy-3-methylchroman (**54**).[18]

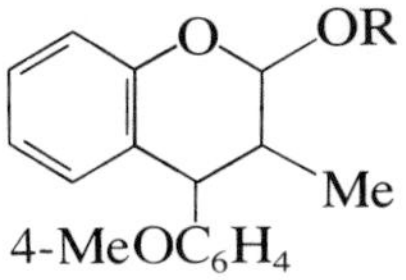

(**53**) R = H
(**54**) R = Me

The majority of known chroman derivatives with one methoxy and two methyl substituents are those containing the methoxy group attached to the benzene ring and a *gem* dimethyl group at the 2-position. These are in general prepared by hydrogenation of the related chromen obtained by reacting the appropriate coumarin with methylmagnesium iodide. They may also be formed by methods related to those used for the preparation of 2,2-dimethylchromans containing a methyl or alkyl group attached to the benzene ring.

Probably the first example was the preparation of 2,2-dimethyl-6-methoxychroman (**56**) by cyclization of 3-chloro-1-(2-hydroxy-5-methoxyphenyl)-3-methylbutane (**55**) in alcoholic potassium acetate. Compound (**55**) was obtained by the reaction between 4-methoxyphenol and isoprene in acetic acid containing hydrogen chloride.[19]

2,2-Dimethyl-5-methoxychroman (**58**) is prepared by the cyclization under acid conditions of 1-(2-hydroxy-6-methoxyphenyl)-3-methylbutan-3-ol (**57**) furnished by the Grignard reaction between dihydro-5-methoxycoumarin and methylmagnesium iodide.[20] It has also been obtained by

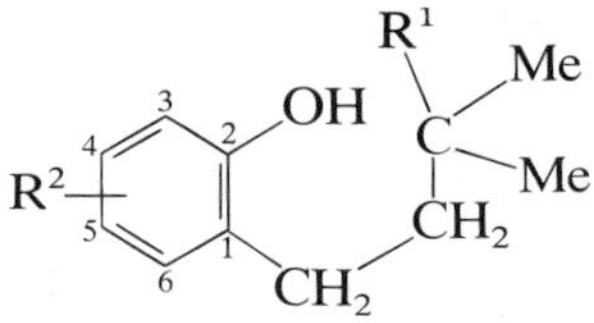

(**55**) $R^1 = Cl, R^2 = 5\text{-OMe}$
(**57**) $R^1 = OH, R^2 = 6\text{-OMe}$

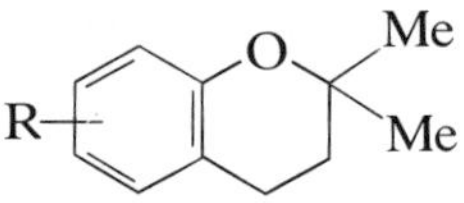

(**56**) R = 6-OMe
(**58**) R = 5-OMe
(**59**) R = 5-OH
(**60**) R = 7-OMe
(**63**) R = 7-OH
(**64**) R = 8-OH
(**65**) R = 8-OMe

boiling 2,2-dimethyl-5-hydroxychroman (**59**) with dimethyl sulfate in acetone in the presence of potassium carbonate.[21]

2,2-Dimethyl-7-methoxychroman (**60**) is obtained by hydrogenation of either 2,2-dimethyl-7-methoxychroman-4-one (**61**) in the presence of copper chromite, or of the related chromen (**62**) in ethanol, using Raney nickel. The first method, which proceeds through the chromen (**62**), gives a yield of 60%, and the second a yield of 80–85%.[22] 2,2-Dimethyl-7-methoxychroman (**60**) may also be prepared by the methylation of the 7-hydroxy compound (**63**) with dimethyl sulfate.[23,24]

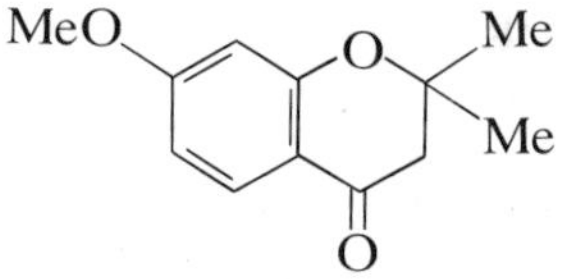

(**61**)

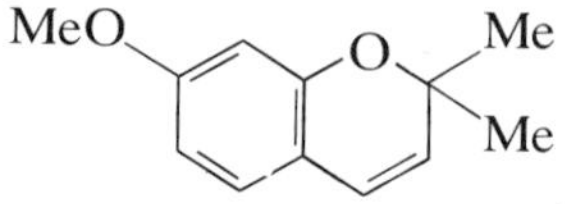

(**62**)

Methylation of 2,2-dimethyl-8-hydroxychroman (**64**) with methyl iodide in acetone containing potassium carbonate affords 2,2-dimethyl-8-methoxychroman (**65**).[25] The alkenylation of 2-methoxyphenol (guaiacol) with isoprene in the presence of phosphoric acid gives ≤70% monoalkylated products that cyclize to give various compounds including 2,2-dimethyl-8-methoxychroman (**65**) obtained from 2-methoxy-6-(3-methylbut-2-enyl)phenol.[26] Similarly 2,4-dimethyl-8-methoxychroman may be formed by the cyclization of 1,2-dimethoxy-3-(pent-2-enyl)benzene, one of the products formed by the alkenylation of 1,2-dimethoxybenzene (veratrol) with penta-1,3-diene (piperylene), in the presence of phosphoric acid or aluminum chloride.[27]

But-3-enyl diphenyl phosphate (**66**) and 3-methylbut-3-enyl diphenyl phosphate (**67**) react with 2-methoxyphenol in the presence of a Lewis acid, under mild conditions, to give 8-methoxy-4-methylchroman (**68**) and 4,4-dimethyl-8-methoxychroman (**69**), respectively.[28] Similarly 2-methylphenol affords the related 8-methyl derivatives.

Dimethyl-7-methoxychromans with one methyl group at either the 6- or 8-position and the other at the 2-position may be obtained by methylation

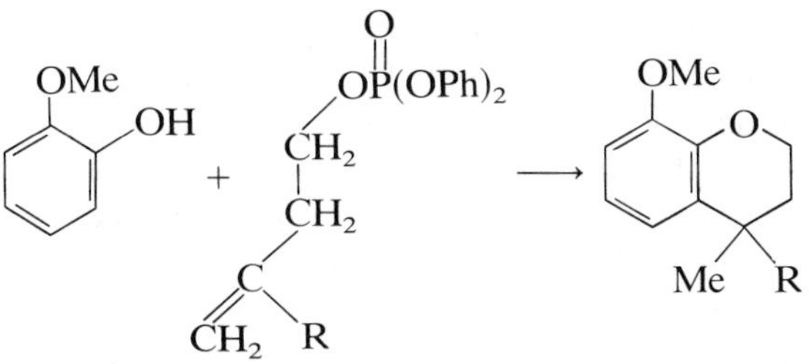

(66) R = H
(67) R = Me

(68) R = H
(69) R = Me

of the related hydroxy compounds, by methods similar to those used for the preparation of 7-methoxychroman.[13]

The acetal (**70**) in benzene on treatment with a slow stream of 1,1-dimethylethene in the presence of boron trifluoride etherate yields 2,2-dimethyl-4-ethoxychroman (**71**).[29]

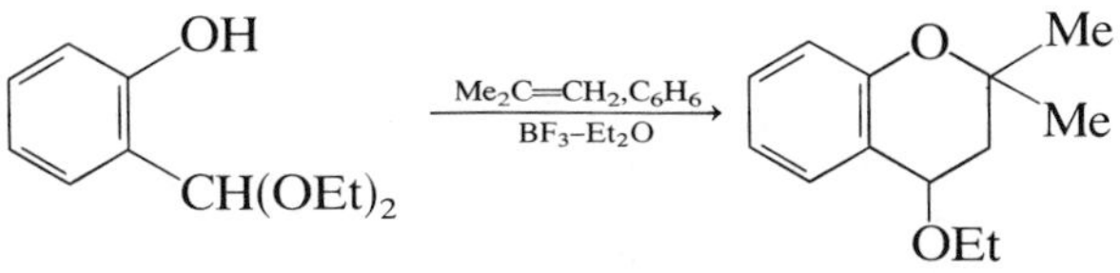

$Me_2C{=}CH_2, C_6H_6$; BF_3–Et_2O

(70) (71)

In Chapter V the syntheses of a number of 3,4-dihalogeno-2,2-dimethyl-6-, 7-, and 8-methoxy-chromans (Section II, 2), some 3-halogeno-4-methoxy- and other 4-alkoxychromans (Sections IV, 3 and IV, 4), and related dihydronaphthopyran derivatives (Section IV, 5) are described. In a similar manner 3-halogeno-2-methoxychromans are obtained from 2,3-dihalogenochromans.[30]

2. Preparation of Dimethoxy- and Dialkoxychromans

The dimethoxychromans and related compounds reported in the literature are in general those which contain two ether groups attached to the benzene ring. An exception is 2,2-dimethoxychroman (**74**), which is obtained on trapping the *o*-quinone methide (**73**), formed by the photochemical decarbonylation of dihydrobenzofuran-2-one (**72**), with 1,1-dimethoxyethene.[31]

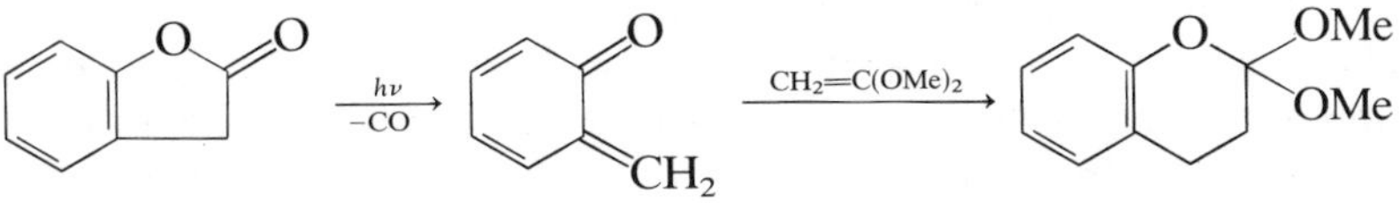

$h\nu$, $-CO$; $CH_2{=}C(OMe)_2$

(72) (73) (74)

Like some of the methoxychromans, 6,7-dimethoxy-2-methylchroman (**75**)[13] and 7,8-dimethoxy-2,2-dimethylchroman (**76**)[32,33] are obtained by the Clemmensen reduction of the related chroman-4-one. Similarly 5,7-dimethoxy-2,2-dimethylchroman-4-one on reduction affords 6,7-dimethoxy-2,2-dimethylchroman (**77**), which has also been prepared by the hydrogenation of 6,7-dimethoxy-2,2-dimethylchrom-3-ene (ageratochromene) using 2% palladium on barium sulfate as catalyst.[34] Ageratochromene was isolated from the essential oil of *Ageratum houstonianum* Mill. 5,8-Dimethoxy-2,2-dimethylchroman (**78**) has been prepared by methylation, using methyl iodide in the presence of potassium carbonate, of 5-hydroxy-8-methoxy-2,2-dimethylchroman (**79**).[33] The latter was originally obtained as a degradation product during investigation of the structure of ichthynone isolated from Jamaican Dogwood, *Piscidia erythrina* L.

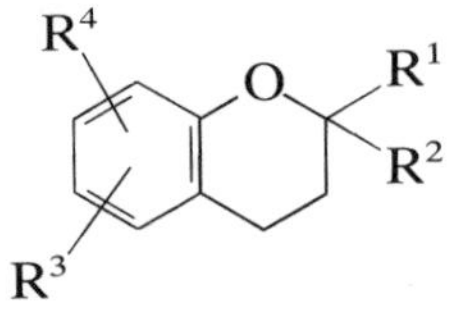

(**75**) $R^1 = H, R^2 = Me, R^3 = 6\text{-OMe}, R^4 = 7\text{-OMe}$
(**76**) $R^1 = R^2 = Me, R^3 = 7\text{-OMe}, R^4 = 8\text{-OMe}$
(**77**) $R^1 = R^2 = Me, R^3 = 6\text{-OMe}, R^4 = 7\text{-OMe}$
(**78**) $R^1 = R^2 = Me, R^3 = 5\text{-OMe}, R^4 = 8\text{-OMe}$
(**79**) $R^1 = R^2 = Me, R^3 = 5\text{-OH}, R^4 = 8\text{-OMe}$

2-Ethoxy-7-methoxy-2-methyl-4-phenylchroman (**80**) is an example of a dialkoxychroman containing a methoxy group attached to the benzene ring and an ethoxy group on the heterocyclic ring. It is prepared by methylation of the related 7-hydroxy compound, obtained originally by condensation of benzylideneacetone and resorcinol in ethanol in the presence of hydrogen chloride.[35]

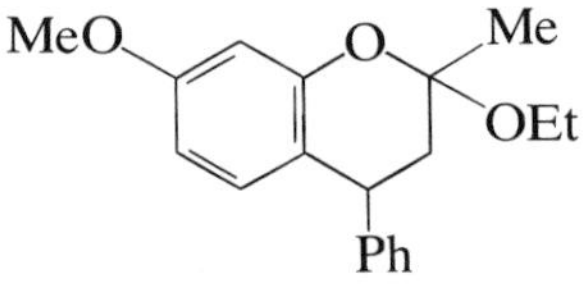

(**80**) 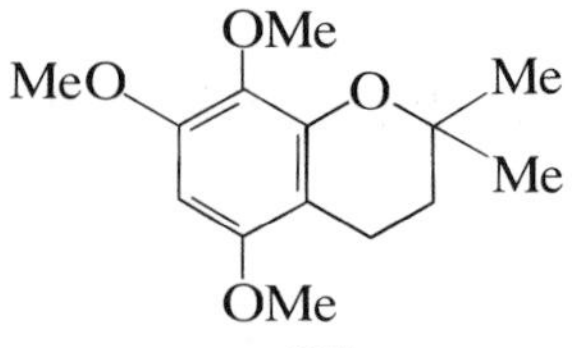 (**81**)

3. Preparation of Miscellaneous Alkoxychromans

2,2-Dimethyl-5,7,8-trimethoxychroman (**81**) was prepared as an intermediate during the synthesis of dihydroevodione, by the catalytic hydrogenation of the parent chromen in methanol in the presence of Raney nickel.[36]

A number of tri- and tetramethyl-6-methoxychromans have been synthesized by methods similar to those used for obtaining tri-, tetra-, and pentamethylchromans. 2,4,4,7-Tetramethylchroman-2-ol (**82**) on boiling with ethanol saturated with hydrogen chloride affords 2-ethoxy-2,4,4,7-tetramethylchroman (**83**).[37] Similar treatment of 2,4,4-trimethylchroman-2-ol (**84**) in methanol yields 2-methoxy-2,4,4-trimethylchroman (**85**), also obtained on boiling 2,4,4-trimethylchrom-2-ene in methanol containing hydrogen chloride.[38]

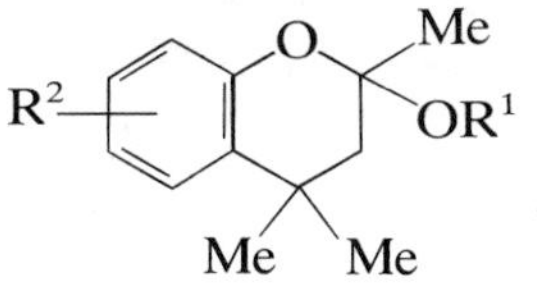

(**82**) $R^1 = H$, $R^2 = 7\text{-Me}$
(**83**) $R^1 = Et$, $R^2 = 7\text{-Me}$
(**84**) $R^1 = R^2 = H$
(**85**) $R^1 = Me$, $R^2 = H$

A few methoxychromans have been prepared with a methoxy group attached to the benzene ring and one or two phenyl groups attached to the heterocyclic ring. In general these contain a *gem*-diphenyl group at the 2-position and are furnished by the hydrogenation of the parent chromen in ethyl acetate using platinic oxide as catalyst.[39] 7-Methoxy-4-phenylchroman (**87**) has been obtained by passing hydrogen chloride into a solution of 1-(2-hydroxy-4-methoxyphenyl)-1-phenylpropan-3-ol (**86**) in ethanol. The starting compound is prepared by reduction of the related coumarin.[40]

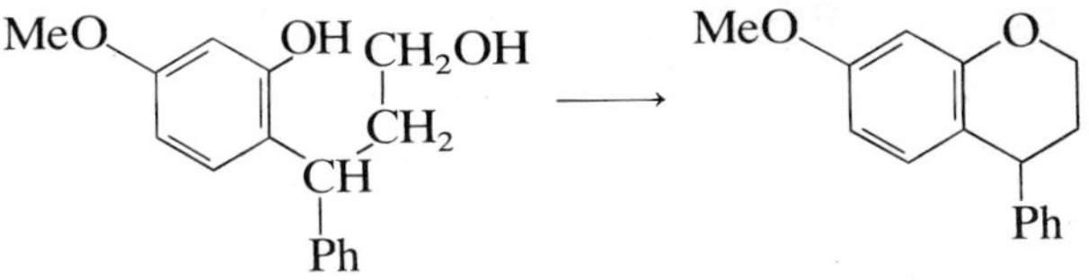

(**86**) (**87**)

The reaction between 1,3-diphenyl-3-(2-hydroxyphenyl)propan-1-one (**88**) and methanol in the presence of hydrogen chloride affords 2,4-diphenyl-2-methoxychroman (**89**). The 2-ethoxy analog can be prepared in a similar manner by using ethanol in place of methanol.[41]

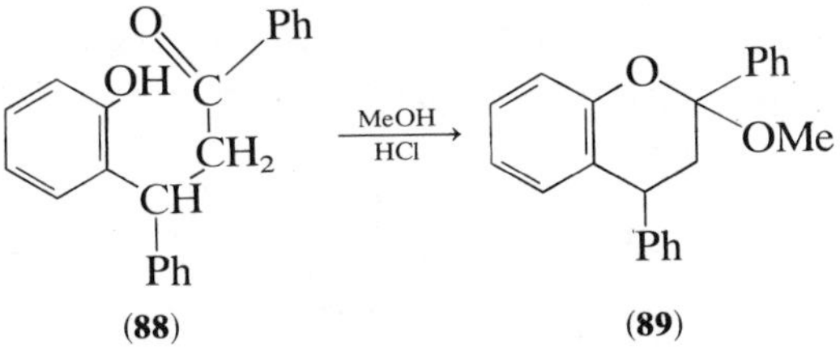

MeOH, HCl

(**88**) (**89**)

1,3-Diphenyl-1*H*-naphtho[2,1-*b*]pyran on boiling with methanol containing hydrogen chloride yields 2,3-dihydro-1,3-diphenyl-3-methoxy-1*H*-naphtho[2,1-*b*]pyran.[42]

$$\text{(90)} \xrightarrow[175^\circ]{EtCOCH{=}CH_2} \text{(91)}$$

(**90**) (**91**)

2-(*N*-Dimethylaminomethyl)cyclohexanone (**90**) when condensed with ethylvinyl ketone at 175° in the presence of a small amount of hydroquinone affords 2-ethoxy-5,6,7,8-tetrahydrochroman (**91**).[43] Some methoxy- and ethoxychromans are listed in Table 2.

III. SPECTRAL PROPERTIES

1. Infrared Spectra

Infrared spectra of the methoxychromans are characterized by absorption bands due to the aromatic ring and the methoxy groups. For example, the ir spectrum of 7-methoxychroman shows bands at 1620, 1585, 1504 (aromatic ring), 1156, 1130, and 1112 cm^{-1} (OMe).[2] The spectra of 5,8- and 7,8-dimethoxy-2,2-dimethylchroman show ν_{max}(KBr) of 1610–1596 and 1490 and ν_{max}($CDCl_3$) at 1610, 1575, and 1496 cm^{-1}, respectively.[33]

2. Ultraviolet Spectra

The uv spectra of a number of alkoxychromans have been reported but have not been discussed systematically. Some of them are listed in Table 1.

3. Nuclear Magnetic Resonance Spectra

The ^{1}H-nmr spectrum of 2-methoxy-2-methylchroman in carbon tetrachloride displays peaks at δ1.47 (singlet, equivalent to 3H), δ1.6–3.1 (complex, equivalent to 4H), δ3.20 (singlet, equivalent to 3H), and δ6.4–7.2 (complex, equivalent to 4H); that of 2,8-dimethyl-2-methoxychroman at δ1.47 (singlet, equivalent to 3H), δ1.6–2.1 (complex, equivalent to 2H), δ2.16 (singlet, equivalent to 3H), δ2.2–3.1 (complex, equivalent to 2H), δ3.18 (singlet, equivalent to 3H), and 6.5–7.0 (complex, equivalent to 3H)[12]; and that of 2,2-dimethoxychroman at δ1.97–2.80 (A_2B_2 pattern, equivalent to 4H), δ3.30 (singlet, methoxy protons, 6H), and δ6.85 (multiplet, aromatic protons, 3H).[31]

5,8-Dimethoxychroman and 7,8-dimethoxychroman were prepared as model compounds during the investigation of the structure of degradation products obtained from ichthynone isolated from Jamaican Dogwood, *P.*

TABLE 1. ULTRAVIOLET DATA OF SOME ALKOXYCHROMANS

Substituents									
2	4	5	6	7	8	Solvent	λ_{max} (nm)	log ε	Ref.
			OMe			EtOH	206	4.15	3
							230	3.79	
							294	3.54	
					OMe	EtOH	210	4.19	3
							225	Shoulder	
							277	3.27	
	Me		OMe				294	3.54	15
	Me			OMe			282	3.50	15
							288	3.46	
Me_2		OMe			OMe	95% EtOH	285	3.04	33
Me_2				OMe	OMe	95% EtOH	274–283	3.50	33

erythrina L., and their ^{1}H nmr spectra reported. The spectrum of 5,8-dimethoxy-2,2-dimethylchroman shows peaks at δ1.37 (dimethyl protons at C-2, 6H), δ1.77 (triplet, $J = 7$, protons at C-3, 2H), δ2.66 (triplet, $J = 7$, protons at C-4, 2H), δ3.77 (methoxy protons at C-8, 3H), δ3.79 (methoxy protons at C-5, 3H), δ6.29 (doublet, $J = 9$ aromatic proton at C-6, 1H), and δ6.63 (doublet, $J = 9$, aromatic proton at C-7, 1H); and that of 7,8-dimethoxy-2,2-dimethylchroman shows peaks at δ1.37 (dimethyl protons at C-2, 6H), δ1.79 (triplet, $J = 7$, protons at C-3, 2H), δ2.74 (triplet, $J = 7$, protons at C-4, 2H), δ3.83 (methoxy protons at C-7 and C-8, 6H), δ6.43 (doublet, $J = 9$, aromatic proton at C-6, 1H), and δ6.76 (doublet, $J = 9$, aromatic proton at C-5, 1H).[33]

4. Mass Spectra

On electron bombardment, methoxychromans with the methoxy group attached to the benzene ring undergo in general the same pattern of fragmentation as shown by chroman (Chapter II, Section III, 5), the *m/e* values of the fragments differing by 30 (OMe). The mass spectra of 6- and 8-methoxychroman show fragments with the following *m/e* values, the corresponding values for chroman being given in parentheses: 164 (134), 149 (119), 136 (106), 121 (91), 108 (78), and 107 (77).

4-Ethoxychroman undergoes two types of fission—one to give chromen by the loss of ethanol and therefore the same fragmentation pattern as chromen, the other the more usual ether fission to give the fragment (**92**) at *m/e* 133, and hence fragments at *m/e* 105 and 77.[3]

(**92**)
m/e 133

m/e 105

$C_6H_5^+$
m/e 77

IV. REACTIONS

1. Demethylation

7-Methoxychroman (**2**) may be demethylated to 7-hydroxychroman either by treatment with ethylmagnesium iodide[1] or by boiling with 48% hydrobromic acid.[6] 8-Methoxychroman (**5**) on boiling with a mixture of acetic acid and 48% hydrobromic acid, and 2,2-dimethyl-5-methoxychroman (**58**) on reacting with hydroiodic acid or aluminum bromide, afford 8-hydroxychroman[44] and 2,2-dimethyl-5-hydroxychroman,[20] respectively.

During an attempt to demethylate the 7-methoxy group of 6-acetyl-7,8-dimethoxy-2,2-dimethylchroman (**93**) using excess boron trichloride in methylene dichloride at 0° for 1 hr, a tertiary chloride, 5-(3-chloro-3-methylbutyl)-2,3,4-trihydroxyacetophenone (**94**) was formed by cleavage of the chroman ring with concomitant demethylation. Similar treatment of 7,8-dimethoxy-2,2-dimethylchroman (**95**) resulted in cleavage of the heterocyclic ring to yield 4-(2,3,4-trihydroxyphenyl)-2-chloro-2-methylbutane (**96**). In contrast, the corresponding 7,8-dihydroxy-2,2-dimethylchroman (**97**) was recovered unchanged even on prolonged treatment with a large excess of boron trichloride.

$\xrightarrow[CH_2Cl_2]{BCl_3,}$

(**93**) $R^1 = Ac, R^2 = R^3 = OMe$
(**95**) $R^1 = H, R^2 = R^3 = OMe$
(**97**) $R^1 = H, R^2 = R^3 = OH$

(**94**) $R^1 = Ac, R^2 = R^3 = R^4 = OH$
(**96**) $R^1 = H, R^2 = R^3 = R^4 = OH$

The products formed can be explained by cleavage of the chroman ring by nucleophilic attack of the chloride ion at C-2 of the chroman ring to give an intermediate phenoxydichloroborane, which can then react further to give an *o*-phenylene chloroboronate hydrolyzed during work-up. The lack of reaction of the 7,8-dihydroxy compound (**97**) is probably due to direct formation of an *o*-phenylene chloroboronate with the catechol group, thus preventing attack at the heterocyclic ring.[45] The above reaction is useful for the synthesis of *o*-isopentenylphenols from 2,2-dimethylchromans.

2. Acylation and Bromination

6-Methoxychroman (**4**) on heating at 100° with boron trifluoride–acetic acid complex demethylates as well as acetylates to give 7-acetyl-6-hydroxychroman (**98**), which on methylation affords 7-acetyl-6-methoxychroman (**99**). This method is superior to acetylation using acetyl chloride and aluminum chloride in carbon disulfide.[5]

The acetylation of 8-methoxychroman (**5**) with boron trifluoride–acetic acid gives 6-acetyl-8-methoxychroman (**100**) as the only product, whereas similar treatment of 8-hydroxychroman affords a mixture of 6-acetyl-8-hydroxy- and 7-acetyl-8-hydroxychroman, (**101**) and (**102**).[7] Some of the acetyl derivatives were formed during attempts to obtain intermediates for the synthesis of dihydropyranochromones.

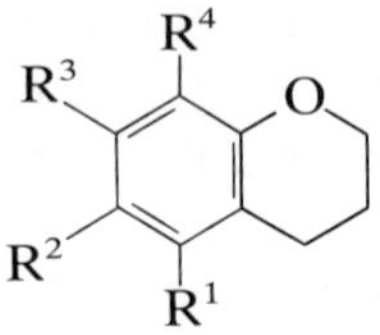

(**98**) $R^1 = R^4 = H, R^2 = OH, R^3 = Ac$
(**99**) $R^1 = R^4 = H, R^2 = OMe, R^3 = Ac$
(**100**) $R^1 = R^3 = H, R^2 = Ac, R^4 = OMe$
(**101**) $R^1 = R^3 = H, R^2 = Ac, R^4 = OH$
(**102**) $R^1 = R^2 = H, R^3 = Ac, R^4 = OH$

A Friedel–Crafts reaction between 2,2-dimethyl-5,7,8-trimethoxychroman (**81**) and acetyl chloride affords a partially crystalline product, which gives a color reaction with ferric chloride. Methylation of the product with dimethyl sulfate in alkali yields 6-acetyl-2,2-dimethyl-5,7,8-trimethoxychroman (dihydroevodione) (**103**).[36]

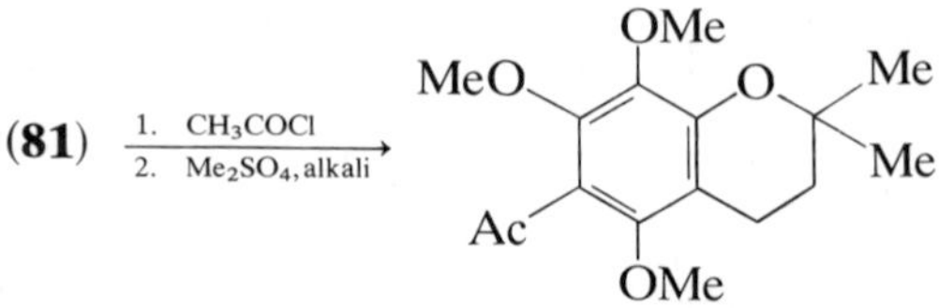

(**81**) $\xrightarrow[\text{2. } Me_2SO_4\text{, alkali}]{\text{1. } CH_3COCl}$ (**103**)

2,2-Dimethyl-7-methoxychroman (**60**) on reacting with dimethylformamide and phosphoryl chloride at 100° yields 2,2-dimethyl-7-methoxychroman-6-carboxaldehyde (**104**),[23] and on treatment with bromine in carbon tetrachloride yields the 6-bromo derivative (**105**).[22]

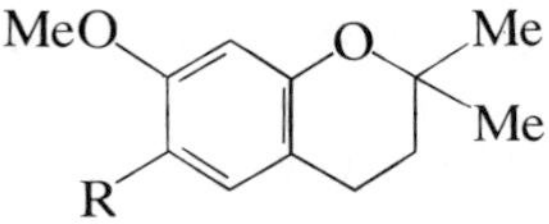

(**104**) R = CHO
(**105**) R = Br

3. Reduction

The Birch reduction of 7-methoxychroman (**2**) in liquid ammonia and *t*-butanol using lithium gave the 5,8-dihydro derivative (**106**). Attempts to purify by chromatography on Florisil yielded 5,8-dihydro-7-methoxychroman (**106**) containing traces of unchanged 7-methoxychroman. The 5,8-dihydro compound (**106**) is partially oxidized in air to the chroman (**2**) and partially hydrolyzed to 4-(3-hydroxypropyl)cyclohexan-1,3-dione in one of its tautomeric forms (**109**). If the crude reduction product is hydrolyzed in methanol containing concentrated hydrochloric acid, chromatography of the mixture of products on Florisil yields 7-methoxychroman (**2**), a ketone believed to be 8*a*-methoxy-4*a*,5,6,7,8,8*a*-hexahydrochroman-7-one (**107**), 4*a*,5,6,7-tetrahydrochroman-7-one (**108**), and a tautomeric form of 4-(3-hydroxypropyl)cyclohexan-1,3-dione (**109**).[2]

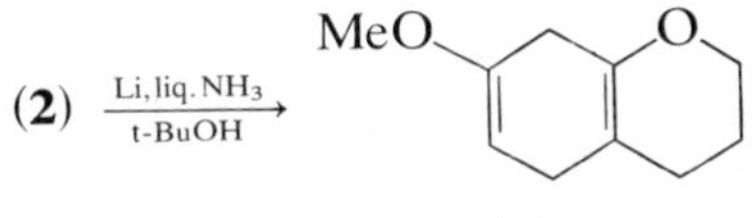

(**2**) $\xrightarrow[\text{t-BuOH}]{\text{Li, liq. NH}_3}$ (**106**)

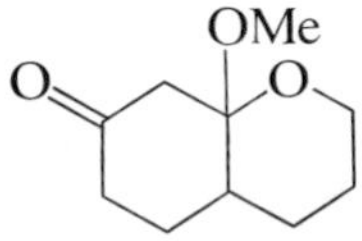

(**107**) 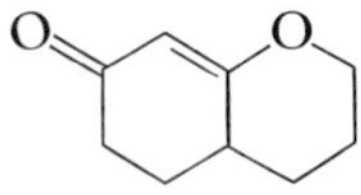 (**108**) 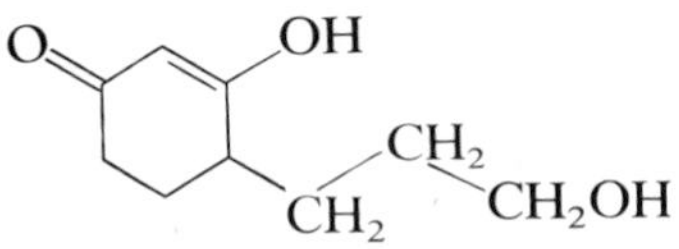 (**109**)

4. Reactions of 2- and 4-Ethoxychromans

2-Ethoxy-2-methylchroman (**26**) on boiling with acetic anhydride affords 2-methylchrom-2-ene (**110**) and is converted on heating with dilute hydrochloric acid into 2-hydroxybenzylacetone (**24**). It is completely stable toward hot alkaline solution.[11]

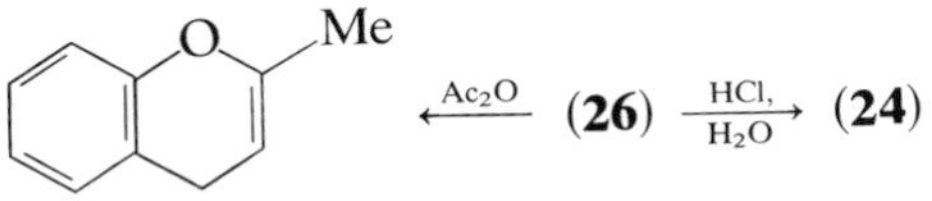

(**110**) $\xleftarrow{Ac_2O}$ (**26**) $\xrightarrow[H_2O]{HCl,}$ (**24**)

No apparent change occurs on boiling 2-ethoxy-8-methyl-6-*t*-butylchroman (**111**) in acidified aqueous ethanol for several hours. However when the solvent is changed to methanol, 2-methoxy-8-methyl-6-*t*-butylchroman (**112**) is recovered, indicating the stability of the chroman ring even after an acid-catalyzed cleavage resulting in acetal exchange. When the reaction is carried out in the presence of an aldehyde trap such as 2,4-dinitrophenylhydrazine, the phenol propionaldehyde derivative (**113**) is formed thus demonstrating a ring opening equilibrium.[17]

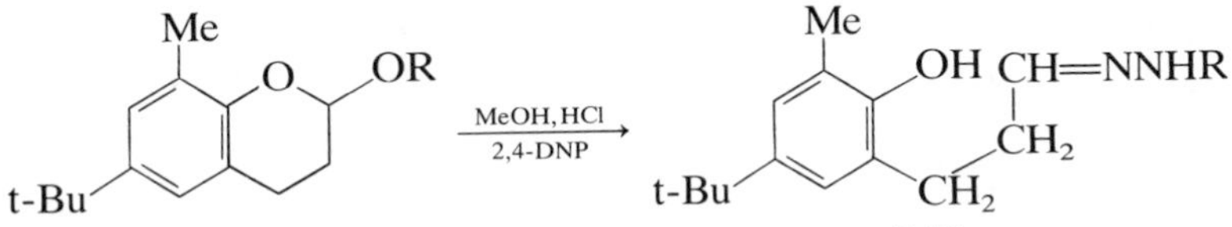

(**111**) R = Et
(**112**) R = Me

$\xrightarrow[\text{2,4-DNP}]{\text{MeOH, HCl}}$ (**113**)

R = 2,4-dinitrophenyl

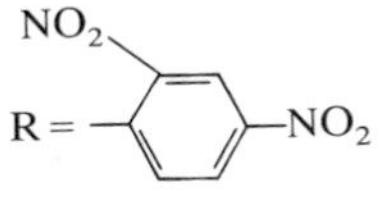

4-Ethoxychroman (**19**) decomposes on heating to give chrom-3-ene (**18**).[9]

V. TABLE OF COMPOUNDS

TABLE 2. ALKOXYCHROMANS

Mol. formula	Substituents 2	2	3	4	5	6	7	8	m.p. or b.p. (mm)	Ref.
$C_{10}H_{12}O_2$						OMe			92 (0.05)	3
									92 (1)	4
									78–82 (0.2)	5
$C_{10}H_{12}O_2$							OMe		273–276 (760)	2
									82–84 (0.05)	6
$C_{10}H_{12}O_2$								OMe	76 (0.01)	3
									92–94 (1.2)	7
$C_{11}H_{14}O_2$	Me	OMe							107 (14)	11
									108–112 (14)	12
$C_{11}H_{14}O_2$	Me					OMe				13
$C_{11}H_{14}O_2$	Me						OMe		130–150 (12)	13
$C_{11}H_{14}O_2$	Me							OMe	51	14
$C_{11}H_{14}O_2$				Me		OMe			100–125 (2)	15
$C_{11}H_{14}O_2$				Me			OMe		90–95 (0.7)	15
$C_{11}H_{14}O_2$	OEt								111.5–112 (12)	9
									108–110 (15)	10
$C_{11}H_{14}O_2$				OEt					93–94 (0.4)	8, 9
										3
$C_{11}H_{14}O_3$	OMe	OMe								31
$C_{12}H_{16}O_2$	Me	Me			OMe				101 (0.8)	20
									114–115 (7)	21
$C_{12}H_{16}O_2$	Me	Me				OMe			47–80 (0.1)	19
$C_{12}H_{16}O_2$	Me	Me					OMe		72 (0.3)	22
									173–174 (30)	23
									102 (1)	24

$C_{12}H_{16}O_2$	Me	Me						OMe	100–102 (2.5)	25
$C_{12}H_{16}O_2$	Me					Me	OMe		53–54	13
$C_{12}H_{16}O_2$	Me						OMe	Me	95–100 (0.5)	13
$C_{12}H_{16}O_2$	Me	OMe						Me	78–85 (ca. 1)	12
$C_{12}H_{16}O_2$					Et			OMe	106–108 (0.5)	7
$C_{12}H_{16}O_2$						Et		OMe	108–110 (0.7)	7
$C_{12}H_{16}O_2$						OMe		Et	105–110 (0.6)	16
$C_{12}H_{16}O_2$	Me	OEt							105 (11)	11
$C_{12}H_{16}O_3$	Me					OMe	OMe		47–48	13
$C_{13}H_{18}O_2$	Me	Me	Me			OMe			50–55 (10^{-6})	19
$C_{13}H_{18}O_2$	Me	Me			Me		OMe		114–115 (1.5–2)	24
$C_{13}H_{18}O_2$	Me	OMe		Me_2					121–124 (17)	38
$C_{13}H_{18}O_2$	Me	Me		OEt					134–137 (16)	29
$C_{13}H_{18}O_3$	Me	Me			OMe		OMe		114 (1)	24
$C_{13}H_{18}O_3$	Me	Me			OMe			OMe	59.5–60	33
$C_{13}H_{18}O_3$	Me	Me				OMe	OMe		60	34
$C_{13}H_{18}O_3$	Me	Me					OMe	OMe	65–70 (0.05)	33
$C_{13}H_{18}O_3$	OEt					OMe		Me		17
$C_{14}H_{18}O_2$	Me	Vinyl		OEt					93 (0.08)	29
$C_{14}H_{20}O_2$	Me	Me				OMe	Me	Me	58.5–59.5	46
$C_{14}H_{20}O_2$	Me				Me	OMe	Me	Me	140–160 (0.2)	13
$C_{14}H_{20}O_2$	Me	OEt		Me_2					126 (15)	35
$C_{14}H_{20}O_4$	Me	Me			OMe		OMe	OMe	145–147 (0.2)	36
$C_{15}H_{20}O_3$	Me	Vinyl		OEt				OMe	110–112 (0.07)	29
$C_{15}H_{22}O_2$	Me	OEt		Me_2			Me		74	37
$C_{15}H_{22}O_2$	OEt				Me	Me	Me	Me	111–113 (17)	43
$C_{16}H_{16}O_2$				Ph			OMe		203–204 (10)	40
$C_{16}H_{24}O_2$	OEt					*t*-Bu		Me	106–107 (0.6)	17
$C_{19}H_{22}O_3$	OEt	Me		Ph			OMe		82	35
$C_{19}H_{30}O_2$	OEt					*t*-Bu		*t*-Bu	134 (0.7)	17

VI. REFERENCES

1. F. Prillinger and H. Schmid, *Monatsh. Chem.*, **72,** 427; *Sitzber. Akad. Wiss. Wien. Math.–Naturw. Klasse*, Abt. IIb, **148,** 125 (1939); *Chem. Abstr.*, **34,** 1017 (1940).
2. R. T. Blickenstaff and I. Y. C. Tao, *Tetrahedron*, **24,** 2495 (1968).
3. B. Willhalm, A. F. Thomas, and F. Gautschi, *Tetrahedron*, **20,** 1185 (1964).
4. L. W. Deady, R. D. Topsom, and J. Vaughan, *J. Chem. Soc.*, 5718 (1965).
5. A. O. Fitton and G. R. Ramage, *J. Chem. Soc.*, 5426 (1963); A. O. Fitton and G. R. Ramage, Brit Patent 1,024,645 (1966); *Chem. Abstr.*, **64,** 17607 (1966); A. H. Wragg and G. P. Ellis, Brit. Patent 1,023,373 (1966); *Chem. Abstr.*, **64,** 19567, (1966); U.S. Patent 3,322,795 (1967); *Chem. Abstr.*, **68,** 49578 (1968).
6. B. Graffe, M.-C. Sacquet, and P. Maitte, *J. Heterocycl. Chem.*, **12,** 247 (1975).
7. R. S. Bramwell and A. O. Fitton, *J. Chem. Soc.*, 3882 (1965).
8. H. Normant and P. Maitte, *C.R. Acad. Sci. Paris*, **234,** 1787 (1952).
9. P. Maitte, *Ann. Chim.* (Paris), **9,** 431 (1954).
10. J. Brugidou and H. Christol, *Bull. Soc. Chim. Fr.*, 1974 (1966).
11. W. Baker and J. Walker, *J. Chem. Soc.*, 646 (1935).
12. L. J. Dolby, C. A. Elliger, S. Esfandiari, and K. S. Marshall, *J. Org. Chem.*, **33,** 4508 (1968).
13. O. Dann, G. Volz, and O. Huber, *Justus Liebigs Ann. Chem.*, **587,** 16 (1954).
14. K. W. Merz and H. Pfäffle, *Arch. Pharm.* (Weinheim), **288,** 86 (1955).
15. B. D. Tilak and Z. Muljiani, *Tetrahedron*, **24,** 949 (1968).
16. A. O. Fitton and G. R. Ramage, *J. Chem. Soc.*, 2481 (1964).
17. D. A. Bolon, *J. Org. Chem.*, **35,** 3666 (1970).
18. H. Nozaki, I. Otani, R. Noyori, and N. Kawanisi, *Tetrahedron*, **24,** 2183 (1968).
19. L. I. Smith and H. E. Ungnade, *J. Org. Chem.*, **4,** 311 (1939).
20. R. Huls, *Bull. Classe Sci. Acad. Roy. Belg.*, **39,** 1064 (1953).
21. H. Fukami and M. Nakajima, *Agr. Biol. Chem.* (Tokyo), **25,** 247 (1961).
22. J. R. Beck, R. Kwok, R. N. Booher, A. C. Brown, L. E. Patterson, P. Pranc, B. Rockey, and A. Pohland, *J. Am. Chem. Soc.*, **90,** 4706 (1968); R. Kwok and A. Pohland, Ger. Offen. 1,933,153 (1970); *Chem. Abstr.*, **72,** 90422 (1970).
23. J. N. Chatterjea, K. D. Banerji, and N. Prasad, *Chem. Ber.*, **96,** 2356 (1963).
24. P. R. Iyer and G. D. Shah, *Indian J. Chem.*, **6,** 227 (1968).
25. M. Hallett and R. Huls, *Bull. Soc. Chim. Belg.*, **61,** 33 (1952).
26. E. A. Vdovtsova, *Zh. Org. Khim.*, **6,** 1238 (1970).
27. E. A. Vdovtsova and G. F. Fedorova, *Zh. Org. Khim.*, **6,** 1230 (1970).
28. Y. Butsugan, H. Tsukamoto, N. Morito, and T. Bito, *Chem. Lett.*, 523 (1976).
29. R. Merten and G. Mueller, *Chem. Ber.*, **97,** 682 (1964).
30. J. Badin and G. Descotes, *Bull. Soc. Chim. Fr.*, 1949 (1970).
31. O. L. Chapman and C. L. McIntosh, *J. Chem. Soc.* (*D*), *Chem. Commun.*, 383 (1971).
32. M. Nakayama, S. Hayashi, M. Tsukayama, T. Horie, and M. Masumura, *Chem. Lett.*, 315 (1972); M. Tsukayama, *Bull. Chem. Soc. Japan*, **48,** 80 (1975).
33. J. S. P. Schwarz, A. I. Cohen, W. D. Ollis, E. A. Kaczka, and L. M. Jackson, *Tetrahedron*, **20,** 1317 (1964).
34. A. R. Alertsen, *Acta Chem. Scand.*, **9,** 1725 (1955); *Acta Polytech. Scand. Chem. Met. Ser.*, **13,** 1 (1961).
35. D. W. Clayton, W. E. Elstow, R. Ghosh, and B. C. Platt, *J. Chem. Soc.*, 581 (1953).
36. R. Huls and S. Brunelle, *Bull. Soc. Chim. Belg.*, **68,** 325 (1959).
37. W. Baker, R. F. Curtis, and J. F. W. McOmie, *J. Chem. Soc.*, 1774 (1952).
38. W. Webster and D. P. Young, *J. Chem. Soc.*, 4785 (1956).
39. D. Abson, K. D. Bartle, J. Bryant, R. Livingstone, and R. B. Watson, *J. Chem. Soc.*, 2978 (1965).
40. P. C. Mitter and P. K. Paul, *J. Indian Chem. Soc.*, **8,** 271 (1931).

41. G. A. Holmberg and J. Axberg, *Acta Chem. Scand.*, **17,** 967 (1963).
42. R. Livingstone, D. Miller, and S. Morris, *J. Chem. Soc.*, 5148 (1960).
43. J. Brugidou and H. Christol, *Bull. Soc. Chim. Fr.*, 1693 (1966).
44. R. Heiss, A. Seyberlich, I. Hammann, and W. Behrenz, Brit. Patent 1,126,140 (1968), *Chem. Abstr.*, **69,** 106557 (1968).
45. R. J. Molyneux, *J. Chem. Soc. (D), Chem. Commun.*, 318 (1974).
46. L. I. Smith and R. W. H. Tess, *J. Am. Chem. Soc.*, **66,** 1523 (1944).

CHAPTER V

Halogenochromans

R. LIVINGSTONE

The Polytechnic, Huddersfield, U.K.

I. INTRODUCTION

The halogenochromans found in the literature are in the main the monohalogeno derivatives which contain the halogen atom attached to the benzene ring and the dihalogeno derivatives with both the halogen atoms attached to the heterocyclic ring. The former are obtained from the appropriate halogeno derivatives of phenol, which by various synthetic routes developed for the preparation of chroman may be converted to the required monohalogenochroman. The latter are formed generally by the addition of halogen to the related chromen, but some have been formed by replacing a hydroxy group attached to C-4 of the related halogenohydrin by halogen. This method has been used for the preparation of some chroman derivatives containing a halogeno substituent at the 4-position.

II. PREPARATION OF HALOGENOCHROMANS

1. Preparation of Monohalogenochromans

The first monohalogenochroman was prepared in 1920 by Rindfusz, Ginnings, and Harnack[1] by the dehydration of 4-bromophenyl 3-hydroxypropyl ether (**1**) to give 6-bromochroman (**2**). They found that the dehydration could be effected more conveniently with phosphorus pentoxide than with zinc chloride. The reaction was carried out by boiling the mixture in an inert solvent, such as toluene or methyl ethyl ketone.[1] A 70% yield of 6-bromochroman (**2**) is obtained by treating the acetate of 2-allyl-4-bromophenol (**3**) with hydrogen bromide in the presence of benzoyl peroxide.[2] 6-Bromo- and 6-chlorochroman, (**2**) and (**4**), may also be prepared by heating 4-bromophenyl 3-chlorophenyl ether (**5**) and 4-chlorophenyl 3-chloropropyl ether (**6**), respectively, with stannic chloride.[3]

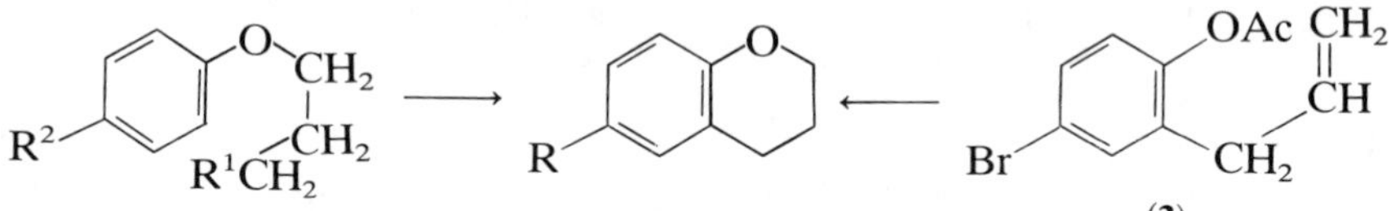

(**1**) $R^1 = OH, R^2 = Br$
(**5**) $R^1 = Cl, R^2 = Br$
(**6**) $R^1 = R^2 = Cl$

(**2**) $R = Br$
(**4**) $R = Cl$
(**7**) $R = H$

(**3**)

Chroman (**7**) on treatment with *N*-bromosuccinimide in carbon tetrachloride in the presence of benzoyl peroxide gives a mixture of 4- and 6-bromochroman (**8**) and (**2**).[4]

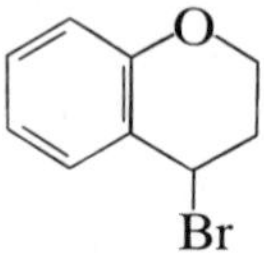

(**8**)

The cyclization of the acetate of 2-(2,3-dibromopropyl)phenol (**9**) with potassium hydroxide yields a mixture which on seeding in light petroleum yields 90% 2-bromomethyl-2,3-dihydrobenzo[*b*]furan (**10**); the mother liquor affords about 10% 3-bromochroman (**11**).[5]

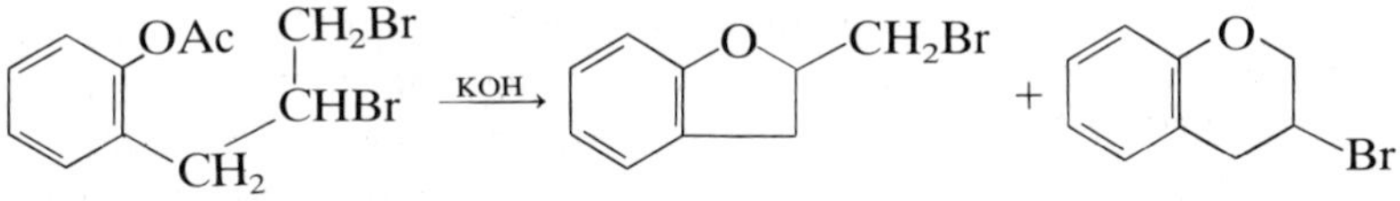

(9) (10) (11)

The smooth decomposition of readily prepared 1,3-diphenoxypropanes in the presence of aluminum chloride to give a phenol and a chroman provides a simple and useful synthesis for 6-halogenochromans (**2**) and (**4**) and 8-halogenochromans (**16**) and (**17**) from 1,3-di(4-halogenophenoxy)- and 1,3-di(2-halogenophenoxy)propanes (**12**) and (**13**) and (**14**) and (**15**), respectively.[6] Pure 6-chloro-, 8-chloro-, and 6-bromochroman (**4**), (**17**), and (**2**) were prepared by this method, but the 8-bromo derivative was not obtained pure. Chroman itself was obtained as a by-product in the preparation of both 6- and 8-bromochroman.

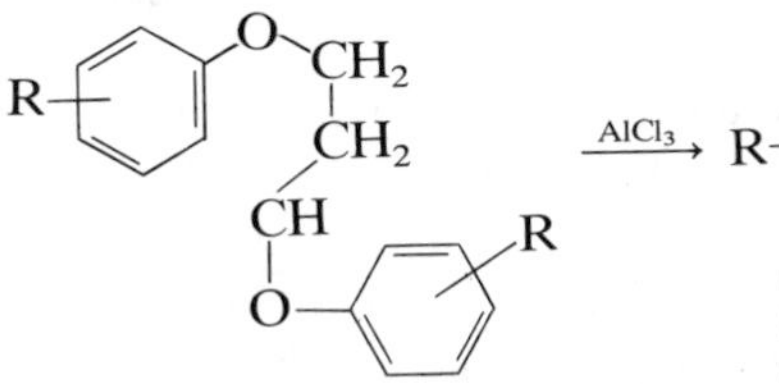

(**12**) R = 4-Br
(**13**) R = 4-Cl
(**14**) R = 2-Br
(**15**) R = 2-Cl
(**18**) R = 3-Cl

(**16**) R = 8-Br
(**17**) R = 8-Cl
(**19**) R = 5-Cl
(**20**) R = 7-Cl

Decomposition of 1,3-di(3-chlorophenoxy)propane (**18**) yielded a mixture of 5- and 7-chlorochromans (**19**) and (**20**) which were separated by preparative gas chromatography. The 7-chloro derivative was eluted in a pure form before the 5-isomer, which was contaminated by some of the 7-isomer. A second elution would probably allow the collection of pure samples, but this was not attempted.[6] 6-Chlorochroman may be obtained by the diazotization of 6-aminochroman, followed by treatment of the diazonium salt with cuprous chloride.[7]

4-Chlorophenyl crotonate (**21**) in hydrogen fluoride at 100° gives 6-chloro-2-methylchroman-4-one (**22**), which undergoes a Clemmensen reduction to afford 6-chloro-2-methylchroman (**23**).[8]

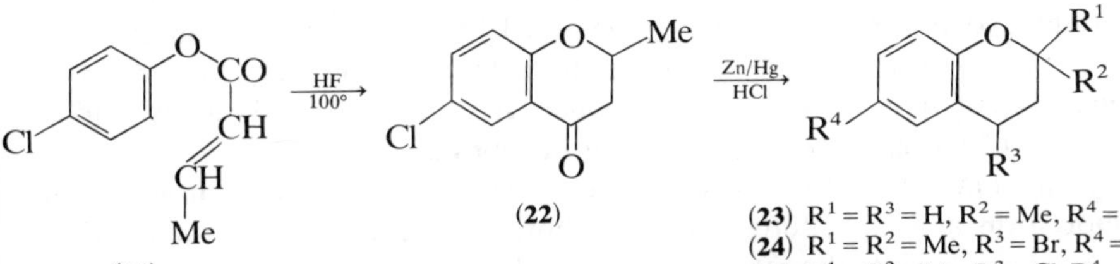

(**21**) (**22**)

(**23**) $R^1 = R^3 = H$, $R^2 = Me$, $R^4 = Cl$
(**24**) $R^1 = R^2 = Me$, $R^3 = Br$, $R^4 = H$
(**25**) $R^1 = R^2 = Me$, $R^3 = Cl$, $R^4 = H$
(**26**) $R^1 = R^2 = Me$, $R^3 = OH$, $R^4 = H$

4-Bromo- and 4-chloro-2,2-dimethylchromans (**24**) and (**25**), containing the halogen atom attached to the heterocyclic ring, have been prepared by stirring 2,2-dimethylchroman-4-ol (**26**) with phosphorus tribromide and phosphorus pentachloride, respectively, in dry benzene at room temperature for several hours. The required monohalogeno derivatives were obtained by distillation.[9]

The reaction between 4-chlorophenol and isoprene on heating in the presence of a metallic phenolate (e.g., Zn, Al, Mg, Ca, Na, Li) yields 6-chloro-2,2-dimethylchroman. 6,8-Dichloro-2,2-dimethylchroman and 6-chloro-2,2,5,7-tetramethylchroman are similarly prepared, while 6-chloro-2,2,3-trimethylchroman may be obtained from 4-chlorophenol and 2,3-dimethylbuta-1,3-diene.[10] The alkenylation of 4-chlorophenol with 4-chloropent-2-ene without a catalyst and in the presence of a small amount of ferric chloride, phosphoric acid, or KU-2 catalyst gives mono- and dialkenylated phenols and their cyclization products, including 6-chloro-2,4-dimethylchroman.[11]

Some monohalogenochromans with more than two methyl substituents have been prepared by direct halogenation or by replacing a hydroxyl group by halogen. For example, 6-bromo-2,2,5,7,8-pentamethylchroman (**27**) is obtained by treating a solution of 2,2,5,7,8-pentamethylchroman in carbon tetrachloride with a solution of bromine in the same solvent[12]; and 4-bromo-2,6,8-trimethylchroman (**28**) is produced by reacting 2,6,8-trimethylchroman-4-ol with phosphorus tribromide in alcohol-free chloroform at 50°.[13] Some monohalogenochromans are listed in Table 5.

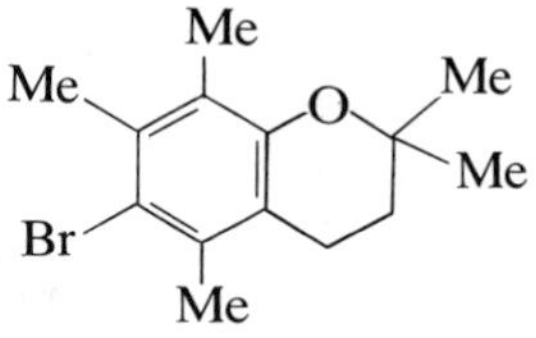

Me Me O Me Me Br Me

(**27**)

Me O Me Me Br

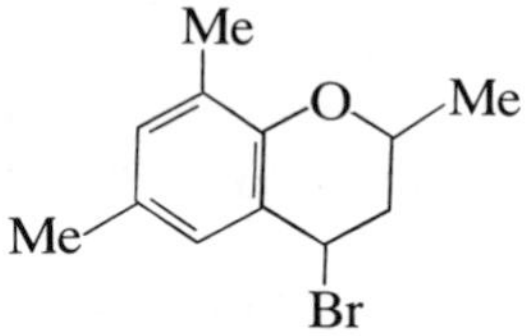

(**28**)

2. Preparation of Dihalogenochromans

trans-3,4-Dibromochroman (**30**) was the first dihalogeno derivative of chroman to be prepared.[5] It was obtained by Maitte[3] by the addition of bromine to chromen (**29**) in carbon tetrachloride. The related *trans*-3,4-dichloro derivative (**31**) is formed when chlorine is passed slowly into a solution of chromen (**29**) in carbon tetrachloride at 0°.[14]

Treatment of 2,2-dimethylchromen (**32**) with 1 mole bromine in chloroform at −15° affords *trans*-3,4-dibromo-2,2-dimethylchroman (**33**)[15] in good yield, while the reaction between chlorine and 2,2-dimethylchroman (**32**) in chloroform gives *trans*-3,4-dichloro-2,2-dimethylchroman (**34**).[16]

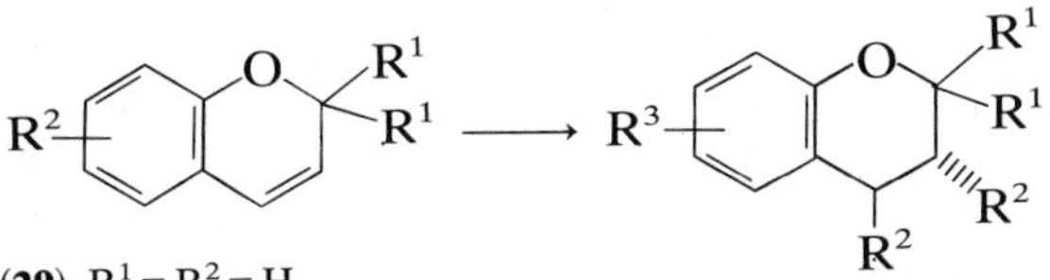

(29) $R^1 = R^2 = H$
(32) $R^1 = Me, R^2 = H$
(39) $R^1 = Me, R^2 = 6\text{-}OMe$
(40) $R^1 = Me, R^2 = 8\text{-}OMe$
(45) $R^1 = Me, R^2 = 7\text{-}OMe$
(47) $R^1 = Me, R^2 = 5\text{-}OMe$
(53) $R^1 = Me, R^2 = 6\text{-}Br$

(30) $R^1 = R^3 = H, R^2 = Br$
(31) $R^1 = R^3 = H, R^2 = Cl$
(33) $R^1 = Me, R^2 = Br, R^3 = H$
(34) $R^1 = Me, R^2 = Cl, R^3 = H$
(37) $R^1 = Ph, R^2 = Br, R^3 = H$
(38) $R^1 = Ph, R^2 = Cl, R^3 = H$
(41) $R^1 = Me, R^2 = Br, R^3 = 6\text{-}OMe$
(42) $R^1 = Me, R^2 = Br, R^3 = 8\text{-}OMe$
(43) $R^1 = Me, R^2 = Cl, R^3 = 6\text{-}OMe$
(44) $R^1 = Me, R^2 = Cl, R^3 = 8\text{-}OMe$
(46) $R^1 = Me, R^2 = Cl, R^3 = 7\text{-}OMe$
(52) $R^1 = Me, R^2 = Br, R^3 = 6\text{-}Br$

cis-3,4-Dichloro-2,2-dimethylchroman (**36**) is obtained by adding freshly distilled thionyl chloride to *cis*-3-chloro-2,2-dimethylchroman-4-ol (**35**) in ether at 0° and then setting aside the mixture for several hours. Removal of the solvent affords a mixture of the *trans*- and *cis*-dichloro derivatives (**34**) and (**36**), which is chromatographed on a neutral alumina column; elution with light petroleum (b.p. <40°) gives 46% trans isomer and 37% cis isomer.[17]

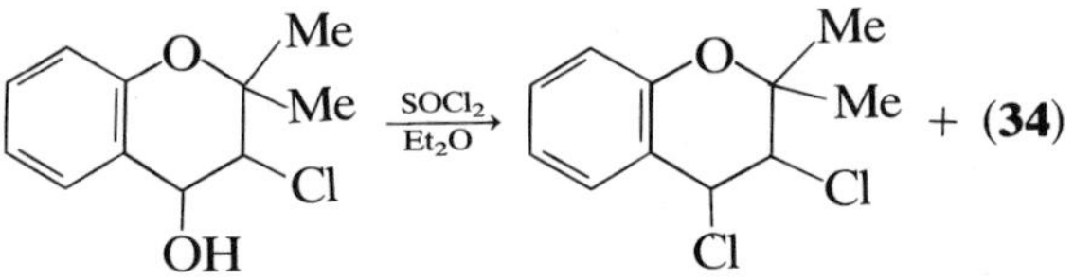

$\xrightarrow[Et_2O]{SOCl_2}$ + (**34**)

(**35**) (**36**)

trans-3,4-Dibromo- and *trans*-3,4-dichloro-2,2-diphenylchroman (**37**) and (**38**)[16] are obtained by methods similar to those used for the preparation of the related 2,2-dimethyl derivatives. It is essential in all the above preparations and those that follow that after the addition of the halogen is completed the solvent is removed at a temperature <40°C.

6-Methoxy- and 8-methoxy-2,2-dimethylchromens (**39**) and (**40**) in chloroform react with both bromine and chlorine to give respectively the expected 3,4-dibromo- and 3,4-dichloro derivatives (**41**) and (**42**), and (**43**) and (**44**). The mother liquor from the crystallization of 3,4-dibromo-8-methoxy-2,2-dimethylchroman (**42**) (from light petroleum b.p. 60–80°) gives an oil on evaporation which when boiled with methanol yields a dibromo dimethoxy compound, probably 3,6-dibromo-4,8-dimethoxy-2,2-dimethylchroman (**54**), indicating the formation of some 3,4,6-tribromo-8-methoxy-2,2-dimethylchroman during the reaction between the chromen (**40**) and bromine.

7-Methoxy-2,2-dimethylchromen (**45**) in chloroform reacts with chlorine at 0° to give a poor yield of 3,4-dichloro-7-methoxy-2,2-dimethylchroman (**46**), which slowly loses hydrogen chloride. The reaction between the 7-methoxychromen (**45**) and bromine results in the evolution of hydrogen bromide, and no recognizable product can be isolated. No 3,4-dihalogeno-5-methoxy-2,2-dimethylchroman was obtained on addition of halogen to 5-methoxy-2,2-dimethylchroman (**47**).[18]

2-Methyl-4-phenylchrom-2-ene (**48**) on treatment with bromine in carbon tetrachloride at 0° in the dark, followed by several hours at room temperature, gives 2,3-dibromo-2-methyl-4-phenylchroman (**49**). When the reaction with bromine is carried out at room temperature, hydrogen bromide is evolved to give a small yield of a solid, which is probably a monobromochromen.[19]

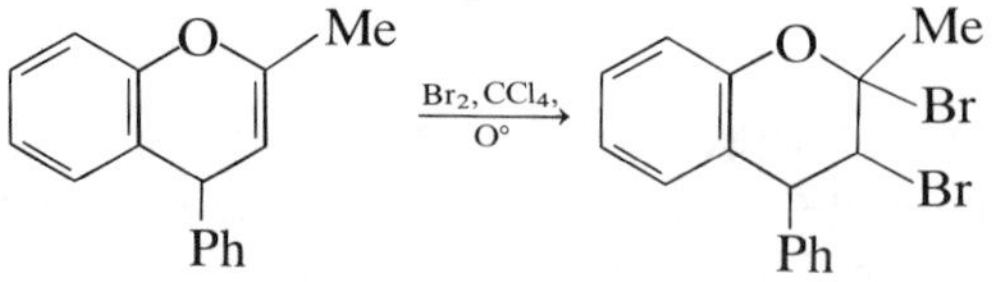

Br_2, CCl_4, 0°

(**48**) (**49**)

Bromination of 2,2,4,4,7-pentamethylchroman (**50**) occurs when it is treated with bromine in chloroform at 50° to give 6,8-dibromo-2,2,4,4,7-pentamethylchroman (**51**).[20] A number of dihalogenochromans are listed in Table 5.

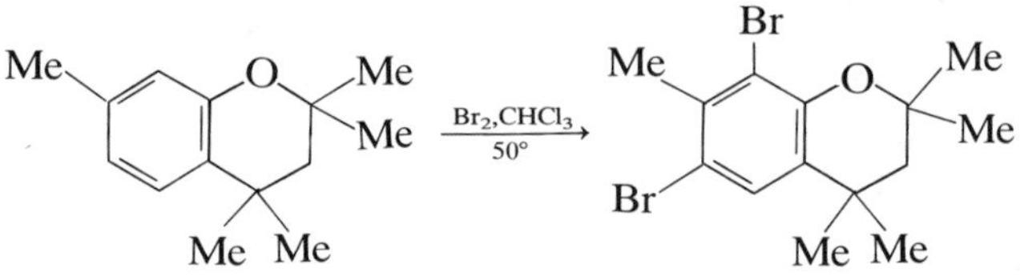

Br_2, $CHCl_3$, 50°

(**50**) (**51**)

3. Preparation of Trihalogenochromans

The trihalogenochromans found in the literature are those containing one halogeno substituent in the benzene ring and two halogeno substituents in the heterocyclic ring. Originally they were obtained as secondary products resulting from attempts to form 3,4-dihalogenochromans by the addition of halogens to certain chromens.

The reaction between 2,2-dimethylchromen (**32**) and 1 mole bromine in carbon disulfide at temperatures between −20° and room temperature affords a mixture of 3,4-dibromo- and 3,4,6-tribromo-2,2-dimethylchromans (**33**) and (**52**). The tribromo derivative (**52**) is prepared by reacting 2,2-dimethylchromen (**32**) or 3,4-dibromo-2,2-dimethylchroman

(**33**) with 2 or 1 mole bromine, respectively in chloroform. A good yield of the tribromo derivative (**52**) is obtained by reacting 6-bromo-2,2-dimethyl chromen (**53**) with bromine in chloroform at −15° and leaving the mixture at room temperature for several hours.[15]

5-Methoxy-2,2-dimethylchromen (**47**) reacts with chlorine in chloroform at 0° to give probably 3,4,6-trichloro-5-methoxy-2,2-dimethylchroman (**55**) in poor yield. It has not been found possible to isolate any 3,4-dichloro-5-methoxy-2,2-dimethylchroman following the above reaction.[18]

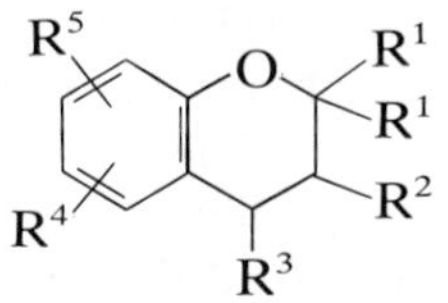

(**54**) $R^1 = Me$, $R^2 = Br$, $R^3 = OMe$, $R^4 = 6\text{-}Br$, $R^5 = 8\text{-}OMe$
(**55**) $R^1 = Me$, $R^2 = R^3 = Cl$, $R^4 = 5\text{-}OMe$, $R^5 = 6\text{-}Cl$

A number of trihalogeno-2,2-dimethylchromans have been prepared by the addition of bromine or chlorine to 5-, 6-, 7-, and 8-chloro-2,2-dimethylchromen[21], generally in chloroform; but the addition of bromine to 7-chloro-2,2-dimethylchromen was carried out in carbon tetrachloride. Some trihalogenochromans are listed in Table 5.

4. Preparation of Monohalogenodihydronaphthopyrans

The reaction between 2,3-dihydro-1*H*-naphtho[2,1-*b*]pyran-1-ol (**56**) and phosphorus tribromide in dry benzene at room temperature affords 1-bromo-2,3-dihydro-1*H*-naphtho[2,1-*b*]pyran (**57**). Under similar conditions treatment of the dihydronaphthopyranol (**56**) with phosphorus pentachloride yields 1-chloro-2,3-dihydro-1*H*-naphtho[2,1-*b*]pyran (**58**).[9] Some mono-halogenodihydronaphthopyrans are listed in Table 6.

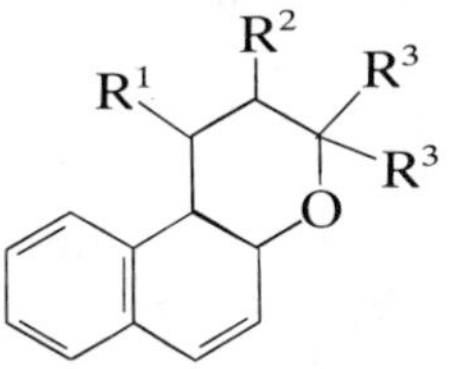

(**56**) $R^1 = OH$, $R^2 = R^3 = H$
(**57**) $R^1 = Br$, $R^2 = R^3 = H$
(**58**) $R^1 = Cl$, $R^2 = R^3 = H$
(**59**) $R^1 = R^2 = Br$, $R^3 = H$
(**61**) $R^1 = R^2 = Cl$, $R^3 = H$
(**64**) $R^1 = OH$, $R^2 = Br$, $R^3 = H$
(**65**) $R^1 = Cl$, $R^2 = Br$, $R^2 = H$
(**67**) $R^1 = R^2 = Br$, $R^3 = Me$

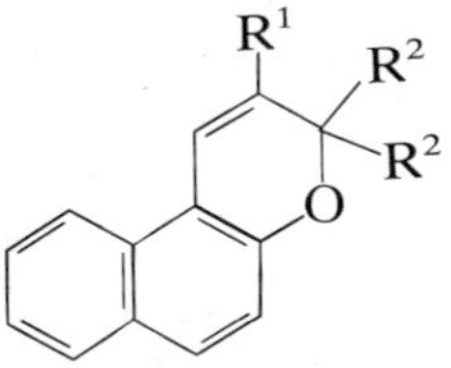

(**60**) $R^1 = R^2 = H$
(**66**) $R^1 = H$, $R^2 = Me$
(**68**) $R^1 = Br$, $R^2 = Me$

5. Preparation of Dihalogenodihydronaphthopyrans

A number of dihalogenodihydronaphthopyrans have been obtained by one of two methods. The first is by the addition of halogen to the related naphthopyran in a solvent such as chloroform, and the second is to replace the hydroxyl group in the corresponding halogenohydrin by halogen. In the latter case the halogenohydrin is obtained from the related halogenoketone.

1,2-Dibromo-1,2-dihydro-3*H*-naphtho[2,1-*b*]pyran (**59**) has been prepared by the addition of bromine to 3*H*-naphtho[2,1-*b*]pyran (**60**) in ether. Similarly 1,2-dichloro-1,2-dihydro-3*H*-naphtho[2,1-*b*]pyran (**61**) is obtained by bubbling chlorine through a solution of the naphthopyran (**60**) in ether (for nomenclature see Chapter II).

A number of 3,4-dihalogeno-3,4-dihydro-2*H*-naphtho[1,2-*b*]pyrans, for example, (**62**), and 1,2-dihalogeno-1,2-dihydro-3*H*-naphtho[2,1-*b*]pyrans, for example, (**59**) and (**61**), are obtained by boiling the appropriate *cis*- or *trans*-3-halogeno-3,4-dihydro-2*H*-naphtho[1,2-*b*]pyran-4-ol or 2-halogeno-1,2-dihydro-3*H*-naphtho[2,1-*b*]pyran-1-ol with either phosphorus tribromide, thionyl chloride, or phosphorus pentachloride in benzene. Boiling 3-bromo-3,4-dihydro-2*H*-naphtho[1,2-*b*]pyran-4-ol (**63**) with phosphorus tribromide in benzene gives 3,4-dibromo-3,4-dihydro-2*H*-naphtho[1,2-*b*]pyran (**62**), and boiling 2-bromo-1,2-dihydro-3*H*-naphtho[2,1-*b*]pyran-1-ol (**64**) with thionyl chloride in benzene yields 2-bromo-1-chloro-1,2-dihydro-3*H*-naphtho[2,1-*b*]pyran (**65**).[22]

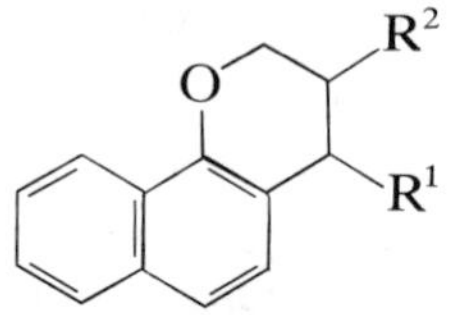

(**62**) $R^1 = R^2 = Br$
(**63**) $R^1 = OH, R^2 = Br$

Addition of bromine to 3,3-dimethyl-3*H*-naphtho[2,1-*b*]pyran (**66**) in carbon disulfide at −15° gives 1,2-dibromo-1,2-dihydro-3,3-dimethyl-3*H*-naphtho[2,1-*b*]pyran (**67**). When the reaction is carried out in chloroform, the substitution product 2-bromo-3,3-dimethyl-3*H*-naphtho[2,1-*b*]pyran (**68**) is formed.[15] A similar effect is observed with 3,3-diphenyl-3*H*-naphtho[2,1-*b*]pyran.[23] Bubbling chlorine through a solution of either 3,3-dimethyl- or 3,3-diphenyl-3*H*-naphtho[2,1-*b*]pyran in chloroform or carbon tetrachloride at 0° yields 1,2-dichloro-1,2-dihydro-3,3-dimethyl-3*H*-naphtho[2,1-*b*]pyran or the corresponding 3,3-diphenyl compound, respectively. Some dihalogenodihydronaphthopyrans are listed in Table 6.

6. Preparation of Trihalogenodihydronaphthopyrans

3,4-Dihydro-2,2-dimethyl-3,4,6-trichloro-2*H*-naphtho[1,2-*b*]pyran (**69**) is prepared in a 70% yield by bubbling chlorine through a solution of

6-chloro-2,2-dimethyl-2*H*-naphtho[1,2-*b*]pyran (**70**) in carbon tetrachloride at 0°. It is also formed when the 6-chloronaphthopyran (**70**) is stirred with sulfuryl chloride in benzene for several hours and may be obtained in high yield by stirring together *cis*- or *trans*-3,6-dichloro-3,4-dihydro-2,2-dimethyl-2*H*-naphtho[1,2-*b*]pyran-4-ol (**71**) and phosphorus pentachloride in benzene.[22]

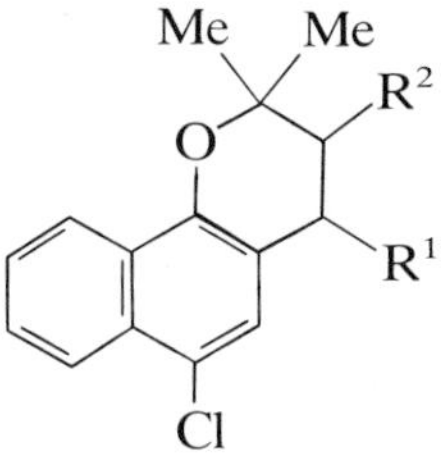

(**69**) $R^1 = R^2 = Cl$
(**71**) $R^1 = OH, R^2 = Cl$
(**73**) $R^1 = R^2 = Br$
(**74**) $R^1 = OH, R^2 = Br$
(**75**) $R^1 = OMe, R^2 = Br$

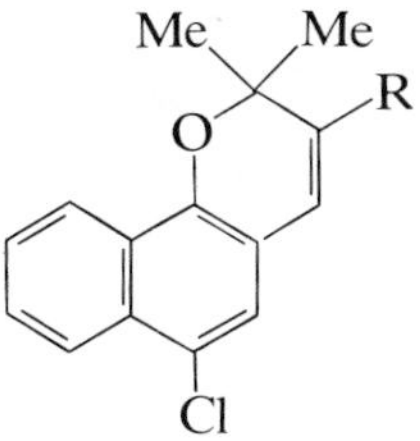

(**70**) R = H
(**72**) R = Br

When bromine is added to 6-chloro-2,2-dimethyl-2*H*-naphtho[1,2-*b*]pyran (**70**) in chloroform or carbon tetrachloride at room temperature, hydrogen bromide is evolved on removing the solvent and 3-bromo-6-chloro-2,2-dimethyl-2*H*-naphtho[1,2-*b*]pyran (**72**) is obtained from the residual gum. Although 6-chloro-3,4-dibromo-3,4-dihydro-2,2-dimethyl-2*H*-naphtho-[1,2-*b*]pyran (**73**) has not been isolated, evidence for its formation is obtained by boiling the gum in aqueous acetone or methanol, which gives 3-bromo-6-chloro-2,2-dimethyl-2*H*-naphtho[1,2-*b*]pyran-4-ol (**74**) or 3-bromo-6-chloro-2,2-dimethyl-4-methoxy-2*H*-naphtho[1,2-*b*]pyran (**75**), respectively. These compounds are obtained by replacing the bromine in the 4-position by either a hydroxyl or methoxy group. Attempts to convert 3-bromo-6-chloro-2,2-dimethyl-2*H*-naphtho[1,2-*b*]pyran-4-ol (**74**) to the dibromo derivative (**73**) by reacting it with phosphorus tribromide failed to give a stable product.[22] Some trihalogenodihydronaphthopyrans are listed in Table 3.

III. PHYSICAL CONSTANTS AND SPECTRA

1. Nuclear Magnetic Resonance Spectra and Conformation of Some Dihalogenochromans and Their Derivatives

The ^{1}H-nmr spectral data of the 3,4-dihalogenochromans (**76a** and **76b**), prepared by the addition of halogen to chromen, and of the related 4-acetoxy-3-halogeno-chromans (**76c** and **76d**) obtained from the corresponding 3-halogeno-chroman-4-ols reveal that their structures are closely related (Table 1). The significant features from the coupling constant data are that $^3J_{CA}$, $^3J_{CB}$, and $^3J_{CX}$ are small (3–4 Hz), $^4J_{BX}$ is ca. 1.5 Hz, and $^4J_{AX}$ is ca. 0.

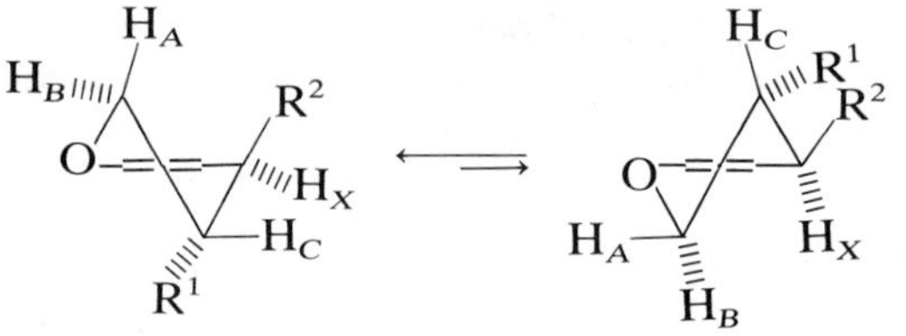

(a) (76) (a[1])

	R^1	R^2
a	Cl	Cl
b	Br	Br
c	Cl	OAc
d	Br	OAc
e	Cl	OH
f	Br	OH
g	Cl	OMe
h	Br	OMe

Dreiding models show that, with the hetero ring in the half-chair conformation and with proton H_X pseudo-equatorial, see (**76a**), protons H_B and H_X lie close to a planar zigzag orientation for which ${}^4J$ coupling is a maximum, and H_A and H_X are unfavorably situated for ${}^4J$ coupling. The alternative "sofa" conformation (**76a**^{1}) does not have the correct geometry for ${}^4J$ coupling, and this structure will not be considered further. The observed coupling constants between the C-2 methylene protons and proton H_C are small (${}^3J_{BC} \approx 2 \times {}^3J_{AC} \approx 3$–$4$ Hz), and application of the Karplus equation ($J_{AB} = 10 \cos^2 \phi$, where ϕ = dihedral angle) indicates that the C-3–H_C bond bisects the C-2 methylene system and that in the proposed half-chair conformation H_C is equatorial. The 3-*ax*-halogeno substituent is antiperiplanar to H_A, and the observation that ${}^3J_{BC} \approx 2 \times {}^3J_{AC}$ is therefore ascribed to the electronegativity effect of the 3-*ax* substituent; no modification to the half-chair geometry of the hetero ring is thus required.

The signal of 2-axial proton H_A lies at lower field than that of the 2-equatorial proton H_B for chromans (**76a**), (**76b**), (**76c**), and (**76d**). This reversal of the normal order may be explained by the fact that in the diaxial conformation the 3-axial halogeno group has a much stronger deshielding effect on proton H_A (dihedral angle $\phi = 180°$) than on proton H_B ($\phi = 60°$). The 3,4-dihalogenochromans and their hydrolysis products exist predominantly in the *trans*-diaxial conformation, (**76a**), (**76b**), (**76e**), and (**76f**), because the corresponding diequatorial conformation is destabilized by dipole–dipole repulsion between the 3-*eq* and 4-*eq* substituents, and by repulsion between the 4-*eq* substituent and the peri hydrogen.

In the case of *cis*-4-acetoxy-3-halogenochromans, (**77a**) and (**77b**), the ^{1}H-nmr coupling constants (${}^3J_{AC} \approx 8$, ${}^3J_{BC} \approx 2.5$ Hz) between proton H_C and the C-2 methylene protons suggest that H_C has an axial orientation and that

TABLE 1. NUCLEAR MAGNETIC RESONANCE DATA OF *cis*- AND *trans*-3,4-DISUBSTITUTED CHROMANS[a]

Substituents			Chemical shift (δ ppm)				Coupling constants (Hz)					Solvent
	3(R^1)	4(R^2)	H_X	H_C	H_A	H_B	$^3J_{CX}$	$^3J_{AC}$	$^3J_{BC}$	$^4J_{BX}$	$^2J_{AB}$	
(76)												
a	Cl	Cl	5.12	4.44	4.65	4.25	3.0	1.5	3.0	1.7	12.1	$CDCl_3$
b	Br	Br	5.18	4.11	4.51	3.83	2.5	1.6	2.5	2.1	12.9	PhH
c	Cl	OAc	5.86	4.55	4.46	4.37	3.2	1.4	3.8	1.3	12.5	Me_2CO
d	Br	OAc	6.21	4.62	4.47	4.38	3.4	1.6	3.9	1.2	12.7	Pyridine
g	Cl	OMe	4.71	4.33	4.29	4.16					12.7	$CDCl_3$
h	Br	OMe	4.33	4.25	4.28	4.23					12.3	$CDCl_3$
(77)												
a	Cl	OAc	6.30	4.97	4.35	4.60	3.8	7.6	2.5	0.9	11.6	DMSO-d_6
b	Br	OAc	6.23	5.01	4.37	4.63	3.8	8.3	2.4	0.9	11.9	DMSO-d_6

[a] Spectra at both 60 and 100M Hz. R^1 and R^2 and H_X, H_C, H_A, and H_B refer to (**76a**) and (**76b**) or (**77a**) and (**77b**). From Reference 19.

it lies outside the angle made by the methylene protons. The $^4J_{BX}$ coupling is still appreciable (0.9 Hz), so that H_X must prefer a pseudoequatorial orientation.

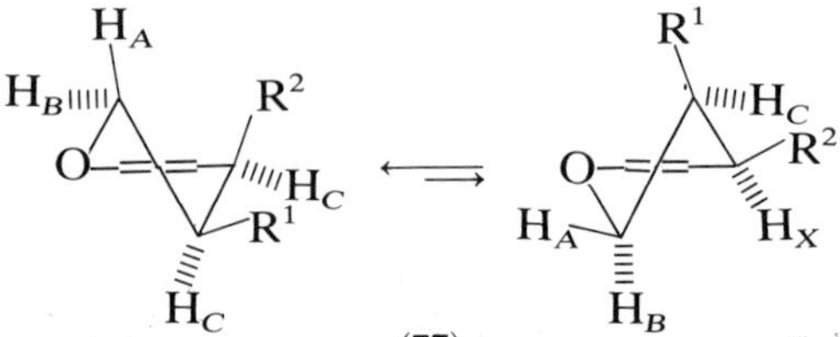

(**77**)

	R^1	R^2	
a	Cl	OAc	**a**^{1}
b	Br	OAC	**b**^{1}

These preferred orientations for H_C and H_X prove that the compounds possess a cis configuration and that the most stable half-chair conformation is that of 4-*ax*-acetoxy-3-*eq*-halogenochroman (**77a**) and (**77b**).[14] By assuming that $^3J_{AaxCax}$ is 10.5 Hz and J_{AeqCeq} is 1.5 Hz (derived from the nmr data obtained by Bolger and co-workers[24] for a number of substituted flavans), the position of the conformational equilibria may be calculated. Thus 68% (ΔG 0.45 kcal/mole) *cis*-4-acetoxy-3-chloro- and 75% (ΔG 0.65 kcal/mole) *cis*-4-acetoxy-3-bromochroman exist in the conformation in which the acetoxy group occupies a pseudoaxial position. The free-energy difference between the two conformations of these cis isomers is probably due to the destabilization of (**77a**^{1}) and (**77b**^{1}) by steric repulsion between the peri hydrogen and the 4-*eq*-acetoxy group and by dipole–dipole repulsion between the 3-*ax*-halogeno substituent and the ring oxygen [there will also be some dipole-dipole stabilization of the other conformations, (**77a**) and (**77b**)]. The signal of the 2-*eq* proton H_B lies at a lower field than that of the 2-*ax* proton H_A; this is the order normally observed in "cyclohexane systems," and it is to be expected, since in the major conformation the 3-*eq*-halogeno group lies between the C-2 geminal protons ($\phi = 60°$), thus shielding both protons equally. The differential shielding is now due to the anisotropy effect of the C-3–C-4 bond.[14] The ^{1}H-nmr spectral data of 3,4-dihalogeno-2,2-dimethylchromans obtained by the addition of halogen to 2,2-dimethylchromen and of 3-halogen-2,2-dimethylchroman-4-ols formed by the hydrolysis of the former compounds are shown in Table 2. The configuration of the substituents at C-3 and C-4 was determined by comparing the observed vicinal coupling constants with those predicted by the Karplus $\cos^2 \phi$ equation for the measured dihedral angles (ϕ) between H_A and H_B (**78**). (The variation of coupling constants with the electronegativity of substituents was neglected.)

Dreiding models of 3,4-dihalogeno-2,2-dimethylchromans show that the heterocyclic ring is distorted from the half-chair cyclohexene analog; although there is no eclipsing strain along the C-2–C-3 bond, strain is present along the C-3–C-4 bond. The values 155° and 85° of the dihedral angle between H_A and H_B in the two conformations (**78a**[1]) and (**78a**) of the *trans*-dihalogeno compound, respectively, are those for the structures which are estimated to have the least combined angle and eclipsing strain in the pyran ring. Because of the flexibility of this ring the C-3–C-4 dihedral angle may be varied without undue strain. The above angles give theoretical coupling constants of 7.4 (**78a**[1]) and ~0 Hz/s (**78a**). These values, when compared with those obtained from nmr spectral data (Table 2), show that in the 3,4-dihalogeno-2,2-dimethylchromans obtained by addition of halogen to 2,2-dimethylchroman and in their hydrolysis products 3-halogeno-2,2-dimethylchroman-4-ols, the most probable configuration is that with the 3- and 4-substituents trans and equatorial (**78a**[1]).

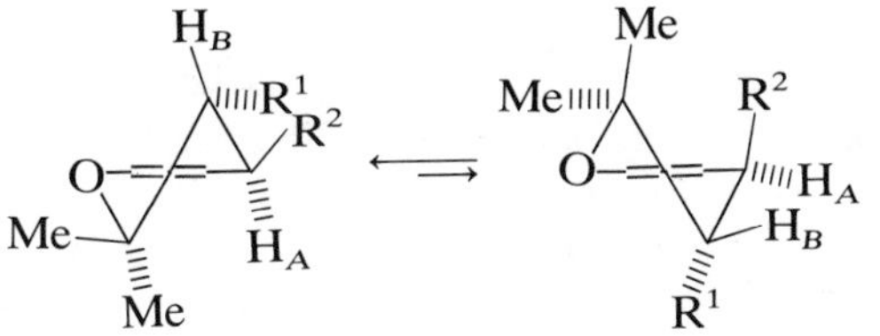

a[1] (**78**) **a**

It is likely that samples of *trans*-3,4-dihalogeno-2,2-dimethylchromans would contain the 3-*ax*, 4-*ax*-dihalogen conformer (**78a**) in amounts that would be undetected by nmr. This conformation (**78a**) is not favored because of the strong repulsion between the 2-*ax*-methyl and the 4-*ax*-halogen. In solution this repulsion energy would be increased, because solvation increases the effective size of the 4-*ax*-halogen, thus making it difficult to predict the equilibrium point in the conformational mixture.[25]

TABLE 2. ^{1}H-NUCLEAR MAGNETIC RESONANCE DATA OF *trans*-3,4-DISUBSTITUTED 2,2-DIMETHYLCHROMANS[a]

Substituents		Chemical shift (δ ppm)		Coupling constants (Hz)
3(R^1)	4(R^2)	H_A	H_B	J_{AB}
Cl	Cl	5.2	4.2	8.5
Cl	OH	4.7	3.9	7.6
Cl	OMe	4.53	4.13	7.8
Br	Br	5.5	4.5	9.7
Br	OH	4.9	4.1	10.8
Br	OMe	4.7	4.19	7.5

[a] Spectra at 60M Hz for ~10% solutions in $CDCl_3$. R^1 and R^2 refer to (**78a**[1]). From Reference 25.

The H^1-nmr spectral data of 3-halogeno-4-methoxy chromans and 2,2-dimethyl-3-halogeno-4-methoxychromans obtained by reacting the respective dihalogeno compounds with methanol show that the most probable configuration in the former case is that with the 3- and 4-substituents trans and axial (**76g**) and (**76h**) (Table 1), and in the latter case they are trans and equatorial (**78a**^{1}) (Table 2).[26]

2. Nuclear Magnetic Resonance Spectra and Conformation of Some Dihalogenodihydronaphthopyrans and Their Derivatives

Dreiding models of 3,4-dihalogeno-3,4-dihydro-2*H*-naphtho[1,2-*b*]pyrans, for example, (**62**), and 1,2-dihalogeno-1,2-dihydro-3*H*-naphtho[2,1-*b*]pyrans, for example, (**59**) and (**61**), show that the heterocyclic ring is distorted from the half-chair cyclohexene analog. Although there is no eclipsing strain along the C-2–C-3 bond in either series, strain is present along the C-3–C-4 bond in the first series and the C-2–C-1 bond in the second series of compounds. The models show that there are four possible structures, two conformations of the *trans*-dihalogeno (**76a**) and (**76a**^{1}) and two of the *cis*-dihalogeno compounds (**77a**) and (**77a**^{1}). They give values of 175°, 65°, 55°, and 55° for the dihedral angle between H_C and H_X in (**76a**), (**76a**^{1}), (**77a**), and (**77a**^{1}), respectively, and are those for the structures which are estimated to have the least combined angle and eclipsing strain in the pyran ring. Because of the flexibility of this ring, the C-3–C-4 dihedral angle in the disubstituted dihydro[1,2-*b*]pyrans and the C-2–C-1 dihedral angle in the disubstituted dihydro[2,1-*b*]pyrans may be varied without undue ring strain. These angles give theoretical coupling constants of 10.5 (**76a**), 1.8 (**76a**^{1}), 3.3 (**77a**), and 3.3 Hz (**77a**^{1}).

The ^1H-nmr spectra of the 3,4-disubstituted dihydronaphtho[1,2-*b*]pyrans and the 1,2-disubstituted dihydronaphtho[2,1-*b*]pyrans, obtained by the methods indicated in Section II, 5, show, on comparing the observed vicinal coupling constants (Table 3) with those predicted by the Karplus $\cos^2 \phi$ equation for the measured dihedral angles (ϕ) between H_C and H_X (the variation of coupling constants with the electronegativity of substituents is neglected), that in all the 3,4-dihalogeno-3,4-dihydro-2*H*-naphtho[1,2-*b*]pyrans and 1,2-dihalogeno-1,2-dihydro-3*H*-naphtho[2,1-*b*]pyrans reported, the most probable configuration is that with the 3- and 4- and 1- and 2-substituents trans and axial (**76a**).[9]

3. Ultraviolet Spectra

The uv absorption spectra of some 3,4-dibromochromans and their derivatives are recorded in Table 4. As would be expected, they are very similar to those of the parent chromans. In the case shown, the introduction

TABLE 3. ^{1}H-NUCLEAR MAGNETIC RESONANCE DATA OF *trans*-DIHALOGENODIHYDRONAPHTHOPYRANS

[1,2-*b*] [2,1-*b*]

	Substituents				Chemical shift (δ ppm)				Coupling constants (Hz)				
Isomer	1(R²)	2(R¹)	3(R¹)	4(R²)	H_X	H_C	H_A	H_B	$^3J_{CX}$	$^3J_{AC}$	$^3J_{BC}$	$^4J_{BX}$	$^4J_{AB}$
[1,2-*b*]			Br	Br	5.62	4.83	5.08	4.5	2.1	1.5	2.1	2.2	12.0
			Br	Cl	5.35	4.68	4.84	4.43	1.8	1.2	1.7	2.4	12.0
			Cl	Cl	5.21	4.38	4.81	4.53	2.5	1.7	1.9	2.0	12.0
			Cl	Br	5.40	4.60	4.97	4.41	2.2	1.4	1.7	2.3	12.0
[2,1-*b*]	Br	Br			5.94	4.87	5.08	4.40	2.2	1.5	2.2	2.3	12.0
	Cl	Br			5.80	4.75	4.93	4.45	2.2	1.5	2.2	2.1	12.0
	Br	Cl			5.75	4.73	5.02	4.33	2.1	1.4	2.1	2.1	12.0

[a] Spectra at 60M Hz for ~10% solution in $CDCl_3$. R^1 and R^2, and H_X, H_C, H_A, and H_B refer to (**76a**). From Reference 9.

TABLE 4. ULTRAVIOLET DATA OF SOME DIBROMOCHROMANS AND THEIR DERIVATIVES

Substituents					
2	3	4	6	λ_{max} (nm)	log ε
Me_2	Br	Br		276	3.40
				284	3.43
Me_2	Br	OH		276	3.08
				284	3.10
Me_2	Br	OMe		277	3.51
				285	3.10
Me_2	Br	Br	Br	297	3.42
				301	3.40
Me_2	Br	OMe	Br	288	3.42
				297	3.40

of the 6-bromo group into the aromatic ring shifts the absorption peaks to the higher wavelength region.[27]

IV. REACTIONS

The reactions of 3,4-dihalogenochromans are mainly concerned with the replacement of the halogen in the 4-position by other groups such as hydroxyl and alkoxy. Since the configuration of the 3,4-dihalogenochromans obtained by the addition of halogen to the appropriate chromen is trans, the resulting derivatives formed by replacement of the 4-halogen atom are also trans; and these, especially the halogenohydrins, can be used to produce other derivatives of chroman. The halogenohydrins, for instance, are easily converted to the corresponding 3,4-epoxychromans, which on reacting with the appropriate reagents, give a wide variety of chroman derivatives. *cis*-Halogenochromanols are obtained by an indirect method, namely, by reduction of the 3-halogenochroman-4-one in methanol with sodium borohydride following halogenation of the chroman-4-one.

Of the few known monohalogenochromans those containing a halogen atom attached to the carbon at positions 3 or 4 partake in elimination reactions to give the parent chromen or in substitution reactions to form, for example, related hydroxy or amino compounds.

1. Reactions of Monohalogenochromans

2,2-Dimethyl-4-halogenochromans (**24**) and (**25**) on boiling in aqueous acetone give 2,2-dimethylchroman-4-ol (**26**). Under similar conditions, 1-bromo- and 1-chloro-2,3-dihydro-1*H*-naphtho[2,1-*b*]pyran, (**57**) and (**58**), afford 2,3-dihydro-1*H*-naphtho[2,1-*b*]pyran-1-ol (**56**).[9]

It is reported that 2,2-dimethylchroman (**79**) on treatment with *N*-bromosuccinimide in boiling carbon tetrachloride, in the presence of benzoyl peroxide, probably gives 4-bromo-2,2-dimethylchroman (**24**). The product resulting from this reaction has not been purified; but the residue obtained following filtration and removal of the solvent has been heated with sodium ethoxide in benzene to give 2,2-dimethylchromen (**32**), indicating that 4-halogenochromans on treatment with sodium ethoxide in benzene dehydrohalogenate to yield the related chromen.[28]

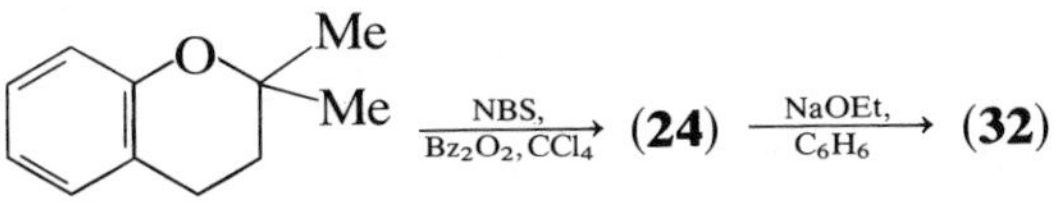

$$(\mathbf{79}) \xrightarrow[\text{Bz}_2\text{O}_2,\ \text{CCl}_4]{\text{NBS,}} (\mathbf{24}) \xrightarrow[\text{C}_6\text{H}_6]{\text{NaOEt,}} (\mathbf{32})$$

(**79**)

4-Bromochroman (**8**) on treatment with an excess of diethylamine is converted into 4-diethylaminochroman (**80**).[3,4] It also reacts with the appropriate Grignard reagent to give 4-ethyl-, 4-propyl-, 4-butyl-, and 4-phenylchroman (**81**).[3]

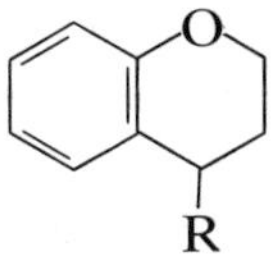

(**80**) R = NEt_2
(**81**) R = Et, Pr, Bu, Ph

3-Bromochroman (**11**) reacts with alcoholic potassium hydroxide and dimethylamine to give chromen (**29**) and 3-dimethylaminochroman (**82**), respectively.[5]

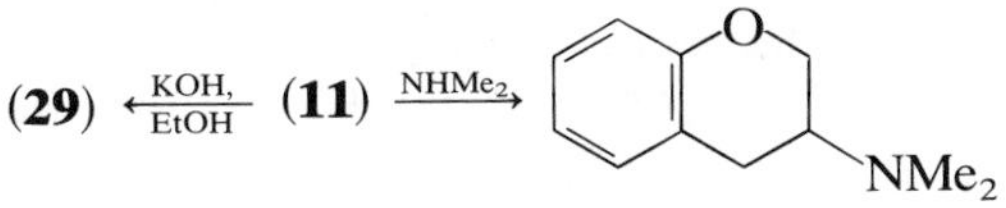

$$(\mathbf{29}) \xleftarrow[\text{EtOH}]{\text{KOH,}} (\mathbf{11}) \xrightarrow{\text{NHMe}_2} (\mathbf{82})$$

(**82**)

2. Hydrolysis of 3,4-Dihalogenochromans

The hydrolysis of *trans*-3,4-dibromo- and *trans*-3,4-dichlorochromans (**30**) and (**31**) using aqueous acetone yields the corresponding *trans*-3-halogenochroman-4-ol (**83**). Evidence for the trans configuration of the halogenohydrins is suggested by the fact that on treatment with potassium hydroxide in ether they yield 3,4-epoxychroman (**84**). It is known that the stereochemical preference for the formation of epoxides by dehydrohalogenation of 1,2-halogenohydrins is that the two interacting groups should be antiperiplanar.[14]

(**30**) or (**31**) $\xrightarrow[Me_2CO]{H_2O,}$ (**83**) R = Br or Cl $\xrightarrow[Et_2O]{KOH,}$ (**84**)

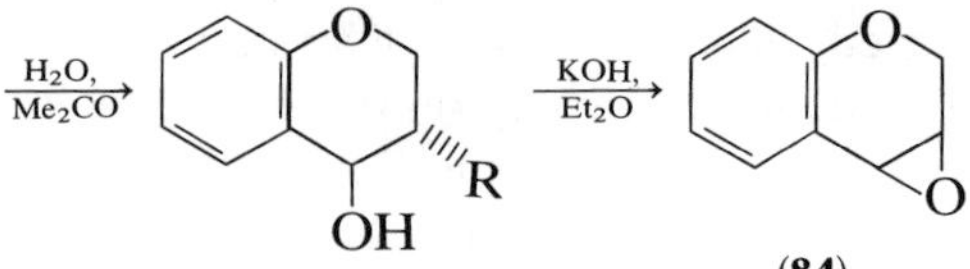

Sodium borohydride reduction of 3-halogenochroman-4-ones (**85**) yields *cis*-3-halogenochroman-4-ols (**86**). The cis configuration is assigned on the basis of the ^{1}H-nmr coupling constants of the acetoxy derivatives (Table 1) and the fact that dehydrohalogenation with potassium hydroxide gives chroman-4-one (**87**).[14] *cis*-3-Bromo-2,6-dimethylchroman-4-ol may be obtained in a similar way. Treatment of the respective bromohydrins with thionyl chloride and phosphorus tribromide affords the corresponding *trans*-bromochloro- and *trans*-dibromochromans.[29]

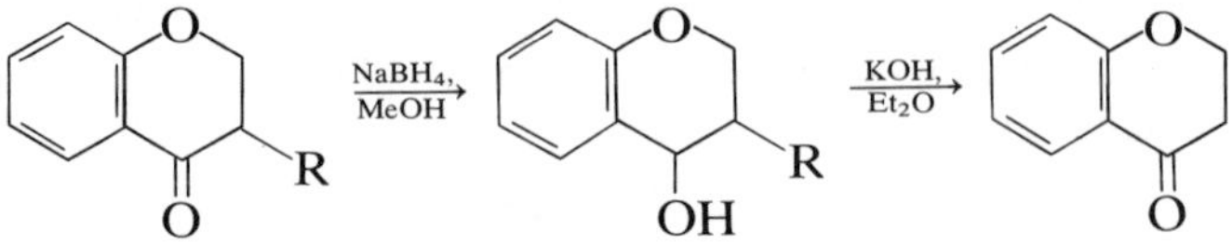

(**85**) R = Br or Cl $\xrightarrow[MeOH]{NaBH_4,}$ (**86**) R = Br or Cl $\xrightarrow[Et_2O]{KOH,}$ (**87**)

The ^{1}H-nmr coupling constants of these compounds show that the hetero ring exists in the half-chair conformation, with both substituents axial in the trans isomer (**76**), and that the most stable conformation of the cis isomer is that in which the 4-substituent is pseudoaxial (**77**). The hydrolysis of the *trans*-3,4-dihalogenochromans (**30**) and (**31**) proceeds with retention of configuration owing to the effect of the neighboring 3-halogeno substituent.

A detailed study, including the kinetic measurements, of the hydrolysis of 3,4-dichloro- and 3,4-dibromo-2,2-dimethylchroman (**78**, $R^1 = R^2 = Cl$ or Br) has been reported. The results show a decrease in the rate coefficients for both halides (**78**, $R^1 = R^2 = Cl$ or Br) on the addition of common ion and an increase on the addition of noncommon ion, thus indicating an S_N1 (or S_N1-like) mechanism. The amount of the *trans*-3-*ax*,4-*ax* conformer (**78a**) (Section III, 1) will be small at equilibrium, but it may hydrolyze more rapidly than the *trans*-3-*eq*,4-*eq* conformer (**78a^1**) (equilibrium would be maintained by the rapid inversion of the heterocyclic ring), in which case it would play an important part in the mechanism.

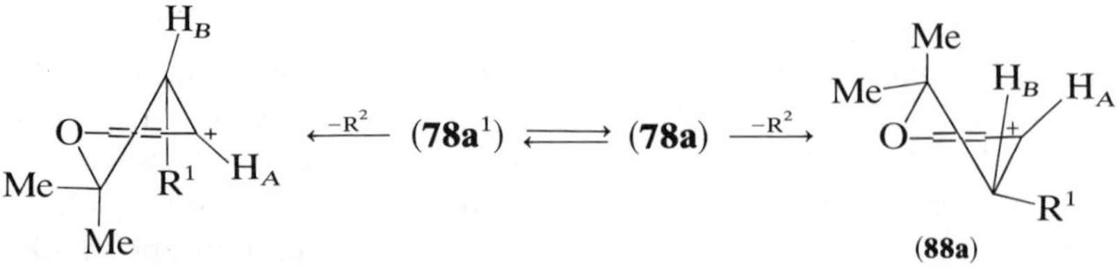

(**88a^1**) $\xleftarrow{-R^2}$ (**78a^1**) $\rightleftharpoons$ (**78a**) $\xrightarrow{-R^2}$ (**88a**)

In the *trans*-3-*ax*,4-*ax*-dihalogeno conformation (**78a**) there is strong repulsion between the 2-*ax*-methyl and 4-*ax*-halogen, which would increase the rate of ionization. The carbonium ion (**88a**) which formed would be shielded from attack on one side, and the major product would be the *cis*-halogenohydrin. Since only *trans*-halogenohydrins have been isolated, to obtain these from the *trans*-3-*ax*,4-*ax*-dihalogeno compounds there would have to be strong neighboring-group participation with both bromine and chlorine. Chlorine generally is a weakly anchimeric-assisting group; and so, if the *trans*-3-*ax*,4-*ax*-dichloro conformer is mechanistically important, then the *cis*-chlorohydrin should be isolated. On this evidence, it is considered that the 3-*ax*,4-*ax*-dihalogeno conformation is not important in the hydrolysis of *trans*-3,4-dihalogeno-2,2-dimethylchromans.

In the *trans*-3-*eq*,4-*eq*-dihalogeno conformation (**78a**^{1}) the repulsion between the 2-*ax*-methyl and the 4-*ax*-hydrogen would increase the rate of ionization equally for both dihalogeno compounds. However, the 2-*ax*-methyl group would shield one side of the carbonium ion, making attack from the same side as the leaving group more favorable. This agrees with the isolation of *trans*-3-*eq*-halogeno-2,2-dimethylchroman-4-*eq*-ols (~98%) from the hydrolyses. The oxygen hetero atom distorts the pyran ring (Section III, 1) from the cyclohexene analog so that there will be considerable eclipsing strain along the C-3–C-4 bond in the solvated compounds. Formation of the transition state (**88a**^{1}) relieves this strain, and the increased tendency to form the carbonium ion is accompanied by rate enhancement.

The salt effects and variation of energy and entropy of activation indicate a similar mechanism for the solvolyses of the dibromo and dichloro compounds, but the results do not distinguish between the chemical S_N1 scheme and schemes involving "internal" and "external" ion pairs.[25]

3. Methanolysis of 3,4-Dihalogenochromans

When 3,4-dihalogenochromans (**30**) and (**31**) are boiled with methanol containing dry hydrogen chloride for a few hours, or with methanol alone for several hours, the chlorine or bromine in the 4-position is replaced by a methoxy group to give the corresponding 3-halogeno-4-methoxychromans (**89**) and (**90**).

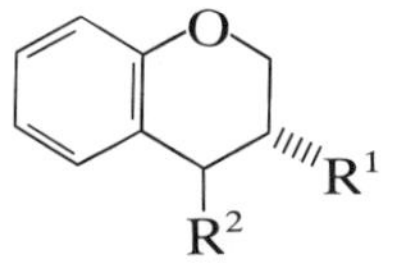

(**89**) $R^1 = Br, R^2 = OMe$
(**90**) $R^1 = Cl, R^2 = OMe$
(**91**) $R^1 = Br, R^2 = OH$
(**92**) $R^1 = Cl, R^2 = OH$

Nuclear magnetic resonance spectral data show that in 3-halogeno-4-methoxychroman the most probable configuration is that with the 3- and 4-substituents trans and axial, but that in 2,2-dimethyl-3-halogeno-4-methoxychroman they are trans and equatorial. The effects of the neighboring phenylene group and halogen and the oxygen hetero atom are the same as for the hydrolysis of 2,2-dimethyl-3,4-dihalogenochromans.

Comparison of the rate coefficients for the methanolysis of 3,4-dihalogenochromans shows that k_1(dibromo)/k_1(dichloro) is approximately twice k_1(bromo)/k_1(chloro) for the methanolysis of some monohalogen compounds, and that for the dihalogeno-2,2-dimethylchromans k_1(dibromo)/k_1(dichloro) is ca. ten times greater than k_1(bromo)/k_1(chloro) (cf. ca. seven times in case of hydrolysis). Along with the explanation already given (Section IV, 2) for such differences, the extra increase in the case of the 2,2-dimethyl compounds may be cited as evidence for the repulsion between the 2-*ax*-methyl and the 4-*ax*-halogen, which would increase the rate of ionization.[26]

3-Halogeno-4-methoxychromans (**89**) and (**90**) may also be obtained by boiling the related 3-halogenochroman-4-ols (**91**) and (**92**) in methanol containing dry hydrogen chloride. 3-Bromo-2,2-dimethyl-4-methoxychroman (**93**) with boiling acetic acid yields 3-bromo-2,2-dimethylchromen (**94**).[15]

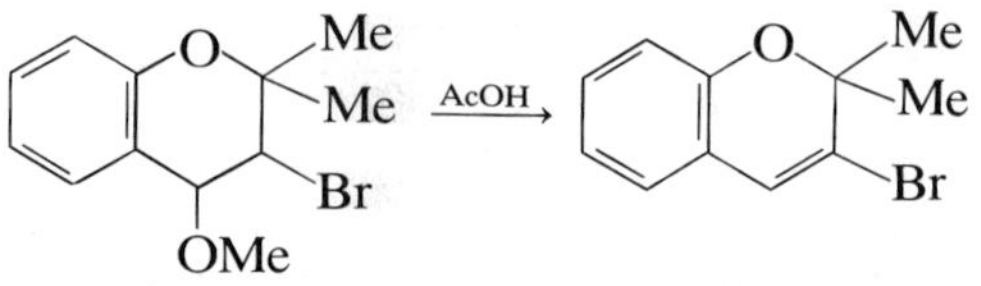

O Me Me Br OMe —AcOH→ O Me Me Br

(**93**) (**94**)

4. Chroman Derivatives Obtained Indirectly From 3,4-Dihalogenochromans

trans-3-Bromo-2,2-dimethylchroman -4-ol (**95**) prepared by the hydrolysis of *trans*-3,4-dibromo-2,2-dimethylchroman (**33**) has also been obtained by the reaction between 2,2-dimethylchromen (**32**) and hypobromous acid. When boiled with acetic acid, the bromohydrin (**95**) dehydrates to give 3-bromo-2,2-dimethylchromen (**94**).

O R^1 R^1 R^2 OH

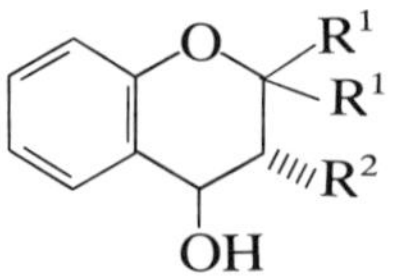

(**95**) $R^1 = Me, R^2 = Br$,
(**96**) $R^1 = Me, R^2 = Cl$,
(**97**) $R^1 = Ph, R^2 = Cl$,

In order to obtain 3-chloro-2,2-dimethyl- and 3-chloro-2,2-diphenylchroman-4-ol, (**96**) and (**97**), it is necessary to treat the appropriate dichloro compounds (**34**) and (**38**) in boiling acetone with a molecular proportion of an aqueous potassium hydroxide solution. Similar treatment of 3,4-dibromo- and 3,4-dichloro-2,2-dimethylchroman results in the formation of the respective 3-halogeno-2,2-dimethylchroman-4-ol.[16]

3,4-Dibromo-2,2-dimethyl-6- and -8-methoxychromans, their dichloro analogs, 3,4-dichloro-7-methoxychroman, and 3,4,6-trichloro-5-methoxychroman react in acetone with an equimolar proportion of aqueous potassium hydroxide to give the corresponding 2,2-dimethyl-3-halogenomethoxychroman-4-ols. 3,4-Dibromo-2,2-dimethyl-8-methoxychroman and 3-bromo-2,2-dimethyl-8-methoxychroman-4-ol on prolonged boiling with methanol yield 3-bromo-4,8-dimethoxy-2,2-dimethylchroman. The corresponding dibromo-6-methoxychroman and 3-bromo-6-methoxychroman-4-ol yield no 3-bromo-4,6-dimethoxy-2,2-dimethylchroman on treatment under similar conditions with methanol or methanolic hydrogen chloride.[18]

A number of derivatives obtained directly and indirectly from di- and trihalogenochromans are listed in Table 7.

5. Dihydronaphthopyran Derivatives Obtained Directly and Indirectly From Dihalogenodihydronaphthopyrans

Many of the reactions of the dihalogenodihydronaphthopyrans are similar to those of the dihalogenochromans. For instance, the hydrolysis of 1,2-dichloro-1,2-dihydro-3,3-dimethyl-3*H*-naphtho[2,1-*b*]pyran (**98**) at room temperature, either with an aqueous acetone solution of potassium hydroxide or with aqueous acetone for several days, gives 2-chloro-1,2-dihydro-3,3-dimethyl-3*H*-naphtho[2,1-*b*]pyran-1-ol (**99**). Treatment of the analogous dibromo compounds (**100**) and (**101**) with aqueous acetone gives the corresponding bromohydrins (**102**) and (**103**).[22]

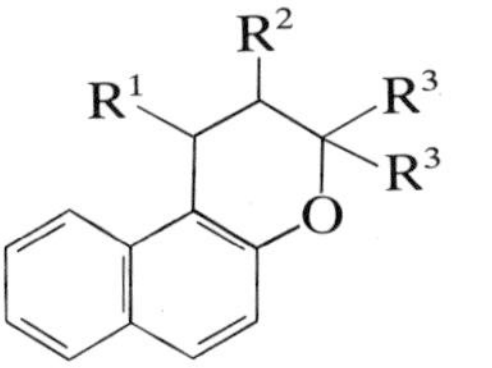

(**98**) $R^1 = R^2 = Cl, R^3 = Me$
(**99**) $R^1 = OH, R^2 = Cl, R^3 = Me$
(**100**) $R^1 = R^2 = Br, R^3 = Me$
(**101**) $R^1 = R^2 = Br, R^3 = Ph$
(**102**) $R^1 = OH, R^2 = Br, R^3 = Me$
(**103**) $R^1 = OH, R^2 = Br, R^3 = Ph$
(**104**) $R^1 = OMe, R^2 = Br, R^3 = Me$

1,2-Dibromo-1,2-dihydro-3,3-dimethyl-3*H*-naphtho[2,1-*b*]pyran (**100**) on boiling with methanol containing dry hydrogen chloride affords the 1-methoxy derivative (**104**), which like the dibromo compound (**100**) on pyrolysis yields 2-bromo-3,3-dimethyl-3*H*-naphtho[2,1-*b*]pyran (**105**).[15] A number of derivatives obtained directly and indirectly from di- and trihalogenodihydronaphthopyrans are listed in Table 8.

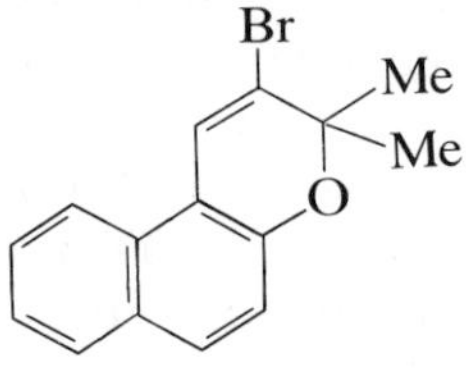

(**105**)

V. TABLES OF COMPOUNDS

TABLE 5. MONO-, DI-, AND TRIHALOGENOCHROMANS[a]

Mol. formula	Substituents							m.p. or b.p. (mm)	Yield (%)	Ref.
	2	3	4	5	6	7	8			
C_9H_8BrClO		Br	Cl					95.5–96.5	57	26
C_9H_8BrClO		Cl	Br					117–118	80	26
$C_9H_8Br_2O$		Br	Br					125.5–126.5		3, 5, 30
								127–128	94	14
$C_9H_8Cl_2O$		Cl	Cl					89-90	43	14
C_9H_9BrO		Br						139–140 (20)		5
C_9H_9BrO					Br			143–144 (18)		1–4
								140–142 (16)	50	6
C_9H_9BrO							Br[b]	100–104(1)	35	6
C_9H_9ClO				Cl[c]				—	54	6
C_9H_9ClO					Cl			123 (15)		3
								86–89 (1)	70	6
								123–125 (16)		7
C_9H_9ClO						Cl[e]		91 (1)	54	6
C_9H_9ClO							Cl	93 (2)	55	6
$C_{11}H_{11}Br_2ClO$		Br	Br	Cl				71–72	95	21
$C_{11}H_{11}Br_2ClO$		Br	Br		Cl			69.5–70.5	92	21

TABLE 5. (*Contd.*)

Mol. formula	Substituents 2	3	4	5	6	7	8	m.p. or b.p. (mm)	Yield (%)	Ref.
$C_{11}H_{11}Br_2ClO$		Br	Br			Cl		65–66	86	21
$C_{11}H_{11}Br_2ClO$		Br	Br				Cl	45	91	21
$C_{11}H_{11}Br_3O$		Br	Br		Br			60	77	15
$C_{11}H_{11}Cl_3O$		Cl	Cl		Cl			58–59	88	21
$C_{11}H_{11}Cl_3O$		Cl	Cl			Cl		65–66	70	21
$C_{11}H_{11}Cl_3O$		Cl	Cl				Cl	25	80	21
$C_{11}H_{12}BrClO$	Me_2	Br	Cl					84–85	64	26
$C_{11}H_{12}BrClO$	Me_2	Cl	Br					53–54	74	26
$C_{11}H_{12}Br_2O$	Me_2	Br	Br					81–82	82	15
$C_{11}H_{12}Cl_2O$	Me_2	Cl	Cl					59–60	62	16
$C_{11}H_{12}Cl_2O$ cis-	Me_2	Cl	Cl					63–64	37	17
$C_{11}H_{13}BrO$	Me_2		Br					70–75(0.5)	39	9
$C_{11}H_{13}ClO$	Me_2		Cl					85–89(1)	51	9
$C_{11}H_{13}ClO$	Me_2				Cl			28–29		10
$C_{12}H_{13}Cl_3O_2$		Cl	Cl	OMe	Cl			136–137		18
$C_{12}H_{14}Br_2O_2$	Me_2	Br	Br		OMe			84–85	67	18
$C_{12}H_{14}Br_2O_2$	Me_2	Br	Br				OMe	93–94	52	18
$C_{12}H_{14}Cl_2O_2$	Me_2	Cl	Cl		OMe			52–53	73	18
$C_{12}H_{14}Cl_2O_2$	Me_2	Cl	Cl			OMe		68–70	36	18
$C_{12}H_{14}Cl_2O_2$	Me_2	Cl	Cl				OMe	90–91	98	18
$C_{21}H_{16}Br_2O$	Ph_2	Br	Br					137–138	89	16
$C_{21}H_{16}Cl_2O$	Ph_2	Cl	Cl					99–100	57.5	16

[a] All dihalogeno compounds are trans, except where indicated.
[b] Not pure.
[c] Mixture of 5- and 7-isomers separated as nitro derivatives.

TABLE 6. MONO-, DI-, AND TRIHALOGENODIHYDRONAPHTHOPYRANS[a]

[1,2-*b*] [2,1-*b*]

Mol. formula	Isomer	Substituents 1	2	3	4	6	m.p.	Reagent	Yield (%)	Ref.
$C_{13}H_{10}BrClO$	[1,2-*b*]			Cl	Br		93–94	PBr_3	45	22
$C_{13}H_{10}BrClO$	[1,2-*b*]			Br	Cl		113–114	$SOCl_2$	78	22
$C_{13}H_{10}BrClO$	[2,1-*b*]	Br	Cl				129–130	PBr_3	66	22
$C_{13}H_{10}BrClO$	[2,1-*b*]	Cl	Br				119–120	$SOCl_2$	81	22
$C_{13}H_{10}Br_2O$	[1,2-*b*]			Br	Br		91–92	PBr_3	70	22
$C_{13}H_{10}Br_2O$	[2,1-*b*]	Br	Br				122–123	PBr_3	70	22
								Br_2[c]	42	22
$C_{13}H_{10}Cl_2O$	[1,2-*b*]			Cl	Cl		91–92	$SOCl_2$	84	22
								PCl_5	81	22
								Cl_2[c]	53	22
$C_{13}H_{10}Cl_2O$	[2,1-*b*]	Cl	Cl				128–129	$SOCl_2$	72	22
								PCl_5	66	22
$C_{13}H_{11}BrO$	[2,1-*b*]	Br					77–78	PBr_3	73	9
$C_{13}H_{11}ClO$	[2,1-*b*]	Cl					67–88[b]	PCl_3	60	9
$C_{15}H_{13}Cl_3O$	[1,2-*b*]		Me_2	Cl	Cl	Cl	106–107	Cl_2	69	22
$C_{15}H_{14}Br_2O$	[2,1-*b*]	Br	Br	Me_2			97–98	Br_2	43	15
$C_{15}H_{14}Cl_2O$	[2,1-*b*]	Cl	Cl	Me_2			117–119	Cl_2	52	23
$C_{25}H_{18}Br_2O$	[2,1-*b*]	Br	Br	Ph_2			118–122	Br_2	85	23
$C_{25}H_{18}Cl_2O$	[2,1-*b*]	Cl	Cl	Ph_2			159–161	Cl_2	75	23

[a] These compounds were prepared from the appropriate halogenohydrin and were crystallized from light petroleum, b.p. 40–60°, unless otherwise stated.
[b] Crystallized from light petroleum, b.p. <40°.
[c] Prepared from 3*H*-naphtho[2,1-*b*]pyran.

TABLE 7. DERIVATIVES OF HALOGENOCHROMANS

Mol. formula	Configuration	Substituents 2	3	4	5	6	7	8	m.p.	Ref.
$C_{10}H_{11}BrO_2$			Br	OMe						26
$C_{10}H_{11}ClO_2$			Cl	OMe						26
$C_{12}H_{14}Br_2O_2$		Me_2	Br	OMe		Br			54	15
$C_{12}H_{15}BrO_2$	trans	Me_2	Br	OMe					74	15
$C_{12}H_{15}BrO_2$		Me_2				Br	OMe		39–41	31
$C_{12}H_{15}ClO_2$	trans	Me_2	Cl	OMe					71–73	26
$C_{13}H_{16}Br_2O_3$		Me_2	Br	OMe		Br		OMe	146–148	18
$C_{13}H_{17}BrO_3$		Me_2	Br	OMe				OMe	88	18

TABLE 8. DERIVATIVES OF DIHYDROHALOGENONAPHTHOPYRANS[22]

Mol. formula	Isomer	Configuration	Substituents 1	2	3	4	6	m.p.
$C_{14}H_{13}BrO_2$	[1,2-*b*]	trans			Br	OMe		122–123
$C_{14}H_{13}BrO_2$	[2,1-*b*]	trans	OMe	Br				162–163
$C_{14}H_{13}ClO_2$	[1,2-*b*]	trans			Cl	OMe		64–65
$C_{14}H_{13}ClO_2$	[2,1-*b*]	trans	OMe	Cl				134–135
$C_{15}H_{15}BrO_2$	[1,2-*b*]	trans			Br	OEt		129–130
$C_{15}H_{15}BrO_2$	[2,1-*b*]	trans	OEt	Br				162–163
$C_{15}H_{15}ClO_2$	[1,2-*b*]	trans			Cl	OEt		65–66
$C_{15}H_{15}ClO_2$	[2,1-*b*]	trans	OEt	Cl				138–139
$C_{16}H_{16}BrClO_2$	[1,2-*b*]			Me_2	Br	OMe	Cl	78–79
$C_{16}H_{16}Cl_2O_2$	[1,2-*b*]			Me_2	Cl	OMe	Cl	142–143

VI. REFERENCES

1. R. E. Rindfusz, P. M. Ginnings, and V. L. Harnack, *J. Am. Chem. Soc.*, **42,** 157 (1920).
2. C. D. Hurd and W. A. Hoffman, *J. Org. Chem.*, **5,** 212 (1940).
3. P. Maitte, *Ann. Chim.* (France), **9,** 431 (1954).
4. H. Normant and P. Maitte, *C.R. Acad. Sci., Paris,* **234,** 1786 (1952).
5. A. Gabert and H. Normant, *C.R. Acad. Sci., Paris,* **235,** 1407 (1952).
6. L. W. Deady, R. D. Topsom, and J. Vaughan, *J. Chem. Soc.*, 5718 (1965).
7. G. Brancaccio, G. Lettieri, and R. Viterbo, *J. Heterocycl. Chem.*, **10,** 623 (1973).
8. O. Dann, G. Volz, and O. Huber, *Justus Liebigs Ann. Chem.*, **587,** 16 (1954).
9. I. A. R. Derrick, R. Livingstone, and B. J. McGreevy, unpublished observations (1971); I. A. R. Derrick, Ph.D. Thesis, Council for National Academic Awards, 1971.
10. Farbenfabriken Bayer A.G. *Br. Patent* 906, 483 (1962), *Chem. Abstr.*, **58,** 6806, (1963).
11. E. A. Vdovtsova, and A. G. Kakharov (USSR). Deposited Publ. 1972, VINITI 4340–72; *Chem. Abstr.*, **85,** 46313 (1976).
12. L. I. Smith, H. H. Hoehn, and H. E. Ungnade, *J. Org. Chem.*, **4,** 351 (1939).
13. Schering A.G. *Br. Patent* 758, 313 (1956); *Chem. Abstr.*, **51,** 9708 (1957).
14. W. D. Cotterill, J. Cottam, and R. Livingstone, *J. Chem. Soc.* (*C*), 1006 (1970).
15. R. Livingstone, D. Miller, and S. Morris, *J. Chem. Soc.*, 3094 (1960); R. Livingstone, *J. Chem. Soc.*, 76 (1962).
16. J. Cottam, R. Livingstone, and S. Morris, *J. Chem. Soc.*, 5266 (1965).
17. W. D. Cotterill, and C. Turner, unpublished observations (1972); C. Turner, Ph.D. Thesis, Council for National Academic Awards, 1972.
18. J. D. Hepworth and R. Livingstone, *J. Chem. Soc.* (*C*), 2013 (1966).
19. R. H. Hall and B. K. Howe, *J. Chem. Soc.*, 2886 (1959).
20. W. Baker, R. F. Curtis, and J. F. W. McOmie. *J. Chem. Soc.*, 76 (1951).
21. J. D. Hepworth, T. K. Jones, and R. Livingstone, unpublished observations, 1974; T. K. Jones, Ph.D. Thesis, Council for National Academic Awards, 1974.
22. M. Iqbal and R. Livingstone, unpublished observations, 1968, 1972; M. Iqbal, M.Phil. Thesis, Council for National Academic Awards, 1968; M. Iqbal, Ph.D. Thesis, Council for National Academic Awards, 1972.
23. J. B. Abbott, C. J. France, R. Livingstone, and D. P. Morrey, *J. Chem. Soc.* (*C*), 1472 (1967).
24. B. J. Bolger, A. Hirwe, K. G. Marathe, E. M. Philbin, M. A. Vickars, and C. P. Lillya, *Tetrahedron*, **22,** 621 (1966).

25. R. Binns, W. D. Cotterill, and R. Livingstone, *J. Chem. Soc.*, 5049 (1965).
26. R. Binns, W. D. Cotterill, I. A. R. Derrick, and R. Livingstone, *J. Chem. Soc., Perkin Trans. II*, 732 (1974).
27. R. Livingstone and S. Morris, unpublished observations, 1960; S. Morris, Ph.D. Thesis, University of London, 1960.
28. G. R. Clemo and N. D. Ghatge, *J. Chem. Soc.*, 4347 (1955).
29. H. Hofmann and G. Salbeck, *Chem. Ber.*, **103,** 2768 (1970).
30. I. Iwai and J. Ide, *Chem. Pharm. Bull.* (Tokyo), **11,** 1042 (1963).
31. J. R. Beck, R. Kwok, R. N. Bocher, A. C. Brown, L. E. Patterson, P. Pranc, B. Rockey, and A. Pohland, *J. Am. Chem. Soc.*, **90,** 4706 (1968); R. Kwok and A. Pohland, Ger. Offen. 1,933,153 (1970); *Chem. Abstr.*, **72,** 90422 (1970).

CHAPTER VI

Nitro- and Aminochromans

I. M. LOCKHART

BOC Limited, London SW19 3UF, U.K.

I. INTRODUCTION

This chapter is concerned with the chemistry of the chromans which contain nitro groups and those with amino or other basic groups in the molecule. In the case of the nitrochromans the nitro group is linked directly to a carbon atom of chroman (**1**), and in the majority of examples the nitro group is substituted in the aromatic ring. A wide variety of chromans with a basic group in the molecule has been described. As in *Chemical Abstracts* up to and including 1971, those compounds with an amino group attached directly to a carbon of chroman are referred to as chromanamines, prefixed with a numeral to indicate the position of the amino group. For example, 6-aminochroman is referred to as 6-chromanamine. The other type of chroman covered in this chapter is that with a basic group in a side chain.

The nitro- and aminochromans are being considered in the same chapter, as their chemistry is frequently intertwined. Reduction of nitrochromans has often been used for the preparation of chromanamines.

(**1**)
CAS Registry No. 254-04-6

The chromanamines provide the bulk of the interest in the groups of compounds discussed. Little work has been reported on 2-chromanamines. Although only a few groups of workers have studied the 3-chromanamines, a large number of compounds have been reported which show interesting pharmacological activity. A variety of 4-chromanamines and their derivatives have been described but the group has not assumed any great importance. The wealth of work that has been reported on chromanamines with the basic substituent in the aromatic ring centers on the tocopheramines, a specialized group of 6-chromanamines having the general formula (**2**). In addition to the 6-amino group, all have a methyl substituent and the 4,8,12-trimethyltridecyl group at position 2. The group is subdivided according to the methyl groups at positions 5, 7, and/or 8, as shown in Table 1.

R^3, R^2, H_2N, R^1, Me, $[(CH_2)_3CHMe]_3Me$

(**2**)

In each of the sections in this chapter the nitrochromans are discussed first; the 2-, 3-, and 4-chromanamines are treated in sequence followed by the chromanamines with the amino substituent in the aromatic ring (Bz-chromanamines). 2,3-Chromandiamines are included with the 2-chromanamines, and the 3,4-chromandiamines with the 3-chromanamines. The sections are concluded with the chemistry of chromans having a basic side chain.

TABLE 1. TOCOPHERAMINE NOMENCLATURE

Tocopheramine (**2**)	R^1	R^2	R^3
α-Tocopheramine	Me	Me	Me
β-Tocopheramine	Me	H	Me
γ-Tocopheramine	H	Me	Me
δ-Tocopheramine	H	H	Me

II. PHYSICAL PROPERTIES AND SPECTRA

1. Nitrochromans

The majority of the nitrochromans that have been isolated are crystalline solids, often with relatively low melting points. The uv spectrum of 7-nitrochroman showed λ_{max} at 237, 279, and 332 nm (ε 10,900, 6700, and 2700) when measured in 95% ethanol.[1]

Infrared data have been recorded for 2,2-dimethyl-6-nitrochroman as a KBr disc[2] and for 2,5,6,7-tetramethyl-2-(4,8,12-trimethyltridecyl)-8-nitrochroman in CCl_4.[3] In the ir spectra of 2-methoxy-6-nitrochroman, 2-ethoxy-6-nitrochroman, and 2-ethoxy-6,8-dinitrochroman, all of which were measured in KBr discs, bands at 1340, 1330, and 1340 cm^{-1}, respectively, were assigned to C–NO_2 bonds.[4]

Data from their nmr spectra have been reported for a number of 6-nitrochromans including 2,2-dimethyl-,[2] 2-methoxy-,[4] 2-ethoxy-,[4] 2-methyl-3-ethyl-,[5] 2-ethyl-3-methyl-,[5] 2-pentyl-,[5] 8-chloro-,[1] and 8-methyl-6-nitrochroman,[1] as well as for 6-nitrochroman itself.[1] Information is also available for 6,8-dinitrochroman[1] and for its 2-ethoxy analog.[4]

No molecular ion appeared in the mass spectrum of 2,5,6,7-tetramethyl-2-(4,8,12-trimethyltridecyl)-8-nitrochroman (MW 473); the 100% peak was at *m/e* 194. The other most significant peaks were *m/e* 459 (27.7%), 234 (23.9%), 177 (26.9%), 164 (24.6%), and 163 (23.7%).[3]

Pertinent information on data available on nitrochromanamines is discussion under the appropriate chromanamine.

2. Chromanamines

A. 2-*Chromanamines*

The majority of 2-chromanamines that have been described are of type (**3**)[6] or 3-halo-2-chromanamines of type (**4**).[7] Both groups are tertiary amines; the former are high-melting solids while the latter have melting points in the region of 100°.

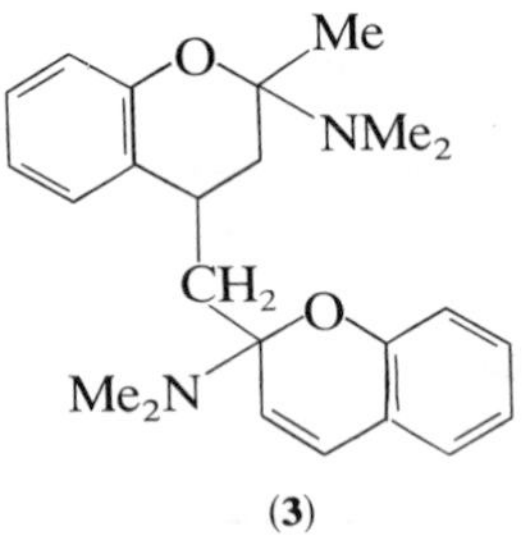

Me
NMe_2
CH_2
Me_2N

(**3**)

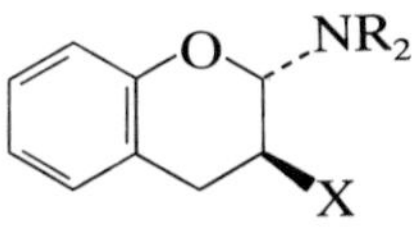

(4) 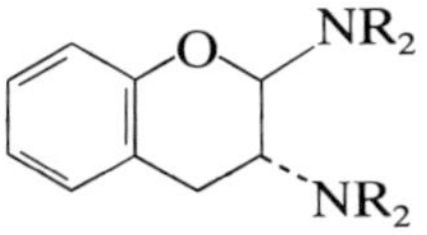 (5)

The ir spectrum of the *N,N*-dimethyl-2-chromanamine derivative (**3**) has been published.[6] Nuclear magnetic resonance data have been reported for 3-halo-2-chromanamines of type (**4**) and for 2,3-chromandiamines of type (**5**) and have been used to confirm the configuration of the products.[7]

B. *3-Chromanamines*

Most of the 3-chromanamines that have been reported have been isolated as crystalline hydrochlorides. The free bases of isomers of 2-methyl-3-chromanamine and their *N,N*-dimethyl analogs have been prepared as liquids which were purified by distillation in vacuo.

Proton magnetic resonance spectra of the cis and trans isomers of 2-methyl-3-chromanamine and its *N,N*-dimethyl analog have been recorded.[8] Interpretation of the results enabled the configurations and preferred conformations to be assigned to the free bases and their hydrochlorides and to the *N*-acetyl and *N*-phthaloyl derivatives of 2-methyl-3-chromanamine.[8] The assignments to the various conformations, (**6**)–(**9**), are listed in Table 2. Similar work has more recently been carried out and the ir and nmr data reported for 3-chromanamine, its hydrochloride, and its *N*-acetyl and *N*-phthaloyl derivatives.[9] The assignment of the preferred conformations between types (**10**) and (**11**) are given in Table 2.

TABLE 2. PREFERRED CONFORMATIONS OF 3-CHROMANAMINES

Chromanamine	Type of conformation	Ref.
3-Chromanamine	(**10**) $R^1 = R^2 = H$	9
Hydrochloride	(**11**) $R^1 = R^2 = H$	9
N-Acetyl	(**10**) $R^1 = H, R^2 = COMe$	9
N-Phthaloyl	(**11**) $R^1R^2 = COC_6H_4CO$	9
cis-2-Methyl-3-chromanamine	(**6**) $R^1 = R^2 = H$	8
Hydrochloride	(**6**) $R^1 = R^2 = H$	8
N-Acetyl	(**6**) $R^1 = H, R^2 = COMe$	8
N-Phthaloyl	(**6**) $R^1R^2 = COC_6H_4CO$	8
trans-2-Methyl-3-chromanamine	(**8**) $R^1 = R^2 = H$	8
Hydrochloride	(**7**) $R^1 = R^2 = H$	8
N-Acetyl	(**7**+**8**) $R^1 = H, R^2 = COMe$[a]	8
N-Phthaloyl	(**8**) $R^1R^2 = COC_6H_4CO$	8
cis-2,*N,N*-Trimethyl-3-chromanamine	(**9**) $R^1 = R^2 = Me$	8
Hydrochloride	(**6**) $R^1 = R^2 = Me$	8
trans-2,*N,N*-Trimethyl-3-chromanamine	(**8**) $R^1 = R^2 = Me$	8
Hydrochloride	(**7**) $R^1 = R^2 = Me$	8

[a]The tentative suggestion was made that conformation (**7**) predominated in the mixture.

(**6**) $R^3 = H, R^4 = Me$
(**7**) $R^3 = Me, R^4 = H$

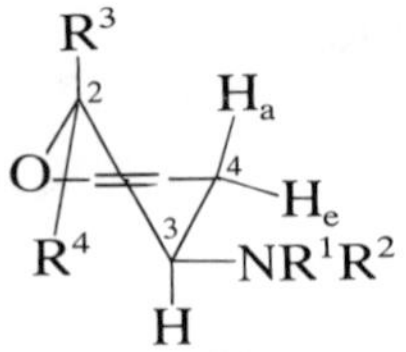

(**8**) $R^3 = H, R^4 = Me$
(**9**) $R^3 = Me, R^4 = H$

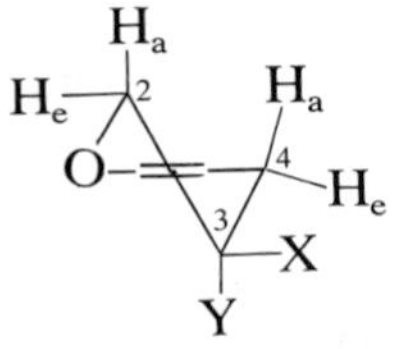

(**10**) $X = NR^1R^2, Y = H_a$
(**11**) $X = H_e, Y = NR^1R^2$

Nuclear magnetic resonance data have been reported for 2-hydroxy-3-chromanamines of type (**12**).[7]

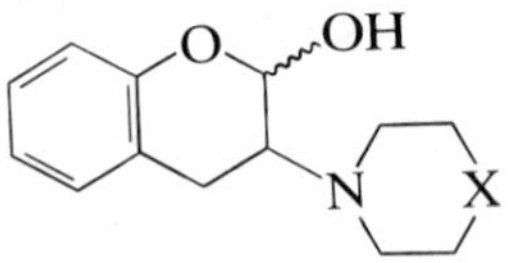

(**12**)

The pK_a values of a number of 2-methyl-3-chromanamines have been determined[8] and are listed in Table 3. Optical data for (+)-R-3-chromanamine hydrochloride and its (−)-S-enantiomer, stereospecifically synthesized as described in a later section (III, 2Bc), were reported as $[\alpha]_D^{25} + 58.7°$ ($c = 0.8H_2O$) and $[\alpha]_D^{25} - 47°$ ($c = 0.83H_2O$), respectively.[10]

TABLE 3. pK_a VALUES OF SOME 2-METHYL-3-CHROMANAMINES

3-Chromanamine	pK_a
cis-2-Methyl	7.61
trans-2-Methyl	7.64
cis-2,*N*,*N*-Trimethyl	6.51
trans-2,*N*,*N*-Trimethyl	7.20

C. 4-*Chromanamines*

The uv spectrum of 6-hydroxy-2,2,5,7,8-pentamethyl-4-chromanamine has been determined in ethanol;[11] the nmr spectrum of 4-chromanamine in CCl_4 has also been reported.[12] pK_a values for a number of 7-methoxy-4-chromanamines are available and are listed in Table 4.[13]

TABLE 4. pK_a VALUES OF SOME 7-METHOXY-4-CHROMANAMINES

4-Chromanamine	pK_a
7-Methoxy	8.62
2-Methyl-7-methoxy	8.65
2-Isopropyl-7-methoxy	8.60
2-*t*-Butyl-7-methoxy	8.55
3-Methyl-7-methoxy	8.60

D. *Bz-Chromanamines*

Ultraviolet, ir, and nmr data for a range of chromanamines with the basic group attached directly to the aromatic ring of the chroman nucleus have been published and the relative references are given in Table 5.

Refractive indices have been reported for tocamin [2-methyl-2-(4,8,12-trimethyltridecyl)-6-chromanamine], for α-, β-, γ-, and δ-tocopheramine, and for several of their *N*-alkyl derivatives.[14,15]

Mass spectra have been recorded for the *N*-acetyl and *N*-trimethylsilyl derivatives of 2,5,6,7-tetramethyl-2-(4,8,12-trimethyltridecyl)-8-chromanamine.[3]

Thermal activation energies have been determined for a number of tocopheramines.[16]

ABLE 5. UV, IR, AND NMR SPECTRAL DATA FOR Bz-CHROMANAMINES

	Ref. to data on		
hromanamine	uv	ir	nmr
-Chromanamine	1		
,6-Chromandiamine (diacetyl deriv.)	1		
,2-Dimethyl-6-chromanamine		2	2
-Methyl-*N*-acetyl-6-chromanamine	1		
ocamin (**2**, $R^1 = R^2 = R^3 = H$)	14, 17		
-Tocopheramine (**2**, $R^1 = R^2 = R^3 = Me$) and *N*-alkyl derivatives	14, 17–18	19	
-Tocopheramine (**2**, $R^1 = R^3 = Me$, $R^2 = H$) and *N*-alkyl derivatives	14, 17		
-Tocopheramine (**2**, $R^1 = H$, $R^2 = R^3 = Me$) and *N*-alkyl derivatives	14, 17		
-Tocopheramine (**2**, $R^1 = R^2 = H$, $R^3 = Me$) and *N*-alkyl derivatives	14, 17		
,5,7,8-Tetramethyl-2-tridecyl-6-chromanamine	19		
-Ethoxy-6-chromanamine		4	4
-Nitro-6-chromanamine hydrochloride	1		
-Nitro-*N*-acetyl-6-chromanamine	1	1	1
-Nitro-6-chromanamine	1		
-Nitro-*N*-acetyl-6-chromanamine	1	1	1
-Nitro-8-methyl-*N*-acetyl-6-chromanamine		1	1
,8-Dinitro-*N*-acetyl-6-chromanamine	1	1	1
-Nitro-*N*-acetyl-6-chromanamine	1	1	1
,7-Chromandiamine (diacetyl deriv.)	1		
,8-Chromandiamine	1		
,8-Chromandiamine (diacetyl deriv.)	1		
-Ethoxy-6,8-chromandiamine		4	4
V-Acetyl-8-chromanamine	1		
-Chloro-8-chromanamine	1		1
-Nitro-8-chromanamine	1		
-Nitro-*N*-acetyl-8-chromanamine	1	1	1
-Nitro-8-chromanamine	1		1
-Nitro-*N*-acetyl-8-chromanamine	1		

3. Chromans with a Basic Side Chain

Little in the way of spectral data is available for the various chromans with a basic side chain. Ultraviolet information has been reported for the quaternary base (**13**)[20] and for the mixture of imine hydrochlorides (**14a**) and (**14b**).[21] Infrared and nmr data have also been recorded for the latter mixture.[21] The 8-dipropylaminomethylchromans (**15a**) and (**15b**) were characterized by their uv, ir, and nmr data, and in the former case mass spectral details were also provided.[22] The position of the OH and NH bands in the ir spectrum of the 2-aminomethylchroman derivative (**16**) has also been reported.[23]

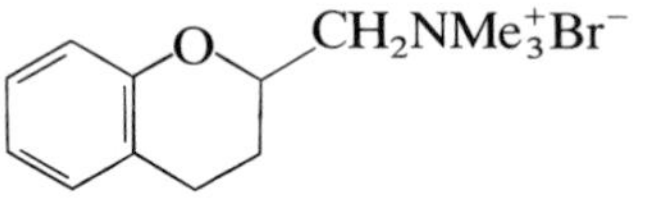

(**13**)

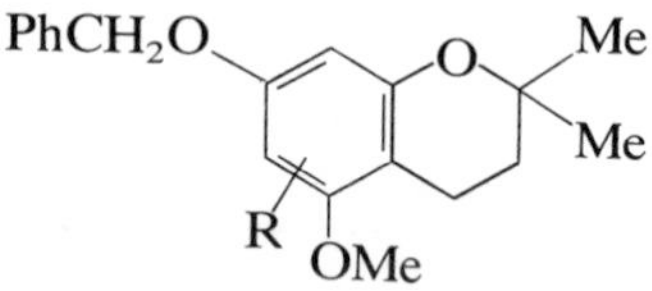

(**14a**) R = 6-CMe=$NH_2^+Cl^-$
(**14b**) R = 8-CMe=$NH_2^+Cl^-$

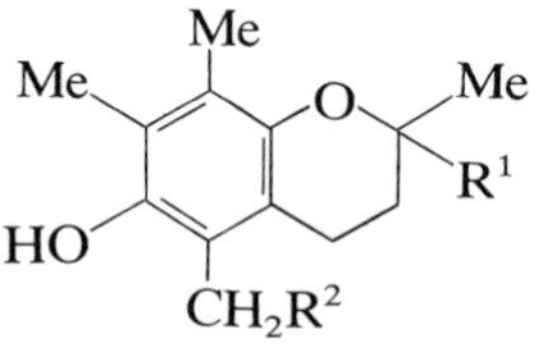

(**15a**) $R^1 = CH_2[CH_2CH{=}CHMeCH_2]_3H$, $R^2 = NPr_2$
(**15b**) $R^1 = CH_2[CH_2CH{=}CHMeCH_2]_8H$, $R^2 = NPr_2$

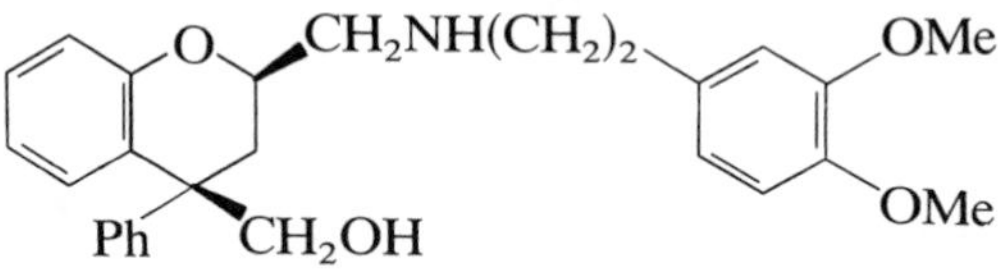

(**16**)

4. Miscellaneous Basic Chromans

The immonium salts (**17a**) and (**17b**) have been characterized by their ir and nmr spectra.[24]

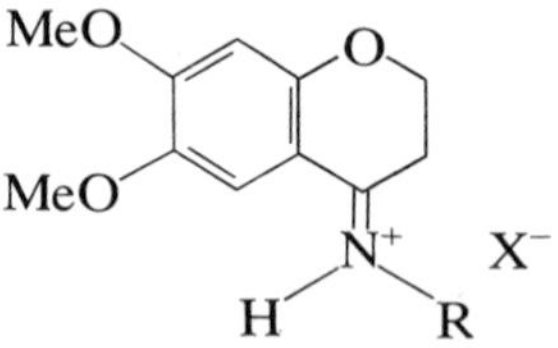

(**17a**) R = H, X = Cl
(**17b**) R = Me, X = I

III. METHODS OF PREPARATION

1. Nitrochromans

2-Methyl-3-nitrochroman (**19**) has been prepared by sodium borohydride reduction of 2-methyl-3-nitro-2*H*-1-benzopyran (**18**),[25] which in its turn was obtained by condensation of salicylaldehyde with 2-chloro-1-nitropropane.[26] With the exception of this example, the nitrochromans that have been described all appear to have the nitro group in the aromatic ring (Bz-nitrochromans), and their preparation is described in the following sections.

(**18**) → (**19**)

A. *Nitration of Chromans*

Nitration of chroman (**1**) with nitric acid affords either the 6-nitro derivative (**20**) or the 6,8-dinitrochroman (**21**) according to the conditions used. Dropwise addition of 60% nitric acid at 15–25° afforded 6-nitrochroman (**20**) as a yellow powder in 35% yield,[27,28] while 6,8-dinitrochroman (**21**) was isolated in 90% yield when chroman was heated with fuming nitric acid at 30–35°.[29] Although a large number of nitrochromans have been prepared as derivatives of chromans,[30–33] their structures have not been specifically identified. The use of concentrated nitric acid with a range of chromans afforded mononitroderivatives, except with 6-methoxy and 8-methoxychroman, from which dinitro compounds were isolated.[32] Nitration in acetic acid has been claimed to give the mononitro-6-methoxychroman,[32] although nitration with concentrated nitric acid in glacial acetic acid has given dinitro-6-methoxychroman[31] and dinitro-8-methoxychroman.[33] Unspecified nitro derivatives of 2,2-dichromanylethers (**22**) have been prepared.[30] By addition of the chroman to 60% nitric acid, the 6-nitro analogs of 2,5-dimethyl- and 7,8-dimethylchroman were prepared in 45% yield.[34] Nitric acid in glacial acetic acid at 50° was used to convert 2,2-dimethylchroman to the 6-nitro derivative in 47% yield.[2] 7-Methoxy-2,2-dimethylchroman was nitrated with a similar mixture, and, although not isolated, subsequent hydrogenation to 7-methoxy-2,2-dimethyl-6-chromanamine confirmed that the 6-nitro derivative had been formed.[35] 2-Methoxy-6-nitrochroman and the corresponding 2-ethoxy analog have been prepared in yields of just over 20% by nitration of the chroman in chloroform with concentrated nitric acid.[4]

A systematic study of the products of nitration of a variety of chromans has been reported;[1] it was demonstrated that the 6- and 8-positions were

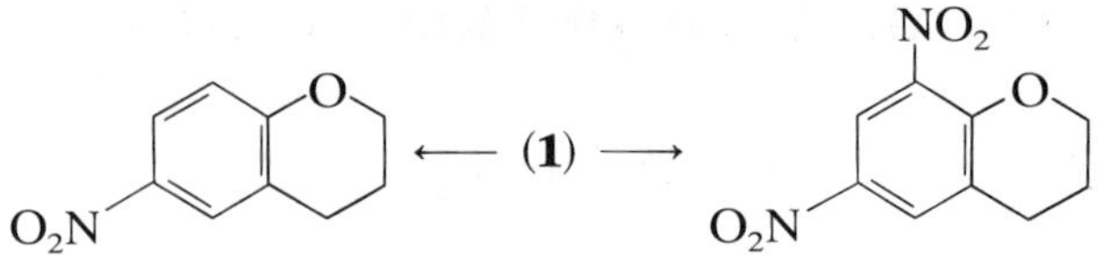

NO_2
O
O
$\longleftarrow$ (**1**) $\longrightarrow$
O_2N
O_2N
(**20**) (**21**)

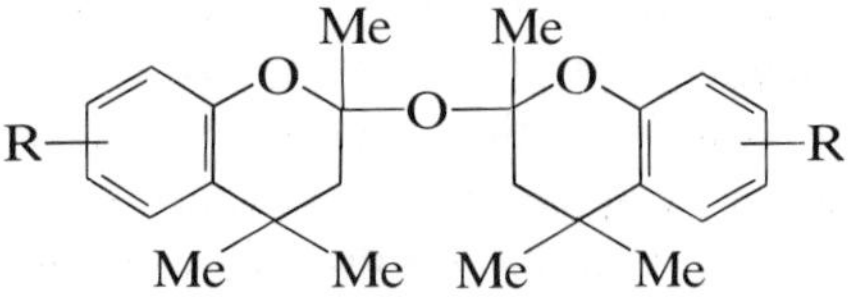

Me Me
O O
R— —O— —R
Me Me Me Me
(**22**)

involved in the nitration of chroman. Nitration of *N*-acetyl-8-chromanamine with fuming nitric acid afforded both the 5- and 6-nitro derivatives, while the 5-, 7-, and 8-nitro derivatives were all formed from *N*-acetyl-6-chromanamine. Further nitration of the 5-nitro derivative did not occur, but both the 7-nitro and 8-nitro derivatives afforded *N*-acetyl-7,8-dinitro-6-chromanamine. Nitration of *N*-acetyl-8-methyl-6-chromanamine with fuming nitric acid yielded the 7-nitro derivative without evidence of the formation of the 5-nitro compound.[1]

Although the majority of direct nitrations have been carried out with nitric acid, when 2,5,6,7-tetramethyl-2-(4,8,12-trimethyltridecyl)chroman (**23**) in acetic anhydride at 0° was treated with cupric nitrate and allowed to warm up to room temperature, the 8-nitro analog (**24**) was obtained as a yellow oil in 38% yield.[3]

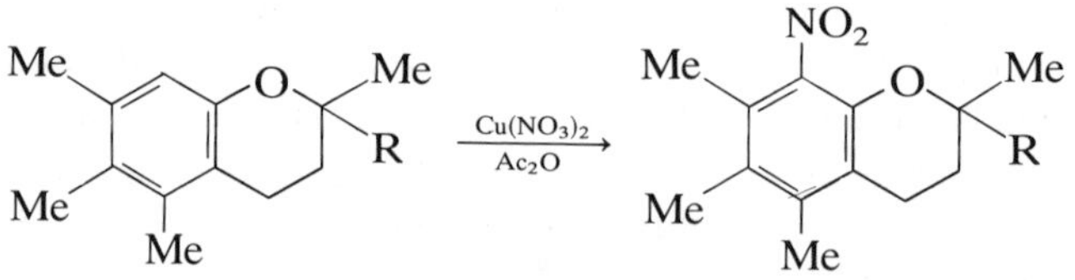

Me O Me $\xrightarrow[Ac_2O]{Cu(NO_3)_2}$ NO_2 Me O Me
Me R Me R
Me Me
(**23**) R = $[(CH_2)_3CHMe]_3Me$ (**24**)

B. *From Nitrophenols*

There are a number of examples of the condensation of 4-nitrophenols (**25**) with an unsaturated compound such as phytol (**26a**)[36–38] or a halogen derivative (**26b**)[38] in the presence of an acidic condensing agent to give 2-methyl-2-(4,8,12-trimethyltridecyl)-6-nitrochromans of type (**27**).

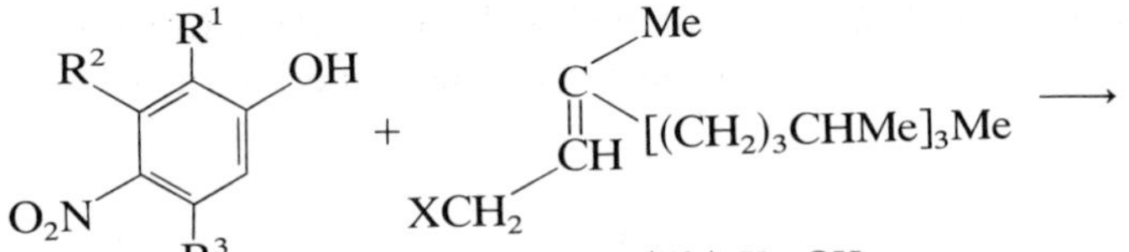

R^1, R^2, OH, O_2N, R^3 (25) + Me, C=CH, $[(CH_2)_3CHMe]_3Me$, XCH_2 ⟶

(26a) X = OH
(26b) X = Br

R^1, R^2, O, Me, $[(CH_2)_3CHMe]_3Me$, O_2N, R^3

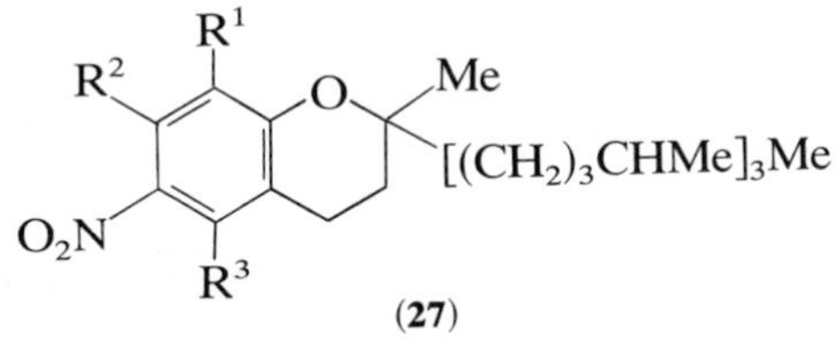

(27)

The reaction of nitrophenols of type (**28**) with an appropriate unsaturated compound (**29**) in the presence of stannic chloride in 1,2-dichloroethane has been used to prepare 6-nitrochromans substituted in the heterocyclic ring (**30**).[5]

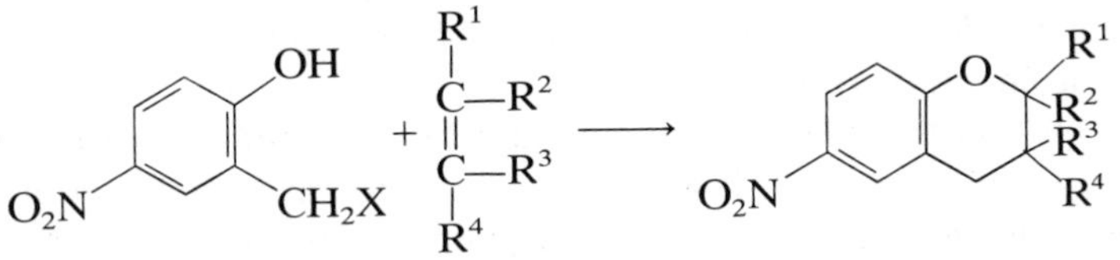

O_2N, OH, CH_2X (28) + R^1, C—R^2, C—R^3, R^4 (29) ⟶ O_2N, O, R^1, R^2, R^3, R^4 (30)

Reaction of a benzyl alcohol derivative (**28**, X = OH) with an unsaturated ether (**29**, R^1 = OEt, R^2 = R^3 = R^4 = H) afforded the nitrochroman (**30**, R^1 = OEt, R^2 = R^3 = R^4 = H) in 60% yield.[4] 6,8-Dinitrochromans with an ethoxy substituent at C-2 were similarly prepared but in lower yield.[4]

2-Phytyl-4-nitrophenol (**28**, X = CH=CMe$[(CH_2)_3CHMe]_3$Me), prepared by condensation of sodium nitromalonaldehyde with phytylacetone in ethanolic alkali, was cyclized to 2-methyl-2-(4,8,12-trimethyltridecyl)-6-nitrochroman (**27**, R^1 = R^2 = R^3 = H) by refluxing in acidic alcohol solution.[39] The 8-methyl analog was similarly prepared.[39]

C. *Miscellaneous*

Cyclization of 3-(4-nitrophenoxy)-3-methylbut-1-yne (**31**) in diethylaniline at 220° afforded 2,2-dimethyl-6-nitro-2*H*-1-benzopyran (**32**).[40] Treatment of the latter with bromine in carbon tetrachloride gave *trans*-3,4-dibromo-2,2-dimethyl-6-nitrochroman (**33**),[40] while when it was dissolved in ethanolic hydrochloric acid and treated with iron powder, the mixture refluxed spontaneously, and the 2,2-dimethyl-6-chromanamine formed was isolated as its *N*-acetyl derivative, presumably through the intermediate 6-nitrochroman (**33**).[41]

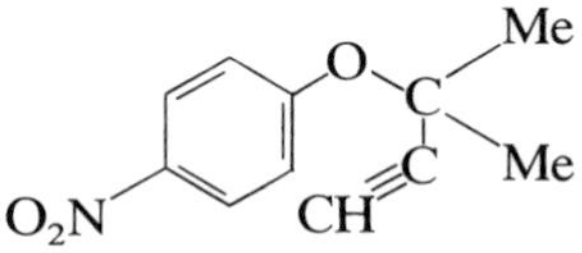

(**31**) 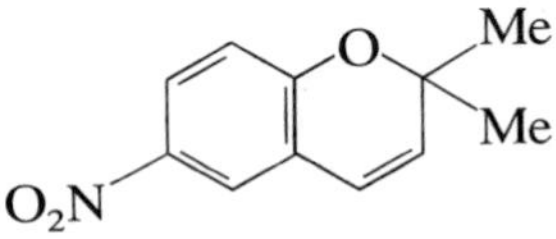 (**32**) 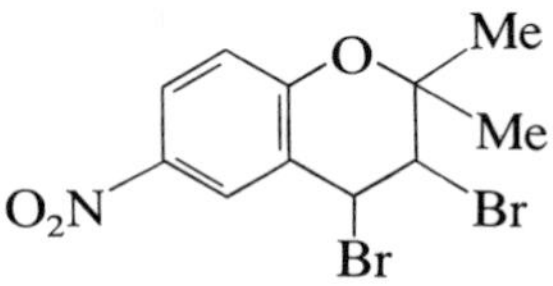 (**33**)

2. Chromanamines

A. *2-Chromanamines*

A number of methods have been used for the preparation of 2-chromanamines. 2,4,4,7-Tetramethyl-2-chromanamine (**34**, $X=NH_2$) was obtained by Hofmann degradation of the amide (**34**, $X=CONH_2$) of 2,4,4,7-tetramethylchroman-2-carboxylic acid.[42]

In studies on the reaction of halohydrins of type (**35**), X = Br or Cl, with secondary amines in ether, it was shown that with one equivalent of base the 2-chromanamine derivative (**4**) was obtained in good yield.[7] Nuclear magnetic resonance studies showed that the products had a *trans*-diequatorial configuration. When the bromohydrin was used, a *gem*-diamino derivative of 2-methyl-2,3-dihydrobenzofuran (**36**) was also formed. 2,3-Chromandiamines of type (**5**) were obtained when a large excess of the amine was used in the reaction.

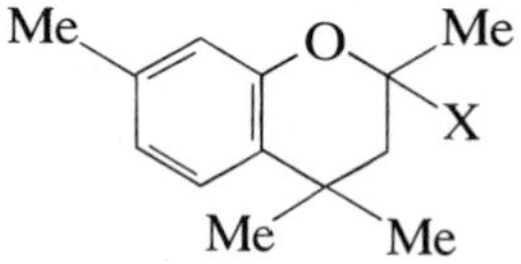

(**34**) 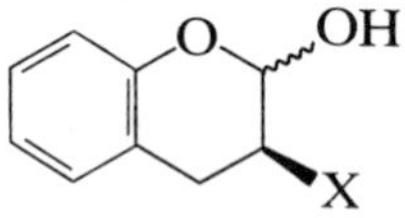 (**35**) 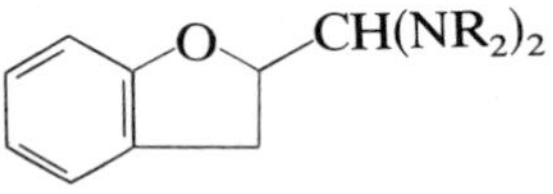 (**36**)

Salicylaldehyde and dry acetone react with dimethylamine in ethanol to give a bis-2-chromanamine having structure (**3**); the mechanism of the reaction has been discussed.[6] On hydrolysis with 2 *N*-hydrochloric acid at room temperature for five days, one of the dimethylamino groups was replaced by hydroxyl.

Pyran-containing fused ring systems have been prepared by refluxing a phenolic Mannich base with an enamine in dioxan.[43] With 1-dimethylaminomethyl-2-naphthol (**37**) and *N*-isobutenylmorpholine (**38**), the 2,3-dihydro-3-morpholino-1*H*-naphtho[2,1-*b*]pyran (**39**) was isolated and characterized.[43]

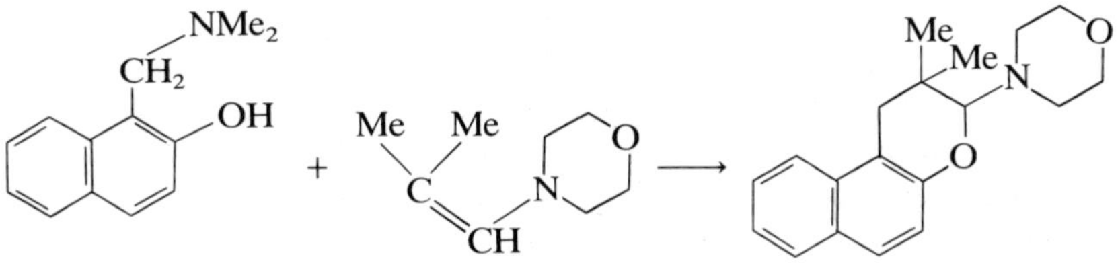

(37) (38) (39)

B. *3-Chromanamines*

The principal methods used to prepare 3-chromanamines have involved reduction. 3-Nitro-2*H*-1-benzopyrans, 3-amino-4-chromanone hydrochlorides, and 3-amino-4-chromanols have been converted to 3-chromanamines using a variety of agents, the method of choice often being determined by the stereochemical configuration required in the product. The preparation of 3-amino-4-chromanone hydrochlorides and 3-amino-4-chromanols has been described in an earlier volume in this series.[44,45]

a. Reduction of 3-Nitro-2*H*-1-benzopyrans. 2-Methyl-3-chromanamine (**40**, $R^1 = Me$, $R^2 = R^3 = H$) hydrochloride was first prepared by hydrogenation of 2-methyl-3-nitro-2*H*-1-benzopyran (**18**) in the presence of Raney nickel,[26] and it melted at 216–217.5°. However when reduction of the nitro compound (**18**) was effected with lithium aluminum hydride in ether and the resulting 2-methyl-3-chromanamine converted to its hydrochloride, it melted at 260–261°.[8,25] The same product was obtained using lithium borohydride in diethylene glycol dimethyl ether or dibutyl ether.[25]

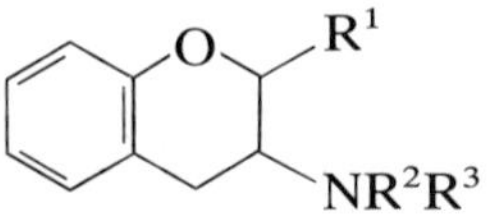

(40) 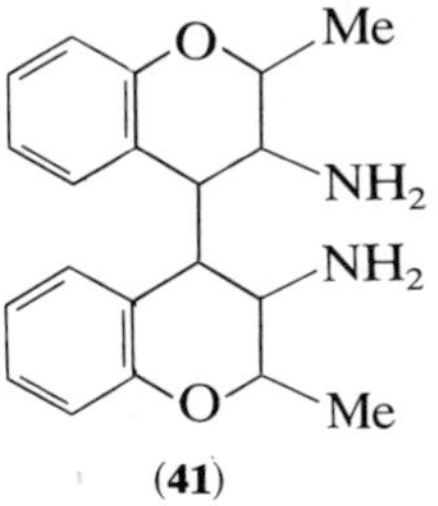 (41)

Conformational studies have established that the higher-melting hydrochloride has the cis configuration (originally designated α) and the lower-melting compound is the trans isomer (originally designated β).[8] By-products have been isolated from both methods of reduction. A substantial quantity of the dimer (**41**) was formed in the catalytic hydrogenation, while with lithium aluminum hydride, some 3-hydroxyamino-2-methylchroman (**40**, $R^1 = Me$, $R^2 = OH$, $R^3 = H$) was formed.[8]

b. Reduction of 3-Nitrochromans. *cis*-2-Methyl-3-chromanamine has been prepared by reduction of 2-methyl-3-nitrochroman with lithium aluminum hydride.[25]

c. REDUCTION OF 3-AMINO-4-CHROMANONE HYDROCHLORIDES. 3-Amino-4-chromanone (**42**, $R^1 = R^2 = H$) hydrochloride has been reduced to 3-chromanamine by catalytic hydrogenation in ethanol–acetic acid–sulfuric acid in the presence of palladium on charcoal.[46] 7-Methoxy-3-chromanamine was prepared similarly. Hydrogenation of 3-acetamido-4-chromanones (**42**, $R^1 = Ac$, $R^2 = H$) in acetic acid in the presence of palladium on charcoal and perchloric acid afforded *N*-acetyl-3-chromanamines (**40**, $R^1 = R^2 = H$, $R^3 = Ac$) which were converted to the corresponding 3-chromanamines by acid hydrolysis.[9,47]

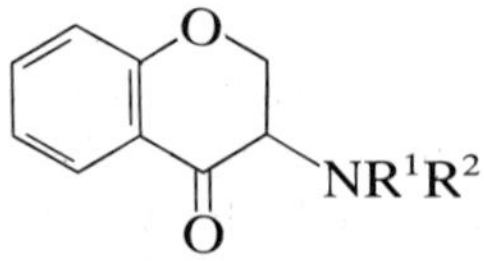

O
NR¹R²
O

(**42**)

The synthesis of optical enantiomers of 3-chromanamine hydrochloride has been elegantly achieved[10] by the sequence of reactions illustrated in Scheme 1. The chromanamines were finally obtained from the corresponding phthalimidochromanones by catalytic hydrogenation followed by removal of the phthaloyl protecting group.

The reduction of 3-amino-4-chromanols, themselves often prepared from 3-amino-4-chromanone hydrochlorides, is described in the next section.

d. REDUCTION OF 3-AMINO-4-CHROMANOLS. Many 3-chromanamines have been prepared by catalytic hydrogenation of 3-amino-4-chromanols (prepared by reduction of 3-amino-4-chromanone hydrochlorides with sodium borohydride or lithium aluminum hydride) in glacial acetic acid–concentrated sulfuric acid in the presence of palladium on charcoal at 60°.[46,48]

e. MISCELLANEOUS METHODS. *N,N*-Dimethyl-3-chromanamine was prepared by reaction of 3-bromochroman with dimethylamine.[49]

2-Hydroxy-3-morpholinochroman (**12**, X = O) and 2-hydroxy-3-piperidinochroman (**12**, $X = CH_2$) have been obtained in good yield by acid hydrolysis of the corresponding 2,3-chromandiamine analogs (**5**, NR_2 = morpholino and **5**, NR_2 = piperidino).[7]

Although 3-chromanones are not among the most easily accessible of compounds,[50] their reductive amination is clearly a potential route to the 3-chromanamines. *N*-Cyclopropyl-3-chromanamine has been prepared by catalytic hydrogenation of a mixture of cyclopropylamine and 3-chromanone in ethanol.[46] 3-Chromanamine hydrochloride has also been prepared by catalytic hydrogenation of 3-chromanone oxime in ethanol in the presence of Raney nickel.[51] Direct halogenation has been used to prepare 3-chromanamines with chloro substituents in the aromatic ring.[46,48]

N-Alkylated 3-chromanamines have been prepared by conventional procedures. *N*-Methylation of the amino group has been effected by the action

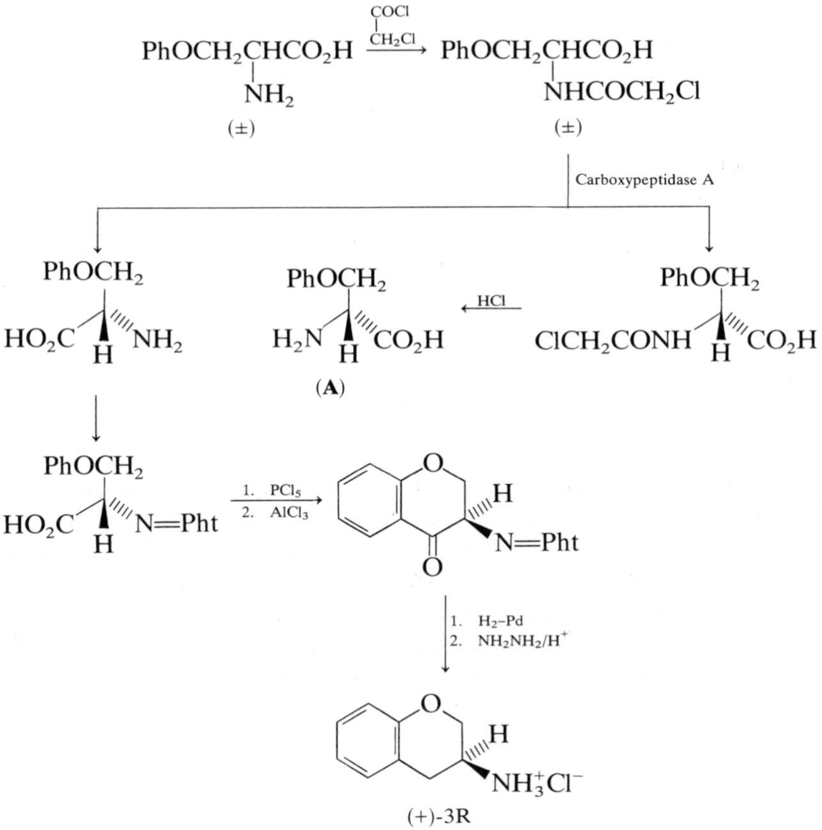

Scheme 1. Synthesis of enantiomers of 3-chromanamine hydrochloride. [The (−)-3S enantiomer was similarly obtained from the amino acid (A).]

of refluxing formic acid and formaldehyde or by catalytic hydrogenation of a mixture of the amine and formaldehyde in ethanol in the presence of palladium on charcoal.[8,46,52] Using acetaldehyde in place of formaldehyde, *N*,*N*-diethyl-3-chromanamines were obtained, while replacement of the aldehyde with acetone afforded a mono-*N*-isopropyl-3-chromanamine.[46] *N*-Monomethyl- and *N*-monoethyl-3-chromanamines have been prepared by lithium aluminum hydride reduction of an appropriate *N*-acylchromanamine.[46,48]

f. 3,4-Chromandiamines. Very little work has been reported on the preparation of 3,4-chromandiamines. Reduction of 3-amino-4-chromanone oxime (**43**) with lithium aluminum hydride afforded 3,4-chromandiamine (**44**), which was isolated as the dihydrochloride.[53] The bisacetyl derivative of 3,4-chromandiamine was prepared as its cis and trans isomers by catalytic hydrogenation of 3-acetamido-4-chromanone oxime in acetic acid in the presence of palladium on charcoal. Hydrolysis of the trans isomer with ethanolic hydrogen chloride, followed by evaporation, and reaction of the residue with phosgene in the presence of potassium carbonate afforded *trans*-2-oxoimidazolido[4,5-*d*]chroman (**45**), which on heating with 0.1 *N* hydrochloric acid afforded *trans*-3,4-chromandiamine dihydrochloride.[54]

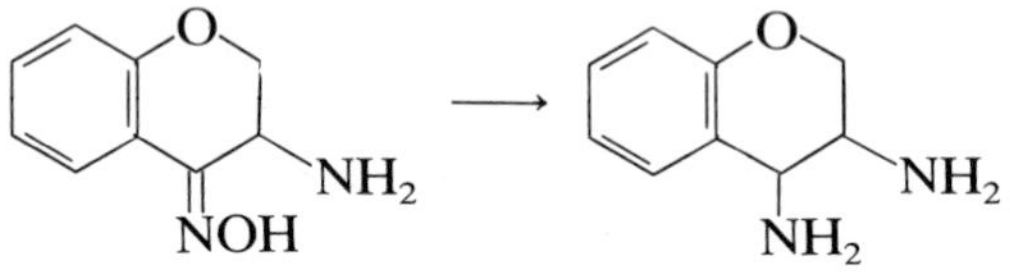

(**43**) (**44**) 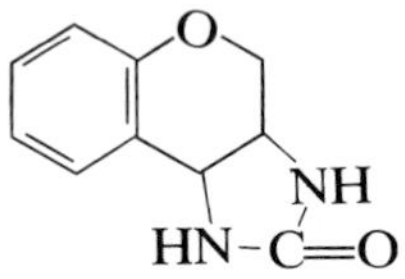 (**45**)

C. *4-Chromanamines*

4-Chromanamines (**46**) have in the majority of cases been prepared by reduction of 4-chromanone oximes. This reaction has been discussed in an earlier volume[44] (Section VI, 15B). Essentially, the use of lithium aluminum hydride leads to the formation of a mixture of the 4-chromanamine and the 2,3,4,5-tetrahydro-1,5-benzoxazepine (**47**). Catalytic hydrogenation affords the chromanamine. Other methods of preparation include reductive amination of 4-chromanones, the action of amines on 4-halochromans, and the action of alkali on chroman-4-carboxamide.

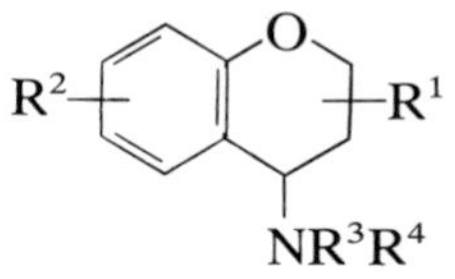

(**46**) 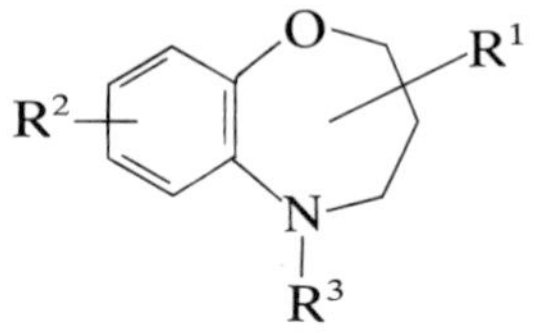 (**47**)

a. Catalytic Hydrogenation of 4-Chromanone Oximes. Examples of the preparation of 4-chromanamines by catalytic hydrogenation of 4-chromanone oximes are listed in Table 6. Yields are normally in the range of 70–85%.

b. Reduction of 4-Chromanone Oximes with Lithium Aluminum Hydride. Reduction of 4-chromanone oximes with lithium aluminum hydride normally affords a mixture of a 4-chromanamine (**46**) with the 2,3,4,5-tetrahydro-1,5-benzoxazepine (**47**). The mixed amines have been separated as the 4-toluenesulfonyl derivatives[57] by recrystallization of the hydrochlorides,[13,58] by utilizing the differences in basicity of the two amines,[11] and by chromatography on silica gel.[12] Various examples of this reduction and the respective yields of the two alternatives amines are given in Table 7. The results are probably too few to make any valid predictive generalizations.

TABLE 6. CATALYTIC HYDROGENATION OF 4-CHROMANONE OXIMES

4-Chromanone oxime	Solvent	Catalyst	Yield of 4-chrom anamine (%)	Ref.
Parent	EtOH	Ni	72	55
2,2-Dimethyl	EtOH	Ni	12	11
7-Methoxy	HOAc + $HClO_4$[a]	Pd–C	70	13
2-Methyl-7-methoxy	HOAc + $HClO_4$[a]	Pd–C	80	13
3-Methyl-7-methoxy	HOAc + $HClO_4$[a]	Pd–C	85	13
2-Isopropyl-7-methoxy	HOAc + $HClO_4$[a]	Pd–C	72	13
2-*t*-Butyl-7-methoxy	HOAc + $HClO_4$[a]	Pd–C	70	13
8-Phenyl	MeOH + HCl	Pd–C		56

[a]The acetic acid used as solvent contained about 0.7% perchloric acid.

However there are indications that substituents in the 2-position lead to a higher proportion of the 4-chromanamine in the mixed product and that this proportion increases with the bulk of the 2-substituent. Findings with the lithium aluminum hydride reduction of flavanone oxime also fit in with this pattern; 4-aminoflavan (**46**, $R^1 = 2\text{-Ph}$, $R^2 = R^3 = R^4 = H$) hydrochloride was isolated in 50% yield and 2-phenyl-2,3,4,5-tetrahydro-1,5-benzoxazepine (**47**, $R^1 = Ph$, $R^2 = R^3 = H$) hydrochloride in 9% yield.[58]

c. Reduction of 4-Chromanone Oximes with sodium amalgam 2,2,5,7,8-Pentamethyl-6-hydroxy-4-chromanone oxime, which was apparently resistant to a number of reducing agents, was converted to

TABLE 7. REDUCTION OF 4-CHROMANONE OXIMES WITH LITHIUM ALUMINUM HYDRIDE

4-Chromanone oxime	Yield (%) of 4-Chromanamine (**46**)	Yield (%) of 1,5-Benzoxazepine (**47**)	Ref.
Parent	16	34	57
	16	60	12
2,2-Dimethyl	57	[a]	11
2,2,6-Trimethyl	42	17.6	11
3-Methyl	23	54	58
2,2,5,7,8-Pentamethyl-6-hydroxy (di-OAc)[b]	[c]	56	11
7-Methoxy	—	70	13
2-Methyl-7-methoxy	10	50	13
3-Methyl-7-methoxy	—	85	13
2-Isopropyl-7-methoxy	50	35	13
2-*t*-Butyl-7-methoxy	80	—	13

[a]Could not be isolated.
[b]The diacetate was used in order to avoid the separation of the insoluble phenate during the lithium aluminum hydride reduction.
[c]Not isolated; sample impure.

2,2,5,7,8-pentamethyl-6-hydroxy-4-chromanamine in 25% yield by the action of sodium amalgam in acetic acid.[11]

d. Reductive Amination of 4-Chromanones. 4-Chromanamines have been prepared by catalytic hydrogenation of a 4-chromanone in the presence of ammonia or a primary or secondary amine. Hydrogenation of 8-phenyl-4-chromanone in methanol in the presence of ethylamine with Raney nickel as a catalyst afforded *N*-ethyl-8-phenyl-4-chromanamine (**46**, $R^1 = R^3 = H$, $R^2 = 8\text{-Ph}$, $R^4 = Et$).[56] Palladium on charcoal and platinum chloride have also been used as the catalyst in similar reductive aminations.[59]

4-Chromanone has been shown to resist imine formation under a variety of conditions.[24] However when 3,4-dimethoxyphenol (**48**) and an excess of vinyl cyanide (**49**) were heated with sodium at 140° for 5 hr, the phenoxyethyl cyanide (**50**) was obtained in 42% yield. Subsequent treatment with ethereal hydrogen chloride formed the imine hydrochloride (**17a**) in almost quantitative yield, which on refluxing with methyl iodide and sodium carbonate in 1,2-dimethoxyethane was converted to 6,7-dimethoxy-4-chromanone-methylimmonium iodide (**17b**).[24]

MeO, MeO, OH + CH_2=CH–C≡N → MeO, MeO, O–CH$_2$CH$_2$–C≡N → (**17a**) → (**17b**)

(**49**) (**50**)

e. Reaction of Amines with 4-Halochromans. *N,N*-Diethyl-4-chromanamine has been prepared by reaction of an excess of diethylamine with 4-bromochroman.[60] 4-Chloro-8-phenylchroman, prepared by the action of thionyl chloride on 8-phenyl-4-chromanol, has been converted to *N,N*-dimethyl-8-phenyl-4-chromanamine (**46**, $R^1 = H$, $R^2 = 8\text{-Ph}$, $R^3 = R^4 = Me$) by the action of dimethylamine at room temperature in a sealed tube.[56] In preparing analogs the dimethylamine has been replaced by diethylamine or ethylamine. Treatment of 2,6,8-trimethyl-4-chromanol with phosphorus tribromide afforded 4-bromo-2,6,8-trimethylchroman, which gave the 4-piperidinochroman (**46**, $R^1 = 2\text{-Me}$, $R^2 = 6,8\text{-diMe}$, $R^3R^4 = —(CH_2)_5—$) on reaction with excess piperidine at room temperature.[56]

f. Hofmann Degradation of Chroman-4-carboxamides. Reaction of 4-phenyl-4-chromancarboxamide with aqueous potassium hypobromite afforded 4-phenyl-4-chromanamine in 64% yield.[61,62]

D. *Bz-Chromanamines*

The majority of chromans with an amino substituent in the aromatic ring are 6-chromanamines. Most of the simpler compounds have been prepared by reduction of the corresponding nitrochromans, usually by catalytic hydrogenation. The other important method, which has been especially used for the synthesis of tocopheramines, normally involves the condensation of an *N*-acyl-4-aminophenol or a 4-nitrophenol with an unsaturated compound in the presence of an acidic catalyst. When the nitrophenol was used, conversion to the chromanamine was finally achieved by hydrogenation.

a. Reduction of Nitrochromans. Catalytic hydrogenation has been the favored method for the conversion of nitrophenols to chromanamines. Catalysts used include Raney nickel,[4,27,28,34] palladium on charcoal,[35,39] and platinum oxide.[1,2,41] It has been claimed that in the hydrogenation of 2-methyl-2-(4,8,12-trimethyltridecyl)-6-nitrochroman over palladium on charcoal, not only was the nitro group reduced to give the 6-chromanamine, but there was some further reduction of the aromatic ring to give the hexahydro-6-chromanamine.[39] Examples of the catalytic hydrogenation of nitrochromans to chromanamines are listed in Table 8.

There are a number of examples of the use of other reducing agents.

TABLE 8. CATALYTIC HYDROGENATION OF NITROCHROMANS

R^1	R^2	Position of NO_2	Catalyst	Ref.
H	H	5	PtO_2/EtOH	1
H	H	6	Ra–Ni/EtOH	27, 28
H	H	7	PtO_2/EtOH	1
2,2-DiMe	H	6	PrO_2/MeOH	41
			PtO_2/C_6H_6–MeOH	2
2-Me-2-(4,8,12-trimethyltridecyl)	H	6	Pd–C/EtOH	39
2-Me-2-(4,8,12-trimethyltridecyl)	8-Me	6	Pd–C/EtOH	39
H	5,8-DiMe	6	RaNi/EtOH	34
H	7,8-DiMe	6	RaNi/EtOH	34
2-OEt	H	6	RaNi/EtOH	4
2-OEt	H	6,8	RaNi/EtOH	4
H	8-MeCONH	6	PtO_2/EtOH	1
H	6-MeCONH	7	PtO_2/EtOH	1

2,5,6,7-Tetramethyl-2-(4,8,12-trimethyltridecyl)-8-chromanamine (**51**, X = NH_2) was obtained in 66% yield when the corresponding nitro compound (**51**, X = NO_2) was refluxed in ethanolic sodium hydroxide in the presence of zinc dust.[3] Addition of sodium sulfide and sulfur to a boiling aqueous ethanolic solution of 6,8-dinitrochroman effected reduction of only the 8-nitro group and afforded 6-nitro-8-chromanamine in 49% yield.[1] 6,8-Chromandiamine was obtained in 84% yield when 6,8-dinitrochroman was treated with tin and hydrochloric acid.[1] 6-Chloro-8-chromanamine and 8-methyl-6-chromanamine were prepared by similar reductions of the appropriate nitro compound.[1]

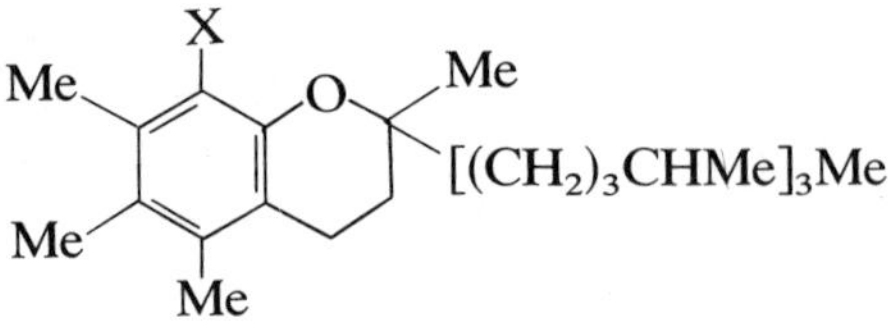

X, Me, Me, Me, O, Me, $[(CH_2)_3CHMe]_3Me$

(**51**)

b. Reaction of Acylaminophenols or Nitrophenols with Unsaturated Compounds. Tocopheramines (**2**) have frequently been prepared by reaction of an appropriately substituted 4-formylaminophenol (**52**) with phytol (**26a**) in the presence of an acidic catalyst such as formic acid. The mixture was normally heated to about 135° for 22 hr. Subsequent removal of the acyl protecting group afforded the tocopheramine (**2**).

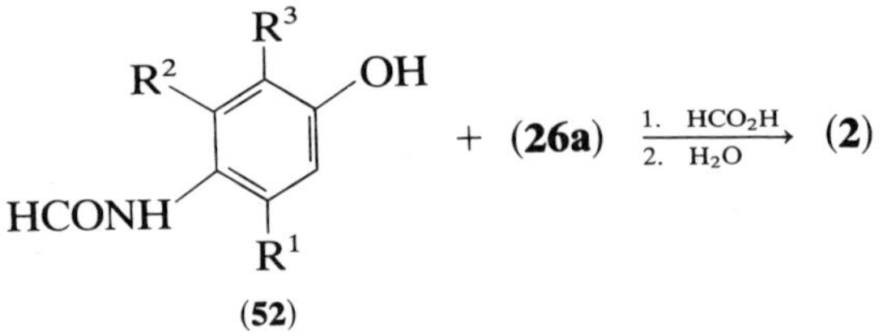

R^3, R^2, OH, HCONH, R^1 + (**26a**) $\xrightarrow[2.\ H_2O]{1.\ HCO_2H}$ (**2**)

(**52**)

α-Tocopheramine (**2**, $R^1 = R^2 = R^3 =$ Me) has been prepared in this way.[17,63–65] Phytol has been replaced by isophytol (**53a**),[14,15,17,66–68] and by changing the substitution of the methyl groups in the aminophenol, other tocopheramines have been synthesized.[17,66,68] (±)-α-Tocopheramine[3,4-^{14}C] has been synthesized from 2,3,5-trimethyl-4-formylaminophenol and isophytol[1,2-^{14}C].[69] The formyl protecting group was hydrolyzed with alcoholic hydrochloric acid. After reaction of 2,3-dimethyl-4-formylaminophenol (**52**, R^1 = H, $R^2 = R^3$ = Me) with isophytol[1,2-^{14}C], the *N*-formyl group was reduced with lithium aluminum hydride in tetrahydrofuran to give (±)-*N*-methyl-γ-tocopheramine[3,4-^{14}C].[69] Tritium-labeled tocopheramines as well as ^{14}C-labeled derivatives have been prepared.[15]

$$HO{-}\underset{\underset{\underset{CH_2}{\|}}{CH}}{\overset{Me}{\overset{|}{C}}}{-}R$$

(**53a**) $R = [(CH_2)_3CHMe]_3Me$
(**53b**) $R = (CH_2)_3CHMe_2$

In other syntheses the alcohol (**26a** or **53a**) has been replaced by an allylic halide such as phytyl bromide (**26b**) or a diene such as phytadiene.[65] Zinc chloride has been used as the acidic catalyst in place of formic acid.[65] In closely related work 2,3-dimethyl-4-formylaminophenol (**52**, $R^1 = H$, $R^2 = R^3 = Me$) was refluxed with dihydrolinalol (**53b**) in the presence of formic acid to give the *N*-formyl-6-chromanamine, which on subsequent reduction with lithium aluminum hydride in THF afforded 2,7,8,*N*-tetramethyl-2-isohexyl-6-chromanamine.[66]

Two routes have been used to prepare the *N*-formyl derivative of 2,2,5,7,8-pentamethyl-6-chromanamine.[70] 2,3,5-Trimethyl-4-formylaminophenol (**52**, $R^1 = R^2 = R^3 = Me$) has either been heated at 100° for 4 hr with isoprene (**54**) and formic acid or has been refluxed with dimethylallyl bromide (**55a**) in benzene in the presence of zinc chloride. Hydrolysis to the 6-chromanamine was effected by refluxing with 25% hydrochloric acid.[70] Reaction of 2,3,5-trimethyl-4-formylaminophenol with 3-methyl-3-tridecyl-1-bromo-2-propene (**55b**) afforded 2,5,7,8-tetramethyl-2-tridecyl-6-chromanamine.[19]

$$CH_2{=}\overset{\overset{Me}{|}}{C}{-}CH{=}CH_2$$

(**54**)

$$BrCH_2{-}CH{=}C(Me)R$$

(**55a**) $R = Me$
(**55b**) $R = (CH_2)_{12}Me$

As an alternative to 4-formylaminophenols, similar reactions have been carried out with 4-nitrophenols. For example, condensation of 2,3,5-trimethyl-4-nitrophenol (**25**, $R^1 = R^2 = R^3 = Me$) with phytol (**26a**) afforded the nitrochroman (**27**, $R^1 = R^2 = R^3 = Me$) which was converted to α-tocopheramine (**2**, $R^1 = R^2 = R^3 = Me$) on reduction.[37] Using dimethyl-4-nitrophenols, the method has been adapted to the preparation of other tocopheramines.[36] Tocamin (**2**, $R^1 = R^2 = R^3 = H$) and δ-tocopheramine (**2**, $R^1 = R^2 = H$, $R^3 = Me$) have been prepared by catalytic hydrogenation of the corresponding nitro compounds in the presence of palladium on charcoal, which in turn had been obtained by cyclization of 2-phytyl-4-nitrophenols[39] (see Section III, 1B).

Other methods that have been used to prepare tocopheramines include the reaction of 2-methyl-2-(4,8,12-trimethyltridecyl)chroman with a reactive diazonium salt such as a nitrobenzenediazonium salt, followed by reduction of the azo compound to give the appropriate tocamin or tocopheramine derivative.[71]

c. MISCELLANEOUS. 7-Methoxy-2,2-dimethyl-5-chromanamine (**57**) has been prepared by the action of sodamide in liquid ammonia on 6-bromo-7-methoxy-2,2-dimethylchroman (**56**);[35,72] the 5-chromanamine was isolated as the hydrochloride.[35]

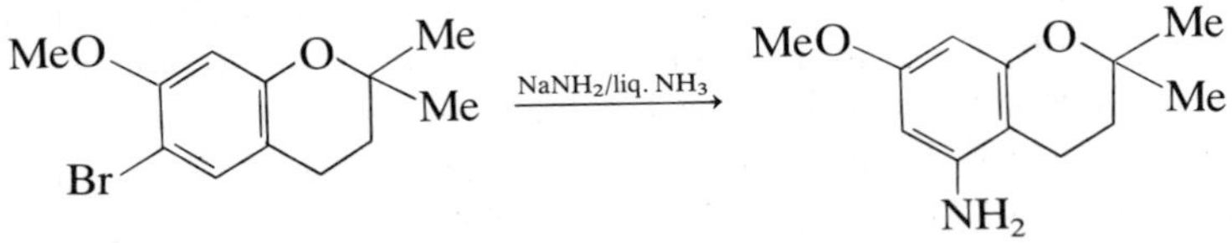

MeO, Br, O, Me, Me —$NaNH_2$/liq. NH_3→ MeO, O, Me, Me, NH_2

(**56**) (**57**)

5-Amino-γ-tocopherol (**58**, $R^1 = R^2 = Me$) and 5-amino-δ-tocopherol (**58**, $R^1 = H$, $R^2 = Me$) have been prepared by hydrogenation of the corresponding 5-nitrosotocopherols in the presence of palladium on calcium carbonate.[73]

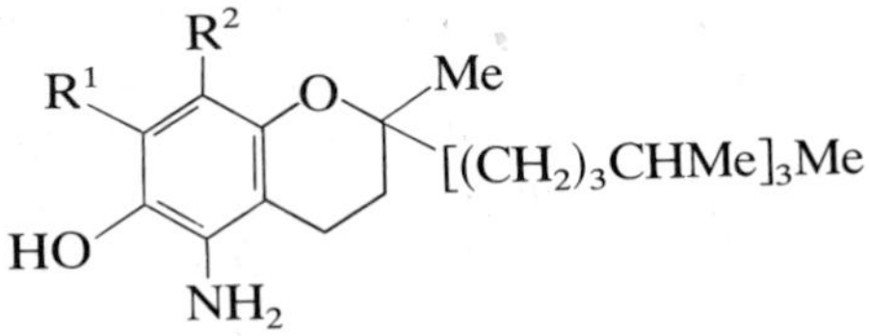

R^2, R^1, O, Me, $[(CH_2)_3CHMe]_3Me$, HO, NH_2

(**58**)

7-Chromanamine was obtained in low yield when 3-aminophenyl 3-bromopropyl ether (**59**) was refluxed in benzene for 8 hr.[74] As was indicated earlier (Section III, 1C), when the 4-nitrophenyl ether (**31**) in ethanol was treated with concentrated hydrochloric acid and iron powder, there was an exothermic reaction, and on work-up the *N*-acetylaminochromene (**60**) was isolated; subsequent catalytic hydrogenation afforded *N*-acetyl-2,2-dimethyl-6-chromanamine (**61**).[41]

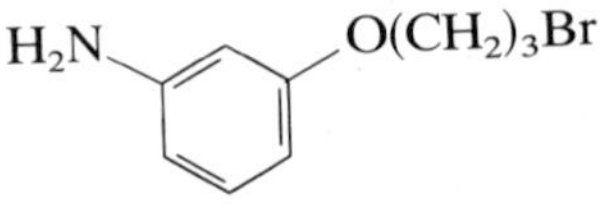

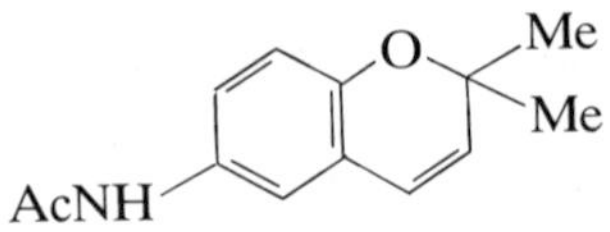

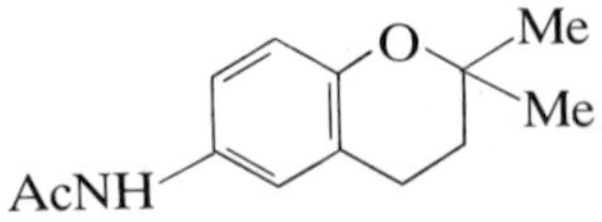

H_2N, $O(CH_2)_3Br$ — O, Me, Me, AcNH — O, Me, Me, AcNH

(**59**) (**60**) (**61**)

3. Chromans with a Basic Side Chain

A. *Chromanalkylamines*

a. 2-AMINOALKYLCHROMANS. 2-Aminomethylchromans (**62**, R = H) have been prepared by reaction of 2-chloromethylchroman with an alkali metal phthalimide in dimethylformamide followed by scission of the phthaloyl group with hydrazine.[75] Alternatively, 2-chloromethylchroman has been heated with phenoxyalkylamines to give an appropriate 2-phenoxyalkylaminomethylchroman. For example, 2-phenoxyethylaminomethylchroman (**62**, $R = (CH_2)_2OPh$) was prepared by heating 2-chloromethylchroman and phenoxyethylamine at 160° for 2 hr. Similar products may be obtained by reaction of 2-aminomethylchroman with a phenoxyalkyl halide.[75]

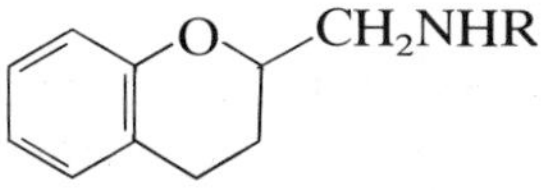

CH_2NHR

(**62**)

Reaction of the cis isomer of the 2-bromomethylchroman-4-carboxylate (**63**) with excess of the homoveratrylamine (**64**) at room temperature was used to prepare the *cis*-homoveratrylaminomethylchroman (**65**, $R = CO_2Me$). Reduction of the free base with lithium aluminum hydride in ether afforded the chroman-4-methanol (**65**, $R = CH_2OH$), which has been converted to its hydrochloride and formate.[23] The latter compound was first isolated as a by-product in the lithium aluminum hydride reduction of the tricyclic amide (**66**, X = O) to the amine (**66**, $X = H_2$). Its formation was a consequence of the use of excess reducing agent. With a controlled amount of lithium aluminum hydride, an 85% yield of the desired amine (**66**, $X = H_2$) was obtained; otherwise 40–50% of the chroman-4-methanol (**65**, $R = CH_2OH$) was formed.[23]

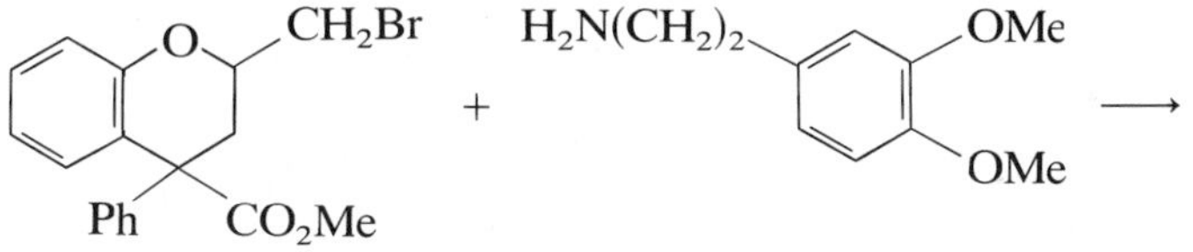

CH_2Br Ph CO_2Me + $H_2N(CH_2)_2$ OMe OMe ⟶

(**63**) (**64**)

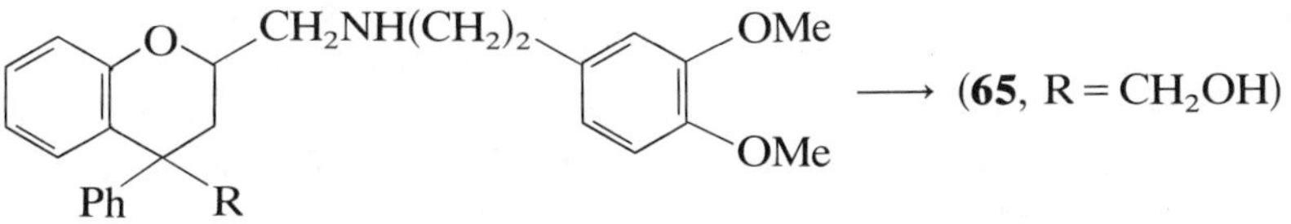

$CH_2NH(CH_2)_2$ OMe OMe Ph R ⟶ (**65**, $R = CH_2OH$)

(**65**, $R = CO_2Me$)

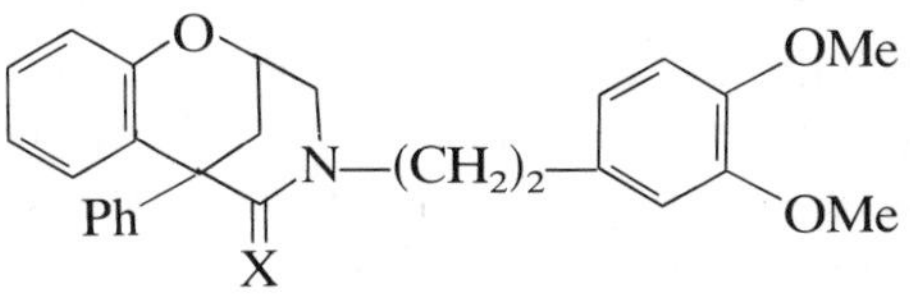

O N—$(CH_2)_2$ OMe OMe Ph X

(**66**)

2-Aminomethylchroman has also been prepared from chroman-2-carboxylic acid. The latter was converted to the acid chloride by the action of thionyl chloride[76] or oxalyl chloride;[20] subsequent reaction with ammonia formed chroman-2-carboxamide which on reduction with lithium aluminum hydride was converted to 2-aminomethylchroman (**62**, R = H).[76] In a similar series of reactions with ammonia replaced with dimethylamine, 2-dimethylaminomethylchroman was prepared; it was converted to the quaternary salt with methyl bromide.[20]

Hydrogenation of 8-methyl-2-dimethylaminomethylchromone (**67**)[77] in the presence of platinum black afforded 2-dimethylaminomethyl-8-methyl-4-chromanol (**68**).[20]

Me, O, CH_2NMe_2, O; H_2 / Pt black →; Me, O, CH_2NMe_2, OH

(**67**) (**68**)

b. 3-AMINOALKYLCHROMANS. Reduction of 2*H*-chromene-3-carboxamide (**69**) with lithium aluminum hydride afforded 3-aminomethylchroman (**70**).[78]

O, $CONH_2$ → O, CH_2NH_2

(**69**) (**70**)

c. 4-AMINOALKYLCHROMANS. A number of 4-phenyl-4-aminoalkyl chromans have been prepared by reduction of the corresponding 4-carboxamides,[79,80] the latter having been prepared by treatment of the acid chloride with ammonia or an appropriate secondary amine. Reduction of 4-phenyl-4-chromancarboxamide (**71**, $R^1 = R^2 = R^3 = H$) with lithium aluminum hydride afforded 4-phenyl-4-aminomethylchroman (**72**, $R^1 = R^2 = R^3 = H$), which was isolated as the hydrochloride in 66% yield.[79] However when the 2,2-dimethyl analog (**71**, $R^1 = Me$, $R^2 = R^3 = H$) was similarly reduced, no basic product but only 2,2-dimethyl-4-phenylchroman was isolated;[80] this same product was obtained in 70% yield when reduction was effected with sodium bis(methoxyethoxy)aluminohydride.[80] However a number of 4-aminomethylchromans of type (**72**, $R^1 = H$)[79] and (**72**, $R^1 = Me$)[80] have been obtained by reduction of the appropriate amide of type (**71**).

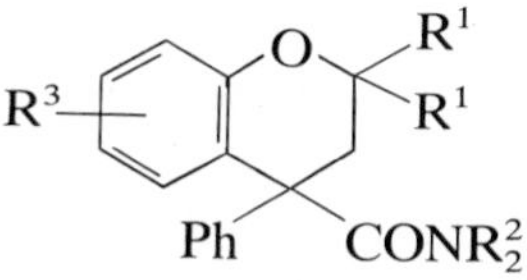

(71)

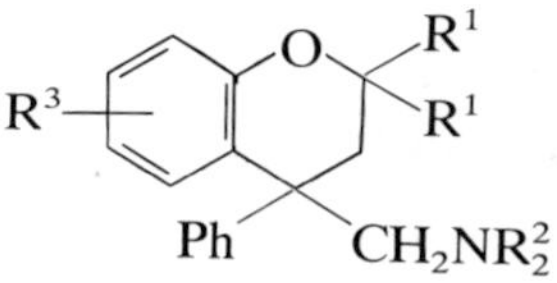

(72)

4-Dimethylaminopropylchromans (**74**) have been prepared by hydrogenation of the analogous chromenes (**73**) in the presence of palladium on charcoal.[81] Ethanol was used as the solvent for the 3-methyl compound (**74**, $R^1 = Me$, $R^2 = H$) and acetic acid for the 6-methyl analog (**74**, $R^1 = H$, $R^2 = Me$). Yields were in the range of 60–70%.

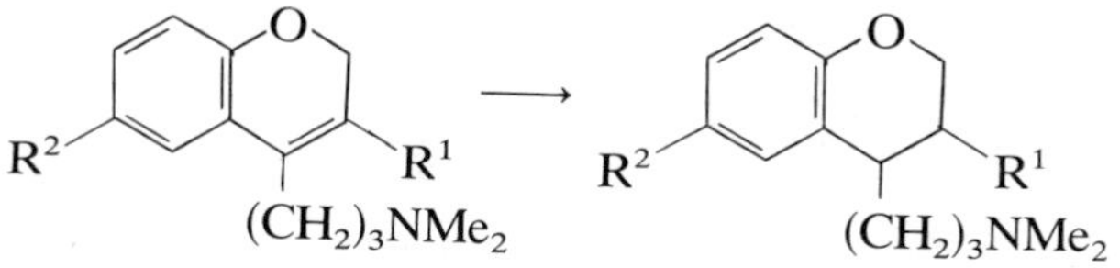

(73)

(74)

d. Bz-Aminoalkylchromans. Refluxing natural vitamin E with morpholine and formalin afforded 5-morpholinomethyl-γ-tocopherol (**75**).[82] When natural γ-tocotrienol was heated with dipropylamine and formalin and the product purified by chromatography on alumina, the 5-dipropylaminomethyl compound (**15a**) was isolated.[22] Similar treatment of plastochromanol-8 afforded the Mannich base (**15b**).[22]

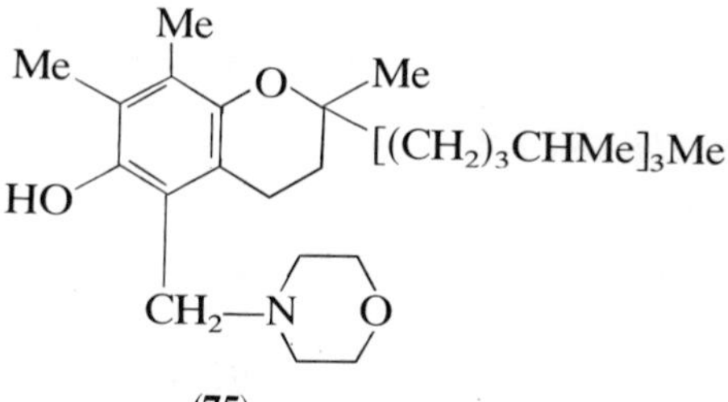

(75)

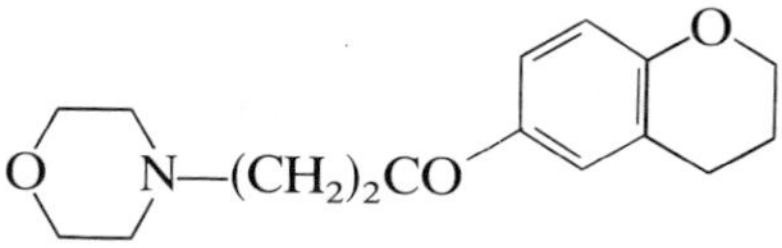

(76)

Another example of the use of the Mannich reaction to prepare basic chromans was provided by the reaction of 6-acetylchroman with piperidine and paraformaldehyde in ethanol to give 6-(3-piperidinopropionyl)chroman (**76**).[27]

Compounds of type (**80**), where R is a heteroaromatic nucleus, were prepared by condensation of chroman-6-carboxylic acid chloride (**78**) with a monosubstituted piperazine (**77**) in the presence of triethylamine in THF to give the acyl derivative (**79**) followed by reduction with lithium aluminum hydride in THF.[83] Alternatively 2-chloropyrimidine was condensed with 6-piperazinylmethylchroman (**80**, R = H) to give the 2-pyrimidine derivative (**80**, R = 2-pyrimidinyl).[83]

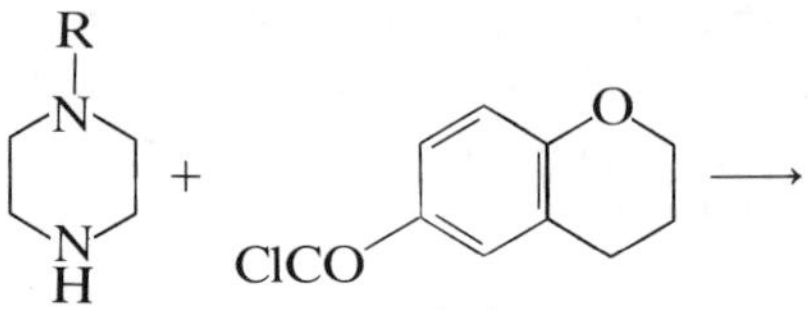

R N N H (**77**) + ClCO (**78**) O →

R—N N—CO O (**79**) → R—N N—CH_2 O (**80**)

When 7-benzyloxy-5-methoxy-2,2-dimethylchroman (**81**) and acetonitrile in ether were added to zinc chloride and hydrogen chloride passed through the cooled mixture, a mixture of the two imine hydrochlorides (**82**) and (**83**) was isolated. The former was the major product.[21]

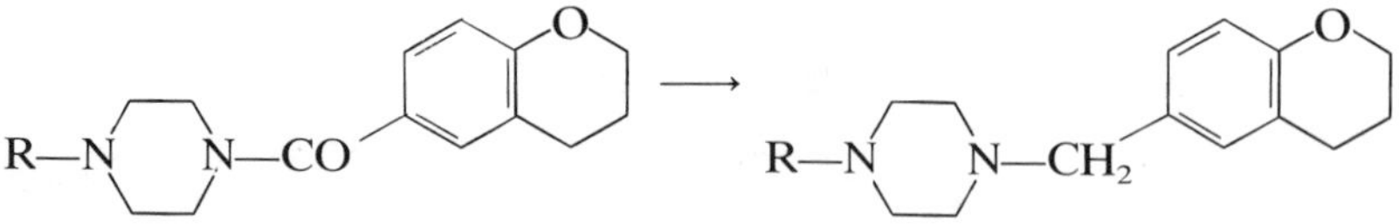

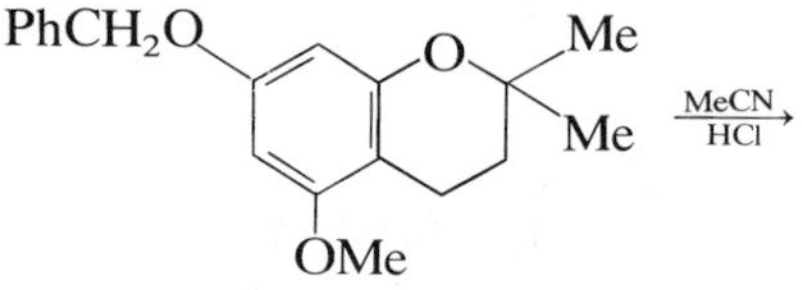

$PhCH_2O$ O Me Me OMe (**81**) $\xrightarrow[\text{HCl}]{\text{MeCN}}$

Me C=$NH_2^+Cl^-$ $PhCH_2O$ O Me Me OMe (**82**) + $PhCH_2O$ O Me Me Me C $Cl^-H_2N^+$ OMe (**83**)

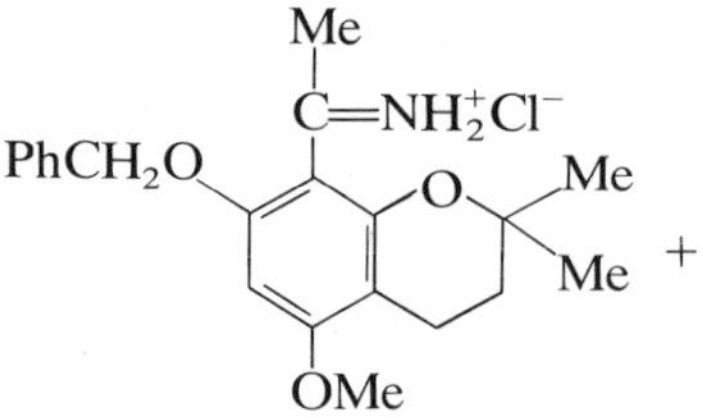

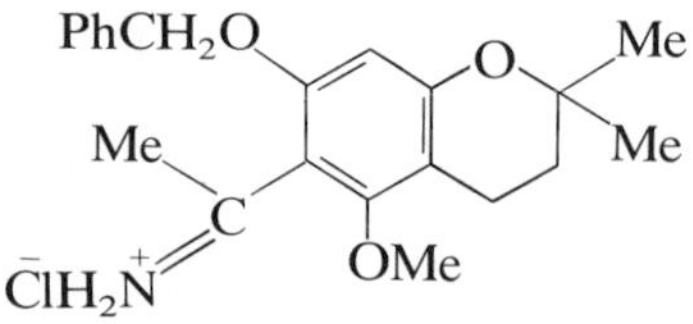

B. *Chromanethanolamines*

A number of 2-chromanethanolamines[84,85] and 6-chromanethanolamines[86–88] have been reported.

a. 2-CHROMANETHANOLAMINES. Chroman-2-carboxylic acid (**84**) was converted to the acid chloride by the action of thionyl chloride or oxalyl chloride. Reaction with diazomethane followed by treatment of the diazomethyl ketone with hydrogen chloride or hydrogen bromide afforded the halomethyl ketone (**85**, X = Cl or Br). Reduction with sodium borohydride afforded the halohydrin (**86a**, X = Cl or Br) which on reaction with an appropriate amine led to the ethanolamine (**86b**).[84,85] Although reaction of the chlorohydrin (**86a**, X = Cl) with an amine proceeds through an intermediate epoxide, the evidence indicated that of the two possible products only the secondary alcohol of type (**86b**) was obtained.[85] In the case of the *t*-butylamino derivative (**86b**, X = NHBu^t), the two possible racemates were separated by fractional crystallization.[84,85] Reaction of compounds of type (**86b**, X = NH_2) with an aldehyde followed by reduction of the product with sodium borohydride afforded secondary amine derivatives.[84]

CO_2H — 1. $SOCl_2$ 2. CH_2N_2 3. HX → $COCH_2X$ — $NaBH_4$ →

(**84**) (**85**)

$CHOHCH_2X$ — HNR^1R^2 → (**86b**) $X = NR^1R^2$

(**86a**) X = halogen

b. 6-CHROMANETHANOLAMINES. 6-Acetylchroman was converted to chroman-6-ylglyoxal (**87**) by oxidation with selenium dioxide in dioxan at 90°. Addition of sodium borohydride to a mixture of the glyoxal (**87**) and isopropylamine in methanol led to the formation of the ethanolamine (**88**).[86–88]

(**87**) R = COCHO
(**88**) R = $CHOHCH_2NHPr^i$

IV. REACTIONS

1. Reactions of Nitrochromans

The most important reaction of the nitrochromans is their reduction to chromanamines; many examples of this have been described in earlier sections. 2-Methyl-3-nitrochroman has been reduced to *cis*-2-methyl-3-chromanamine either with lithium aluminum hydride in ether under reflux for 4 hr or with lithium borohydride in THF for 50 hr.[25]

A variety of examples of the importance of this reduction in the preparation of bz-chromanamines (Section III, 2Da and b) have been reported.[1–4,27,28,34–37,39] The conversion has been effected by catalytic hydrogenation,[1,2,4,27,28,34,39] and by the action of zinc dust in alkali,[3] iron powder in ethanolic hydrochloric acid,[41] and tin in hydrochloric acid.[1] The last reagent was used to prepare 6,8-chromandiamine from 6,8-dinitrochroman,[1] but it has been reported that an attempted reduction of the same compound with zinc and hydrochloric acid was unsuccessful.[29] As an example of selective reduction, 6-nitro-8-chromanamine was prepared by adding an aqueous solution of sodium sulfide nonahydrate and sulfur in water to a boiling solution of 6,8-dinitrochroman in aqueous ethanol.[1]

Although this reaction is not specifically related to the nitro group, refluxing *trans*-3,4-dibromo-2,2-dimethyl-6-nitrochroman with aqueous acetone for 20 hr afforded *trans*-3-bromo-2,2-dimethyl-6-nitro-4-chromanol. On stirring the latter with potassium hydroxide pellets in ether, 3,4-epoxy-2,2-dimethyl-6-nitrochroman (**89**) was obtained.[40]

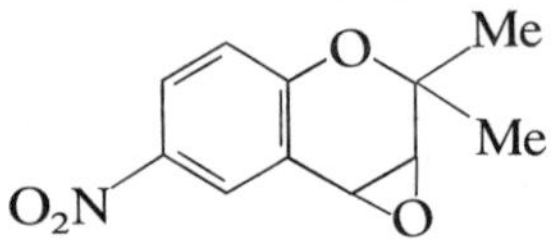

O Me Me O_2N O

(**89**)

2. Reactions of Chromanamines

The chromanamines fall into two broad groups. Those having an amino group in position 2, 3, or 4 behave as would be expected of alicyclic amines; 5-, 6-, 7-, and 8-chromanamines are typical aromatic amines. These generalizations are also true of amino substituents in the dihydronaphthopyrans.

The chromanamines form salts with acids and derivatives such as acetates. The free amino group is alkylated by conventional techniques.

A. *2-Chromanamines*

The action of nitrous acid on 2,4,4,7-tetramethyl-2-chromanamine (**34**, $X=NH_2$) gave a 10% yield of 2,4,4,7-tetramethyl-2-chromanol (**34**, X = OH).[42] 2-Chromanols have also been obtained from 2-chromanamines of type (**3**). The action of 50% acetic acid at 20° replaced both amino groups with hydroxyl producing a diol of type (**90**) but with 2 *N* hydrochloric acid, only one of the dimethylamino groups was replaced but its identity was not apparently elucidated.[6] With methyl iodide in ethyl acetate, the base (**3**) gave a monomethiodide.[6]

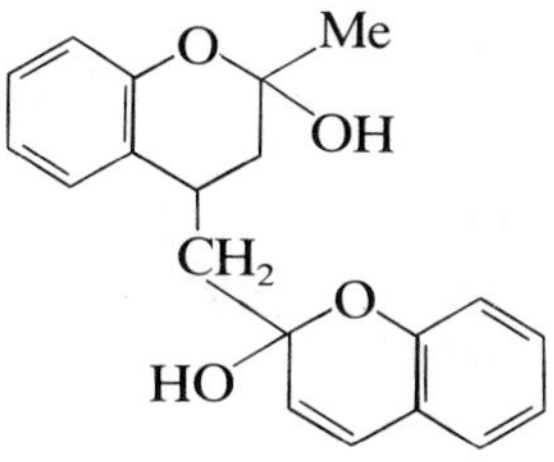

Me OH CH2 O HO

(**90**)

Hydrolysis of 2,3-chromandiamines of type (**5**) with 5% hydrochloric acid afforded a 2-hydroxy-3-chromanamine.[7] On heating the diamine (**5**, $R_2 =$ —$(CH_2)_5$—), the basic benzofuran (**91**) was obtained and converted to the dihydro derivative (**92**) by catalytic hydrogenation.[7]

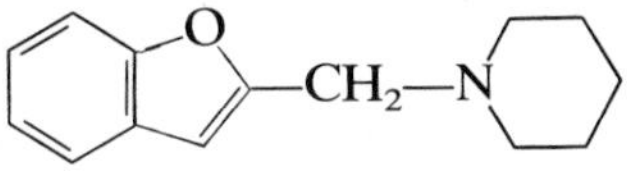

(**91**) 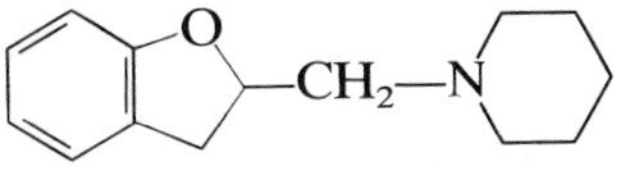 (**92**)

B. *3-Chromanamines*

3-Chromanamines behave as typical alicyclic amines and have usually been isolated as their hydrochlorides. They readily form derivatives such as acetates, and the amino group may be alkylated by standard methods (Section III, 2Be). The aromatic ring is readily halogenated by the action of bromine or chlorine.

On refluxing with cyanamide in ethanol, 3-chromanamine afforded 3-guanidinochroman (**40**, $R^1 = R^2 = H$, $R^3 = C(:NH)NH_2$).[51]

As has been mentioned earlier (Section III, 2Bf), 3,4-chromandiamine hydrochloride reacts with phosgene in the presence of aqueous potassium carbonate to give the oxoimidazolidochroman (**45**). On heating with 0.1 *N* hydrochloric acid, the trans isomer was reconverted to the *trans*-3,4-chromandiamine, whereas attempted hydrolysis of the cis isomer left the tricyclic compound unchanged.[54]

C. *4-Chromanamines*

Most of the 4-chromanamines have been isolated as their hydrochlorides, and like the 3-chromanamines they are acylated and alkylated by standard procedures. Acetamido derivatives have been prepared from the hydrochlorides by the action of acetic anhydride in the presence of pyridine.[13]

Reaction of chloroacetyl chloride with 4-chromanamine (**46**, $R^1 = R^2 = R^3 = R^4 = H$) hydrochloride afforded the 4-chloroacetylaminochroman (**93**,

$n = 1$) which was converted to the 4-dialkylaminoacetylamino (**94**, $n = 1$) hydrochloride on refluxing with the appropriate amine in toluene.[55] Subsequent reduction with lithium aluminum hydride led to the 4-dialkylaminoethylaminochroman (**95**, $n = 1$).[55] In a similar sequence of reactions, 4-(3-chloropropionylamino)chromans (**93**, $n = 2$) were prepared by the action of 3-chloropropionyl chloride on the 4-chromanamine in the presence of alkali.[55–89] Subsequent reaction with an appropriate amine afforded the corresponding dialkylaminopropionylchroman (**94**, $n = 2$) hydrochloride,[55,89] which was reduced to the 4-dialkylaminopropylaminochroman (**95**, $n = 2$) dihydrochloride by the action of lithium aluminum hydride.

NHX

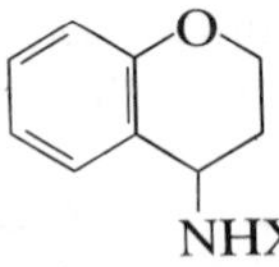

(**93**) $X = CO(CH_2)_{n'}Cl$
(**94**) $X = CO(CH_2)_nNR_2$
(**95**) $X = (CH_2)_{n+1}NR_2$

When a mixture of 4-chromanamine and acrylonitrile was heated over a free flame, 4-(cyanoethyl)aminochroman hydrochloride was obtained in 75% yield.[55]

Reaction of 4-chromanamine with ethyl chloroformate afforded 4-ethoxycarbonylaminochroman (**46**, $R^1 = R^2 = R^3 = H$, $R^4 = CO_2Et$), which on reduction with lithium aluminum hydride was converted to *N*-methyl-4-chromanamine (**46**, $R^1 = R^2 = R^3 = H$, $R^4 = Me$). Treatment of the latter with propargyl bromide in acetone led to the propyne derivative (**46**, $R^1 = R^2 = H$, $R^3 = CH_2C{\vdots}CH$).[90]

Cleavage of the oxazolidine derivative (**96**) by catalytic hydrogenation and treatment with nitrous acid afforded the nitroso compound (**97**). Subsequent cyclization with thionyl chloride and triethylamine gave the tricyclic product (**98**).[91]

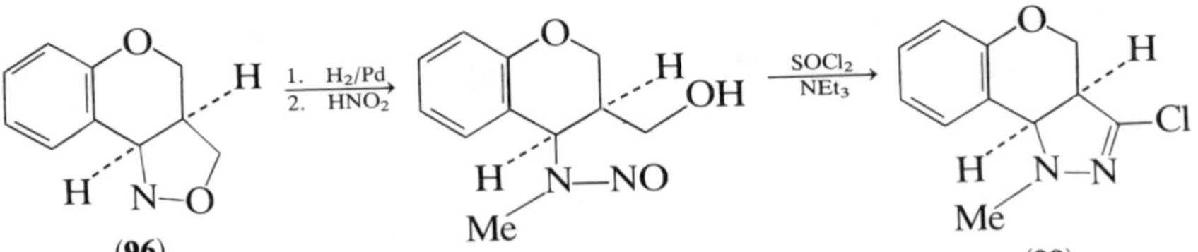

H N–O H 1. H_2/Pd 2. HNO_2 H OH H N–NO Me $SOCl_2$ NEt_3 H Cl H N–N Me

(**96**) (**97**) (**98**)

D. *Bz*-Chromanamines

The 5-, 6-, 7-, and 8-chromanamines and the related chromandiamines behave like typical aromatic amines. They form salts such as hydrochlorides and picrates with acids and may be *N*-alkylated[17,68] and *N*-acylated[17,41,66,68,92] by conventional techniques. Monoalkyl derivatives have been prepared by reduction of the acyl derivative, usually with lithium aluminum hydride,[15,17,66,68,92] or by alkylation of the *N*-acyl compound followed by hydrolysis.[67] Alkylation of *N*-acyl compounds followed by lithium aluminum hydride reduction of the amide has been used to prepare unsymmetrical di-*N*-alkylated derivatives. *N*-Formyltocopheramines have been obtained as intermediates in the preparation of the tocopheramines (Section III, 2Db). Treatment of 6-chromanamine with chloroacetyl chloride in acetic acid afforded the 6-chloroacetamidochroman which on subsequent reaction with diethylamine was converted to 6-diethylaminoacetamidochroman (**99**).[27,28] 6-Piperidinoacetylaminochroman was similarly prepared.[27,28]

The bz-chromanamines may be diazotized and the diazonium salts used as synthetic intermediates. For example, reaction of cuprous chloride in concentrated hydrochloric acid with diazonium salts prepared from bz-chromanamines has been used to prepare the corresponding bz-chlorochromans.[1] Deamination of 5-nitro-6-chromanamine has been effected by addition of hypophosphorous acid to a solution of the diazonium salt;[1] the reaction has been used with a number of other nitrochromanamines. Hydrolysis of the diazonium salt affords the corresponding phenol, a method that is illustrated by the preparation of 8-hydroxy-2,5,6,7-tetramethyl-2-(4,8,12-trimethyltridecyl)chroman from the 8-chromanamine analog.[3] 6-Hydroxy-2,2,5,7,8-pentamethylchroman,[70] 6-hydroxy-5,8-dimethylchroman and its 7,8-dimethyl analog,[34] and (±)-α-tocopherol[64] have all been prepared similarly.

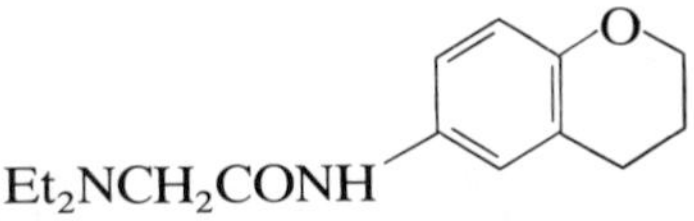

Et_2NCH_2CONH–(chroman)

(**99**)

Several nitro derivatives of *N*-acetyl-bz-chromanamines have been prepared by direct nitration; the 5- and 6-nitro derivatives of *N*-acetyl-8-chromanamine were separated by fractional crystallization from benzene.[1] Crystallization and solvent fractionation have been used to separate the *N*-acetyl derivatives of 5-, 7-, and 8-nitro-6-chromanamines.[1] Nitration of *N*-acetyl-6-chromanamine was used to prepare the 7,8-dinitro compound.[1]

Reaction of 2,2-dimethyl-6-chromanamine (**100**) in THF with the dihydrofuranone (**101**) (prepared in situ from diethyl malonate, sodium hydride, and chloracetyl chloride in tetrahydrofuran) afforded the ester (**102**), which on heating in diphenyl ether at 255° for 5 min gave a mixture of the two tetracyclic compounds (**103**) and (**104**).[2]

Oxidation of α-tocopheramine (**2**, $R^1 = R^2 = R^3 = Me$) hydrochloride with ferric chloride in methanolic hydrogen chloride afforded the yellow

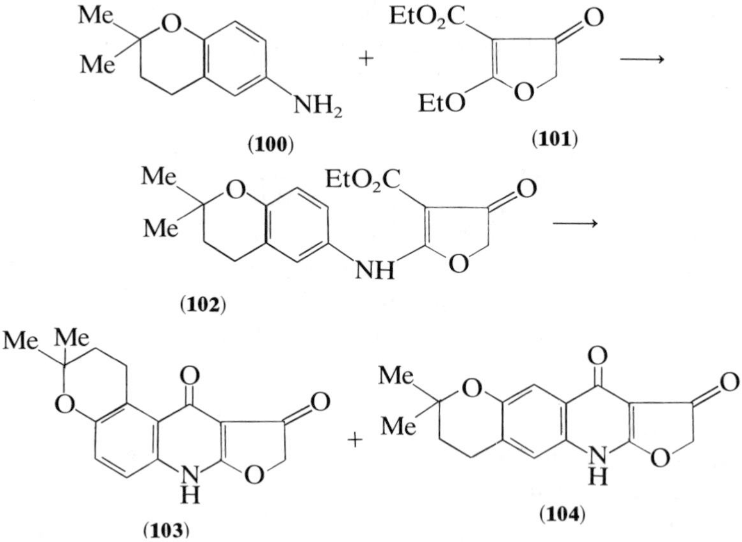

Me, Me, O, NH_2 + EtO_2C, O, EtO, O →

(100) (101)

Me, Me, O, EtO_2C, O, NH, O →

(102)

Me Me, O, O, O, N H, O + Me, Me, O, O, O, N H, O

(103) (104)

tocopherylquinone (**105**).[63] The orange-red azochroman (**106**) was obtained on oxidation of α-tocopheramine (**2**, $R^1 = R^2 = R^3 = Me$) with alkaline ferricyanide. When dissolved in dioxan and treated with stannous chloride in hydrochloric acid, it was reconverted to α-tocopheramine.[18] Oxidation of tocamine (**2**, $R^1 = R^2 = R^3 = H$) in ethanol with silver acetate afforded a yellow product whose properties were consistent with it being the nitroso compound (**107**).[39]

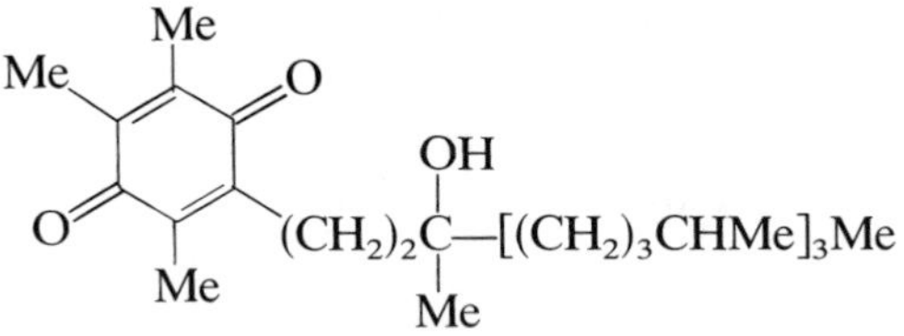

Me, Me, O, O, Me, OH, $(CH_2)_2C$—$[(CH_2)_3CHMe]_3Me$, Me

(105)

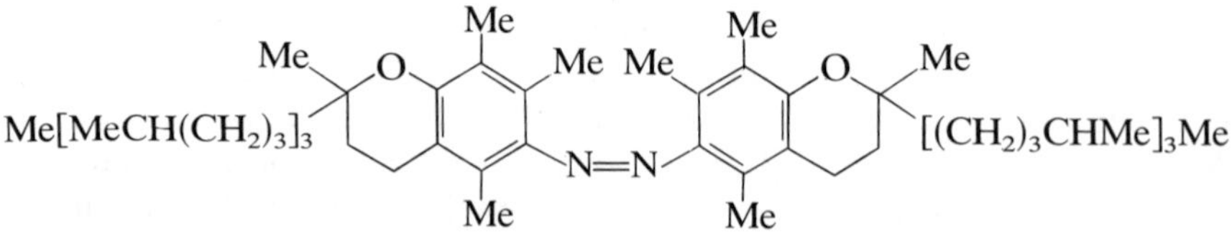

Me, Me, O, Me Me, Me, O, Me, $Me[MeCH(CH_2)_3]_3$, N=N, $[(CH_2)_3CHMe]_3Me$, Me, Me

(106)

O, Me, $[(CH_2)_3CHMe]_3Me$, ON, OEt

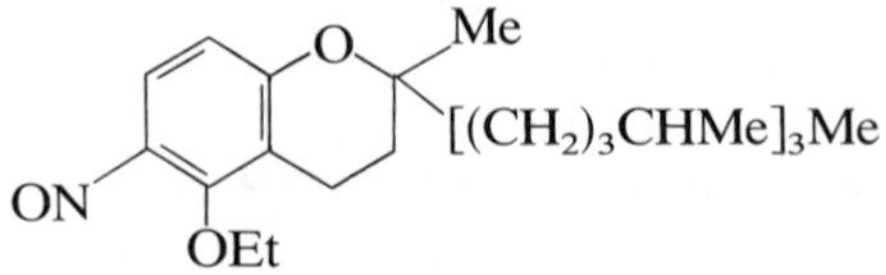

(107)

6-Chromanamines react with isocyanates and isothiocyanates in benzene to give chroman-6-yl ureas and thioureas of type (**108**) where X = O or S and R = alkyl or aryl.[92]

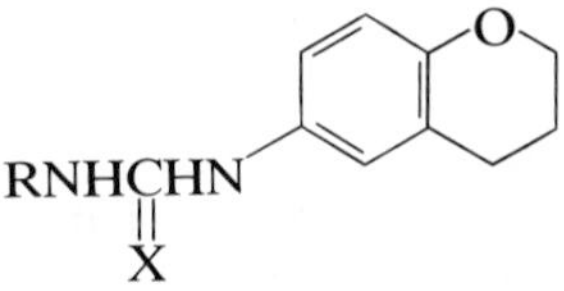

RNHCHN
||
X

(**108**)

When 7-methoxy-2,2-dimethyl-5-chromanamine (**109**) was refluxed with 2-iodobenzoic acid in the presence of potassium carbonate, metallic copper, and cuprous carbonate in pentanol, the secondary amine (**110**) was formed.[72] The same product has been prepared using 2-bromobenzoic acid.[35] Subsequent heating in polyphosphoric acid at 90° for 2 hr afforded the tetracyclic compound (**111**),[35,72] which on methylation gave dihydro-acronycine.[35]

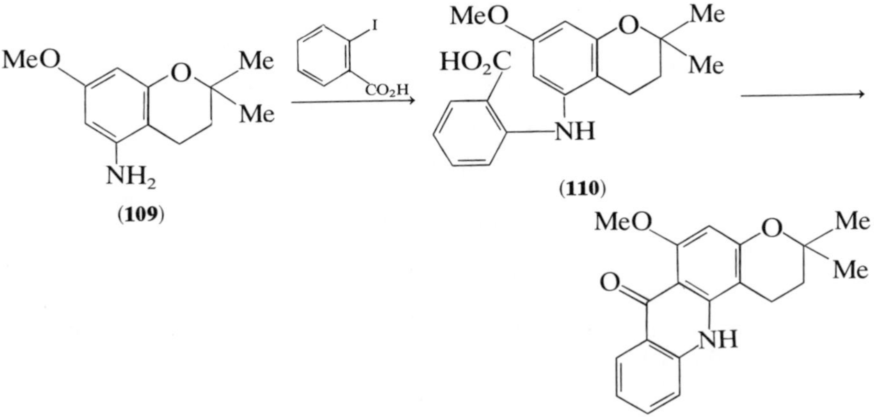

MeO O Me Me NH_2 (**109**)

I CO_2H

MeO O Me Me HO_2C NH (**110**)

MeO O Me Me O NH (**111**)

3. Chromans with a Basic Side Chain

Analogs of 2-phenoxyalkylaminomethylchroman (**62**, R = $(CH_2)_nOPh$) have been prepared by heating a 2-aminomethylchroman with an appropriate phenoxyalkyl halide.[75] When 2-aminomethylchroman was heated on the steam bath with 3,5-dimethyl-1-guanylpyrazole sulfate in water, bis(2-guanidinomethylchroman) (**62**, R = C(:NH)NH_2) sulfate was obtained.[76] Bis(3-guanidinomethylchroman) (**112**) sulfate has been similarly prepared in 55% yield by heating 3-aminomethylchroman with 3,5-dimethyl-1-guanylpyrazole sulfate in water at 90° for 16 hr.[78]

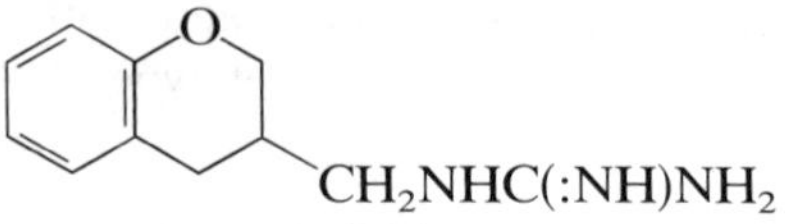

$CH_2NHC(:NH)NH_2$

(**112**)

Reduction has been used to remove the basic group of 5-aminomethylchromans. Hydrogenation of the 5-dipropylaminomethyl compounds (**15a**) and (**15b**) in the presence of palladium chloride afforded the corresponding 5-methyl compounds (**113**; $n = 3$ and 8, respectively).[22] When 5-morpholinomethyl-γ-tocopherol (**75**) was refluxed with zinc dust and sodium acetate in the presence of acetic acid and acetic anhydride for 1.5 hr, 5-acetoxymethyl-γ-tocopheryl acetate (**114**) was isolated.[82]

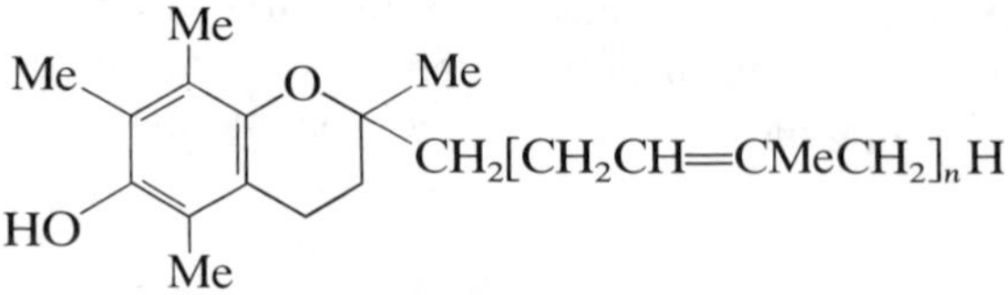

Me, Me, O, Me, HO, Me, $CH_2[CH_2CH{=}CMeCH_2]_nH$

(**113**)

Me, Me, O, Me, $MeCO_2$, CH_2OCOMe, $[(CH_2)_3CHMe]_3Me$

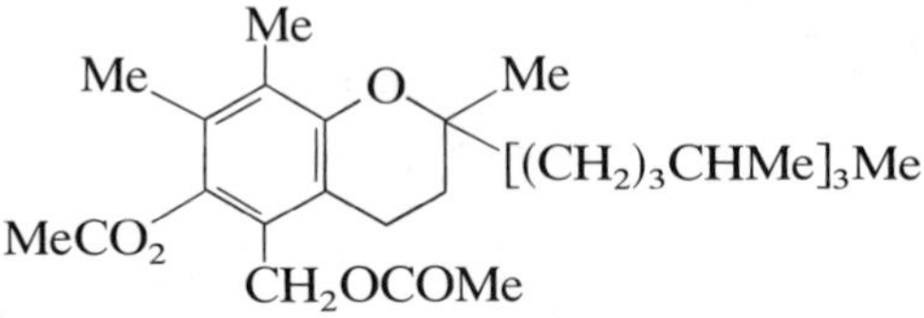

(**114**)

Hydrolysis of the mixture of imine hydrochlorides (**82**) and (**83**) by vigorous boiling with water overnight afforded a product from which 8-acetyl-7-benzyloxy-5-methoxy-2,2-dimethylchroman was isolated.[21] When the imine hydrochloride (**17a**) was refluxed with ethanol in the presence of sodium carbonate for 2 hr, the benzopyranopyridine (**115**) was isolated in 61% yield and characterized.[24]

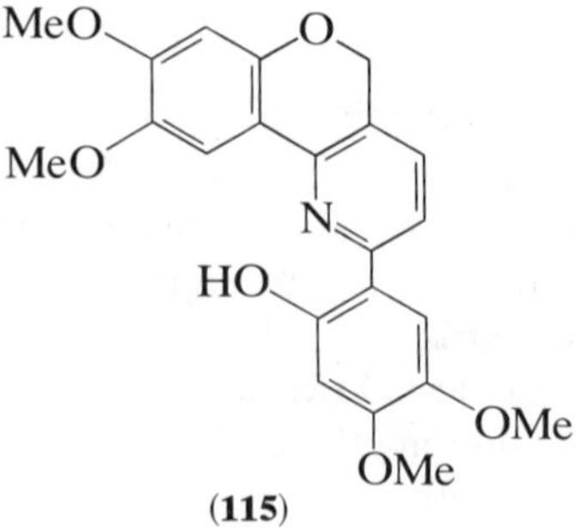

MeO, O, MeO, N, HO, OMe, OMe

(**115**)

V. BIOLOGICAL AND PHARMACOLOGICAL ACTIVITY

There has been little evidence of biological activity among the nitrochromans. It has however been claimed that 2,5,7-trimethyl- and 2,5,7,8-tetramethyl-2-(4,8,12-trimethyltridecyl)-6-nitrochromans exhibit vitamin E-like activity.[38]

Among the chromanamines, most interest centers on the 3-chromanamines and on the bz-chromanamines, where the 6-chromanamines have been extensively studied. These groups, together with chromans with a basic side chain, are discussed in the following sections. No biological or pharmacological activity has apparently been reported for 2-chromanamines. 4-Chromanamines have been prepared in work aimed at the treatment of depression and angina,[93] and 4-phenyl-4-chromanamine has been synthesized for pharmacological studies.[61] As yet no useful activity has apparently been found among the 4-chromanamines.

1. 3-Chromanamines

3-Chromanamines show a variety of pharmacological properties, and the series illustrates the very profound changes that can occur as a result of relatively small modifications of chemical structure. *cis*-2,*N*,*N*-Trimethyl-3-chromanamine hydrochloride, also known as the α-isomer, reduces the hyperirritability of rats that have been lesioned in the septal area of the brain.[52] The compound suppressed this hyperirritability without causing generalized depression associated with tranquilizers and sedative hypnotic drugs in the same test procedure. The compound also suppressed the mouse-killing instinct of male Sprague–Dawley rats at dose levels that produce no other obvious changes in the behavior of the caged rat.[52] A number of other pharmacological properties[52] led to the conclusion that the compound possessed both major tranquilizing and antidepressant properties.[94] It has been shown by quantitative pharmacoelectroencephalography to produce significant effects on human brain function. It has been predicted that the clinical effects will be similar to those seen after administration of major tranquilizers such as chlorpromazine. The neuroleptic effect would be expected to have an early onset (1 hr); but 6 hr after administration, a central stimulatory profile is expected.[94] *cis*-2-Methyl-3-chromanamine hydrochloride (also known as the α-isomer) exhibits the characteristics of an antidepressant in a number of ways. It has an antitetrabenazine effect in rats, it reduces reserpine-induced ptosis in mice, and suppresses the mouse-killer instinct in rats.[25]

Although a number of 3-chromanamines have been shown to cause central excitation in rats as shown by the increase in locomotor activity determined with jiggle cages, when some analogs with two alkyl groups in the aromatic ring were studied, many were shown to have potent central stimulant activity when examined by this technique.[48]

A series of 13 3-chromanamines, particularly with *N*-substituents or with methyl or methoxy substituents in the 6- or 7-position, have been compared with amphetamine in pharmacological and biochemical studies.[95] The biochemical results showed that they were amphetaminelike. 3-Chromanamine and its 6-methyl and 7-chloro analogs increased the level of serotonin and decreased the level of tyrosine in the rat brain. Rat brain levels of tryptophan, dopamine, and noradrenaline were also measured. All of the compounds studied antagonized reserpine-induced ptosis in mice. Experiments in rats showed that the compounds behaved as psychostimulants, and, with the exception of *N*-methyl-3-chromanamine, antidepressive potential was absent.[95] Stereochemical comparisons of the molecular skeleton of the 3-chromanamines with that of LSD confirmed that these compounds have psychotomimetic potential.[95]

2. Bz-Chromanamines

Of the chromanamines with a basic substituent in the aromatic nucleus, it is only the 6-chromanamines, especially the tocopheramines (**116**), whose biological properties have been studied to any great extent. The most important areas of interest concern the vitamin E activity and their useful properties as antioxidants, especially for the stabilization of food products.

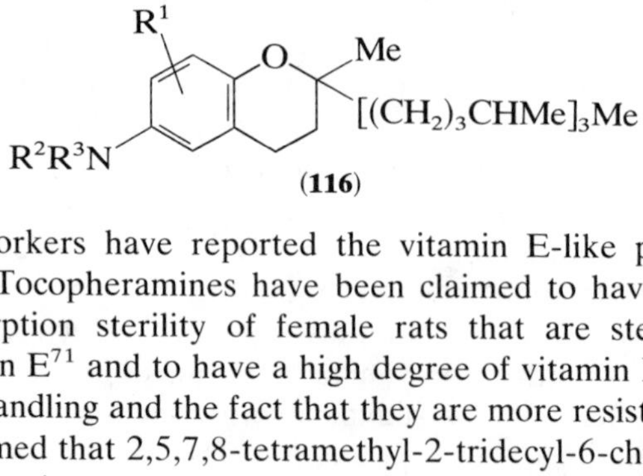

R^1 O Me $[(CH_2)_3CHMe]_3Me$ R^2R^3N

(**116**)

Numerous workers have reported the vitamin E-like properties of tocopheramines. Tocopheramines have been claimed to have a curing effect upon the resorption sterility of female rats that are sterile for lack of sufficient vitamin E[71] and to have a high degree of vitamin E activity.[65] Due to the ease of handling and the fact that they are more resistant to oxidation, it has been claimed that 2,5,7,8-tetramethyl-2-tridecyl-6-chromanamine and the *N*-acyl derivatives are preferable to both natural and synthetic vitamin E.[19] In two different bioassays with the chick, (±)-α-tocopheramine (**116**, $R^1 = $ 5,7,8-triMe, $R^2 = R^3 = H$), (±)-*N*-methyl-β-tocopheramine (**116**, $R^1 = $ 5,8-diMe, $R^2 = Me$, $R^3 = H$), and (±)-*N*-methyl-γ-tocopheramine ($R^1 = $ 7,8-diMe, $R^2 = Me$, $R^3 = H$) had vitamin E activity as great as (±)-α-tocopherol on a molar basis. (±)-β-Tocopheramine (**116**, $R^1 = $ 5,8-diMe, $R^2 = R^3 = H$) and (±)-γ-tocopheramine (**116**, $R^1 = $ 7,8-diMe, $R^2 = R^3 = H$) were considerably less active.[96] The essential equivalence of the vitamin E activity of (±)-*N*-methyl-β-tocopheramine (**116**, $R^1 = $ 5,8-diMe, $R^2 = Me$,

$R^3 = H$) and (±)-α-tocopherol has been confirmed in the rat gestation-resorption bioassay.[97] (±)-α-Tocopheramine (**116**, $R^1 = 5,7,8$-triMe, $R^2 = R^3 = H$), (±)-*N*-methyl-γ-tocopheramine (**116**, $R^1 = 7,8$-diMe, $R^2 = Me$, $R^3 = H$), and (±)-*N,N*-dimethyl-γ-tocopheramine (**116**, $R^1 = 7,8$-diMe, $R^2 = R^3 = Me$) were comparable with (±)-α-tocopheryl acetate in preventing vitamin E deficiency-induced muscular dystrophy in chicks.[98] Tocopheramine hydrochloride showed antidystrophy activity in rabbits.[99]

Studies of the metabolism of some tocopherols and tocopheramines demonstrated that they were converted to γ-lactones via tocopherylquinone; their distribution and storage in various organs showed that retention paralleled biological activity.[100] (±)-α-Tocopheramine (**116**, $R^1 = 5,7,8$-triMe, $R^2 = R^3 = H$) and (±)-*N*-methyl-γ-tocopheramine (**116**, $R^1 = 7,8$-diMe, $R^2 = Me$, $R^3 = H$), both of which show high oral vitamin E activity in rats, have been shown to be resorbed and retained in the tissues to the same extent as (±)-α-tocopheryl acetate and were degraded to the same metabolic end products.[69]

(±)-α-Tocopheramine[3,4-^{14}C] (**116**, $R^1 = 5,7,8$-triMe, $R^2 = R^3 = H$) and (±)-*N*-methyl-γ-tocopheramine[methyl-^{3}H] (**116**, $R^1 = 7,8$-diMe, $R^2 = Me$, $R^3 = H$) were among compounds fed to patients with cannulated thoracic ducts. Radioactivity in the lymph lipids was measured. The highest recovery occurred after administration of the latter, and this was followed by the former compound. Lower recoveries were obtained with labeled (±)-α-tocopheryl acetate and (±)-α-tocopherol. It was suggested that low intestinal absorption may result in a vitamin E deficiency when the diet contains high amounts of polyunsaturated fatty acids.[101]

Morphological change in the rat thyroid gland during vitamin E deficiency was partially or completely normalized by treatment with α-tocopherol or 6-chromanamines of type (**117**) where *x* is an odd number. The normalization occurred with α-tocopheramine (**117**, $x = 3$) and with *N*-methyl-γ-tocopheramine (**116**, $R^1 = 7,8$-diMe, $R^2 = Me$, $R^3 = H$) but did not occur with compounds of type (**117**) where *x* was an even number such as 2 or 4.[102]

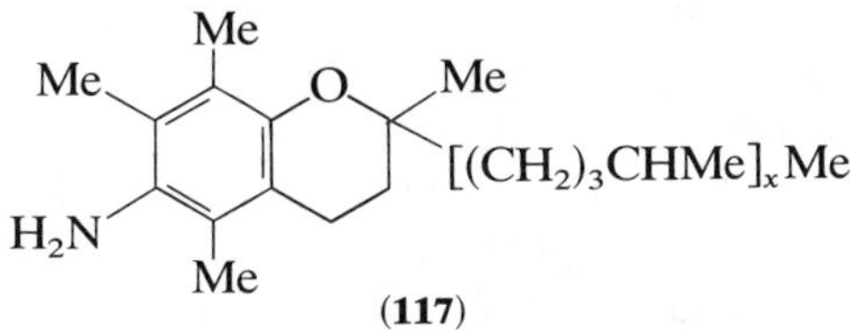

Me; Me; O; Me; $[(CH_2)_3CHMe]_xMe$; H_2N; Me

(**117**)

It has been claimed that α-tocopheramine (**116**, $R^1 = 5,7,8$-triMe, $R^2 = R^3 = H$), *N*-methyl-β-tocopheramine (**116**, $R^1 = 5,8$-diMe, $R^2 = Me$, $R^3 = H$), *N,N*-dimethyl-δ-tocopheramine (**116**, $R^1 = 8$-Me, $R^2 = R^3 = H$), and a number of other tocopheramines are very effective antioxidants for the stabilization of easily oxidized products used for human consumption; they are nontoxic.[66,103–106] Their use has been claimed in vitamin and cosmetic preparations as well as in foodstuffs.[103]

As lipid antioxidants, the tocopheramines were no more effective than the natural tocopherols. It was shown that there was no correlation between in vitro antioxidant activity and in vivo biological activity for overcoming the symptoms of vitamin E deficiency.[107]

The increase in the level of long-chain and unsaturated fatty acids of erythrocyte lipids of vitamin E-deficient rats was shown to be reversed by administration of antioxidants such as *N*-methyl-γ-tocopheramine (**116**, $R^1 = 7,8$-diMe, $R^2 = Me$, $R^3 = H$). The alteration of the fatty acid pattern occurred outside the erythrocyte prior to incorporation.[108]

Inhibition of adenosine triphosphatase from membranes by γ-tocopheramine has been studied.[109]

The biological activity of the tocopheramines in the hemolysis test is similar to that of the corresponding tocopherols. Methylation of the amino group had variable effects. Monomethylation of (±)-α-tocopheramine (**116**, $R^1 = 5,7,8$-triMe) reduced the activity some tenfold while similar treatment of (±)-β- and (±)-γ-tocopheramines (**116**, $R^1 = 5,8$-diMe and $R^1 = 7,8$-diMe, respectively) increased the activity fivefold.[14]

Of other bz-chromanamines it has been claimed that salts of 6-diethylaminoacetamidochroman and 6-piperidinoacetamidochroman have local anaesthetic and hypotensive properties.[27,28]

3. Chromans with a Basic Side Chain

A number of patents have claimed biological activity for chromans with a basic side chain. 2-Phenoxyalkylaminomethylchromans (type **62**, R = phenoxyalkyl) have been claimed to be useful as hypotensives and as adrenolytic and adrenergic blocking agents.[75] 2-Guanidinomethylchroman (**62**, $R = C(:NH)NH_2$) and its salts[76] and the sulfate of 3-guanidinomethylchroman (**112**)[78] are said to behave as antihypertensive agents. A number of analogs of 4-aminomethyl-4-phenylchroman (type **72**) have been claimed to be analgesics, antispasmodics, local anaesthetics, and hypotensive agents.[79] 6-Piperazinomethylchromans (type **80**) are included in a wide range of chroman and furan derivatives said to have possible use in the treatment of peripheral vascular disorders, Parkinson's disease, hypertension, and in the prevention of pregnancy.[83] It is not clear whether any of these claims are substantiated by true clinical efficacy.

1-(Chroman-2-yl)-2-aminoethanols (type **86b**), which contain structural features of pronethalol (**118**) and propanolol (**119**), have been tested as β-adrenergic blocking agents.[84,85]

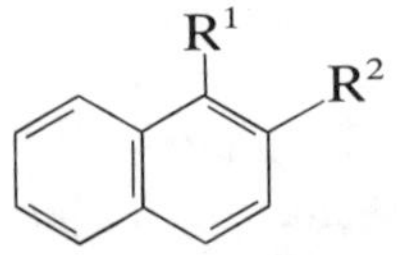

(**118**) $R^1 = H$, $R^2 = CHOHCH_2NHPr^i$
(**119**) $R^1 = OCH_2CHOHCH_2NHPr^i$, $R^2 = H$

The 6-chromanyl analog (**88**) has also been shown to have a pronethalol level of activity;[86–88] this was expected as the group replacing the naphthalene nucleus was of about the same size and had no functionality.[88]

VI. TABLES OF COMPOUNDS

The various nitrochromans that have been prepared are listed in Table 9. In a number of instances, the actual position of the nitro group has not been specifically defined, and this is indicated by a superscript in the table.

Chromanamines that have been reported are listed in Tables 10 through 14; aminodihydronaphthopyrans are indicated as "benzo" bracketed between the two positions of the chroman nucleus which form the point of fusion with the benzene ring. 2,3-Chromandiamines are included with 2-chromanamines and 3,4-chromandiamines with the 3-chromanamines. Chromans with a basic side chain are all listed together in Table 15. Where there is more than one reference to a compound, only a selected m.p. or b.p. is normally given.

TABLE 9. NITROCHROMANS

Mol. formula	Substituents 2	3	4	5	6	7	8	m.p. or b.p. (mm)	Ref.
$C_9H_8BrNO_3$					Br		NO_2[a]	102	32
$C_9H_8BrNO_3$					NO_2[a]		Br	104	32
$C_9H_8ClNO_3$				Cl	NO_2[a]			90	32
$C_9H_8ClNO_3$					Cl		NO_2[a]	64	32
$C_9H_8ClNO_3$					Cl		NO_2	—	1
$C_9H_8ClNO_3$					NO_2[a]	Cl		96	32
$C_9H_8ClNO_3$					NO_2[a]		Cl	112–113	32
$C_9H_8ClNO_3$					NO_2		Cl	99–101	1
$C_9H_8N_2O_5$					NO_2		NO_2	141	29
$C_9H_9NO_3$					NO_2			104	1, 27, 28, 32
$C_9H_9NO_3$						NO_2		90–92	1
$C_{10}H_{10}N_2O_6$					OMe	Di-NO_2[b]		142–143	32
$C_{10}H_{10}N_2O_6$					OMe	Di-NO_2[b]		150–151	31
$C_{10}H_{10}N_2O_6$					DiNO_2[b]		OMe	173	32
$C_{10}H_{10}N_2O_6$					DiNO_2[b]		OMe	178–179	33
$C_{10}H_{11}NO_3$	Me	NO_2						82–85	25
$C_{10}H_{11}NO_3$				Me	NO_2[a]			99–100	32
$C_{10}H_{11}NO_3$					NO_2[a]	Me		74	32
$C_{10}H_{11}NO_3$					NO_2		Me	63–64	1, 32
$C_{10}H_{11}NO_4$	OMe				NO_2			94	4
$C_{10}H_{11}NO_4$					OMe		NO_2[a]	80–81	32

$C_{11}H_{11}Br_2NO_3$	Me, Me	Br	Br		NO_2			130–134	40
$C_{11}H_{12}N_2O_6$	OEt				NO_2		NO_2	69–70	4
$C_{11}H_{13}NO_3$	Me, Me				NO_2			106	2, 41
$C_{11}H_{13}NO_3$	Me			Me	NO_2			72–74	34
								160–165 (2.5)	
$C_{11}H_{13}NO_3$					NO_2	Me	Me	52–53	34
								157–158 (1.2)	
$C_{11}H_{13}NO_4$	OEt				NO_2			52	4
								170 (1.0)	
$C_{12}H_{15}NO_3$	Et	Me			NO_2			120 (0.001)	5
$C_{12}H_{15}NO_3$	Me	Et			NO_2			120 (0.001)	5
$C_{14}H_{19}NO_3$	C_5H_{11}				NO_2			142 (0.001)	5
$C_{15}H_{13}NO_3$					NO_2[a]		Ph	117–118	32
$C_{26}H_{30}N_4O_{11}$	Tetranitro derivative of compound (**22**, R = 6-Me)							167	30
$C_{26}H_{30}N_4O_{11}$	Tetranitro derivative of compound (**22**, R = 7-Me)							145	30
$C_{26}H_{43}NO_3$	Me, TMTD[c]				NO_2			—	39
$C_{27}H_{45}NO_3$	Me, TMTD[c]				NO_2		Me	—	39
$C_{28}H_{34}N_4O_{11}$	Tetranitro derivative of compound (**22**, R = 6, 8-diMe)							155	30
$C_{28}H_{47}NO_3$	Me, TMTD[c]			Me	NO_2	Me		147	38
$C_{29}H_{49}NO_3$	Me, TMTD[c]			Me	NO_2	Me	Me	153	37, 38
$C_{29}H_{49}NO_3$	Me, TMTD[c]			Me	Me	Me	NO_2	Oil	3
$C_{32}H_{44}N_2O_7$	Dinitro derivative of compound (**22**, R = 8-Me, ?-Pr^i)							185	30
$C_{32}H_{44}N_2O_7$	Dinitro derivative of compound (**22**, R = 5-Me, 8-Pr^i)							201	30

[a] Mononitro compound in which the position of the nitro group has not actually been defined.
[b] Dinitro compound in which the positions of the nitro groups have not actually been defined.
[c] TMTD = $[(CH_2)_3CHMe]_3Me$.

TABLE 10. 2-CHROMANAMINES AND 2,3-CHROMANDIAMINES

	Substituents								
Mol. formula	2	3	4	5	6	7	8	m.p.	Ref.
$C_{13}H_{16}BrNO_2$	Mor[a]	Br						117	7
$C_{13}H_{16}ClNO_2$	Mor[a]	Cl						116	7
$C_{13}H_{17}NO$	Me, NH_2		Me, Me			Me		76	42
	N-Acetyl							134	42
$C_{14}H_{18}ClNO$	Pip[b]	Cl						64	7
$C_{16}H_{16}BrNO$	N(Ph)Me	Br						105	7
$C_{16}H_{16}ClNO$	N(Ph)Me	Cl						92	7
$C_{17}H_{24}N_2O_3$	Mor[a]	Mor[a]						179	7
$C_{19}H_{23}NO_2$	Mor[a]	Me, Me		Benzo				154–156	43
$C_{19}H_{28}N_2O$	Pip[b]	Pip[b]						—	7

[a] Mor = *N*-morpholino.
[b] Pip = *N*-piperidino.

TABLE 11. MISCELLANEOUS 2-CHROMANAMINES

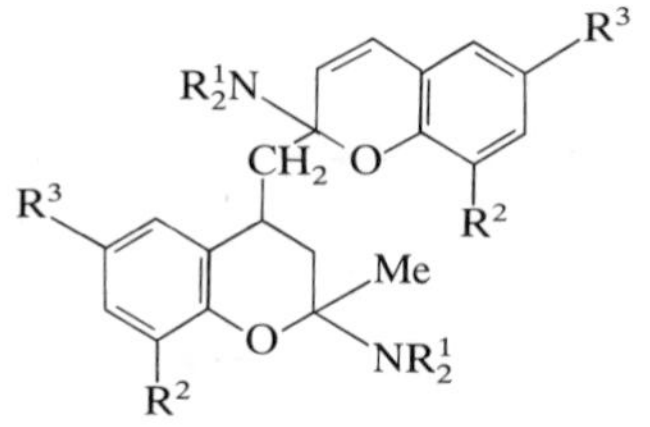

Mol. formula	R^1	R^2	R^3	m.p.	Ref.
$C_{24}H_{26}Br_4N_2O_2$	Me	Br	Br	217–218	6
$C_{24}H_{26}Cl_4N_2O_2$	Me	Cl	Cl	216	6
$C_{24}H_{28}Cl_2N_2O_2$	Me	H	Cl	215	6
$C_{24}H_{30}N_2O_2$	Me	H	H	213	6
	Monoiodomethylate			188–189d	6
$C_{26}H_{34}N_2O_2$	Me	H	Me	221	6
$C_{28}H_{34}Br_4N_2O_2$	Et	Br	Br	192	6
$C_{28}H_{34}Cl_4N_2O_2$	Et	Cl	Cl	193	6
$C_{28}H_{36}Cl_2N_2O_2$	Et	H	Cl	180–181	6
$C_{28}H_{38}N_2O_2$	Et	H	H	182–183	6
$C_{30}H_{42}N_2O_2$	Et	H	Me	185–186	6

TABLE 12. 3-CHROMANAMINES AND 3,4-CHROMANDIAMINES

Mol. formula	Substituents 2	3	4	5	6	7	8	m.p. or b.p. (mm)	Ref.
$C_9H_9Cl_2NO$		NH_2			Cl	Cl		—	—
	Hydrochloride							—	47
$C_9H_{10}ClNO$		NH_2			Cl			—	—
	Hydrochloride							190	9, 47
$C_9H_{10}ClNO$		NH_2				Cl		—	—
	Hydrochloride							156	9, 47, 95
$C_9H_{11}NO$		NH_2						—	—
	Hydrochloride							208	9, 46, 47, 51
	Hydrochloride, (+)3R enantiomer[a]							210	10, 95
	Hydrochloride, (−)3S enantiomer[b]							215	10, 95
	N-Phthaloyl							180	9
$C_9H_{11}NO_2$		NH_2			OH			—	—
	Hydrochloride monohydrate							263	9
$C_9H_{11}NO_2$		NH_2				OH		—	—
	Hydrobromide[c]							258	9
$C_9H_{12}N_2O$		NH_2	NH_2					—	—
	Dihydrochloride							294–296d	53
								295–300d[d]	54
	N,N-Diacetyl							216[e]	54
	N,N-Diacetyl							253[d]	54
$C_{10}H_{13}NO$		NHMe						—	—
	Hydrochloride							118	9, 47, 95
$C_{10}H_{13}NO$	Me	NH_2						85–86 (0.8)[f]	8, 25, 52
	Hydrochloride							260–261	52
	N-Acetyl							112–112.5	8
	N-Phthaloyl							148–149	8
$C_{10}H_{13}NO$	Me	NH_2						101–105 (4.0)[g]	8, 26
	Hydrochloride							217–218	8, 26
	Hydrochloride $0.25H_2O$							214	9, 47

TABLE 12. (*Contd.*)

Mol. formula	Substituents 2	3	4	5	6	7	8	m.p. or b.p. (mm)	Ref.
$C_{10}H_{13}NO$ (*Cont'd*)	*N*-Acetyl							96–97	8
	N-Phthaloyl							158–160	8
$C_{10}H_{13}NO$		NH_2			Me			—	—
	Hydrochloride							218	9, 46, 47, 95
$C_{10}H_{13}NO$		NH_2				Me		—	—
	Hydrochloride							230	9, 47, 95
	N-Benzoyl							172	9
$C_{10}H_{13}NO_2$	Me	NHOH						175–177	8
$C_{10}H_{13}NO_2$		NH_2			OMe			—	—
	Hydrochloride							202	9, 46, 47, 95
	N-Benzoyl							283	9
$C_{10}H_{13}NO_2$		NH_2				OMe		—	—
	Hydrochloride							206	9, 46, 47, 95
$C_{10}H_{13}N_3O$		$NHC(:NH)NH_2$						—	51
	Hydrochloride							—	51
$C_{11}H_{15}NO$		NMe_2						125–126 (15)	49
	Hydrochloride							204	9, 46, 47, 49, 95
	Acid oxalate							166	49
$C_{11}H_{15}NO$		NHEt						—	—
	Hydrochloride							184	9, 47, 95
$C_{11}H_{15}NO$		NH_2		Me		Me		—	—
	Hydrochloride							207.5	48
$C_{11}H_{15}NO$		NH_2			Me	Me		—	—
	Hydrochloride							270–279d	48
								260–265	9, 47, 95
$C_{11}H_{15}NO$		NH_2			Me		Me	—	—
	Hydrochloride $0.5H_2O$							285–287	48
$C_{11}H_{15}NO$		NH_2				Me	Me	—	—
	Hydrochloride							244–246	48

$C_{12}H_{15}NO$		NH ◁					—	—
	Hydrochloride						163–164	46
$C_{12}H_{17}NO$	Me	NMe_2					90–96 (0.9)[f]	8, 52
	Hydrochloride						217–218[f]	8, 52, 94
$C_{12}H_{17}NO$	Me	NMe_2					—	—
	Hydrochloride $0.5H_2O$						195–197[g]	8
$C_{12}H_{17}NO$		NMe_2		Me			—	—
	Hydrochloride						200	9
$C_{12}H_{17}NO$		$NHCHMe_2$					—	—
	Hydrochloride						218	9, 47
							231–233	46
$C_{12}H_{17}NO$		NHEt		Me			—	—
	Hydrochloride						200–202	9, 95
$C_{12}H_{17}NO$		NHEt			Me		—	—
	Hydrochloride						228–230	9, 95
$C_{12}H_{17}NO$		NHMe	Me		Me		—	—
	Hydrochloride						194–195	48
$C_{12}H_{17}NO$		NHMe		Me	Me		—	—
	Hydrochloride						234–235	48
$C_{12}H_{17}NO$		NHMe		Me		Me	—	—
	Hydrochloride						219–220	48
$C_{12}H_{17}NO$	Me	NH_2	Me		Me		—	—
	Hydrochloride						243–245	48
$C_{12}H_{17}NO$	Me	NH_2			Me	Me	—	—
	Hydrochloride						240–241	48
$C_{12}H_{17}NO_2$		NMe_2		OMe			—	—
	Hydrochloride						187–188	46
	Hydrochloride $0.5H_2O$						182	9
$C_{12}H_{17}NO_2$		NHEt		OMe			—	—
	Hydrochloride						170	9
$C_{13}H_{13}NO$		NH_2	Benzo	Benzo			—	—
	Hydrochloride $0.25H_2O$						256.5–257.5	46
$C_{13}H_{18}ClNO$		NH_2	Me	Cl		$CHMe_2$	—	—
	Hydrochloride $0.5H_2O$						223–224	48

TABLE 12 (*Contd.*)

Mol. formula	Substituents 2	3	4	5	6	7	8	m.p. or b.p. (mm)	Ref.
$C_{13}H_{17}NO_3$	OH	Mor[h]						192	7
$C_{13}H_{19}NO$		NMe_2		Me		Me		—	—
	Hydrochloride							243–245	48
$C_{13}H_{19}NO$	Me	$NHCHMe_2$						—	—
	Hydrochloride							279[f]	46
$C_{13}H_{19}NO$		NHEt		Me		Me		—	—
	Hydrochloride							256–257	48
$C_{13}H_{19}NO$		NH_2		Me			$CHMe_2$	—	—
	Hydrochloride							267–269	48
$C_{14}H_{19}NO_2$	OH	Pip[i]						181	7
$C_{14}H_{21}NO$	Me	NEt_2						—	—
	Hydrochloride							162–164	46
$C_{14}H_{21}NO$	Me	NMe_2		Me		Me		—	—
	Hydrochloride							229–230	48
$C_{14}H_{21}NO$	Me	NMe_2				Me	Me	—	—
	Hydrochloride							207–209	48
$C_{14}H_{21}NO$		N(Bu)Me						—	—
	Hydrochloride							131–132	46

$C_{15}H_{17}NO$		NMe_2	Benzo				—	—
	Hydrochloride $0.25H_2O$						252–253	46
$C_{15}H_{23}NO$		NMe_2	Me			$CHMe_2$	—	—
	Hydrochloride						187–189	48
$C_{16}H_{17}NO$		$NHCH_2Ph$					—	47
$C_{17}H_{19}NO$		$NHCH_2Ph$		Me			—	47
$C_{17}H_{19}NO$		$NHCH_2Ph$			Me		—	—
	Hydrochloride						210	9
$C_{17}H_{19}NO_2$		$NHCH_2Ph$		OMe			—	47
$C_{17}H_{19}NO_2$		$NH(CH_2)_2OPh$					—	—
	Hydrochloride						208–210	9, 47, 95
$C_{26}H_{36}N_2O_2$	Me	NHCHMe [i]					—	—
	Dihydrochloride dihydrate						>270d	8

[a] $[\alpha]_D^{25} + 58.7°$ ($c = 0.8H_2O$).[10] [b] $[\alpha]_D^{25} - 47°$ ($c = 0.83H_2O$).[10] [c] Analyzed as dihydrobromide. [d] Trans isomer. [e] Cis isomer. [f] Cis isomer, also referred to as the α-isomer. [g] Trans isomer, also referred to as the β-isomer. [h] Mor = *N*-morpholino. [i] Pip = *N*-piperidino. [j] The substituent at C-4 is

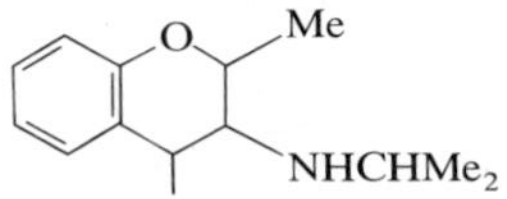

TABLE 13. 4-CHROMANAMINES

Mol. formula	Substituents 2	3	4	5	6	7	8	m.p. or b.p. (mm)	Ref.
$C_9H_{11}NO$			NH_2					103.5 (4.0)	12, 55, 57
	Hydrochloride							231–232	55
	4-Toluenesulfonate							147.5–148	57
$C_{10}H_{13}NO$			NHMe					—	—
	Hydrochloride							178	90, 93
$C_{10}H_{13}NO$		Me	NH_2					—	—
	Hydrochloride							185–187	58
$C_{10}H_{13}NO_2$			NH_2			OMe		—	—
	Hydrochloride							206–207	13
	N-acetyl							170–171	13
$C_{11}H_{12}ClNO_2$			$NHCOCH_2Cl$					163–164	55
$C_{11}H_{14}ClNO_3$			$:\overset{+}{N}H_2Cl^-$		OMe	OMe		234	24
$C_{11}H_{14}N_2O_3$		CH_2OH	N(Me)NO					84	91
$C_{11}H_{15}NO$			NMe_2					—	—
	Hydrochloride							216–218	55
	Methyl 4-toluenesulfonate							130–131	55
$C_{11}H_{15}NO$	Me, Me		NH_2					—	—
	Hydrochloride							>330	11
	N-Benzoyl							149–150	11
$C_{11}H_{15}NO_2$	Me		NH_2			OMe		—	—
	Hydrochloride							204–205	13
	N-Acetyl							184–185	13
$C_{11}H_{15}NO_2$		Me	NH_2			OMe		—	—
	Hydrochloride							205–206	13
	N-Acetyl							150–151	13
$C_{12}H_{14}ClNO_2$			$NHCO(CH_2)_2Cl$					114–115	55
$C_{12}H_{14}N_2O$			$NH(CH_2)_2CN$					—	—
	Hydrochloride							213–214	55
$C_{12}H_{15}NO_3$			$NHCO_2Et$					130	90

$C_{12}H_{16}INO_3$			$:NHMeI^-$		OMe	OMe		189	24
$C_{12}H_{17}NO$	Me, Me		NH_2		Me			—	—
	Hydrochloride							264–265	11
$C_{12}H_{18}N_2O$			$NH(CH_2)_3NH_2$					—	—
	Dihydrochloride							181–184	55
$C_{13}H_{15}NO$			$N(Me)CH_2C{:}CH$					Oil	90, 93
	Hydrochloride							176	90, 93
$C_{13}H_{16}ClNO_2$		Me	$NHCO(CH_2)_2Cl$					126–127.5	89
$C_{13}H_{19}NO$			NEt_2					143.5 (16)	60
$C_{13}H_{19}NO_2$	$CHMe_2$		NH_2			OMe		—	—
	Hydrochloride							195–196	13
	N-Acetyl							151–152	13
$C_{14}H_{21}NO$	Me		NMe_2		Me		Me	—	—
	Hydrochloride							225	56
$C_{14}H_{21}NO_2$	Me, Me		NH_2	Me	OH	Me	Me	—	—
	Hydrochloride							126–127	11
$C_{14}H_{21}NO_2$	Bu^t		NH_2			OMe		—	—
	Hydrochloride							230–231	13
	N-Acetyl							177–178	13
$C_{15}H_{15}NO$			Ph, NH_2					154–155 (1.5)	61, 62
	Hydrochloride							229–230	61, 62
$C_{15}H_{15}NO$		NH_2					Ph	—	—
	Hydrochloride							233–235	56
$C_{15}H_{22}N_2O_2$			$NHCOCH_2NEt_2$					—	—
	Hydrochloride							147–148	55
	Ethiodide							152–155	55
	Methyl 4-toluenesulfonate							125–126	55
$C_{15}H_{24}N_2O$			$NH(CH_2)_2NEt_2$					—	—
	Dihydrochloride							[a]	55
$C_{16}N_{22}N_2O_2$			$NHCOCH_2Pip$[b]					139–141.5	55
	Hydrochloride							197–199	55
	Methyl 4-toluenesulfonate							174–177	55
$C_{16}H_{22}N_2O_3$			$NHCO(CH_2)_2Mor$[c]					128–130	55
	Hydrochloride							203–204	55
	Ethiodide							204–205	55

TABLE 13. (*Contd.*)

Mol formula	Substituents 2	3	4	5	6	7	8	m.p. or b.p. (mm)	Ref.
$C_{16}H_{24}N_2O_2$			$NHCO(CH_2)_2NEt_2$					—	—
	Hydrochloride							133–135	55
	Ethiodide							182–184	55
	Methyl 4-toluenesulfonate							115–116	55
$C_{16}H_{26}N_2O$			$NH(CH_2)_3NEt_2$					—	—
	Dihydrochloride							199–200.5	55
$C_{17}H_{19}NO$			NMe_2				Ph	—	—
	Hydrochloride							230–231.5	56
$C_{17}H_{19}NO$			NHEt				Ph	—	—
	Hydrochloride							185–187	56
	Tartrate							119d	56
$C_{17}H_{24}N_2O_2$			$NHCO(CH_2)_2Pip$[b]					136–137.5	55
	Hydrochloride							146–148	55
	Ethiodide							200–202	55
$C_{17}H_{24}N_2O_3$		Me	$NHCO(CH_2)_2Mor$[c]					—	—
	Hydrochloride							202.5–203	89
$C_{17}H_{25}NO$	Me		Pip[b]		Me		Me	—	—
	Hydrochloride							237–238	56
$C_{19}H_{21}NO$			Pyrr[d]				Ph	—	—
	Hydrochloride							234–237	56
$C_{19}H_{21}NO_2$			Mor[c]				Ph	—	—
	Hydrochloride							206–208.5	56
$C_{19}H_{23}NO$			NEt_2				Ph	—	—
	Hydrochloride							210–212	56

[a] Very hygroscopic. [b] Pip = *N*-piperidino. [c] Mor = *N*-morpholino. [d] Pyrr = *N*-pyrrolidino.

TABLE 14. Bz-CHROMANAMINES

Mol. formula	Substituents 2	3	4	5	6	7	8	m.p. or b.p. (mm)	Ref.
$C_9H_9N_3O_5$					NH_2	NO_2	NO_2	—	—
	N-Acetyl							218–219	1
$C_9H_{10}ClNO$					Cl		NH_2	114–118 (0.001)	1
								47–49	1
$C_9H_{10}N_2O_3$				NO_2	NH_2			62–64	1
	Hydrochloride							160–163d	1
	N-Acetyl							177–180	1
$C_9H_{10}N_2O_3$					NH_2	NO_2		—	—
	N-Acetyl							139–142	1
$C_9H_{10}N_2O_3$					NH_2		NO_2	—	—
	N-Acetyl							188–191	1
$C_9H_{10}N_2O_3$				NO_2			NH_2	79–81	1
	N-Acetyl							159–161	1
$C_9H_{10}N_2O_3$					NO_2		NH_2	84–85	1
	N-Acetyl							182–184	1
$C_9H_{11}NO$				NH_2				98–103 (0.001)	1
$C_9H_{11}NO$					NH_2			74	27, 28
	Picrate							203	27
	N-Acetyl							118	27
$C_9H_{11}NO$						NH_2		140–142 (7.0)	74
	Hydrochloride							158–160d	74
	Chloroplatinate							224–225d	74
	Picrate							182–183d	1, 74
	N-Acetyl							Oil	74
	Benzenesulfonamide							148–148.5	1, 74
$C_9H_{11}NO$							NH_2	—	—
	N-Acetyl							103–105	1
$C_9H_{12}N_2O$				NH_2	NH_2			—	—
	N,N-Diacetyl							238–240	1

TABLE 14. (*Contd.*)

Mol. formula	Substituents 2	3	4	5	6	7	8	m.p. or b.p. (mm)	Ref.
$C_9H_{12}N_2O$					NH_2	NH_2		—	—
	N,N-Diacetyl							241–243	1
$C_9H_{12}N_2O$					NH_2		NH_2	62–63	1
	8-*N*-Acetyl							143–144	1
	N,N-Diacetyl							197–198	1
$C_{10}H_{12}N_2O_3$					NH_2	NO_2	Me	—	—
	N-Acetyl							152–153	1
$C_{10}H_{13}NO$					NHMe			115 (0.001)	92
	Picrate							135–137	92
$C_{10}H_{13}NO$					NH_2			100–105 (0.001)	1
	N-Acetyl							139–140	1
$C_{11}H_{12}ClNO_2$					$NHCOCH_2Cl$			125	27, 28
$C_{11}H_{15}NO$	Me, Me				NH_2			74	2
	N-Acetyl							110–111.5	41
$C_{11}H_{15}NO$				Me	NH_2		Me	—	—
	Hydrochloride							290d	34
$C_{11}H_{15}NO$					NH_2	Me	Me	—	—
	Hydrochloride							300–301d	34
$C_{11}H_{15}NO_2$	OEt				NH_2			120 (0.5)	4
	Hydrochloride							Decomposes	4
$C_{11}H_{16}N_2O_2$	OEt				NH_2		NH_2	Oil	4
$C_{12}H_{17}NO_2$	Me, Me			NH_2		OMe		67–68	35, 72
	Hydrochloride							255–258	35
$C_{12}H_{17}NO_2$	Me, Me				NH_2	OMe		89–90	35
$C_{12}H_{17}NO_2$	Me, Me				NH_2		OMe	—	110
$C_{14}H_{20}N_2O_2$					$NHCONH(CH_2)_3Me$			125–126	92
$C_{14}H_{21}NO$	Me, Me			Me	NH_2	Me	Me	41	70
	Hydrochloride							276	70
	N-Formyl							158	70

$C_{14}H_{21}NO_2$	Me, Me		NHMe	Me	Me	62–64	66, 67
$C_{15}H_{22}N_2O_2$			$NHCOCH_2NEt_2$			180–185 (0.3)	27
						63	27
	Hydrochloride					163	27, 28
	Picrate					201	27
	Ethobromide					188	27, 28
$C_{16}H_{15}ClN_2O_2$			$NHCONHC_6H_4$-4-Cl			226–227	92
$C_{16}H_{15}ClN_2O_2$		Cl	NHCONHPh			204–206	92
$C_{16}H_{15}N_3O_4$			$NHCONHC_6H_4$-3-NO_2			236–238	92
$C_{16}H_{16}N_2O_2$			NHCONHPh			209–210	92
$C_{16}H_{16}N_2OS$			NHCSNHPh			125–126	92
$C_{16}H_{22}N_2O_2$			$NHCOCH_2Pip$[a]			190–195 (0.5)	27
	Hydrochloride					225	27, 28
	Picrate					217	27
$C_{17}H_{18}N_2O_2$			NMeCONHPh			146–148	92
$C_{17}H_{18}N_2O_2$		Me	NHCONHPh			204	92
$C_{17}H_{18}N_2O_3$			$NHCONHC_6H_4$-2-OMe			175–176	92
$C_{17}H_{18}N_2OS$		Me	NHCSNHPh			146–148	92
$C_{17}H_{24}N_2O_2$			$NHCO(CH_2)_2Pip$[a]			—	—
	Hydrochloride					202	27
$C_{18}H_{20}N_2O_3$			$NHCONHC_6H_4$-4-OEt			190–192	92
$C_{19}H_{31}NO$	Me, $(CH_2)_3CHMe_2$	Me	NH_2	Me	Me	—	102
$C_{19}H_{31}NO$	Me, $CH_2(Me)CH(CH_2)_2Me$		NHMe	Me	Me	131–133 (0.01)	15, 66, 67
$C_{19}H_{31}NO$	Me, $CH_2(Me)CH(CH_2)_2Me$	Me	NH_2	Me	Me	172–175 (0.02)	15
$C_{20}H_{33}NO$	Me, $CH_2(Me)CH(CH_2)_2Me$	Me	NHMe	Me	Me	149–157 (0.05)	15, 66, 67
$C_{24}H_{41}NO$	Me, $[(CH_2)_3CHMe]_2Me$	Me	NH_2	Me	Me	—	102
$C_{26}H_{45}NO$ (Tocamin)	Me, TMTD[b]		NH_2			204–208 (0.04)	14–16, 39
	Hydrochloride					153–155	39
$C_{26}H_{45}NO$	Me, $(CH_2)_{12}Me$	Me	NH_2	Me	Me	238 (0.2)	19
						53	
	N-Formyl					235–242 (0.15)	19
						97	
$C_{27}H_{47}NO$	Me, TMTD[b]		NHMe			186–189 (0.007)	66

TABLE 14. (*Contd.*)

Mol. formula	Substituents 2	3	4	5	6	7	8	m.p. or b.p. (mm)	Ref.
$C_{27}H_{47}NO$ (δ-Tocopheramine)	Me, TMTD[b]				NH_2		Me	186–190 (0.007)	14, 98
	Hydrochloride							—	39
$C_{27}H_{47}NO_2$	Me, TMTD[b]			NH_2	OH		Me	—	73
$C_{28}H_{49}NO$	Me, TMTD[b]				NHMe		Me	189–190 (0.005)	14, 15, 17, 66–68
$C_{28}H_{49}NO$	Me, TMTD[b]			Me	NH_2	Me		209–211 (0.08)	14, 15, 98
	Hydrochloride							140–142	65
$C_{28}H_{49}NO$ (β-Tocopheramine)	Me, TMTD[b]			Me	NH_2		Me	200–203 (0.01)	14, 15, 96
$C_{28}H_{49}NO$ (γ-Tocopheramine)	Me, TMTD[b]				NH_2	Me	Me	207–210 (0.03)	14, 15, 96, 104, 107
	N-Formyl							233 (0.01)	17, 68
$C_{28}H_{49}NO$	Me, TMTD[b]			Me		NH_2	Me	—	73
$C_{28}H_{49}NO_2$	Me, TMTD[b]			NH_2	OH	Me	Me	—	73
$C_{29}H_{51}NO$	Me, TMTD[b]				NMe_2		Me	193–195 (0.05)	14, 15, 17, 66, 68, 103, 104
$C_{29}H_{51}NO$	Me, TMTD[b]				NHEt		Me	—	68
$C_{29}H_{51}NO$	Me, TMTD[b]			Me	NHMe	Me		195–200 (0.08)	14, 15, 66, 67
$C_{29}H_{51}NO$	Me, TMTD[b]			Me	NHMe		Me	210–213 (0.03)	14–17, 66–68, 96, 97, 103, 104
$C_{29}H_{51}NO$	Me, TMTD[b]				NHMe	Me	Me	190–195 (0.01)	14–17, 66–69, 96, 98, 100–103, 107, 108, 111, 112
	$3,4\text{-}^{14}C$							—	69

$C_{29}H_{51}NO$ (α-Tocopheramine)	Me, TMTD[b]	Me	NH_2	Me	Me	214–218 (0.015)	14–19, 37, 63, 64, 66, 68, 69, 96, 98, 100–109, 112–114
	Hydrochloride					158–160	63, 65, 99
	Oxalate					153–154	63
	N-Acetyl					235–240 (0.08)	17, 68
	3,4-^{14}C					—	69
$C_{29}H_{51}NO$	Me, TMTD[b]	Me	Me	Me	NH_2	—	3
$C_{30}H_{53}NO$	Me, TMTD[b]	Me	NMe_2		Me	—	68
$C_{30}H_{53}NO$	Me, TMTD[b]		NMe_2	Me	Me	183–186 (0.03)	14, 15, 68, 107
$C_{30}H_{53}NO$	Me, TMTD[b]	Me	NHEt		Me	—	68
$C_{30}H_{53}NO$	Me, TMTD[b]		NHEt	Me	Me	195–197 (0.05)	14–17, 66–68, 104
$C_{30}H_{53}NO$	Me, TMTD[b]	Me	NHMe	Me	Me	200–202 (0.06)	14, 15, 67, 68
$C_{31}H_{55}NO$	Me, TMTD[b]		NEt_2		Me	—	68
$C_{31}H_{55}NO$	Me, TMTD[b]	Me	NMe_2	Me	Me	200–205 (0.02)	14–17, 66, 68, 107
$C_{31}H_{55}NO$	Me, TMTD[b]		$NHPr^i$	Me	Me	—	68
$C_{31}H_{55}NO$	Me, TMTD[b]	Me	NHEt	Me	Me	211–214 (0.01)	14, 15, 17, 66–68
$C_{32}H_{57}NO$	Me, TMTD[b]	Me	NEt_2		Me	—	68
$C_{32}H_{57}NO$	Me, TMTD[b]		NEt_2	Me	Me	—	68
$C_{33}H_{59}NO$	Me, TMTD[b]	Me	NEt_2	Me	Me	—	68
$C_{34}H_{61}NO$	Me, $[(CH_2)_3CHMe]_4Me$	Me	NH_2	Me	Me	—	102
$C_{39}H_{71}NO$	Me, $[(CH_2)_3CHMe]_5Me$	Me	NH_2	Me	Me	—	102

[a] Pip = *N*-piperidino. [b] TMTD = $[(CH_2)_3CHMe]_3Me$.

TABLE 15. CHROMANS WITH A BASIC SIDE CHAIN

Mol. formula	Substituents 2	3	4	5	6	7	8	m.p. or b.p. (mm)	Ref.
$C_{10}H_{12}ClNO$	CH_2NH_2				Cl			—	75
$C_{10}H_{13}NO$	CH_2NH_2							173–182 (1.3)	75, 76
$C_{10}H_{13}NO$		CH_2NH_2						66–68 (0.04)	78
$C_{11}H_{15}NO$	CH_2NH_2						Me	—	75
$C_{11}H_{15}NO_2$	CH_2NH_2					OMe		—	75
$C_{11}H_{15}NO_2$	$CHOHCH_2NH_2$							—	—
	Hydrochloride							226–228	84, 85
$C_{11}H_{15}N_3O$	$CH_2NHC(:NH)NH_2$							—	—
	Sulfate							216–220	76
$C_{11}H_{15}N_3O$		$CH_2NHC(:NH)NH_2$						110	78
	Sulfate							270–280	78
$C_{12}H_{17}NO$	CH_2NMe_2							140 (13.0)	20
	Methobromide							199–202	20
$C_{12}H_{17}NO$	CH_2NH_2					Me	Me	—	75
$C_{13}H_{19}NO_2$	CH_2NMe_2		OH				Me	102.5–104	20
$C_{14}H_{19}NO_2$	$CHOHCH_2NHCH_2CH:CH_2$							96–97	85
$C_{14}H_{21}NO_2$	$CHOHCH_2NHCHMe_2$							—	—
	Hydrochloride							175–176	84
	Hydrochloride $0.25H_2O$							171–173	85
$C_{14}H_{21}NO_2$					$CHOHCH_2NHPr^i$			105–106	86–88
$C_{15}H_{21}NO_3$	$CHOHCH_2Mor$[a]							—	—
	Acid oxalate							120–121	85
$C_{15}H_{22}BrNO_2$	$CHOHCH_2NHBu^t$				Br			108–109	84, 85
$C_{15}H_{23}NO$		Me	$(CH_2)_3NMe_2$					112–120 (0.65–0.8)	81
$C_{15}H_{23}NO$			$(CH_2)_3NMe_2$		Me			—	—
	Hydrochloride							145–147	81
$C_{15}H_{23}NO$	$CHOHCH_2NEt_2$							—	—
	Picrate							123–124	85
$C_{15}H_{23}NO_2$[b]	$CHOHCH_2NHBu^t$							108–109	84, 85
	Hydrochloride							248–249	84, 85

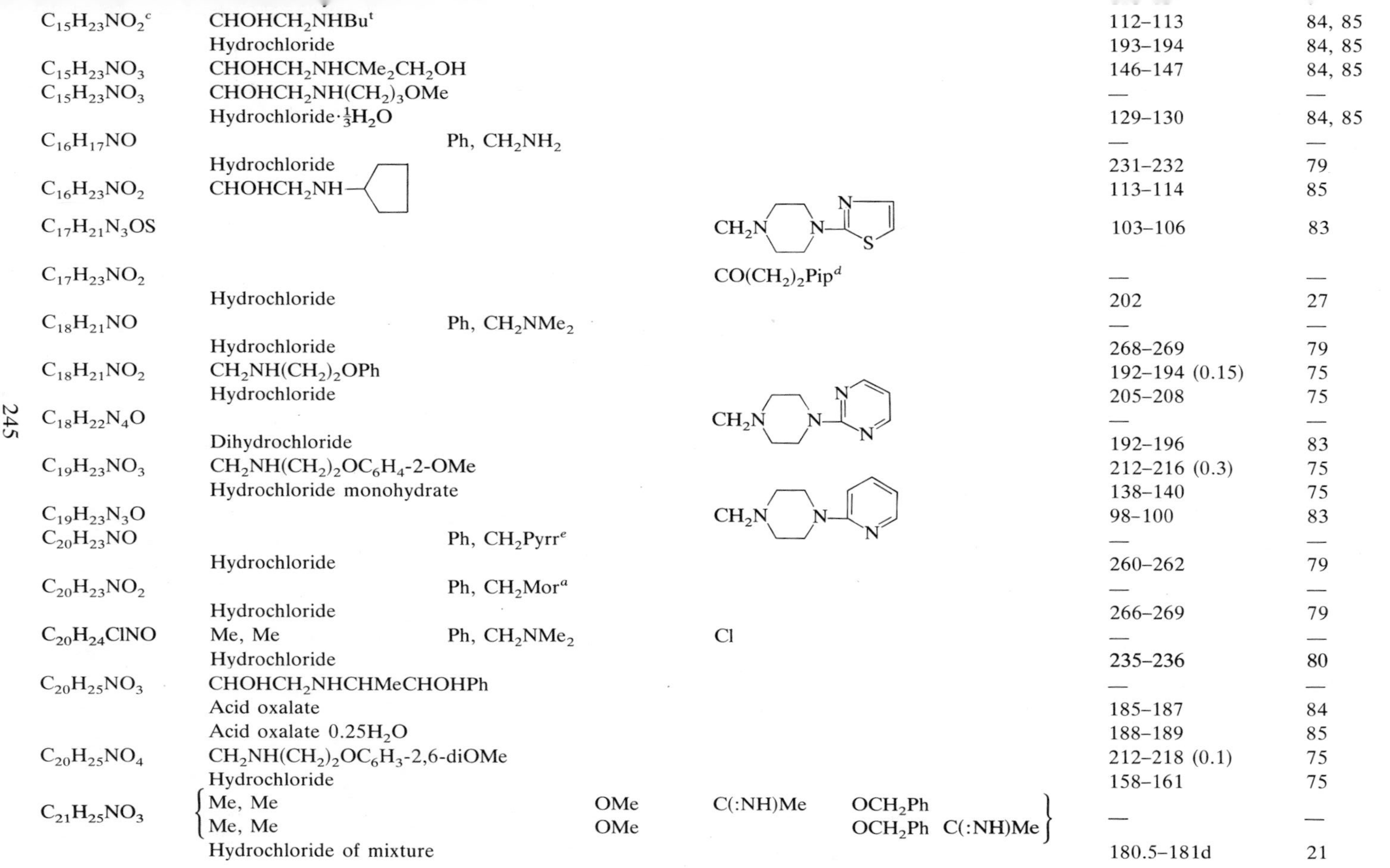

$C_{15}H_{23}NO_2$[c]	$CHOHCH_2NHBu^t$						112–113	84, 85
	Hydrochloride						193–194	84, 85
$C_{15}H_{23}NO_3$	$CHOHCH_2NHCMe_2CH_2OH$						146–147	84, 85
$C_{15}H_{23}NO_3$	$CHOHCH_2NH(CH_2)_3OMe$						—	—
	Hydrochloride·$\frac{1}{3}H_2O$						129–130	84, 85
$C_{16}H_{17}NO$		Ph, CH_2NH_2					—	—
	Hydrochloride						231–232	79
$C_{16}H_{23}NO_2$	$CHOHCH_2NH$–(cyclopentyl)						113–114	85
$C_{17}H_{21}N_3OS$				CH_2N N–(thiazol-2-yl)			103–106	83
$C_{17}H_{23}NO_2$				$CO(CH_2)_2Pip$[d]			—	—
	Hydrochloride						202	27
$C_{18}H_{21}NO$		Ph, CH_2NMe_2					—	—
	Hydrochloride						268–269	79
$C_{18}H_{21}NO_2$	$CH_2NH(CH_2)_2OPh$						192–194 (0.15)	75
	Hydrochloride						205–208	75
$C_{18}H_{22}N_4O$				CH_2N N–(pyrimidin-2-yl)			—	—
	Dihydrochloride						192–196	83
$C_{19}H_{23}NO_3$	$CH_2NH(CH_2)_2OC_6H_4$-2-OMe						212–216 (0.3)	75
	Hydrochloride monohydrate						138–140	75
$C_{19}H_{23}N_3O$				CH_2N N–(pyridin-2-yl)			98–100	83
$C_{20}H_{23}NO$		Ph, CH_2Pyrr[e]					—	—
	Hydrochloride						260–262	79
$C_{20}H_{23}NO_2$		Ph, CH_2Mor[a]					—	—
	Hydrochloride						266–269	79
$C_{20}H_{24}ClNO$	Me, Me	Ph, CH_2NMe_2		Cl			—	—
	Hydrochloride						235–236	80
$C_{20}H_{25}NO_3$	$CHOHCH_2NHCHMeCHOHPh$						—	—
	Acid oxalate						185–187	84
	Acid oxalate $0.25H_2O$						188–189	85
$C_{20}H_{25}NO_4$	$CH_2NH(CH_2)_2OC_6H_3$-2,6-diOMe						212–218 (0.1)	75
	Hydrochloride						158–161	75
$C_{21}H_{25}NO_3$	{ Me, Me		OMe	C(:NH)Me	OCH_2Ph	}	—	—
	{ Me, Me		OMe		OCH_2Ph	C(:NH)Me }		
	Hydrochloride of mixture						180.5–181d	21

TABLE 15. (*Contd.*)

Mol. formula	Substituents 2	3	4	5	6	7	8	m.p. or b.p. (mm)	Ref.
$C_{21}H_{27}NO_2$	$CHOHCH_2NHCHMe(CH_2)_2Ph$							—	—
	Hydrochloride							192–194	84, 85
$C_{21}H_{27}NO_2$	Me, Me		Ph, CH_2NMe_2		OMe			—	—
	Hydrochloride monohydrate							121–123	80
$C_{21}H_{27}NO_3$	$CHOHCH_2NHCMe_2CH_2OPh$							—	—
	Hydrochloride $0.5H_2O$							116–117	84, 85
$C_{21}H_{27}NO_4$	$CH_2NH(CH_2)_2OC_6H_3$-2,6-diOMe						Me	224–226 (0.3)	75
	Hydrochloride							153–154	75
$C_{22}H_{26}ClNO$	Me, Me		Ph, CH_2Pyrr^e		Cl			—	—
	Hydrochloride							248–251	80
$C_{22}H_{27}NO$			Ph, CH_2NH–C_6H_{11}					—	—
	Oxalate							216–218	79
$C_{22}H_{28}ClNO_2$	$CHOHCH_2NHCMe_2(CH_2)_2C_6H_4$-4-Cl							—	—
	Hydrochloride $0.5H_2O$							153–155	84, 85
$C_{22}H_{29}NO_4$	$CH_2NH(CH_2)_2OC_6H_3$-2,6-diOMe					Me	Me	230–234 (0.3)	75
	Hydrochloride							144–146	75
$C_{23}H_{29}ClN_2O$	Me, Me		Ph, $CH_2N(CH_2CH_2)_2NMe$		Cl			—	—
	Hydrochloride							244–245	80
$C_{23}H_{29}NO_2$	Me, Me		Ph, CH_2Pyrr^e		OMe			—	—
	Hydrochloride							232–234	80
$C_{23}H_{30}N_2O$	Me, Me		Ph, $CH_2N(CH_2CH_2)_2NMe$					—	—
	Dihydrochloride							235–255	80
$C_{27}H_{31}NO_4$	$CH_2NH(CH_2)_2C_6H_3$-3,4-diOMe								
			Ph, CH_2OH					108–109	23
	Hydrochloride							233–235d	23
	Formate							156–157	23
$C_{33}H_{57}NO_3$	Me, TMTD[f]			CH_2Mor^a	OH	Me	Me	35–36	82
$C_{35}H_{57}NO_2$	Me, $CH_2[CH_2CH{:}CMeCH_2]_3H$			CH_2NPr_2	OH	Me	Me	Oil	22
$C_{60}H_{97}NO_2$	Me, $CH_2[CH_2CH{:}CMeCH_2]_8H$			CH_2NPr_2	OH	Me	Me	—	22

VII. REFERENCES

1. G. Brancaccio, G. Lettieri, and R. Viterbo, *J. Heterocycl. Chem.*, **10,** 623 (1973).
2. T. R. Govindachari, S. Prabhakar, V. N. Ramachandran, and B. R. Pai, *Indian J. Chem.*, **9,** 1031 (1971).
3. J. Vance and R. Bentley, *Bioorg. Chem.*, **1,** 345 (1971).
4. G. Descotes, J. C. Martin, and N. Mathicolonis, *Bull. Soc. Chim. Fr.*, 1077 (1972).
5. R. R. Schmidt, *Tetrahedron Lett.*, 5279 (1969).
6. R. Kuhn, H. R. Hensel, and D. Weiser, *Justus Liebigs Ann. Chem.*, **611,** 83 (1958).
7. G. Descotes and D. Missos, *Bull. Soc. Chim. Fr.*, 696 (1972).
8. H. Booth, D. Huckle, and I. M. Lockhart, *J. Chem. Soc., Perkin Trans. II*, 227 (1973).
9. N. Sarda, A. Grouiller, and H. Pacheco, *Eur. J. Med. Chem.*, **11,** 251 (1976).
10. N. Sarda, A. Grouiller, and H. Pacheco, *Tetrahedron Lett.*, 271 (1976).
11. N. V. Dudykina, L. M. Meshcheryakova, and V. A. Zagorevskii, *Chem. Heterocycl. Comp.*, 324 (1969).
12. K. Nagarajan, C. L. Kulkarni, and A. Venkateswarlu, *Indian J. Chem.*, **12,** 247 (1974).
13. S. Ito, *Bull. Chem. Soc. Japan*, **43,** 1824 (1970).
14. U. Schwieter, R. Tamm, H. Weiser, and O. Wiss, *Helv. Chim. Acta*, **49,** 2297 (1966).
15. R. Ruegg, H. Mayer, P. Schudel, U. Schwieter, R. Tamm, and O. Isler, *Wiss. Veroeff. Deut. Ges. Ernaehr.*, **16,** 14 (1967); *Chem. Abstr.*, **68,** 105392 (1968).
16. W. Boguth, G. Herrmann, and H. Niemann, *Experientia*, **24,** 232 (1968).
17. Hoffmann–La Roche A.-G., Neth. Appl. 6,414,649 (1965); *Chem. Abstr.*, **64,** 707 (1966).
18. W. Boguth and R. Hackel, *Z. Physiol. Chem.*, **342,** 172 (1965).
19. R. Behnisch and W. Salzer, U.S. Patent 2,359,461 (1944); *Chem. Abstr.*, **39,** 1256 (1945).
20. E. R. Clark and S. G. Williams, *J. Chem. Soc.* (*B*), 859 (1967).
21. F. N. Lahey and R. V. Stick, *Austr. J. Chem.*, **26,** 2291 (1973).
22. H. Mayer, J. Metzger, and O. Isler, *Helv. Chim. Acta*, **50,** 1376 (1967).
23. H. E. Zaugg, R. W. DeNet, and E. T. Kimura, *J. Med. Chem.*, **5,** 430 (1962).
24. F. M. Dean and K. B. Hindley, *Tetrahedron Lett.*, 1445 (1972).
25. I. M. Lockhart, Brit. Patent 1,168,228 (1969); *Chem. Abstr.*, **72,** 31618 (1970).
26. G. B. Bachmann and H. A. Levine, *J. Am. Chem. Soc.*, **70,** 599 (1948).
27. V. Hach, *Collect. Czech. Chem. Commun.*, **24,** 3136 (1959).
28. V. Hach, Czech. Patent 91,157 (1959); *Chem. Abstr.*, **55,** 3618 (1961).
29. G. Chatelus, *Ann. Chim.* (Paris), **4,** 505 (1949).
30. J. B. Niederl, *J. Am. Chem. Soc.*, **51,** 2426 (1929).
31. A. O. Fitton and G. R. Ramage, *J. Chem. Soc.*, 5426 (1963).
32. L. W. Deady, R. D. Topsom, and J. Vaughan, *J. Chem. Soc.*, 5718 (1965).
33. P. S. Bramwell and A. O. Fitton, *J. Chem. Soc.*, 3882 (1965).
34. J. L. G. Nilsson and H. Selander, *Acta Chem. Scand.*, **24,** 2885 (1970).
35. J. R. Beck, R. Kwok, R. N. Booher, A. C. Brown, L. E. Patterson, P. Pranc, B. Rockey, and A. Pohland, *J. Am. Chem. Soc.*, **90,** 4706 (1968).
36. O. Hromatka, Ger. Patent 707,956 (1941); *Chem. Abstr.*, **37,** 2889 (1943).
37. O. Hromatka, Ger. Patent 741,688 (1943); *Chem. Abstr.*, **39,** 2768 (1945).
38. O. Hromatka, U.S. Patent 2,343,773 (1944); *Chem. Abstr.*, **38,** 3421 (1944).
39. R. A. Stein, *J. Med. Chem.*, **10,** 162 (1967).
40. E. A. Watts, Brit. Patent 1,495,526 (1977), Ger. Offen. 2,523,281 (1975); *Chem. Abstr.*, **84,** 121650 (1976).
41. J. A. Miller and H. C. S. Wood, Brit. Patent 1,121,307 (1968); *Chem. Abstr.*, **69,** 96471 (1968).
42. W. Baker, R. F. Curtis, and J. F. W. McOmie, *J. Chem. Soc.*, 1774 (1952).

43. M. Von Strandtmann, M. P. Cohen, and J. Shavel, *Tetrahedron Lett.*, 3103 (1965).
44. I. M. Lockhart, in *Chromenes, Chromanones, and Chromones*, G. P. Ellis, Ed., John Wiley, New York, 1977, Chap. V.
45. I. M. Lockhart, in *Chromenes, Chromanones, and Chromones*, G. P. Ellis, Ed., John Wiley, New York, 1977, Chap. III.
46. I. M. Lockhart, D. Huckle, and M. Wright, unpublished observations.
47. N. Sarda, A. Grouiller, and H. Pacheco, *C.R. Acad. Sci., Paris, Ser. C*, **279,** 281 (1974).
48. I. M. Lockhart and S. A. Foard, *J. Med. Chem.*, **15,** 863 (1972).
49. A. Gabert and H. Normant, *C.R. Acad. Sci. Paris*, **235,** 1407 (1952).
50. I. M. Lockhart, in *Chromenes, Chromanones, and Chromones*, G. P. Ellis, Ed., John Wiley, New York, 1977, Chap. IV.
51. CIBA Limited, Fr. Patent 1,465,838 (1967); *Chem. Abstr.*, **68,** 68884 (1968).
52. I. M. Lockhart, Brit. Patent 1,151,474 (1968); *Chem. Abstr.*, **71,** 101710 (1969).
53. N. V. Dudykina and V. A. Zagorevskii, *J. Org. Chem. USSR*, **2,** 2179 (1966).
54. I. Isaka, K. Kubo, M. Takashima, and M. Murakami, *Yakugaku Zasshi*, **87,** 1556 (1967); *Chem. Abstr.*, **69,** 27175 (1968).
55. V. A. Zagorevskii and N. V. Dudykina, *J. Gen. Chem. USSR*, **32,** 3856 (1962).
56. Schering A.-G., Brit. Patent 758,313 (1956); *Chem. Abstr.*, **51,** 9708 (1957).
57. V. A. Zagorevskii and N. V. Dudykina, *J. Gen. Chem. USSR*, **33,** 317 (1963).
58. N. V. Dudykina and V. A. Zagorevskii, *Sintez Prirodn. Soedin., Analogov i Fragmentov. Akad. Nauk SSSR, Otd. Obshch. Tekhn. Khim.*, 134 (1965), *Chem. Abstr.*, **65,** 683 (1966).
59. I. M. Lockhart, unpublished observations.
60. P. Maitte, *Ann. Chim.* (Paris), **9,** 431 (1954).
61. H. E. Zaugg, R. W. DeNet, and R. J. Michaels, Belg. Patent 618,529 (1962); *Chem. Abstr.*, **59,** 7500 (1963).
62. H. E. Zaugg, R. W. DeNet, and R. J. Michaels, *J. Org. Chem.*, **26,** 4821 (1966).
63. L. I. Smith, W. B. Renfrow, and J. W. Opie, *J. Am. Chem. Soc.*, **64,** 1082 (1942).
64. O. Hromatka, Ger. Patent 703,957 (1941); *Chem. Abstr.*, **36,** 1144 (1942).
65. O. Hromatka, U.S. Patent 2,358,286-7 (1944); *Chem. Abstr.*, **39,** 1513 (1945).
66. U. Schwieter, R. Tamm, and W. Schlegel, Swiss Patent 463,533 (1968); *Chem. Abstr.*, **71,** 49771 (1969).
67. U. Schwieter, R. Tamm, and W. Schlegel, Swiss Patent 463,534 (1968); *Chem. Abstr.*, **71,** 81579 (1969).
68. Hoffmann–La Roche A.-G., Brit. Patent 1,043,486 (1966); Neth. Appl. 6,414,649 (1965); *Chem. Abstr.*, **64,** 707 (1966).
69. U. Gloor, J. Wuersch, U. Schwieter, and O. Wiss, *Helv. Chim. Acta*, **49,** 2303 (1966).
70. O. Hromatka, U.S. Patent 2,354,317 (1944); *Chem. Abstr.*, **39,** 3538 (1945).
71. R. Behnisch, U.S. Patent 2,334,743 (1943); *Chem. Abstr.*, **38,** 2795 (1944).
72. R. Kwok and A. Pohland, Ger. Offen. 1,933,153 (1970); *Chem. Abstr.*, **72,** 90422 (1970).
73. M. L. Quaife, U.S. Patent 2,499,778 (1950); *Chem. Abstr.*, **44,** 4376 (1950).
74. W. C. Wilson and R. Adams, *J. Am. Chem. Soc.*, **45,** 528 (1923).
75. P. N. Green and M. Shapero, Brit. Patent 1,296,431 (1972), *Chem. Abstr.*, **78,** 111124 (1973).
76. J. Augstein, A. M. Monro, and T. I. Wrigley, Brit. Patent 1,004,468 (1965); *Chem. Abstr.*, **63,** 18036 (1965); J. Augstein, A. M. Monro, G. W. H. Potter, and P. Scholfield, *J. Med. Chem.*, **11,** 844 (1968).
77. J. Schmutz, R. Hirt, F. Kunzle, E. Eichenberger, and H. Lauener, *Helv. Chim. Acta*, **36,** 620 (1953).
78. J. Augstein and A. M. Monro, Brit. Patent 1,043,857 (1966); *Chem. Abstr.*, **65,** 18562 (1966).
79. H. E. Zaugg, R. W. DeNet, and R. J. Michaels, U.S. Patent 3,156,688 (1964); *Chem.*

Abstr., **62,** 2764 (1965).
80. H. E. Zaugg, J. E. Leonard, R. W. DeNet, and D. L. Arendsen, *J. Heterocycl. Chem.*, **11,** 797 (1974).
81. P. J. Hattersley, I. M. Lockhart, and M. Wright, *J. Chem. Soc. (C)*, 217 (1969).
82. T. Nakamura and K. Miyao, Jap. Patent 5,936 (1966); *Chem. Abstr.*, **65,** 5498 (1966).
83. G. Regnier, R. Canevari, M. Laubie, and J.-C. Poignant, Brit. Patent 1,407,552 (1975); Ger. Offen. 2,415,082 (1974); *Chem. Abstr.*, **82,** 4307 (1975).
84. Imperial Chemical Industries, Neth. Appl. 6,500,863 (1965); *Chem. Abstr.*, **64,** 3493 (1966).
85. R. Howe, B. S. Rao, and M. S. Chodnekar, *J. Med. Chem.*, **13,** 169 (1970).
86. Imperial Chemical Industries, Belg. Patent 633,973 (1963); *Chem. Abstr.*, **61,** 673 (1964).
87. Imperial Chemical Industries, Neth. Appl. 294,296 (1965); *Chem. Abstr.*, **64,** 2101 (1966).
88. M. S. Chodnekar, A. F. Crowther, W. Hepworth, R. Howe, B. J. McLoughlin, A. Mitchell, B. S. Rao, R. P. Slatcher, L. H. Smith, and M. A. Stevens, *J. Med. Chem.*, **15,** 49 (1972).
89. N. V. Dudykina and V. A. Zagorevskii, *Khim. Geterotsikl. Soedin.*, 250 (1967); *Chem. Abstr.*, **68,** 114579 (1968).
90. J. R. Boissier and R. Ratouis, Fr. Patent 1,584,755 (1970); *Chem. Abstr.*, **74,** 111911 (1971).
91. W. Oppolzer, *Tetrahedron Lett.*, 3091 (1970).
92. G. Lettieri, G. Brancaccio, A. Larrizza, and R. Viterbo, *J. Med. Chem.*, **13,** 584 (1970).
93. J. R. Boissier and R. Ratouis, *Fr. M* 7499 (1970); *Chem. Abstr.*, **75,** 35743 (1971).
94. T. M. Itil, J. Marasa, A. Bigelow, and B. Saletu, *Curr. Ther. Res. Clin. Exp.*, **16,** 80 (1974); *Chem. Abstr.*, **81,** 33512 (1974).
95. N. Sarda, J. Legehand, H. Vial, A. Ramirez, and H. Pacheco, *Eur. J. Med. Chem.*, **11,** 257 (1976).
96. J. G. Bieri and E. L. Prival, *Biochemistry*, **6,** 2153 (1967).
97. J. G. Bieri and K. E. Mason, *J. Nutr.*, **96,** 192 (1968).
98. E. Soendergaard and H. Dam, *Z. Ernachrungswiss.*, **10,** 71 (1970); *Chem. Abstr.*, **74,** 10904 (1971).
99. M. C. Farber, A. E. Milman, and A. T. Milhorat, *Z. Physiol. Chem.*, **295,** 318 (1953).
100. U. Gloor, F. Weber, and O. Wiss, *Wiss. Veroeff. Deut. Ges. Ernaehr.*, **16,** 66 (1967); *Chem. Abstr.*, **67,** 106398 (1967).
101. R. Blomstrand and L. Forsgren, *Int. Z. Vitaminforsch.*, **38,** 328 (1968); *Chem. Abstr.*, **70,** 9647 (1969).
102. S. Blaehser and W. Boguth, *Int. Z. Vitaminforsch.*, **39,** 163 (1969); *Chem. Abstr.*, **71,** 79530 (1969).
103. Hoffmann–La Roche, A.-G., Brit. Patent 1,043,487 (1966); *Chem. Abstr.*, **65,** 19229 (1966).
104. U. Schwieter, W. Schlegel, and R. Tamm, Ger. Patent 1,235,918 (1967); *Chem. Abstr.*, **69,** 21909 (1968).
105. Hoffmann–La Roche A.-G., Brit. Patent 1,193,027 (1970); *Chem. Abstr.*, **73,** 54790 (1970).
106. G. Pongracz, *Int. J. Vitam. Nutr. Res.*, **43,** 517 (1973); *Chem. Abstr.*, **80,** 144456 (1974).
107. H. S. Olcott, J. Van der Veen, and T. Koide, *Nippon Suisan. Gakkaishi*, **36,** 844 (1970); *Chem. Abstr.*, **73,** 129671 (1970).
108. R. Patzelt–Wenczler, *Int. J. Vitam. Nutr. Res.*, **42,** 285 (1972); *Chem. Abstr.*, **77,** 135033 (1972).
109. K. Kawai, M. Nakao, T. Nakao, and G. Katsui, *Am. J. Clin. Nutr.*, **27,** 987 (1974); *Chem. Abstr.*, **81,** 165384 (1974).
110. A. Meidell, *Med. Norsk Farm. Selskap*, **27,** 101 (1965), *Chem. Abstr.*, **63,** 18008 (1965).

111. P. A. Murphy, S. J. Lin, H. S. Olcott, and J. J. Windle, *Lipids*, **11,** 296 (1976).
112. W. Boguth, R. Patzelt–Wenczler, and R. Ropges, *Int. J. Vitam. Nutr. Res.*, **41,** 21 (1971); *Chem. Abstr.*, **75,** 4083 (1971).
113. J. Baraud, R. Bernhard, C. Cassagne, and L. Genevois, *Qual. Plant Mater. Veg.*, **17,** 217 (1969); *Chem. Abstr.*, **72,** 712 (1970).
114. J. J. Gilbert, *Physiol. Zool.*, **48,** 404 (1975).

CHAPTER VII

Bz-Oxochromans

I. M. LOCKHART

BOC Limited, London SW19 3UF, U.K.

1. INTRODUCTION

The compounds discussed in this chapter are derived from chromans and have one or more carbonyl groups substituted in the homocyclic ring. As an inevitable consequence of the substitution, the aromatic nature is reduced and a variety of compounds are possible depending upon the degree of saturation of the aromatic ring of the chroman system. The number of different types of compound is still further increased in the series derived from the various dihydronaphthopyrans according to which ring carries the carbonyl substituent. With a multiplicity of possibilities, the field covered in this chapter is very fragmented and gives little scope for generalizations.

In line with the practice adopted by *Chemical Abstracts* up to and including 1971, the oxochromans will be referred to as 5-, 6-, and 7-chromanones with appropriate qualification to indicate the degree of saturation of the homocyclic ring. There do not appear to be any examples of 8-chromanones. Dioxo derivatives are represented by 5,6-, 5,7-, and 5,8-chromandiones. Chromanones with a carbonyl group in the heterocyclic ring have, with one or two exceptions, been excluded. (These compounds are discussed in Volume 31 of this series entitled *Chromenes, Chromanones, and Chromones*.) The numbering used in the four fully saturated ring systems is indicated in structures (**1**) through (**4**).

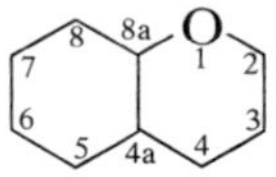

(**1**) 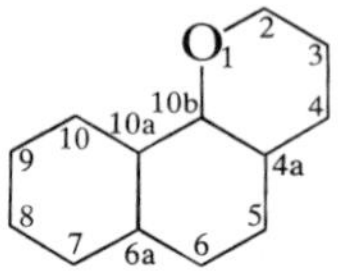 (**2**)

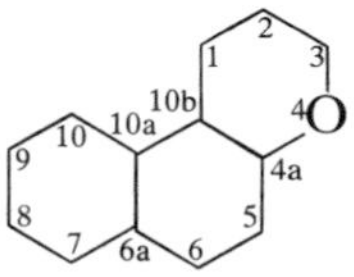

(**3**) 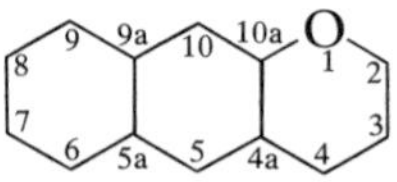 (**4**)

The interest in many of the compounds derives from the fact that they have been isolated from natural sources or were prepared in the course of structural investigations on natural products. α-Lapachone (**5**) and β-lapachone (**6**), the two most widely studied compounds in this group, have been isolated from a number of South American and other woods. These two compounds were first isolated and their structures assigned at the end of the nineteenth century.[1–3]

Chromans with a side-chain oxo substituent are also included in this review. This group effectively comprises the chromans having an acyl or aroyl substituent in the aromatic part of the molecule. Acylchromanones are considered under the appropriate bz-chromanone.

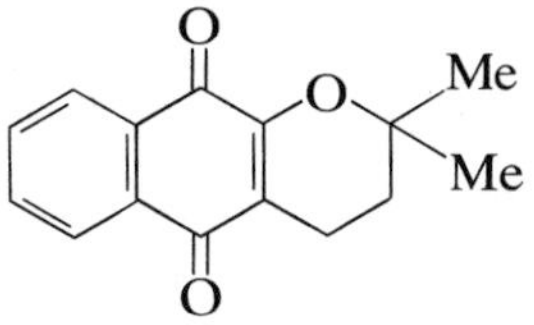

(5) 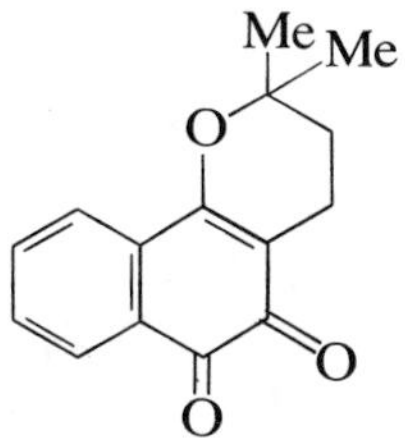 (6)

II. Bz-OXOCHROMANS IN NATURAL PRODUCT CHEMISTRY

A substantial amount of the interest in bz-oxochromans derives from either their occurrence as natural products or their formation in the course of structural studies on naturally occurring substances. Both aspects are discussed in this section.

1. 5-Chromanones

In investigations of the structure of the antibiotic narbomycin, acid hydrolysis of dihydronarbomycin afforded, among other products, crystalline anhydrodihydroaglycon E. On heating with alkali under pressure, the latter product afforded a number of substances that included the 5-chromanone (**7**).[4] The formation of an oxime, the presence of bands in the infrared spectrum at 1717 (strong) and 1672 cm^{-1} (weak), and the formation of acetic and propionic acids on Kuhn–Roth oxidation confirmed the assignment of the 5-chromanone structure.[4]

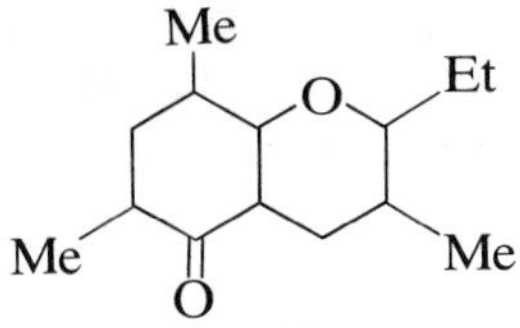

(7)

In studies of hop constituents, it was demonstrated that colupulone (**8**) afforded the dimethylallyl-5-chromanone (**9**) together with the 7-chromanone analog (**10**) on oxidation with sodium persulfate[5,6] and further 5-chromanone (**11**).[6] The latter compound (**11**), which may be a short-lived intermediate in the acid degradation of colupulone, was characterized as a pale-yellow solid. Structures were derived from the uv, ir, nmr, and mass spectra.[6] More detailed studies of the mass spectrum of the chromanone (**9**) showed that it initially loses an alkenyl side chain to form a highly conjugated ion (**12**) (*m*/*e* 331) and a fully aromatic ion (**13**) (*m*/*e* 332).[7]

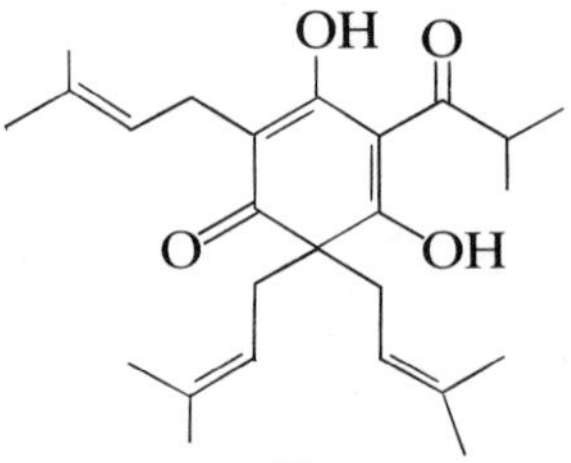

(**8**)

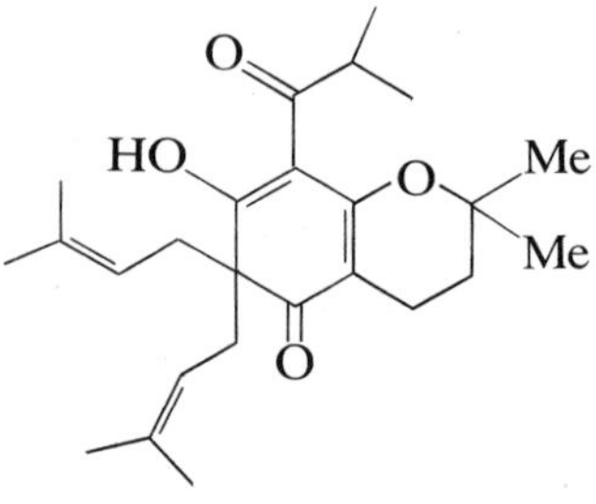

(**9**)

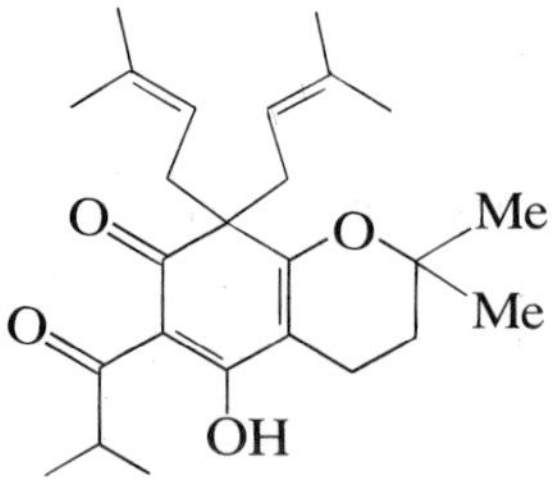

(**10**)

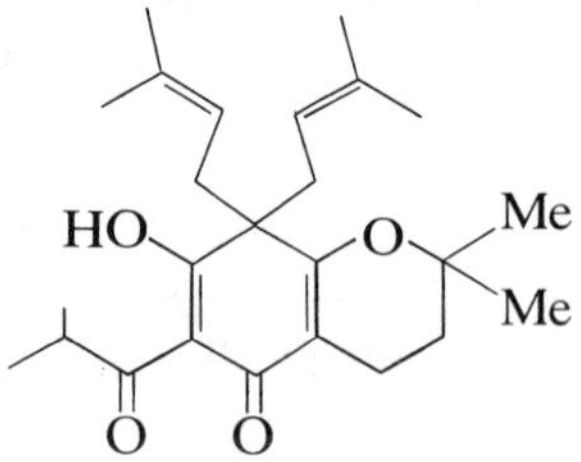

(**11**)

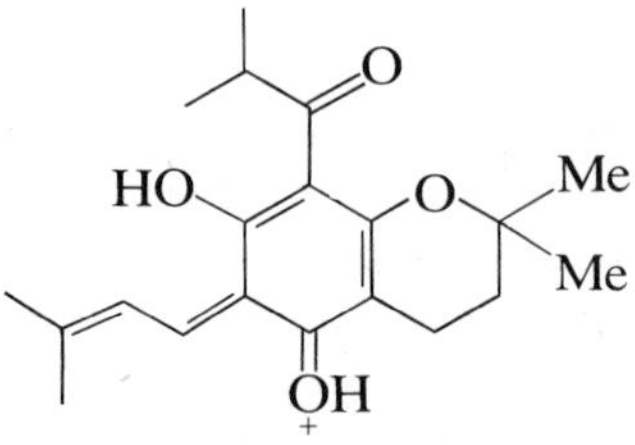

(**12**)

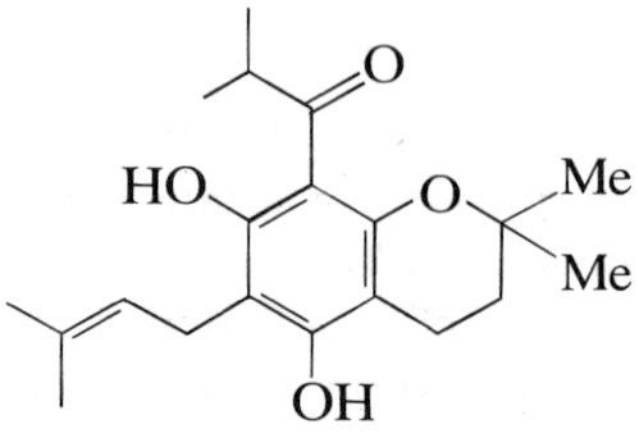

(**13**)

The pentaketide (**14**), diplodialide-D, was isolated from *Diplodia pinea*.[8] On treatment with potassium hydroxide in aqueous methanol at room temperature followed by acidification, the α,β-unsaturated ketone (**15**) was formed and its structure assigned from the ir, uv, nmr, and mass spectra. On subsequent refluxing with methanolic hydrogen chloride, the phenol (**16**) was identified.[8]

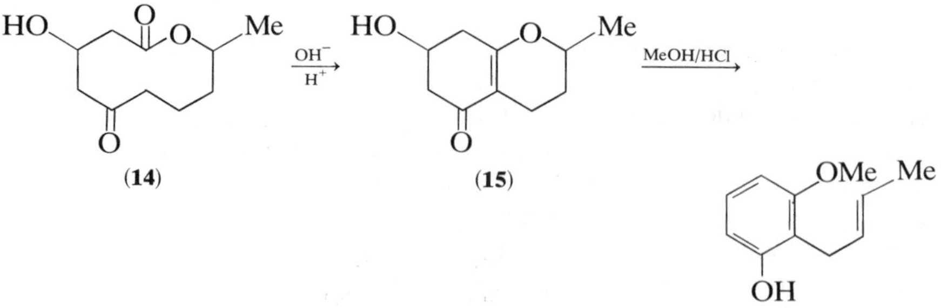

(**14**) $\xrightarrow[H^+]{OH^-}$ (**15**) $\xrightarrow{MeOH/HCl}$ (**16**)

2. 6-Chromanones

6-Chromanones have achieved little significance in natural product chemistry. However the use of ir and mass spectra and deuterium labeling have shown that a dimer of α-tocopherol isolated from rat liver had the structure (**17**).[9]

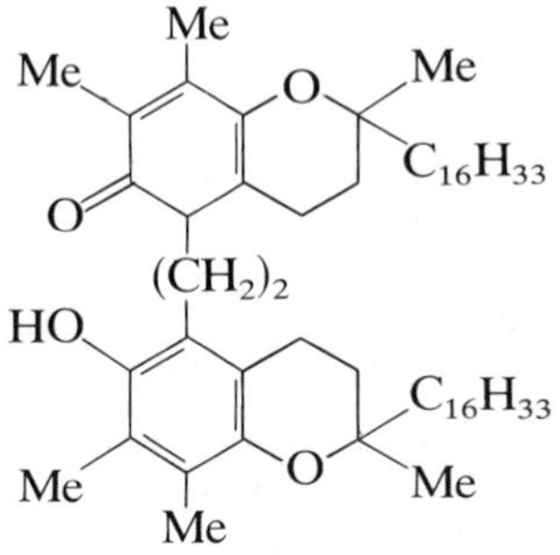

(**17**)

The fact that α-tocopherol is readily oxidized to α-tocopherone, coupled with the ease of the reverse reduction, has led to a discussion of possible mechanisms whereby α-tocopherols could participate catalytically in the biological transfer of electrons.[10]

3. 7-Chromanones

7-Chromanones and related compounds have been isolated and identified among volatile constituents of tobaccos.[11–14] The fully saturated 7-chromanone (**18**), isolated from Burley tobacco,[11,13] was synthesized and shown to impart a camphorlike odor to tobacco, especially during smoking.[15]

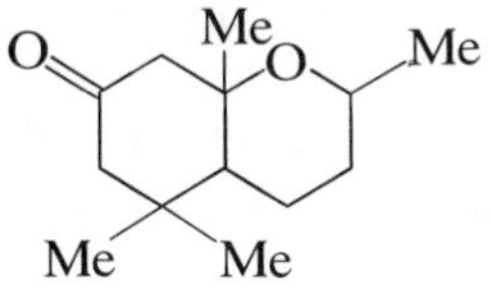

(**18**)

The formation of the 7-chromanone (**10**) on oxidation of colupulone has been referred to earlier (see Section II, 1).

4. Bz-Chromandiones

Tocored (**19**, $R=[(CH_2)_3CHMe]_3Me$)[16] and tocopurple (**20**)[17] have been identified as oxidation products of (±)-α-tocopherol. Interest in such products is in no small part due to the fact that the role of α-tocopherol in muscle metabolism and as an antioxidant in vegetable oils is undoubtedly related to its oxidation.[17] Antioxidant properties of 2,2,7,8-tetramethyl-5,6-chromandione (**19**, R = Me), tocored (**19**, R = $[(CH_2)_3CHMe]_3Me$), and tocopurple (**20**) have been studied.

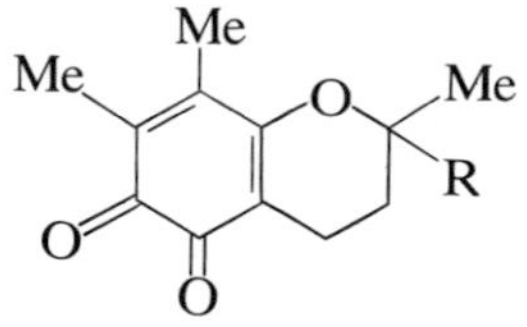

(**19**) 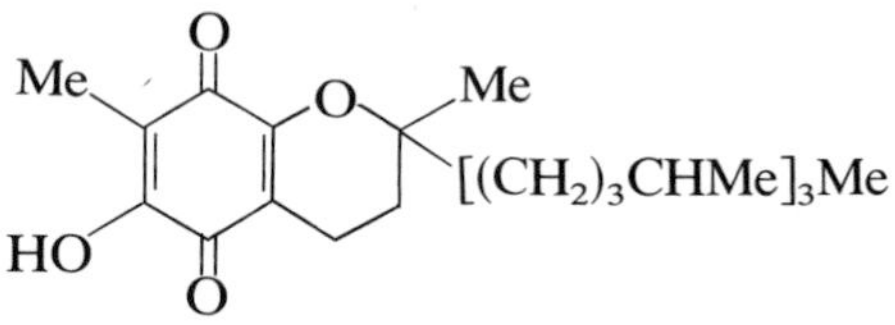 (**20**)

The tetramethylchromandione (**19**, R = Me) and tocored were included in a group of α-tocopherol derivatives evaluated for their ability to protect β-carotene in corn oil from aerial oxidation.[18] In studies on their antioxidative activity in lard, tocored (**19,** R = $C_{16}H_{33}$) was comparable with α-tocopherol, while tocopurple (**20**) was inactive at low concentrations but moderately active at high concentrations.[19] Formation of tocored has been demonstrated during the refining of soya bean oil.[20] In autoxidizing soya bean oil, oxidation of γ-tocopherol to a dimeric oxidation product (5-γ-tocopheroxy-γ-tocopherol) occurred more readily in the presence of tocored. However in a competitive reaction, γ-tocopherol had an inhibiting effect on the subsequent oxidation of the dimer which is believed to be an intermediate in the conversion of γ-tocopherol to tocored.[21] The quantity of tocored in crude soya bean oil increases with the moisture level of the soybeans while the tocopherol decreases.[22] Thus tocored is derived from tocopherol; it was further concluded that part of the tocored remains in a colorless form in deodorized oil but changes to tocored again in color-reverted soybean salad oil.[22]

The tetramethyl-5,6-chromandione (**19**, R = Me) and 2,2,6,7-tetramethyl-5,8-chromandione were tested for their antihemolytic effect on blood from vitamin E-deficient rats.[23] The former compound was among a group studied as inhibitors of cyclic AMP [adenosine 3′,5′-(hydrogen phosphate)] and cyclic GMP [guanosine cyclic-3′,5′-(hydrogen phosphate)] phosphodiesterases; compound (**19**) was shown to be a potent inhibitor of cyclic AMP phosphodiesterase in vitro.[24] However it was shown that single intraperitoneal injections of the compound did not significantly increase the levels of cyclic AMP in any of the histol regions of the glandular stomach of the rat.[25]

In the course of fermentation of soil cultures, biotransformation of the 6-chromanol (**21**) occurred to form the 5,6-chromandione (**22**) in 5.1% yield.[26]

Apparently the only recorded 5,7-chromandione derivatives are xanthochymol (**23**) and isoxanthochymol (**24**), two compounds that were isolated from the fruits of *Garcinia xanthochymus.*[27]

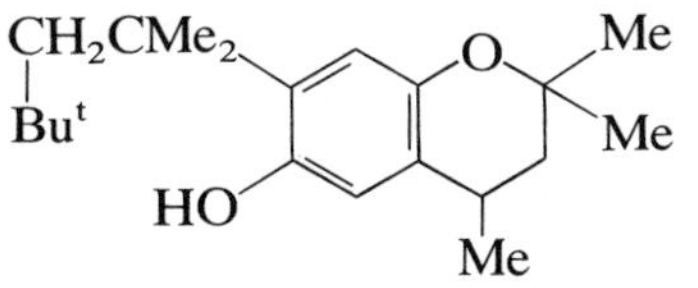

(21) 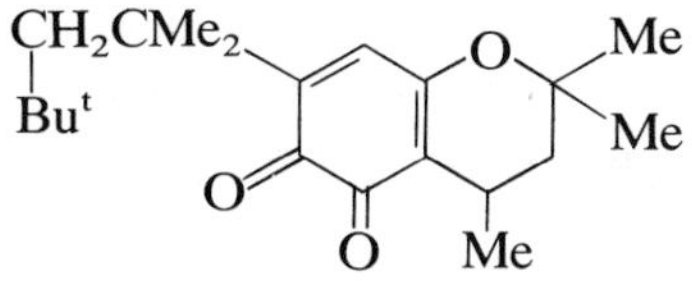 **(22)**

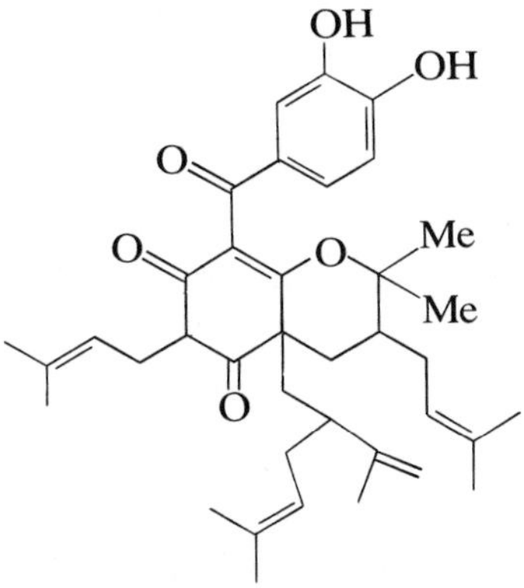

(23) 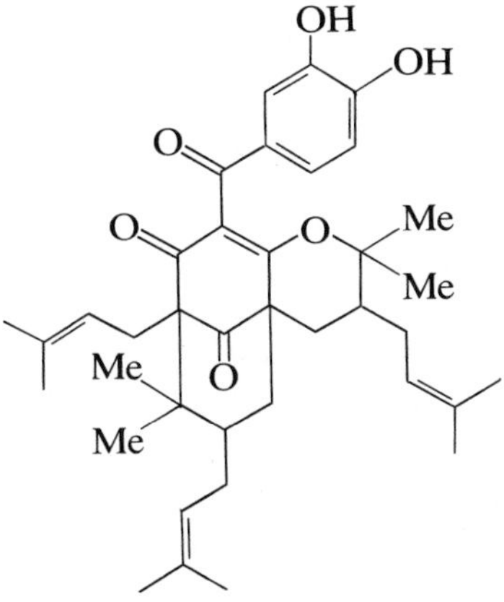 **(24)**

Degradation studies on flavoglaucin, a pigment of *Aspergillus glaucus*, led to the isolation of 2,2-dimethyl-6-hydroxy-7-heptyl-5,8-chromandione.[28,29]

5. Oxo Derivatives of Dihydronaphthopyrans

Of the dihydronaphthopyrans with keto substitution in the naphthalene nucleus, the two most widely studied natural products are α-lapachone (**5**) and β-lapachone (**6**). Work with these compounds started in the late nineteenth century,[1,2] and their structures were eventually established.[3]

β-Lapachone has been isolated by solvent extraction of *Tabebuia avellanedae.*[30,31] Both α- and β-lapachones have been obtained from the heartwood of *T. avellanedae,*[32,33] *T. chrysantha,*[34] and *T. guayacan.*[35,36] β-Lapachone has also been identified in the neutral fraction of extracts of the heartwood of *Phyllarthron comorense* that had been separated by chromatography over deactivated alumina.[37]

Methanol extraction of *Catalpa ovata* wood afforded α-lapachone and a whole range of substituted analogs which were separated by careful chromatography on a variety of supports. The substituted α-lapachones included the 9-hydroxy-,[38,39] 9-methoxy-,[38,39] 4-hydroxy-,[39] 4,9-dihydroxy,[38,39] and 4-oxo-α-lapachones.[39] Infrared, uv, and nmr spectral data were used in the elucidation of the structures.

Feeding studies with tritium-labeled 2-methoxycarbonyl-4-hydroxy-1-tetralone and *C. ovata* indicated that 2-succinoylbenzoic acid, itself derived from shikimic acid, was biosynthesized to the diketone (**25**) through the enol form of 2-carboxy-4-hydroxy-1-tetralone and 2-carboxy-4-oxo-1-tetralone. The diketone (**25**) was transformed into 4,9-dihydroxy-α-lapachone and other metabolites.[40]

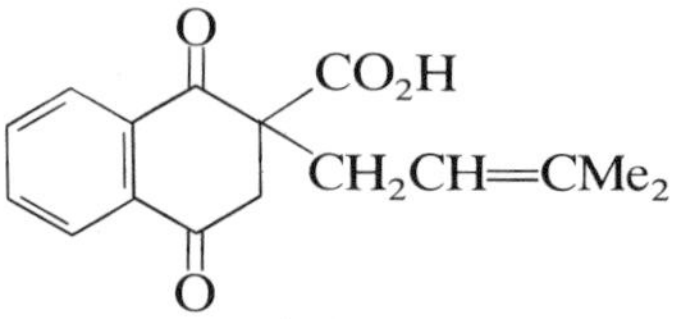

(**25**)

β-Lapachone shows significant activity against some Gram-positive organisms such as *Bacillus subtilis* and *Staphylococcus aureus.*[32] The antimicrobial spectrum of β-lapachone is similar to that of lapachol.[41] Antitumor and toxicological properties of β-lapachone have also been reported.[42] Injected intraperitoneally, the LD_{50} in rats was 80 mg/kg. Oral administration of 9 mg/kg produced death in six cases out of ten, with loss of weight, anorexia, and diarrhea.[42] There was inhibition of Yoshida sarcoma and Walker 256 carcinoma at 7 mg/kg; α-lapachone was not inhibitory.[42] α-Lapachone has also been tested for antibiotic and antineoplastic activity, both in vitro and in vivo.[43]

α-Lapachone and its 4-hydroxy analog have been isolated from *Zeyhera tuberculosa*; the structure of the latter compound was derived from ir, uv, nmr, and mass spectral data.[44]

An orange pigment, stenocarpoquinone-A, isolated from the wood of *Stenocarpus salignus* R.Br, has been identified as (R)-3′-hydroxy-β-lapachone (**26**, R = OH).[45] The structure was elucidated from spectral data, and an interpretation of the mass spectrum was published. The pigment afforded an acetate (**26**, R = OAc). In the synthesis of the racemic product, bromination of lapachol afforded the 3-bromo compound (**26**, R = Br).[3,45] Subsequent alkaline cleavage of the heterocyclic ring led to a glycol which gave the 3-hydroxy derivative (**26**, R = OH) on cyclization in refluxing ethanolic hydrogen chloride. The configuration of the natural product was shown to be represented as (**27**).[45]

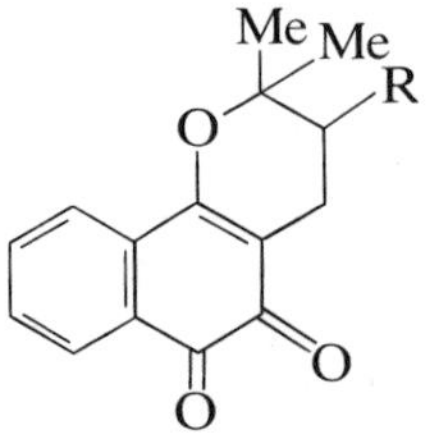

(**26**) 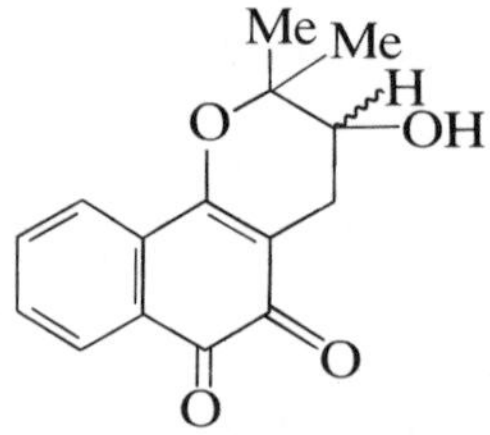 (**27**)

Studies of antibacterial pigments isolated from deep cultures of *Gnomonia erythrostoma* have led to the structures indicated for erythrostominone (**28**,

$R^1 = CH_2COMe$, $R^2 = R^3 = R^4 = OH$, $R^5 = Me$),[46,47] deoxyerythrostominone (**28**, $R^1 = CH_2COMe$, $R^2 = H$, $R^3 = R^4 = OH$, $R^5 = Me$),[46,47] deoxyerythrostominol (**28**, $R^1 = CH_2CHOHMe$, $R^2 = H$, $R^3 = R^4 = OH$, $R^5 = Me$),[46–48] and epierythrostominol (**28**, $R^1 = CH_2CHOHMe$, $R^2 = R^3 = R^4 = OH$, $R^5 = Me$).[48] Total synthesis of the 5,10-dione (**29**, $R^1 = O$, $R^2 = R^4 = H$, $R^3 = OH$) was used to rigorously establish the structure of the ring system in these pigments.[49]

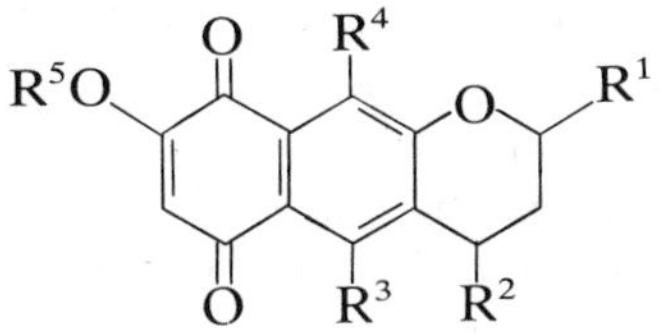

(**28**)

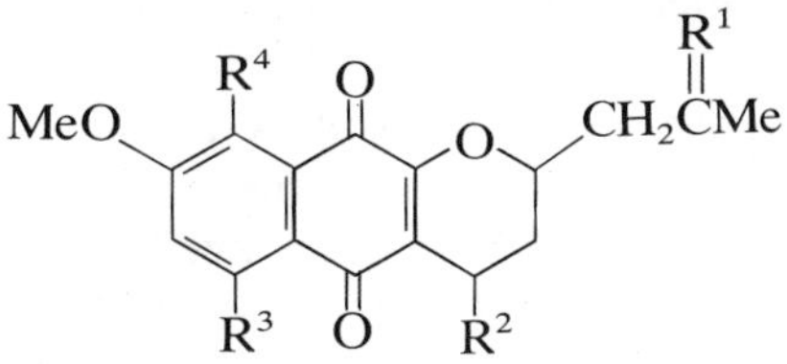

(**29**)

Erythrostominone was converted to deoxyerythrostominone by catalytic hydrogenation in the presence of palladium on charcoal until two molecular equivalents of hydrogen had been taken up.[47] Prolonged hydrogenation afforded a mixture of the 5,10-diketone (**29**, $R^1 = O$, $R^2 = R^4 = H$, $R^3 = OH$) and a 6,9-diketone (**28**, $R^1 = CH_2COMe$, $R^2 = R^3 = H$, $R^4 = OH$, $R^5 = Me$; and/or **28**, $R^1 = CH_2COMe$, $R^2 = R^4 = H$, $R^3 = OH$, $R^5 = Me$).[47,49] Methylation of the 5,10-diketone with methyl iodide and silver oxide gave the dimethoxy compound (**29**, $R^1 = O$, $R^2 = R^4 = H$, $R^3 = OMe$).[49] Acetylation of the mixture and separation of the products by preparative layer chromatography led to the appropriate *O*-acetates.[47] Acetylation of erythrostominone with acetic anhydride in pyridine gave a mixture of isomeric triacetates (**28**, $R^1 = CH_2COMe$, $R^2 = R^3 = R^4 = OAc$, $R^5 = Me$; and **29**, $R^1 = O$, $R^2 = R^3 = R^4 = OAc$). Diacetates (**28**, $R^1 = CH_2COMe$, $R^2 = H$, $R^3 = R^4 = OAc$, $R^5 = Me$; and **29**, $R^1 = O$, $R^2 = H$, $R^3 = R^4 = OAc$) were similarly obtained from deoxyerythrostominone.[47]

Oxidation of erythrostominone with acetic anhydride and dimethyl sulfoxide afforded the triketone (**30**).[47] The trihydroxy compound (**28**, $R^1 = CH_2COMe$, $R^2 = R^5 = H$, $R^3 = R^4 = OH$) was formed on demethylation of deoxyerythrostominone with hydrobromic acid in acetic acid.[47]

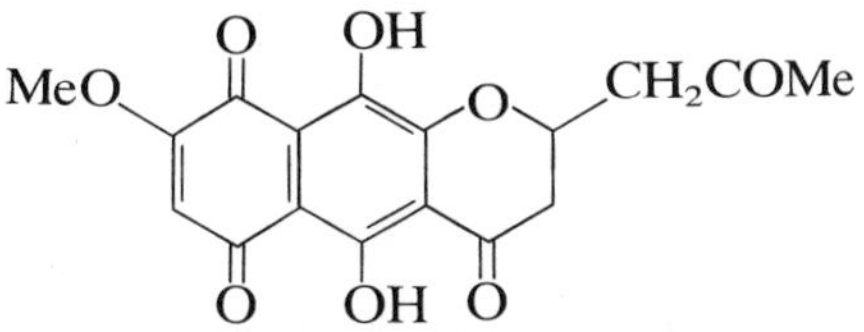

(**30**)

Two main products were obtained on sodium borohydride reduction of deoxyerythrostominone. One of these, the alcohol (**28,** $R^1 = CH_2CHOHMe$,

$R^2 = H$, $R^3 = R^4 = OH$, $R^5 = Me$), on acetylation gave a mixture of triacetates (**29**, $R^1 = H$, OAc, $R^2 = H$, $R^3 = R^4 = OAc$) and (**28**, $R^1 = CH_2CHOAcMe$, $R^2 = H$, $R^3 = R^4 = OAc$, $R^5 = Me$). The other was either the 5-hydroxy compound (**28**, $R^1 = CH_2CHOH$, $R^2 = R^4 = H$, $R^3 = OH$, $R^5 = Me$) or the 10-hydroxy analog (**28**, $R^1 = CH_2CHOH$, $R^2 = R^3 = H$, $R^4 = OH$, $R^5 = Me$), or a mixture of the two.[47] Reduction of erythrostominone with sodium borohydride afforded epierythrostominol and its 2′-epimer.[48] Hydrogenation of erythrostominol gave the 2′-epimer of deoxyerythrostominol. When the products of hydrogenation of erythrostominol were separated by preparative layer chromatography, in addition to the monohydroxy compounds (**28**, $R^1 = CH_2CHOHMe$, $R^2 = R^4 = H$, $R^3 = OH$, $R^5 = Me$; and/or **28**, $R^1 = CH_2CHOHMe$, $R^2 = R^3 = H$, $R^4 = OH$, $R^5 = Me$), a small amount of the monohydroxyquinone (**29**, $R^1 = H,OH$, $R^2 = R^4 = H$, $R^3 = OH$) was identified in the product.[48]

Among other 5,10-dione derivatives of 2,3-dihydro-4*H*-naphtho[2,3-*b*]-pyran, compound (**31**) has been shown to have prothrombin activity.[50] Cryptosporin, a metabolite of the fungus *Cryptosporium pinicola* Linder, has the structure and absolute configuration (**32**).[51] On oxidation with periodate, it was converted to a mixture of the two aldehydonaphthoquinones [**33**, R = OH; and **33**, R = OCHMeCH(OH)OMe]. Cryptosporin formed a triacetate and a tribenzoate. In the presence of 4-toluenesulphonic acid, the tetracyclic compound (**34**) was formed.[51]

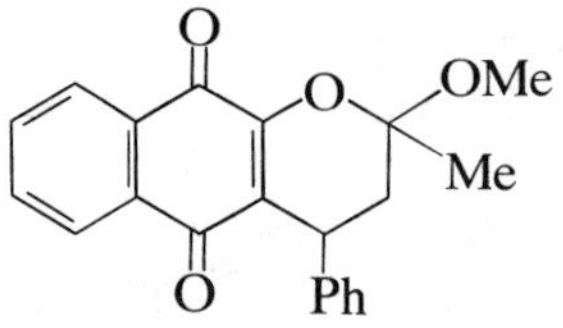

O O OMe Me Ph

(**31**)

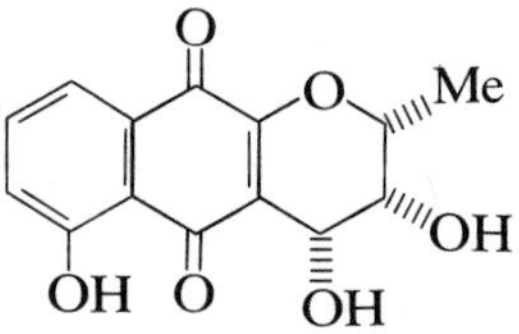

O O Me OH OH O OH

(**32**)

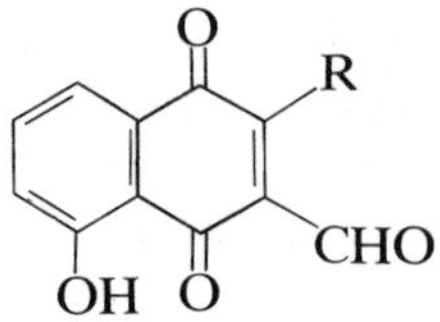

O R CHO OH O

(**33**)

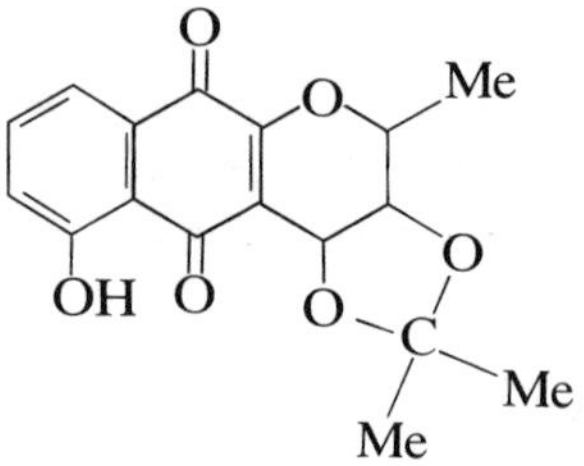

O O Me O OH O O C Me Me

(**34**)

6. Bz-Acylchromans

Among exocyclic ketones, dihydroevodionol is 7-hydroxy-5-methoxy-6-acetyl-2,2-dimethylchroman.[52]

III. METHODS OF PREPARATION

1. 5-Chromanones

5-Chromanones have been prepared by cyclization of compounds such as 2-(3,3-dimethylallyl)-1,3-cyclohexanedione (**35**, $R^1 = Me$, $R^2 = H$). Heating with phosphorus pentoxide for 1.5 hr at 120° afforded a 77% yield of the 2,2-dimethyl-5-chromanone (**36**, $R^1 = Me$, $R^2 = H$).[53,54] Similar cyclizations have been carried out with phosphoric acid at 100° for 18 hr.[55]

(**35**) (**36**)

Michael addition of methyl vinyl ketone (**38**) to 1,3-cyclohexanedione (**37**) in the presence of potassium carbonate followed by treatment of the crude product with diazomethane gave some of the 5-chromanone (**36**, $R^1 = OMe$, $R^2 = H$), in addition to the expected uncyclized compound (**39**, R = Me), which was the major product. The two compounds were separated by gas–liquid chromatography.[56] When the dione (**37**) and the ketone (**38**) were allowed to react in water in an inert atmosphere for 24 hr, the cyclohexene (**39**, R = H) was formed. Subsequent reduction with sodium borohydride in methanol or catalytic hydrogenation in the presence of Raney nickel afforded the 2-methyl-5-chromanone (**36**, $R^1 = R^2 = H$).[57,58]

(**37**) + (**38**) —1. K_2CO_3; 2. CH_2N_2→ (**39**; R = Me) + (**36**, $R^1 = OMe$, $R^2 = H$)

(**39**, R = H) ⟶ (**36**, $R^1 = R^2 = H$)

Although alkenylation of 1,3-cyclohexanedione with allyl bromide in alkali in the presence of copper powder gave the expected 2,2-diallyl derivative, alkenylation with 3,3-dimethylallyl bromide afforded a mixture of products that included two 5-chromanone derivatives (**36**, $R^1 = Me$, $R^2 = H$) and (**36**, $R^1 = Me$, $R^2 = CH_2CH{=}CMe_2$) along with the expected 2,2-dialkenylated 1,3-cyclohexanedione and the two benzofuranone derivatives (**40**, $R^1 = Me$, $R^2 = H$) and (**40**, $R^1 = H$, $R^2 = Me$).[59]

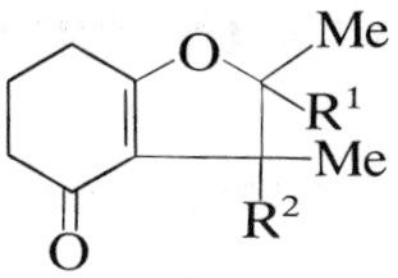

O Me R¹ Me R² O

(**40**)

Stirring the 2-tosyloxyaldehyde (**41**) in dimethyl sulfoxide with the potassium salt of dimedone (**42**) at 20° to 50° for 4 hr afforded the 5-chromanone (**43**) in 44% yield.[60]

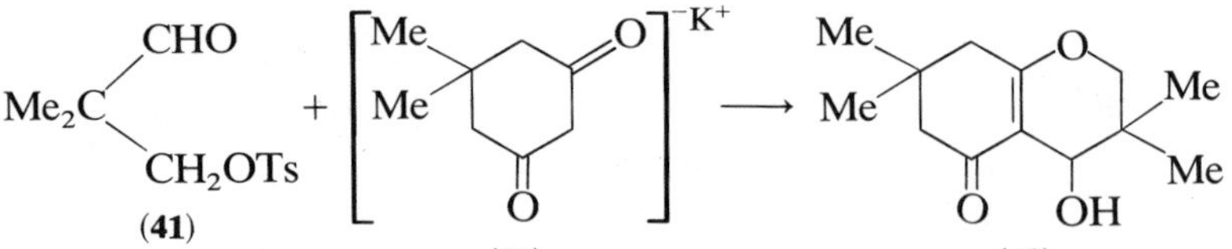

CHO Me_2C CH_2OTs (**41**) + [Me Me O O]$^{-}K^{+}$ (**42**) → Me Me O Me Me O OH (**43**)

In work on the structure of dambonitol, a dialdehyde of presumed structure (**44**) reacted with dimedone in the presence of piperidine to give a product that was believed to be the 5-chromanone (**45**).[61]

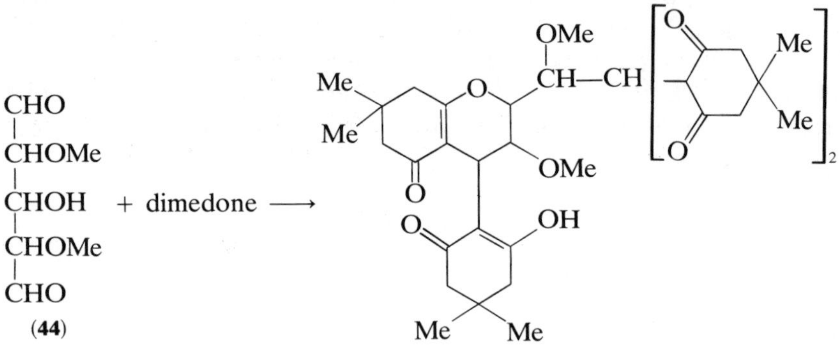

CHO–CHOMe–CHOH–CHOMe–CHO (**44**) + dimedone → Me Me O O OMe CH—CH [O O Me Me]$_2$ OMe O OH Me Me (**45**)

Oxidation of the *trans*-hydroxylactone (**46**) with chromic acid in acetone afforded the *trans*-5-chromanone (**47**). The corresponding cis isomer was similarly prepared.[62]

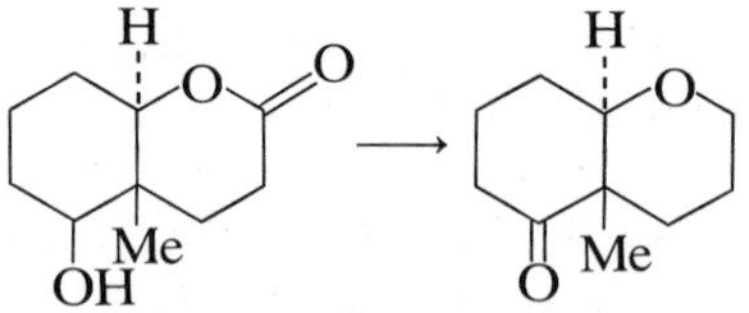

H O O Me OH (**46**) → H O Me O (**47**)

2. 6-Chromanones

The majority of 6-chromanones have been prepared by oxidation of the corresponding chromanols, in particular from tocopherols. Oxidation of α-tocopherol with ferric chloride and 2,2′-bipyridine in ethanol afforded the quinone acetal (**48**, R = Et),[63,64] although the structure was originally incorrectly assigned.[63] In variations of the conditions, 2,2′-bipyridine has been replaced by 1,10-phenanthroline, and oxidation has also been carried out with ferric chloride in ethanol.[65] The mechanism of the reaction has been discussed.[65] A whole range of 8a-hydroxy-α-tocopherones has been prepared in yields ranging from 28 to 78% by oxidation of tocopherols with benzoyl peroxide in anhydrous alcohols.[10] This contrasted with substitution on the 5-methyl group that occurred when benzoyl peroxide oxidation was carried out in hydrocarbon solvents.[66] Efforts to produce a tertiary butoxy analog were unsuccessful.[10] Bromine, diacyl peroxides, a mixture of cyclohexane peroxides, as well as ferric chloride–2,2′-bipyridine are among other oxidizing agents that have been used to prepare 8a-alkoxy-α-tocopherones.[10]

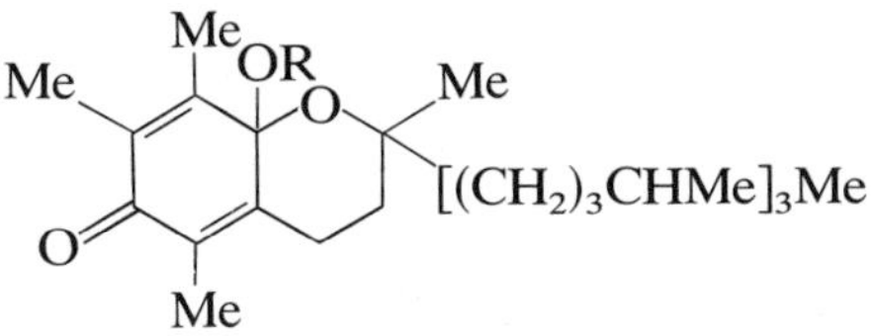

(**48**) 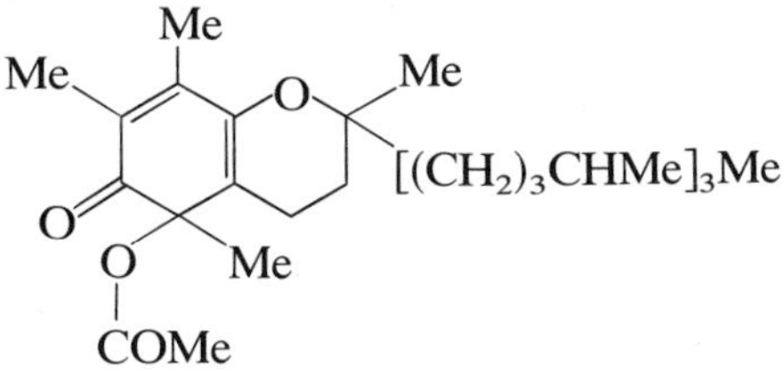 (**49**)

8a-Hydroxy-α-tocopherone (**48**, R = H) has been prepared by oxidation of α-tocopherol with tetrachloro-2-quinone in aqueous acetonitrile, or with *N*-bromosuccinimide in aqueous buffered acetonitrile.[67–69] With ferric chloride in ethanol, 8a-ethoxy-α-tocopherone (**48**, R = Et) was obtained.[68] When α-tocopherol in acetonitrile was treated with tetramethylammonium acetate followed by *N*-bromosuccinimide, a mixture of the *O*-acetates (**48**, R = Ac) and (**49**) was isolated.[68]

Oxidation of the fully saturated 6-hydroxychroman (**50**) with chromium trioxide in acetic acid afforded the ketone (**51**).[70,71]

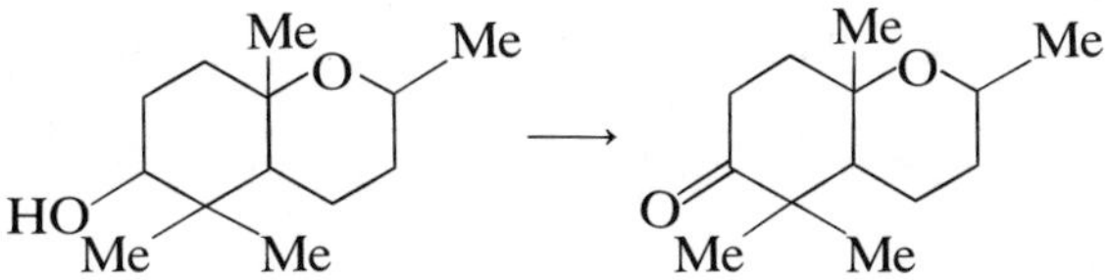

(**50**) → (**51**)

Irradiation of α-tocopherol in isooctane led to a main product that probably had the structure (**52**, $R=C_{16}H_{33}$).[72] The same product has been postulated as an intermediate in the acid isomerization of tocoquinone (**53**, $R=C_{16}H_{33}$) in the presence of 1,1-diphenylethylene to give the dipyran (**54**, $R=C_{16}H_{33}$). Similar results were obtained on isomerization of the analog (**53**, R=Me) to the tricyclic product (**54**, R=Me).[73]

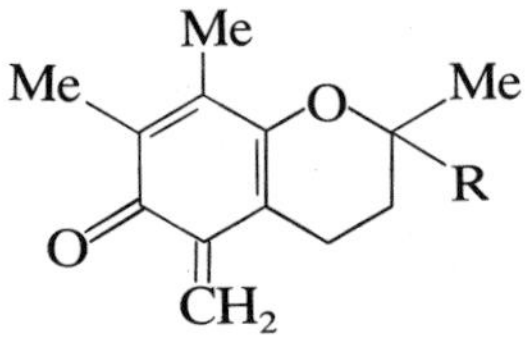

(**52**) 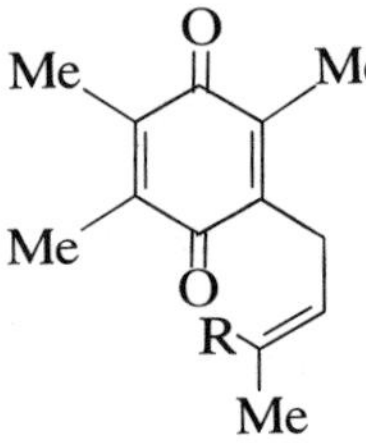 (**53**) $\xrightarrow[H^+]{Ph_2C{=}CH_2}$ (**54**)

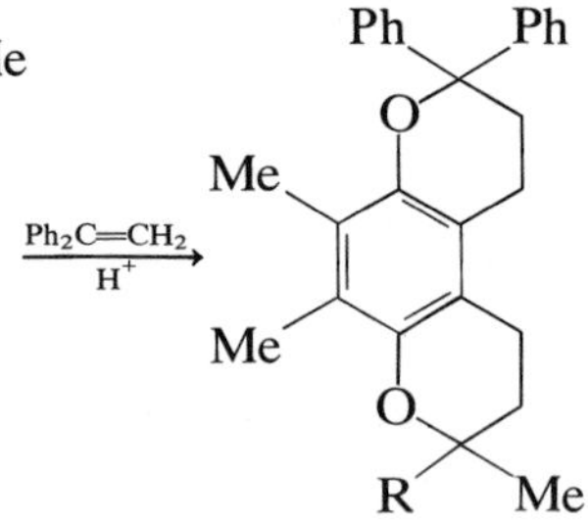

Analogous reactions occurred with phylloquinones (**55**, R=Me or $C_{16}H_{33}$), where the intermediate (**56**) was postulated.[73]

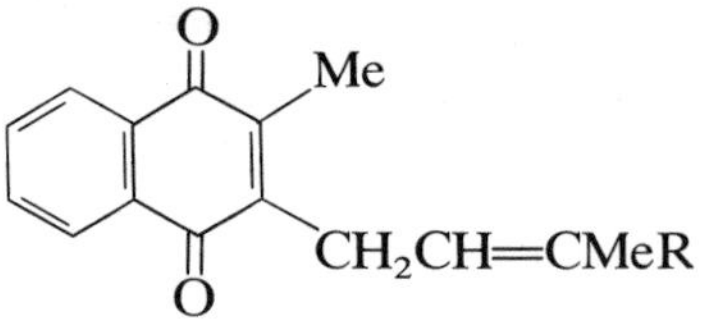

$CH_2CH{=}CMeR$

(**55**) 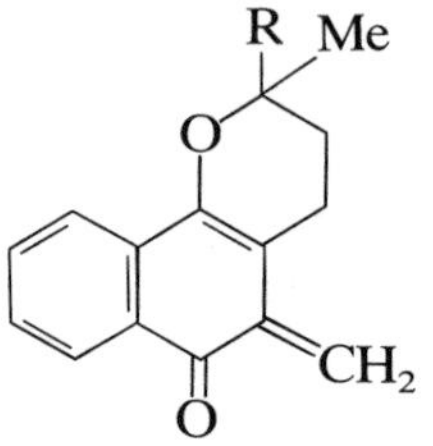 (**56**)

Evidence has been reported for the transient formation of the naphthopyran-6-one (**56**, $R=C_{16}H_{33}$) in acid-catalyzed reactions of vitamin $K_{1(20)}$[74] and in the perchloric acid isomerization of vitamin $K_{1(20)}$ and in its condensation with styrene.[75]

In the course of electrochemical studies of the behavior of α-tocopherol in aqueous acetonitrile, the 6-chromanol (**57**) was converted to the carbonium ion (**58**) on the electrode surface by electrochemical oxidation.[69]

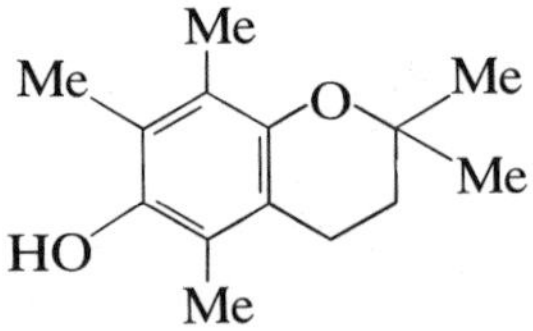

(**57**) $\underset{\text{redn.}}{\overset{\text{oxdn.}}{\rightleftharpoons}}$ (**58**) $+ H^+ + 2e$

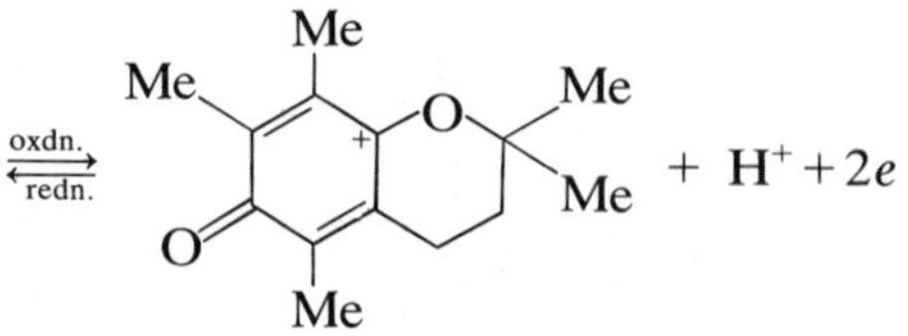

3. 7-Chromanones

7-Chromanones have usually been obtained by specific methods, and few generalizations can be made.

Acid-catalyzed rearrangement of the alcohols (**59**, R = H or Me) with 2.5% perchloric acid in acetic acid afforded a mixture of the corresponding cyclohexenone (**60**) and 7-chromanone (**61**). Prolonging the reaction time to four days resulted in exclusive formation of what was believed to be the 7-chromanone (**61**) in yields of about 70%.[76]

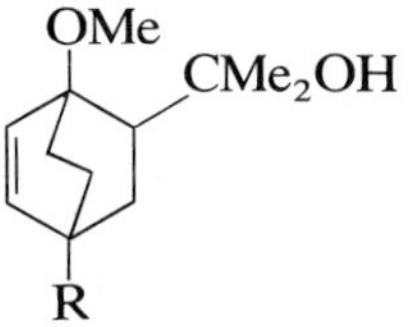

OMe, CMe_2OH, R (**59**)   O, R, $CH_2CH{=}CMe_2$ (**60**)   O, O, Me, Me, R (**61**)

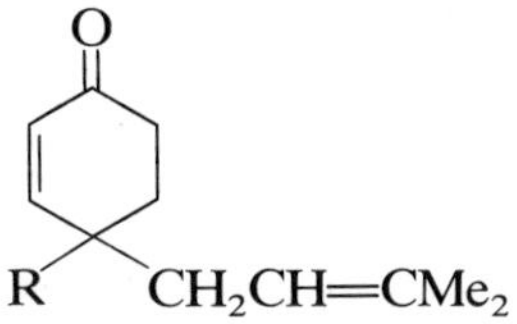

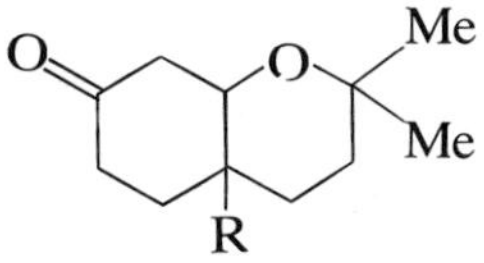

Reduction of 7-methoxychroman with lithium in liquid ammonia and *t*-butanol gave the dihydrochroman (**62**). Replacement of *t*-butanol by methanol, 2-propanol, or *n*-butanol had no significant effect on the products formed. Subsequent treatment with methanolic hydrogen chloride afforded a mixture of the three products (**63**), (**64**), and (**65**), which were separated by chromatography on Florosil.[77] Reduction of the ester (**66**, R = CO_2Me) with lithium aluminum hydride gave the triol (**66**, R = CH_2OH), which on hydrogenation in the presence of rhodium on alumina afforded a mixture of the chromanone (**64**) and the diol (**65**). The same mixture was obtained on acid hydrolysis of the chromanone (**63**) in the absence of methanol. The 7-chromanone (**64**) was shown to exist in two crystalline foms having mp 49–50° and 57–58°.[77]

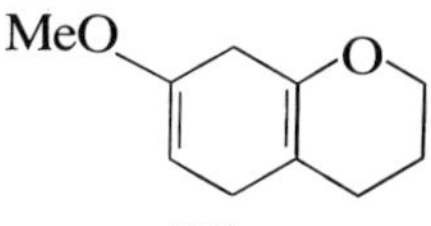

MeO, O (**62**)   O, OMe, O (**63**)

O, O (**64**)   O, OH, CH_2OH (**65**)

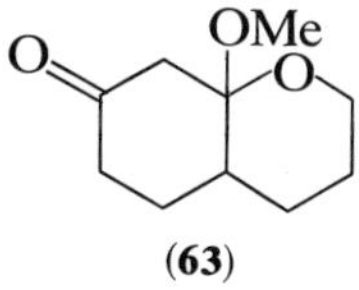

HO, OH, R (**66**)

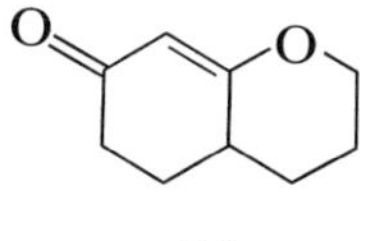

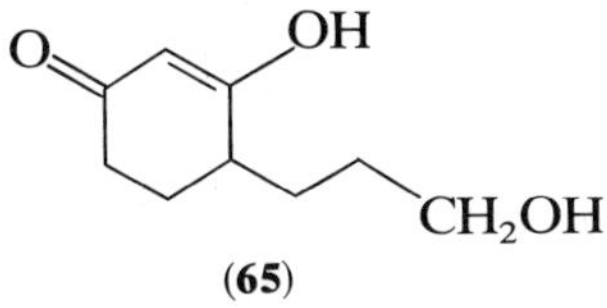

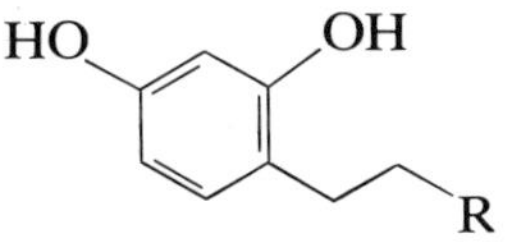

In the course of the synthesis of (±)-lycoramine, the monoketal (**67**) was reduced with lithium aluminum hydride in tetrahydrofuran; subsequent treatment with oxalic acid in boiling aqueous ethanol afforded the 7-chromanone derivative (**68**) in 74% yield.[78,79]

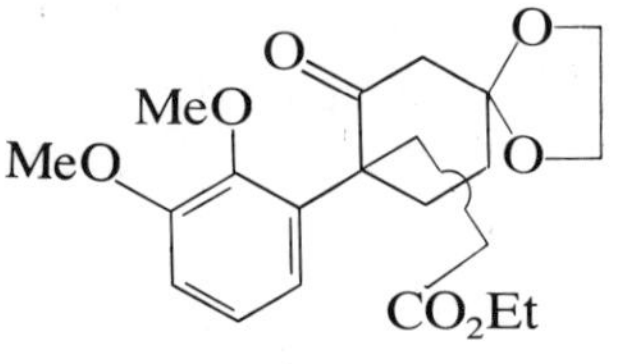

MeO MeO O O O CO_2Et

(**67**)

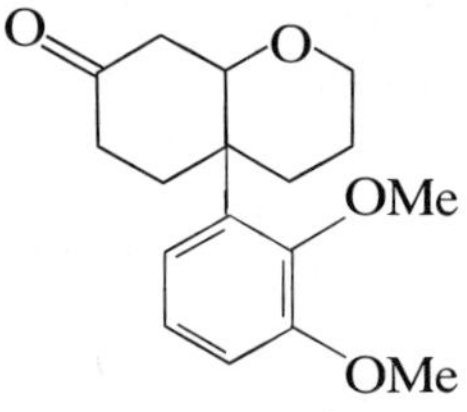

O O OMe OMe

(**68**)

In the course of work on the stereoselective synthesis of (±)-juvabione, it was demonstrated that oxidation of the diol (**69**) with manganese dioxide in chloroform afforded a mixture of the two 7-chromanones (**70**) and (**71**). The isomers, which differed in the configuration of the side chain, were separated by chromatography on silica gel. The pmr spectra showed that there was cis fusion of the two rings in both products.[80,81] Under similar conditions, the 7-chromanone (**74**) was obtained in 32% yield from the diol (**72**), together with the ketol (**73**); the latter rearranged spontaneously to the chromanone (**74**).[81]

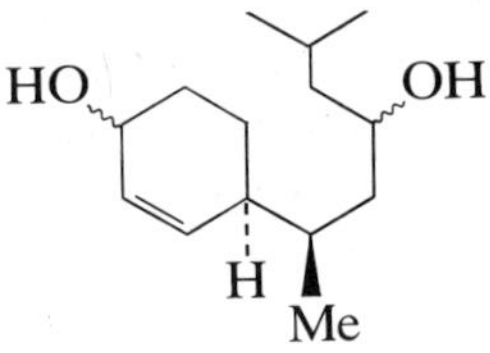

HO OH H Me

(**69**)

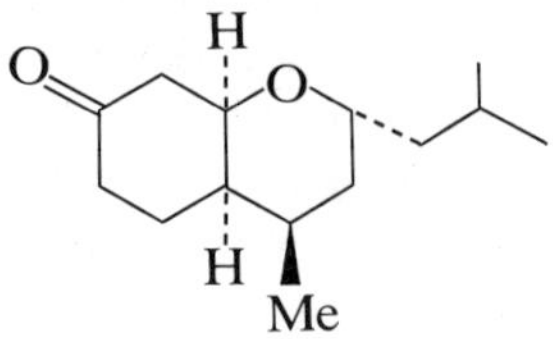

O H O H Me

(**70**)

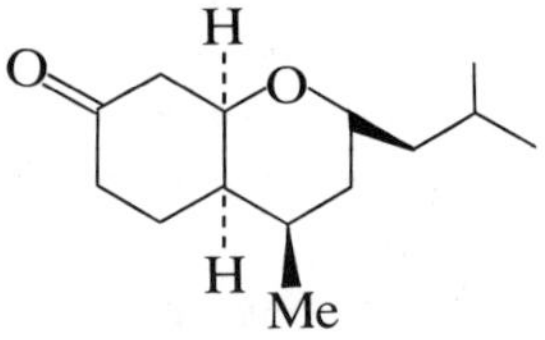

O H O H Me

(**71**)

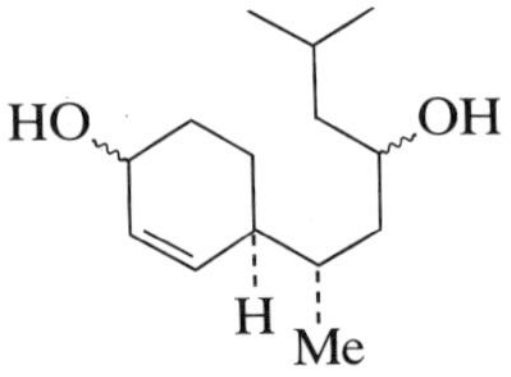

HO OH H Me

(**72**)

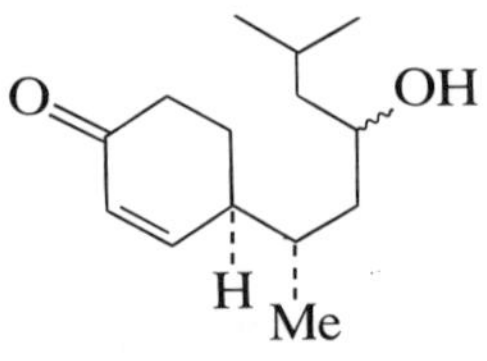

O OH H Me

(**73**)

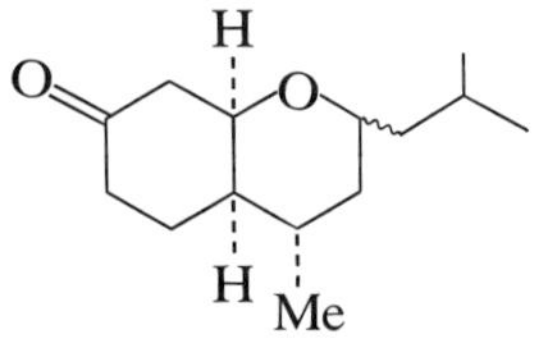

O H O H Me

(**74**)

Cyclization of the ketal dialcohol (**75**) under mild acidic conditions gave the very stable 7-chromanone (**76**).[82]

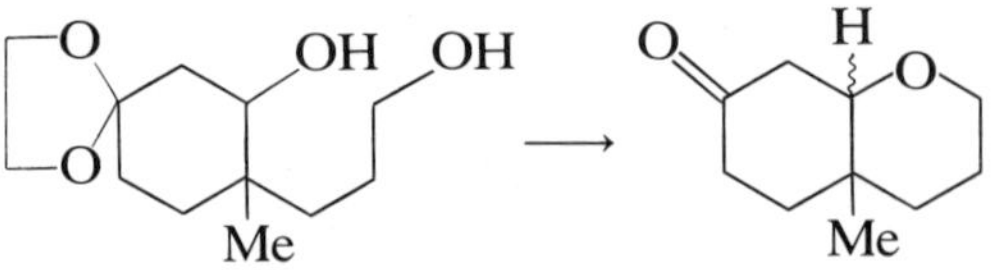

O O OH OH Me → O H O Me

(**75**) (**76**)

4. Bz-Chromandiones

The majority of chromandiones with the keto groups in the homocyclic ring have been prepared by oxidation processes. Oxidation of α-tocopherol in methanol with ferric chloride at 50° afforded four products which were separated by column chromatography on zinc carbonate. One of these, a yellow band comprising 42% by weight of the oxidized products, was identified as tocored (**19**, $R=C_{16}H_{33}$).[16] It has also been prepared by heating ($\pm$)-α-tocopherol with nitric acid on the steam bath in either methanol[16] or ethanol.[22,83] A purple band from the column afforded α-tocopurple (**20**) whose structure was elucidated;[17] other workers have used the same method of preparation.[83]

Tocored (**19**, $R=C_{16}H_{33}$) has also been obtained by oxidation of γ-tocopherol in benzene with benzoyl peroxide; α-tocoquinone was formed on similar treatment of α-tocopherol.[84]

Prolonged oxidation of 2,2,5,7,8-pentamethyl-6-hydroxychroman with silver nitrate or nitric acid afforded brilliant red solutions from which a red crystalline product was isolated. The structure (**19**, R = Me) was established from uv spectral data and from the fact that it formed a yellow phenazine with *o*-phenylenediamine and a tetramethylphenazine with tetramethyl-*o*-phenylenediamine. Alcoholic solutions of the phenazines showed a pale-green fluorescence.[85] The same compound had been isolated previously[86,87] but had been incorrectly assigned the structure 2,2,6,7-tetramethyl-5,8-chromandione.[86] Later work supported the *o*-quinone formulation.[88] In order to establish which methyl group was eliminated, 2,2,7,8-tetramethyl-6-hydroxychroman was oxidized with nitric acid and the tricyclic compound (**77**) with silver nitrate. Both gave excellent yields of the same red compound of mp 109–110° and confirmed the structure (**19**, R = Me).[85] The α-, β-, and γ-tocopherols all gave deep-red compounds when oxidized with nitric acid in alcoholic solution, but these were obtained as oils that failed to crystallize.[85]

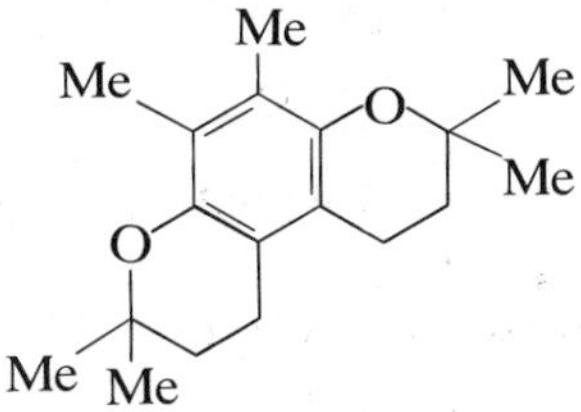

Me Me Me Me O O Me Me

(**77**)

The red 2,7,8-trimethyl-5,6-chromandione isomerized in hydrochloric acid at 80° to the yellow 2,6,7-trimethyl-5,8-chromandione. The former gave a phenazine, whereas the latter did not react with phenylenediamine under the same conditions.[88] The 2,2,7,8-tetramethyl analog similarly isomerized to the 2,2,6,7-tetramethyl-5,8-chromandione.[88]

2,2,7-Trimethyl-6-hydroxy-5,8-chromandione has been obtained by oxidation of either 2,2,7,8-tetramethyl-6-hydroxychroman or of the 2,2,5,7-tetramethyl analog with ferric chloride in methanol.[17]

Oxidation of the alcohol (**78**, R = Me) with chromic acid in aqueous acetic acid afforded two products. The 5,8-chromandione (**80**), possibly formed through the intermediate (**79**), was obtained in 9.4% yield, and the carboxylic acid (**78**, R = CO_2H), in 11.8% yield.[89]

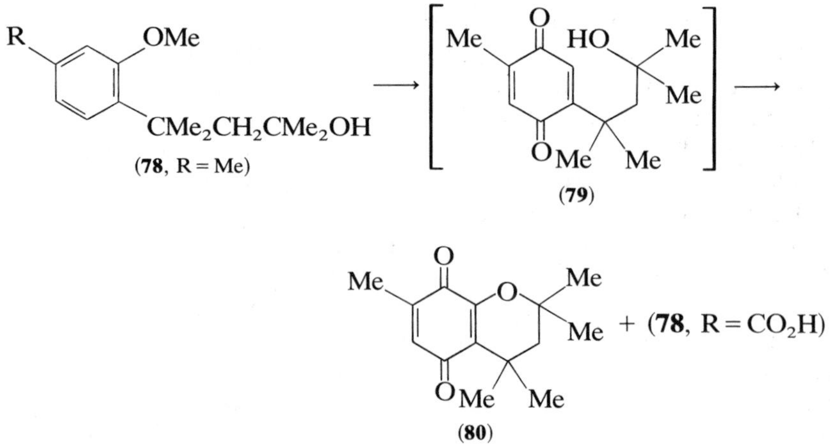

(**78**, R = Me) → [(**79**)] → (**80**) + (**78**, R = CO_2H)

2,2-Dimethyl-7-heptyl-6-hydroxy-5,8-chromandione was obtained by heating 2-hydroxy-3-heptyl-5-anilino-6-isopentenyl-*p*-benzoquinone on the steam bath with aqueous acetic acid containing sulfuric acid.[28,29]

5. Oxo Derivatives of Dihydronaphthopyrans

The action of sulfuric acid on lapachol (**81**, R^1 = Me, R^2 = H), the coloring matter of lapacho and other tropical woods, afforded red β-lapachone (**6**)[2,90] whereas with hydrochloric acid the isomeric yellow α-lapachone (**5**) was obtained.[2,91] With nitric acid a mixture of α- and β-lapachones was formed, with the latter predominating.[2] Although the structures proposed in the earlier work[2] proved to be incorrect, those subsequently assigned have stood the test of time. A number of syntheses of lapachol have been reported.[90,92,93] On treatment of lapachol with aluminum chloride in ether, β-lapachone was obtained without concomitant formation of the α-isomer.[94] On heating lapachol (obtained by extraction of seeds of *Tecoma impetiginosa* Mart) in sealed ampoules at 140–160° for 3 hr, α- and β-lapachones were among the products separated by thin-layer chromatography.[41]

3-Hydroxy-β-lapachone (**83**, $R^1 = R^2$ = Me, R^3 = OH) was identified among the reaction products when dihydroxyhydrolapachol (**82**, R^1 = Me,

$R^2 = OH$) was dissolved in sulfuric acid.[3,93] However on boiling with acetic acid to which a small quantity of sulfuric acid had been added, 3-acetoxy-α-lapachone (**84**, $R^1 = R^2 = Me$, $R^3 = OAc$) was isolated as pale-yellow needles which were converted to the 3-hydroxy-α-lapachone (**84**, $R^1 = R^2 = Me$, $R^3 = OH$) by the action of dilute sulfuric acid. In addition to removing the acetyl group, reaction with dilute sodium hydroxide effected ring cleavage.[3] Hydroxyhydrolapachol (**82**, $R^1 = Me$, $R^2 = H$) is readily converted to lapachones by the action of mineral acids; sulfuric acid gave β-lapachone, and hydrochloric acid gave the α-isomer.[95] α- and β-Lapachone were readily separated by treating an ethanolic solution with saturated sodium bisulfite followed by water; α-lapachone precipitated, while addition of sodium carbonate to the filtrate afforded β-lapachone.[95] In an analogous preparation, the 3-methyl analog of β-lapachone (**83**, $R^1 = R^2 = H$, $R^3 = Me$) was obtained as orange crystals on treatment of the hydroxyquinone (**82**, $R^1 = H$, $R^2 = Me$) with sulfuric acid.[96] It was also obtained by the action of dilute hydrochloric acid over one to two days. It formed an azine on boiling with *o*-phenylenediamine in glacial acetic acid; and on heating with hydrochloric acid at 55–65° for 1.5 hr, the β-lapachone analog was converted to the 5,10-diketone (**84**, $R^1 = R^2 = H$, $R^3 = Me$). On boiling with alkali, the tricyclic quinones were converted to the open-chain compound (**82**, $R^1 = H$, $R^2 = Me$).[96] The β-lapachone analog (**83**, $R^1 = R^2 = H$, $R^3 = Me$) has also been prepared by hydrogenation of the naphthopyran (**85**) in ethanol in the presence of Adams catalyst.[97]

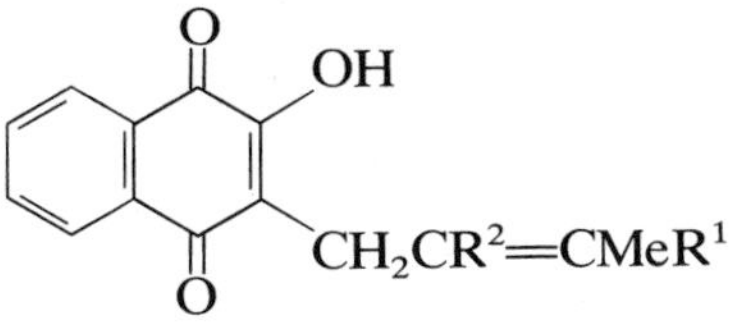

O, OH, $CH_2CR^2{=}CMeR^1$, O

(**81**)

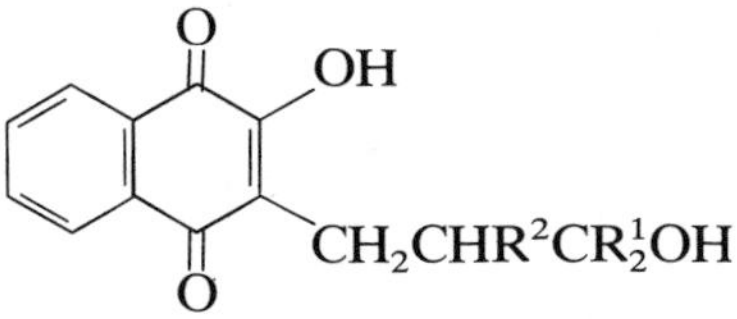

O, OH, $CH_2CHR^2CR^1_2OH$, O

(**82**)

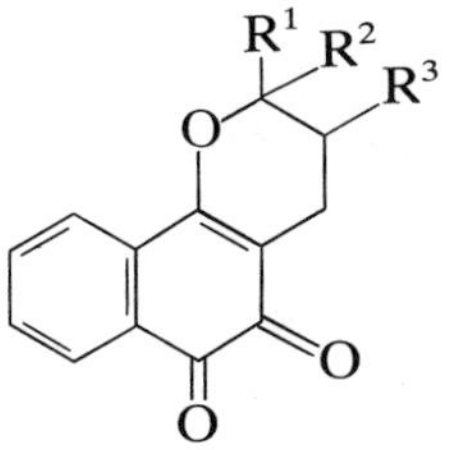

R^1, R^2, R^3, O, O, O

(**83**)

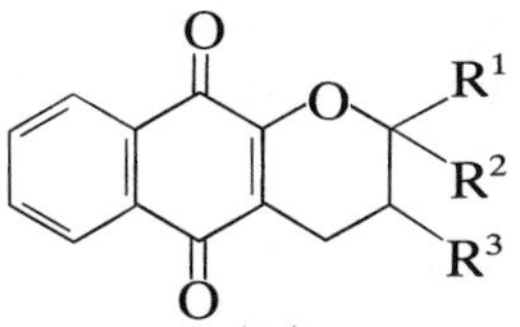

O, O, R^1, R^2, R^3, O

(**84**)

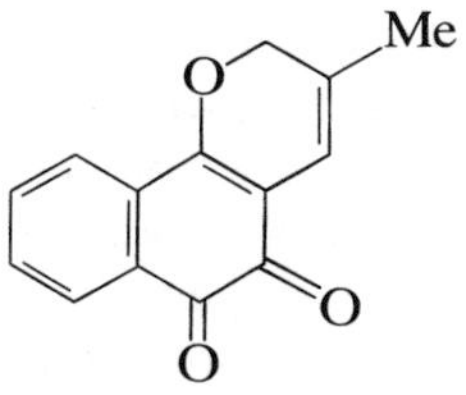

O, Me, O, O

(**85**)

Other lapachone analogs have also been prepared by the action of acids on compounds of the lapachol type (**81**). The action of sulfuric acid on 2-(3-methylallyl)-3-hydroxy-1,4-naphthoquinone (**81**, $R^1 = R^2 = H$) afforded equal parts of orange and yellow isomers which were separated as their bisulfite compounds and shown to be the 2-methyl analogs of the lapachones (**83**, $R^1 = Me$, $R^2 = R^3 = H$) and (**84**, $R^1 = Me$, $R^2 = R^3 = H$), respectively.[92] As would be expected, treatment of the trimethylallylnaphthoquinone (**81**, $R^1 = R^2 = Me$) with cold sulfuric acid for 20 hr gave the *o*-quinone (**83**, $R^1 = R^2 = R^3 = Me$), whereas on heating in a mixture of glacial acetic acid and hydrochloric acid at 100° for 75 min the *p*-quinone (**84**, $R^1 = R^2 = R^3 = Me$) was obtained.[98]

9-Hydroxy-α-lapachone has been obtained by catalytic hydrogenation of 9-hydroxy-α-dehydrolapachone in ethanol in the presence of palladium on barium sulfate.[99]

The parent quinone (**84**, $R^1 = R^2 = R^3 = H$) was prepared in 31% yield by reaction of the chlorophenol (**86**, R = H) with 4-chlorothiophenol in ethanolic sodium hydroxide.[100] When cyclization of the chlorophenol (**86**, R = H) was effected by refluxing with diethanolamine in ethanol, both the quinone (**83**, $R^1 = R^2 = R^3 = H$) and (**84**, $R^1 = R^2 = R^3 = H$) were formed, the former in 10% yield.[101] On heating with hydrochloric acid for 0.5 hr, the former quinone was converted to the latter. In a similar cyclization of the *O*-acetate (**86**, R = Ac), the yield of the *o*-quinone (**83**, $R^1 = R^2 = R^3 = H$) was increased to 40% at the expense of the isomer (**84**, $R^1 = R^2 = R^3 = H$).[101]

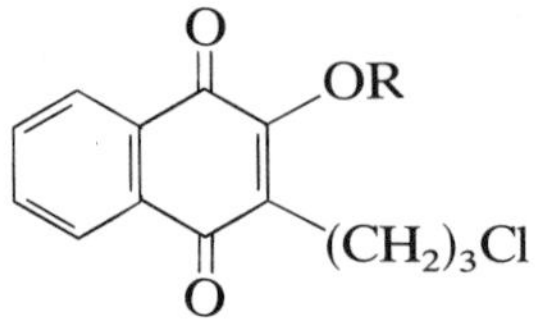

O
OR
$(CH_2)_3Cl$
O

(**86**)

The α-lapachone analog (**84**, $R^1 = R^2 = Me$, $R^3 = Ph$) has been obtained by photochemical reaction of 2-isopropoxy-1,4-naphthoquinone with styrene; the yield was 41%.[102]

The three naphtho[2,1-*b*]pyran-6-one derivatives (**87a**) through (**87c**) were prepared in the course of structural studies on the diterpene borjatriol, a derivative of manoyl oxide isolated from *Sideritis mugronensis.*[103] Oxidation of a 6-hydroxy compound was adopted in each case. The 3-methyl compound (**87a**) was obtained by oxidation with Jones's reagent in acetone, while chromium trioxide in pyridine was adopted for the preparation of compound (**87c**). The 3-carboxylic acid (**87b**) was prepared by oxidation of the 6-hydroxy-3-aldehyde with Jones's reagent.[103]

The hydroxynaphthopyranone (**88**) has been isolated from fermentation extracts of *Sporormia affinis.* The dione (**89**), which was obtained on careful

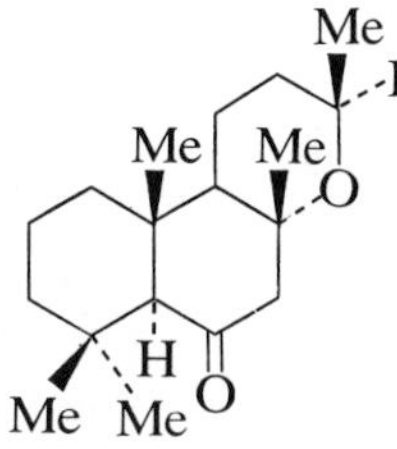

(**87a**) R = Me
(**87b**) R = CO_2H
(**87c**) R =

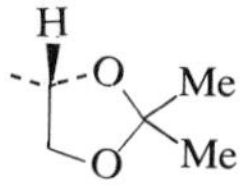

oxidation with Jones's reagent, rearranged in acetone in the presence of sodium hydroxide to give the isomer (**90**).[104]

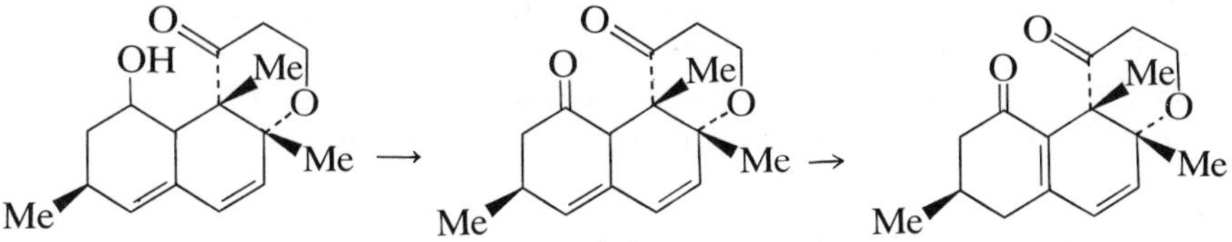

(**88**) (**89**) (**90**)

The 9-oxo compound (**91**, R = C≡CH) was prepared by oxidation of the corresponding hydroxy compound (itself obtained from 2-oxomanoyl oxide) in acetone with chromic acid–sulfuric acid; oxidative dimerization occurred on refluxing with cupric acetate and pyridine.[105] 2-Oxodihydromanoyl oxide (**91**, R = Et) has been prepared by hydrogenation of 2-oxomanoyl oxide (**91**, R = CH=CH_2) which had been extracted from the wood oil of *Dacrydium colensoi.*[106] Reduction to the alcohol followed by dehydration, or by chlorination and dehydrochlorination, led to the olefin (**92**, X = H_2). Epoxidation of the olefin followed by reduction of the product with lithium aluminum hydride afforded an alcohol that was oxidized to the ketone (**93**). Oxidation of the olefin (**92**, X = H_2) with selenium dioxide in acetic acid gave an allylic acetate that was hydrolyzed and oxidized to the unsaturated ketone (**92**, X = O) and finally hydrogenated to the 10-ketone (**94**).[106]

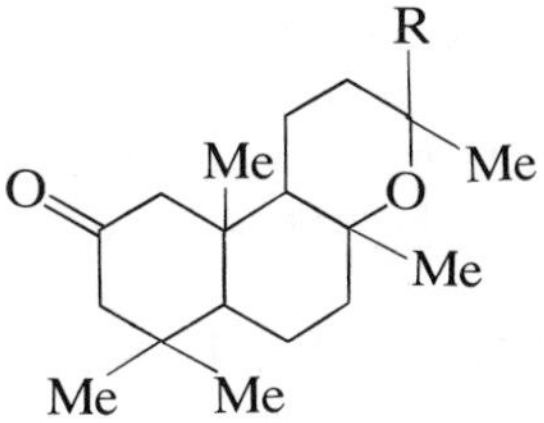

(**91**) 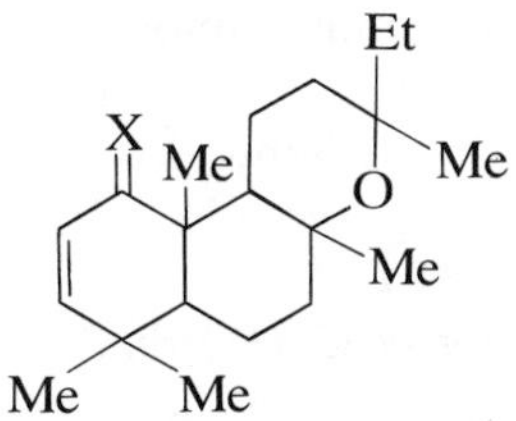 (**92**)

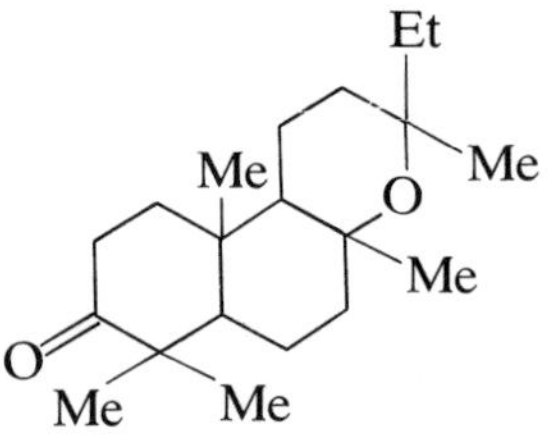

(93) 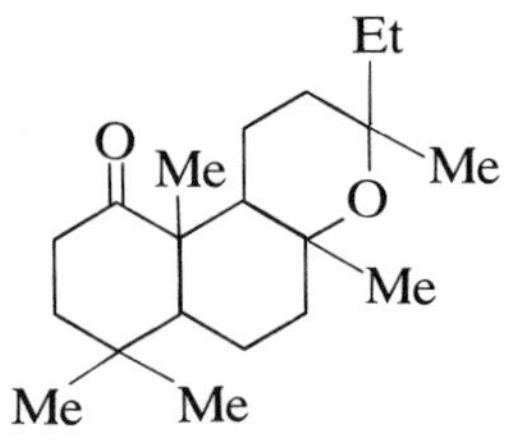 (94)

Deuteration of dihydrooxomanoyl oxide with potassium deuteroxide in deuterium oxide afforded a tetradeuterated product. Only hydrogen atoms α to the carbonyl exchanged and the mass spectrum confirmed the assigned structure (**91**, $R = CH{=}CH_2$) of oxomanoyl oxide.[107]

Reaction of 7-hydroxynonen-3-one (**96**) with 2-methylcyclohexane-1,3-dione (**95**) in toluene in the presence of pyridine and a trace of quinol afforded the naphthopyranone (**97**, $R^1 = Et$, $R^2 = Me$).[108] The diethyl analog (**97**, $R^1 = R^2 = Et$) was similarly prepared. Hydrogenation in toluene in the presence of 5% palladium on charcoal afforded the partially reduced derivatives (**98**), which when dissolved in acetone and treated with 1 *N* sulfuric acid at 25° were converted to hydroxy compounds (**99**).[108]

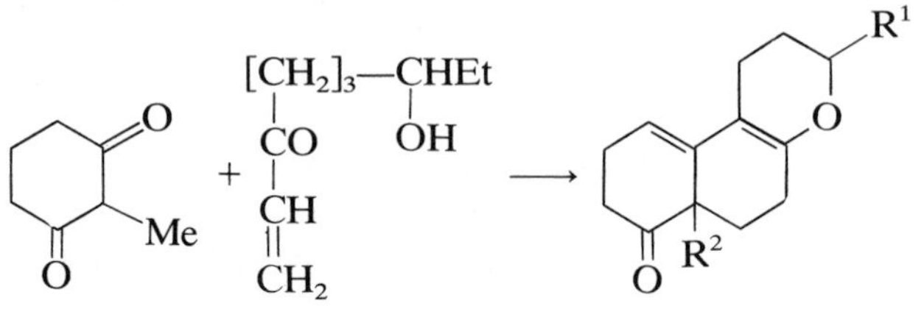

(95) (96) (97) $R^1 = Et$, $R^2 = Me$

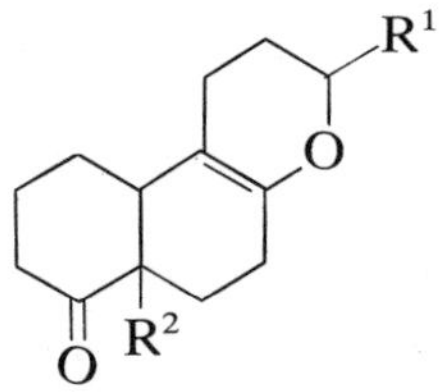

(98) 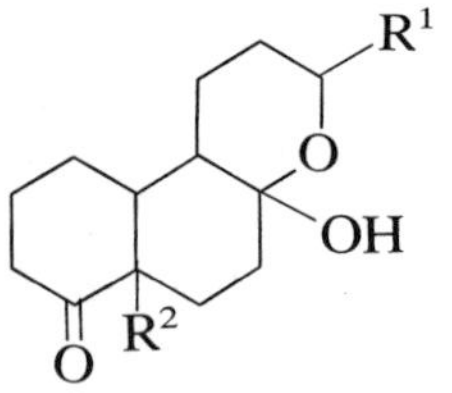 (99)

A general preparation of pyran-containing fused ring systems involves the reaction of equimolar amounts of a phenolic Mannich base with an enamine in dioxan.[109–111] A relevant application is the synthesis of the 2-morpholino-3,3-dimethylnaphthopyranone (**102**) from 3-dimethylaminomethyl-2-hydroxy-1,4-naphthoquinone (**100**, $R = CH_2NMe_2$) and *N*-isobutenylmorpholine (**101**).[109,111]

The addition product (**100**, R = H) from 2-hydroxy-1,4-naphthoquinone and benzylidenacetone has been converted to the naphthopyranone (**31**) by refluxing with methanolic hydrogen chloride.[112]

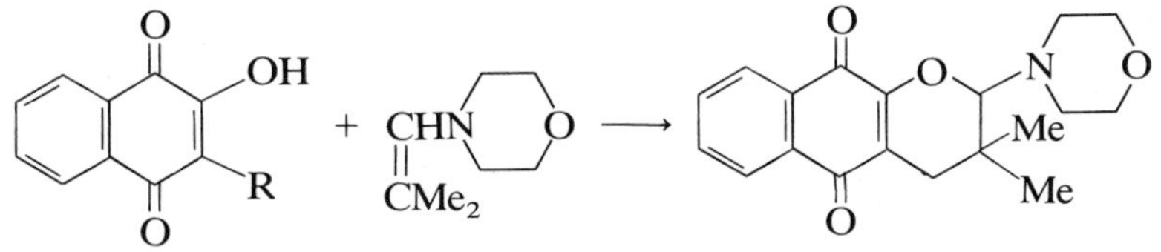

(100) R = CH_2NMe_2)   (101)   (102)

In work related to the structure of various natural products, the trihydroxynaphthopyrandione (**28**, $R^1 = Me$, $R^2 = R^5 = H$, $R^3 = R^4 = OH$) was prepared by catalytic hydrogenation of the corresponding chromene analog, a pigment isolated from the sea urchin *Echinothrix diadema.*[113] In the course of work on products isolated from the roots of *Conospermum teretifolium* R.Br., the chroman (**103**, R = H) was converted to the 6-propionylchroman (**103**, R = COEt) by treatment with propionyl chloride and stannic chloride. On cyclization in refluxing sodium ethoxide in ethanol followed by aerial oxidation, the naphthopyranone (**104**, R = Me) was obtained and demethylated to the dihydroxy analog (**104**, R = H).[114] The *p*-quinone (**29**, $R^1 = O$, $R^2 = R^4 = H$, $R^3 = OMe$) was prepared by cyclization of the chloroketone (**105**) in refluxing sodium ethoxide in ethanol.[49]

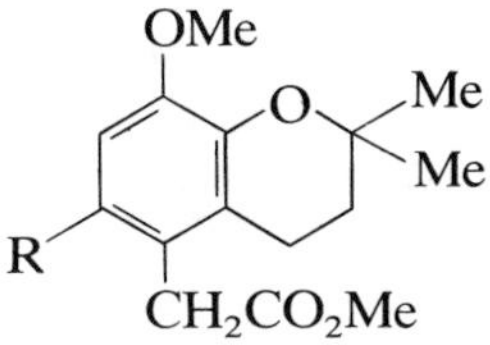

(**103**) 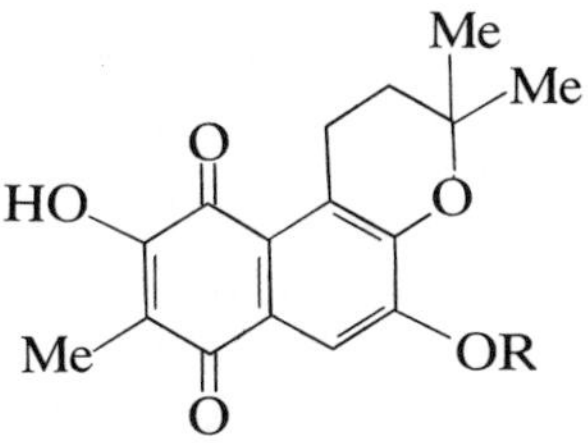   (**104**)

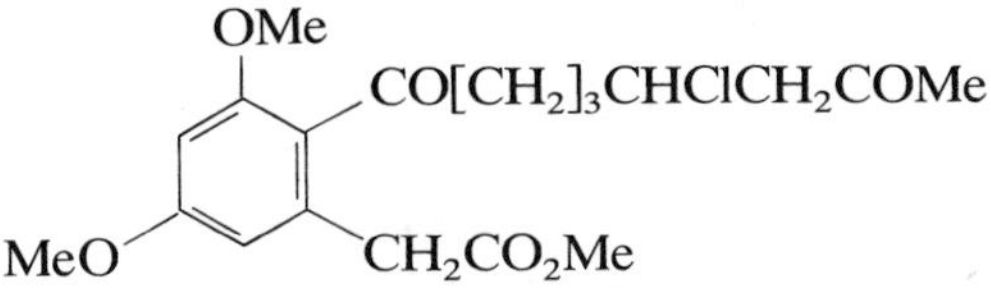

(**105**)

6. Bz-Acylchromans and Related Compounds

The majority of bz-acylchromans have been prepared by the Friedel–Crafts reaction of a chroman with an acid chloride in the presence of aluminum chloride. Substitution normally occurred in the position para to the ether group to give 6-acylchromans. 6-Benzoylchroman and 6-(3,4-dimethoxybenzoyl)chroman were prepared by the action of benzoyl chloride and veratroyl chloride, respectively, in the presence of aluminum chloride.[115] Similarly, reaction of 2,2-dimethylchroman[116] or its 5,7-dihydroxy analog[117] with acetyl chloride in nitrobenzene afforded the 6-acetyl derivatives exclusively. A variety of chroman-6-yl ketones has been similarly prepared using carbon disulfide as solvent;[118–121] yields ranged from about 65%[118] to 80% of theory.[119] The reaction has also been carried out in ether,[122] and acetic anhydride has been employed in place of acetyl chloride.[123] Methyl 2,2-dimethyl-8-methoxy-6-propionylchroman-5-yl acetate (**106**, $R^1 = 2,2\text{-diMe}$, $R^2 = 5\text{-}CH_2CO_2Me\text{-}6\text{-COEt-}8\text{-OMe}$) has been obtained on reaction of the appropriate chroman with propionyl chloride in the presence of stannic chloride.[114]

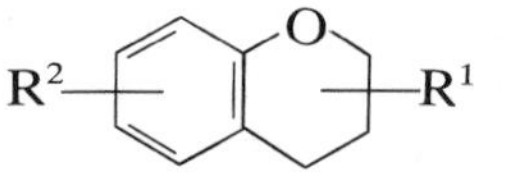

(**106**)

In contrast to the Friedel–Crafts reaction, when 5,7-dihydroxy-2,2-dimethylchroman was condensed with acetonitrile under the conditions of the Hoesch reaction, the crude product afforded a mixture of the 6-acetyl (9%) and the 8-acetyl-5,7-dihydroxy-2,2-dimethylchromans (53%), which were separated by solvent extraction using benzene and methanol.[52,117] With 3-phenylpropionitrile under similar conditions, a mixture of the 6- and 8-(3-phenylpropionitrile) derivatives of 5,7-dihydroxy-2,2-dimethylchroman was separated and characterized.[117]

Heating 7-acetoxychroman with aluminum chloride afforded 6-acetyl-7-hydroxychroman in 62% yield.[123] Only a poor yield of 7-acetyl-6-hydroxychroman was obtained from a similar Fries rearrangement of 6-acetoxychroman;[124] a more satisfactory preparation involved treatment of 6-methoxychroman with boron trifluoride–acetic acid complex.[124–127] Similar treatment of 8-methoxychroman gave the 6-acetyl derivative; but with 8-hydroxychroman at a higher temperature, 7-acetyl-8-hydroxychroman was isolated in 41% yield together with a small quantity of the 6-acetyl analog.[128,129]

Examples have been reported of the preparation of acetylchromans by direct cyclization. 8-Acetyl-6-methoxychroman was formed by the action of sulfuric acid on 2-hydroxy-3-(3-hydroxypropyl)-5-methoxyacetophenone.[130] Reaction of the diacetate of 1,3-butanediol with phenol in the presence of aluminum chloride gave a product that has been described as the 5-acetyl[131] or 8-acetyl[132] derivative of 4-methylchroman.

6-Acetyl-7,8-dimethoxy-2,2-dimethylchroman was obtained by hydrogenation of the corresponding chromene.[133]

Methylation of 8-acetyl-5,7-dihydroxy-2,2-dimethylchroman in acetone with methyl iodide in the presence of potassium carbonate afforded the 7-hydroxy-5-methoxy analog.[117] Similar treatment of the 6-acetyl compound (**106**, $R^1 = 2,2\text{-diMe}$, $R^2 = 5,7\text{-diOH-}6\text{-Ac}$) gave 6-acetyl-2,2-dimethyl-5-hydroxy-7-methoxychroman; while with methyl sulfate in the presence of potassium carbonate, 6-acetyl-5,7-dimethoxy-2,2-dimethylchroman was formed.[52] Methylation of 5,7-dihydroxy-2,2-dimethyl-8-(3-phenylpropionyl)chroman (**106**, $R^1 = 2,2$-diMe, $R^2 = 5,7$-diOH-8-$CO(CH_2)_2Ph$) gave a mixture of the 5-hydroxy-7-methoxy and the 7-hydroxy-5-methoxy analogs; the dimethoxy analog was also prepared. With benzyl bromide, 5,7-dihydroxy-2,2-dimethyl-8-(3-phenylpropionyl)chroman affords the 5-benzyloxy-7-hydroxy compound.[117]

Reaction of 8-acetyl-5,7-dihydroxy-2,2-dimethylchroman with benzaldehyde in ethanol in the presence of alkali gave the 8-cinnamoyl analog (**106**, $R^1 = 2,2$-diMe, $R^2 = 5,7$-diOH-8-COCH=CHPh); conversion to 5,7-dihydroxy-2,2-dimethyl-8-(3-phenylpropionyl)chroman was effected by catalytic hydrogenation.[117] The benzylidene derivatives of 6-acetyl-2,2-dimethyl-5-hydroxy-7-methoxychroman and its 5,7-dimethoxy analog have been hydrogenated in the presence of platinum oxide to give the corresponding 6-(3-phenylpropionyl) compounds.[52]

Addition of sodium to a solution of 7-acetyl-6-hydroxychroman in dry ethyl acetate afforded 7-acetoacetyl-6-hydroxychroman (**106**; $R^1 = H$, $R^2 = 6$-OH-7-$COCH_2COMe$) while the keto ester (**106**, $R^1 = H$, $R^2 = 6$-OH-7-$COCH_2COCO_2Et$) was obtained on reaction with diethyl oxalate.[124–127]

IV. PHYSICAL PROPERTIES AND SPECTRA

The majority of the compounds discussed in this chapter have been isolated as crystalline solids. While chromanones are normally white, many of the chromandione derivatives have a quinonoid structure which imparts a yellow or even red color. The optical density at 465 nm has in fact been utilized for the quantitative estimation of tocored (**19**, $R = C_{16}H_{33}$) in ethanol.[22]

Details have been published of the thin-layer chromatography of 2,2,7,8-tetramethyl-5,6-chromandione (**19**, R = Me) and of 6-hydroxy-2,2,7-trimethyl-5,8-chromandione on silica gel. R_F values in chloroform, benzene, and cyclohexane–tetrahydrofuran were reported and the compounds detected by spraying with either sulfuric acid (brown) or potassium ferricyanide (blue); spraying with silver nitrate did not reveal the former but gave a grey spot with the latter.[134] Methods have been reported for the detection of quinones, including α- and β-lapachones, both on paper chromatograms and by thin-layer chromatography using silica gel plates.[135]

Spectral data have been reported for a number of bz-chromanone derivatives; the nature and source of such information are indicated in Table 1. Compounds for which optical rotations have been recorded are also listed. Many of these compounds are natural products or derivatives prepared in the course of synthesis or structural studies. Optical rotatory dispersion studies have been reported for the fully saturated naphtho[2,1-*b*]pyran-6-one derivatives (**87a**) and (**87b**);[103] rotatory dispersion curves have been published for the 8-, 9-, and 10-ketones (**93**), (**91**, R = Et), and (**94**).[106]

TABLE 1. SPECTRAL DATA AND OPTICAL ROTATIONS OF Bz-OXOCHROMANS AND RELATED COMPOUNDS

Oxochroman	Ref. to data on				
	nmr	ir	uv	ms	$[\alpha]_D$
5-Chromanones					
(**7**)		4			
Type (**36**)	56, 59	56, 59		56, 59	
(**43**)	60	60			
(**47**) cis and trans		64			
6-Chromanones					
Type (**48**)		10, 67, 68	10, 67, 68		
(**51**)	70	70			
7-Chromanones					
Type (**61**)	76	76		76	
(**63**) and (**64**)	77	77			
(**68**)	79	79			
(**70**), (**71**), and (**74**)	81	81			
5,6-Chromandiones					
Tocored (**19**, $R = C_{16}H_{33}$)		22	16, 22, 83		
(**22**)	26			26	
5,7-Chromandiones					
(**23**) and (**24**)	27	27	27	27	
5,8-Chromandiones					
2,2,7,8-tetramethyl			85		
6-Hydroxy-2,2,7-trimethyl		17	17		
Tocopurple (**20**)	17	17	17		
2,3-Dihydro-1*H*-naphtho[2,1-*b*]pyranones					
(**87a–c**) and Me ester of (**87b**)	103	103			103
(**89**) and (**90**)	104	104	104	104	104
(**91**) R = C≡CH and dimer, R = Et		106	105		105
(**91**, R = CH═CH$_2$)-d_4				107	
(**92**, X═O)			106	106	
(**93**) and (**94**)		106			106
2,3-Dihydro-4*H*-naphtho[1,2-*b*]pyranones					
β-Lapachone (**6**)	30, 94	94, 138, 139	85, 139	136	
7-Hydroxy-β-lapachone	39, 99	39, 99	39, 99		
Type (**83**)		101	136		
Type (**26**)	45	45	45	45	45
2,3-Dihydro-4*H*-naphtho[2,3-*b*]pyranones					
α-Lapachone (**5**) (and analogs)	38, 39	38, 39, 139	38, 39, 139	136	39
Type (**28**)	46–48	46–48	46–48, 113	46, 48	46, 47
Type (**29**)	47, 49	47, 49	47, 49	47, 49	49
(**30**)		47	47	47	
(**32**)	51	51	51	51	51
Type (**84**)	102	102	137	102	
6-Acetylchromans	133	133	122, 133, 140	141	
8-Acylchromans			142		

Explanations have been offered for the major ions in the mass spectra of α- and β-lapachones.[136] Ionization constants of β-lapachone and of the 3-methyl lapachone analogs (**83** and **84**, $R^1 = R^2 = H$, $R^3 = Me$) have been measured spectrophotometrically in aqueous sulfuric acid.[137] The oxidation–reduction potential of β-lapachone has been discussed.[138] Studies on the structure of xanthochymol (**23**) involved an X-ray crystallographic analysis of the di-4-bromobenzenesulfonate of isoxanthochymol (**24**).[27]

V. REACTIONS

1. General Reactions of Bz-Oxochromans

The bz-oxochromans and the bz-acylchromans discussed in this chapter exhibit the expected reactions of ketones. Oximes, semicarbazones, and 2,4-dinitrophenylhydrazones have been prepared as crystalline derivatives of a number of 5-chromanones. The 7-chromanone (**61**, R = Me) has been converted to its semicarbazone,[76] and α- and β-lapachones gave monooximes on reaction with hydroxylamine hydrochloride.[143] Oximes, semicarbazones, and 2,4-dinitrophenylhydrazones have also been described for a number of bz-acylchromans. On reaction with benzaldehyde, acetylchromans afforded benzylidene derivatives.[52,117,122]

A number of examples of the reduction of the ketones to alcohols have been described. α-Tocopherones are readily reduced to the α-tocopherols by the action of ascorbic acid or sodium borohydride at room temperature.[10] In related studies, it was shown that in slightly acidic media, 8*a*-hydroxy-α-tocopherone (**48**, R = H) was reduced to α-tocopherol by the action of ascorbic acid, while if the mixture was strongly acidic, ring-opening occurred to give an α-tocopherylquinone of type (**108**, $R = C_{16}H_{33}$); the latter reaction was faster than the rate of dehydration, and no reduction occurred.[69] Treatment of 8*a*-alkoxy-α-tocopherones with dilute aqueous hydrochloric acid afforded α-tocopheryl-*p*-quinone;[10] the same product was obtained on decomposition of 8*a*-hydroxy-α-tocopherone (**48**, R = H) in petroleum ether in alkaline or acidic media.[67] The kinetics and mechanism of the conversion of 8*a*-hydroxy-α-tocopherones (**107**) to the tocopherylquinones (**108**) has been extensively studied using double-potential step chronoamperometry with the pentamethyl compound (**107**, R = Me) as a model. ^{18}O-Isotope studies provided further evidence that the acid catalysis involved proton transfer from the acid to the heterocyclic oxygen followed by removal of a proton from the 8*a*-hydroxyl by water. In base catalysis, the base removed the proton from the hydroxyl group and a proton was transferred to the 1-oxygen from the solvent.[144]

Reduction of the fully saturated 6-chromanone (**51**) to the corresponding 6-chromanol (**50**) has been effected with lithium in ammonia.[70,71] Addition of zinc dust to a refluxing mixture of the 5,8-chromandione (**80**) in acetic

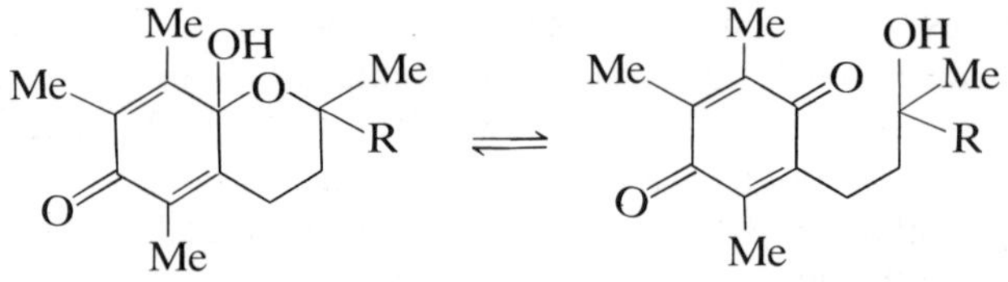

(**107**) ⇌ (**108**)

anhydride–glacial acetic acid afforded 5,8-diacetoxy-2,2,4,4,7-pentamethylchroman.[89] Reductive acetylation of α- and β-lapachones by the action of acetic anhydride in the presence of zinc dust and sodium acetate resulted in the formation of the diacetyl derivatives of the corresponding hydroquinones, (**109**) and (**110**).[91]

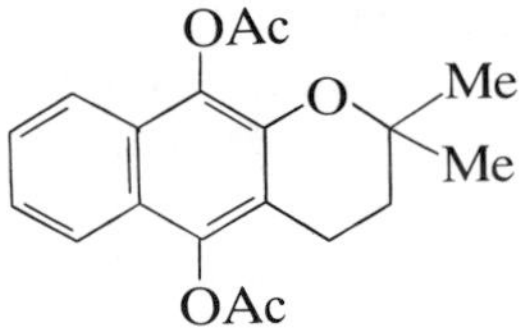

(**109**) 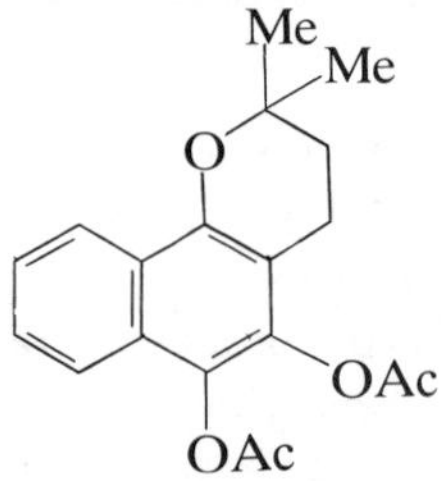 (**110**)

Sodium borohydride reduction of the naphthopyran-6-one (**87c**) gave the 6-hydroxy analog, while the Meerwein–Ponndorf reduction of the carboxylic acid (**87b**) followed by treatment with diazomethane afforded the methyl 6-hydroxy-3-carboxylate (**111**), which was characterized as its *O*-acetate.[103] 3,7,8-Trimethyl-1-tetralone (**112**) was formed on heating the diketone (**90**) with 10% palladium on charcoal in vacuo at 250° for 5 min.[104]

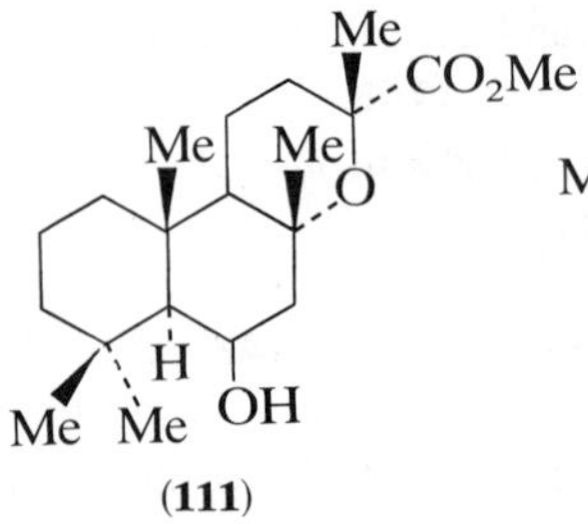

(**111**) 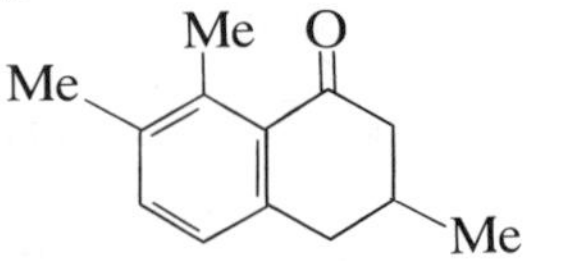 (**112**)

Bz-Ethyl chromans have been prepared by Clemmensen reduction of appropriate acetylchromans.[119,123,124,128,130,132]

α-Lapachones and β-lapachones have been reduced with phosphorus and hydriodic acid to give α- and β-lapachans which were assigned the structures (**113**) and (**114**), respectively.[3]

Dehydrogenation of the tetrahydro-2,2-dimethyl-5-chromanone (**36**, $R^1 =$ Me, $R^2 = H$) over platinum on charcoal afforded 2,2-dimethyl-5-chromanol.[54]

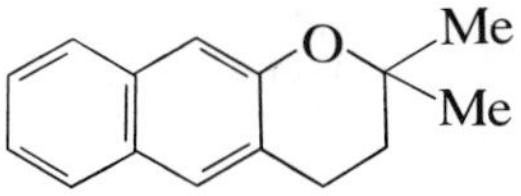

(113) 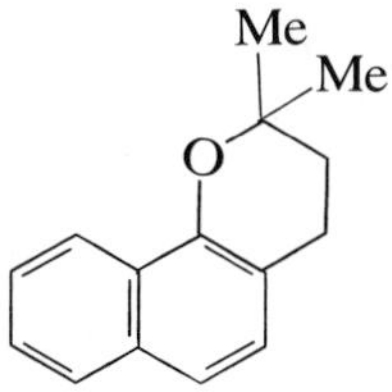 (114)

2. Bz-Chromanones

When the 2-methyl-5-chromanone (**36**, $R^1 = R^2 = H$) was allowed to react with vinylmagnesium chloride at −30°, an unstable 5-vinylchromanol was formed which reacted with cyclopentane-1,3-dione to give the chroman (**115**). The latter compound has been used as an intermediate in the synthesis of 4-oxasteroids.[58]

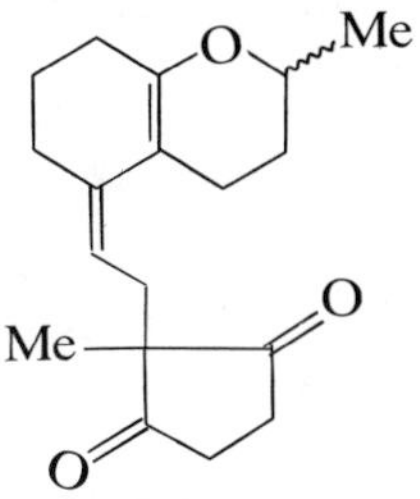

(115)

The 7-chromanones (**70**) and (**71**) have been converted to cyanhydrins (by the action of potassium cyanide in acetic acid), which formed α-hydroxy-esters on methanolysis. On dehydration with phosphoryl chloride in pyridine, a mixture of the unsaturated esters (**116**) and (**117**) was obtained.[80,81]

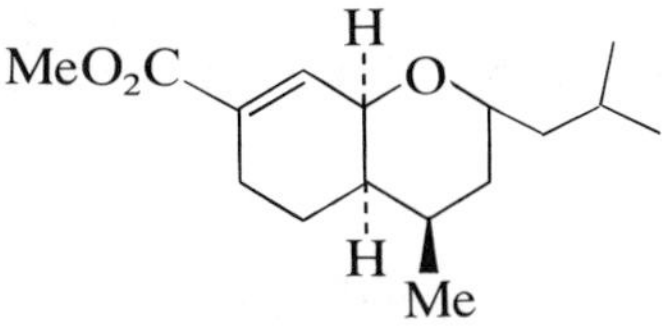

(116) 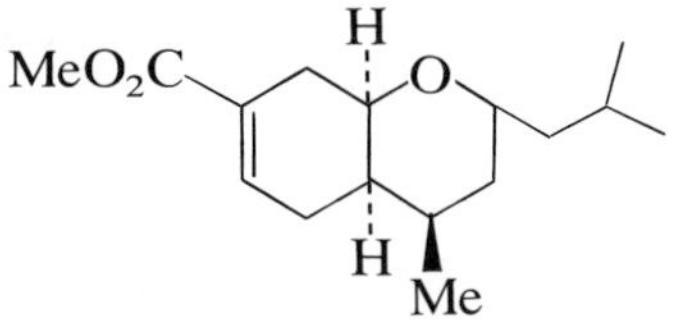 (117)

Treatment of the 4*a*-(2,3-dimethoxyphenyl)-7-chromanone (**68**) with constant boiling hydriodic acid afforded the tricyclic compound (**118**) in 58% yield.[78,79]

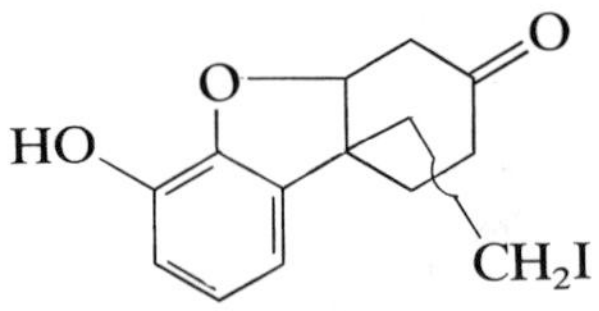

(118)

3. Bz-Chromandiones

The 5,6- and 5,8-chromandiones behave as quinones. Tocored (**19**, R = $C_{16}H_{33}$), a 5,6-chromandione, afforded a phenazine on reaction with *o*-phenylenediamine in glacial acetic acid.[16] The 7,8-dimethyl-5,6-chromandiones give yellow phenazines[85,88] which show a pale-green fluorescence in ethanolic solution.[85] Not surprisingly, 6,7-dimethyl-5,8-chromandiones failed to react with *o*-phenylene diamine.[88]

On refluxing with 6 *N* hydrochloric acid, tocored (**19**, R = $C_{16}H_{33}$) isomerized to the *p*-quinone (**119**).[16] A dihydroacetate was obtained as white needles when a mixture of tocored, pyridine, acetic anhydride, and powdered zinc was stirred for 1 hr.[16]

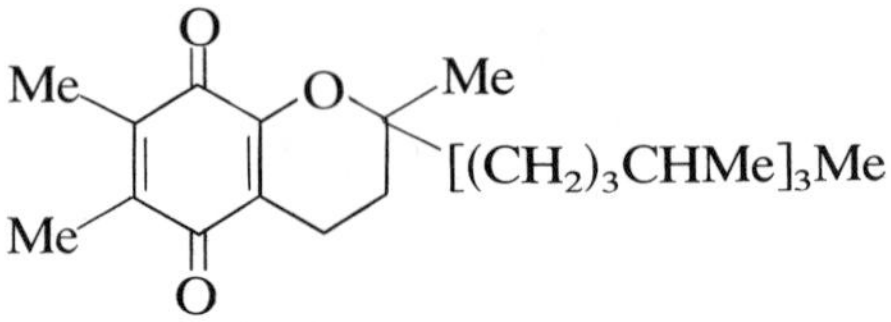

(119)

The interaction of tocored and model compounds with unsaturated fatty esters has been studied. It had been suggested that tocored in crude soybean oil was not removed in conventional refining procedures but that some remained in a colorless form which assumed a red color on atmospheric oxidation. It was shown that changes in the absorption of tocored at 460 nm occur on heating with unsaturated fatty acids such as methyl linoleate and soybean oil fatty acid methyl esters, but not with methyl palmitate. More detailed studies were made using 2,7,8-trimethyl-5,6-chromandione as a model.[83]

4. Oxo Derivatives of Dihydronaphthopyrans

There is a quantitative isomerization of α-lapachone to β-lapachone in sulfuric acid, and the reverse reaction occurs in hydrochloric acid.[2] Ultraviolet spectral studies and measurements of the ionization constants in sulfuric acid have been used in seeking an explanation of the interconversion of the lapachones. In view of the instability of α-lapachone in sulfuric acid, the 3-methyl analogs (**84**, $R^1 = R^2 = H$, $R^3 = Me$) and (**83**, $R^1 = R^2 = H$, $R^3 = Me$) were employed in these studies. The uv spectra of these compounds and of hydrolapachol [**100**, $R = (CH_2)_2CHMe_2$] were very similar in sulfuric acid, in contrast to those in ethanol. This was interpreted as indicating that addition of a proton occurs in sulfuric acid to give the ions (**120**) and (**121**), which together with the cation of hydrolapachol, are examples of the resonant ion (**122**). Spectrophotometric measurement of the ionization constants of β-lapachone (**6**) and of the 3-methyl compounds (**84**) and (**83**) ($R^1 = R^2 = H$, $R^3 = Me$) in aqueous sulfuric acid showed that β-lapachone, an *o*-quinone, is exceptionally basic (pK_a −3.45). The basic strength of the *p*-quinone (**84**, $R^1 = R^2 = H$, $R^3 = Me$) was only $\frac{1}{800}$ of that of the *o*-isomer (**83**, $R^1 = R^2 = H$, $R^3 = Me$). In concentrated sulfuric acid, which converts both isomers to their conjugate acids, the *o*-quinone cation predominates. In hydrochloric acid, which has a higher water content, ionization of the *o*-quinone is insufficient to transform the equilibrium from the side of the unchanged *p*-quinone.[137]

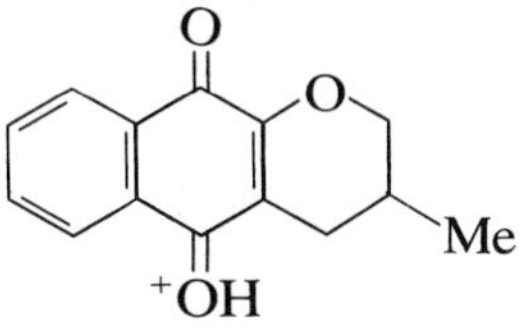

O, O, Me, $^+$OH

(**120**)

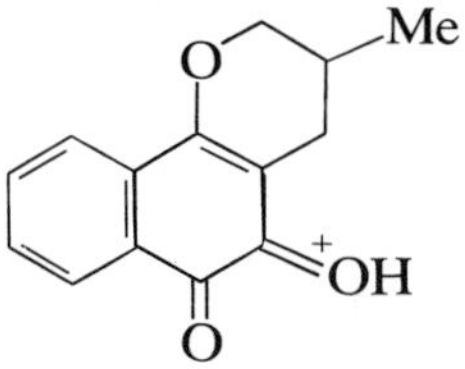

Me, O, $^+$OH, O

(**121**)

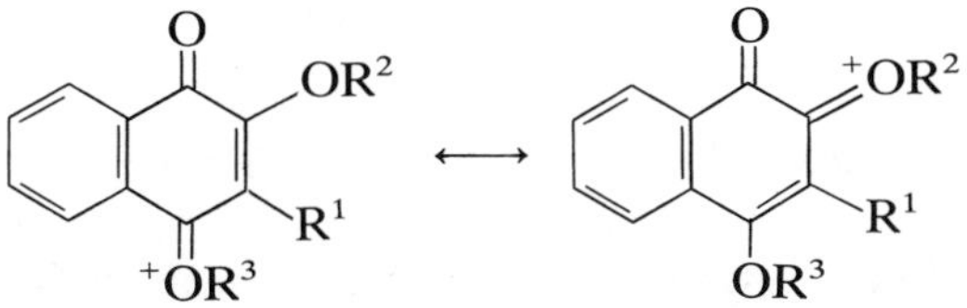

O, OR^2, R^1, $^+OR^3$ ⟷ O, $^+OR^2$, R^1, OR^3

(**122**)

In a similar reaction, 9-hydroxy-α-lapachone has been converted to 7-hydroxy-β-lapachone by the action of concentrated sulfuric acid.[39,99] Methylation of the latter compound with methyl iodide afforded the 7-methoxy analog.[99]

Dehydrogenation of α- and β-lapachones with 2,3-dichloro-5,6-dicyano-1,4-benzoquinone afforded the corresponding dehydrolapachones (**123**) and (**124**), respectively.[139] On addition of dilute sodium hydroxide to 3-bromo-β-lapachone, the dehydrolapachone (**124**) was formed in small quantity in addition to dihydroxyhydrolapachol (**82**, $R^1 = Me$, $R^2 = OH$).[2]

The 2-methyl compound (**83**, $R^1 = Me$, $R^2 = R^3 = H$) dissolves in warm alkali to give a claret-red solution, no doubt due to ring cleavage,[92] as has been demonstrated for β-lapachone.[2]

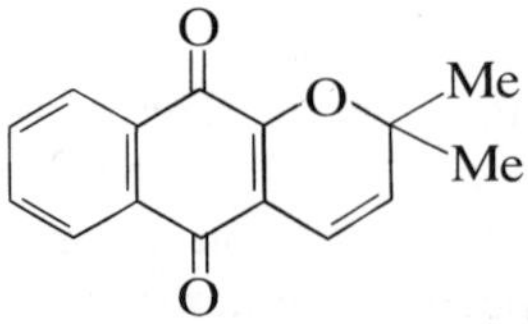

(**123**) 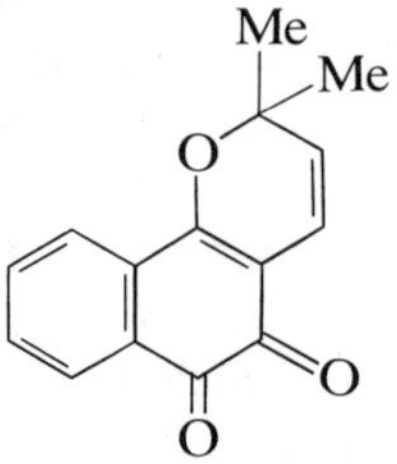 (**124**)

The reactions of erythrostominone (**28**, $R^1 = CH_2COMe$, $R^2 = R^3 = R^4 =$ OH, $R^5 = Me$) in acid solution have been studied. Erythrostominone was unaffected by glacial acetic acid, but with toluene-4-sulfonic acid in benzene and sulfuric acid it gave an unsaturated cyclic acetal (**125**) and/or its $\Delta^{2'}$-isomer. Refluxing with ethanolic hydrogen chloride led to the naphthoquinone (**126**).[46,47]

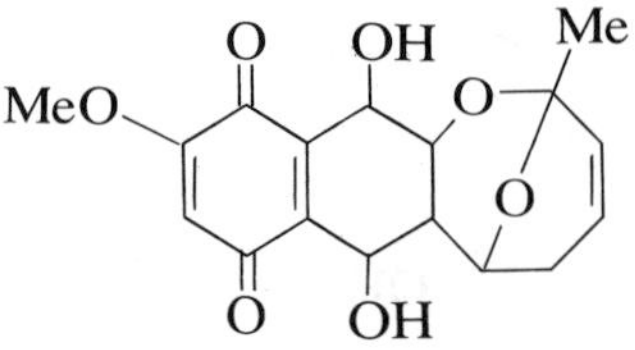

(**125**) 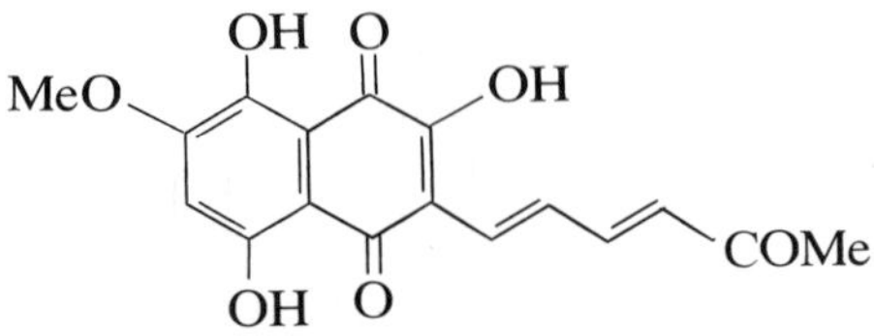 (**126**)

5. Bz-Acylchromans

On oxidation of 6-acetylchroman with selenium dioxide in acetic acid at 90°, a glyoxal was formed which was mixed with isopropylamine in methanol and treated with sodium borohydride to give 1-chroman-6-yl-2-isopropylaminoethanol (**127**, $R = CHOHCH_2NHCHMe_2$).[121]

The acetylchromans form Mannich bases; 6-(2-piperidinopropionyl)-chroman [**127**, $R = CO(CH_2)_2N(CH_2)_5$] hydrochloride was prepared in 68% yield by reaction of 6-acetylchroman with piperidine and paraformaldehyde in ethanol at 5° for 48 hr.[120]

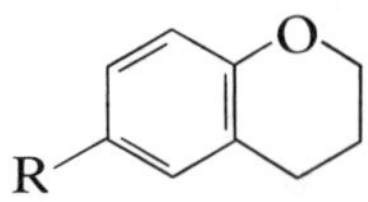

(**127**) 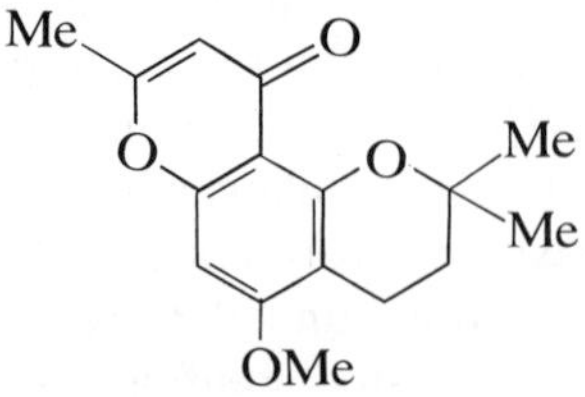 (**128**)

When the oxime of 6-acetylchroman was subjected to a Beckmann rearrangement by heating with polyphosphoric acid at 100–110° for 7.5 min, 6-acetylaminochroman was formed.[120]

On treatment of 8-acetyl-2,2-dimethyl-7-hydroxy-5-methoxychroman in ethyl acetate with sodium on the steam bath, a crude diketone was obtained which cyclized in boiling ethanolic hydrogen chloride to give the chromanopyrone (**128**).[117]

As has been mentioned earlier (Section III, 5), the 6-propionylchroman (**103**, R = COEt) has been cyclized to the naphthopyran (**104**, R = Me) with sodium ethoxide in refluxing ethanol followed by aerial oxidation.[114]

VI. TABLES OF COMPOUNDS

Bz-Chromanones and bz-chromandiones that have been prepared or reported are listed in Table 2. Oxo derivatives of 2,3-dihydro-4*H*-naphtho-[1,2-*b*]pyrans are shown in Table 3, and oxo derivatives of 2,3-dihydro-1*H*-naphtho[2,1-*b*]pyrans are listed in Table 4. The ring systems are numbered as shown in structures (**1**), (**2**), and (**3**), respectively. The extent of saturation of the aromatic nucleus is indicated by the number of substitutents at each position. The absence of substituents at the angular positions is denoted by a dash (—). Where no substituent is indicated, its significance is described in the first footnote to each of these tables.

2,3-Dihydro-4*H*-naphtho[2,3-*b*]pyran-5,10-diones are listed in Table 5, and 2,3-dihydro-4*H*-naphtho[2,3-*b*]pyran-6,9-diones are shown in Table 6. In both cases the ring system is numbered as shown in structure (**4**). In these tables the only substituents indicated are those where there is a difference from the parent dione. Similarly in Table 7, which provides data on bz-acylchromans and related compounds, the only substituents shown are those where there is a difference from chroman itself.

TABLE 2. Bz-CHROMANONES AND Bz-CHROMANDIONES

Mol. formula	Substituents[a] 2	3	4	4a	5	6	7	8	8a	mp or bp (mm)	Ref.
$C_9H_{12}O_2$				H			=O	H	—	—	77
$C_{10}H_{14}O_2$	H, Me			—	=O				—	29–30	57, 58
$C_{10}H_{14}O_3$	H, Me			—	=O		H, OH		—	Oil	8
$C_{10}H_{16}O_2$				Me	=O				H	100–105 (0.04)[b]	62
$C_{10}H_{16}O_2$				Me	=O				H	70–75 (0.045)[c]	62
$C_{10}H_{16}O_2$				Me			=O		H	42.5	82
$C_{10}H_{16}O_3$				H			=O		OMe	—	77
$C_{11}H_{16}O_2$	Me_2			—	=O				—	93–94 (2.0)	53, 54, 59
	2,4-DNP									240–241	53
	Semicarbazone									219	53
$C_{11}H_{16}O_3$	Me, OMe			—	=O				—	—	56
$C_{11}H_{18}O_2$	Me_2			H			=O		H	70–72 140–142 (20)	76
$C_{12}H_{14}O_3$	H, Me			—	=O	=O	Me	Me	—	—	88
$C_{12}H_{14}O_3$	H, Me			—	=O	Me	Me	=O	—	132	88
$C_{12}H_{14}O_4$	Me_2			—	=O	OH	Me	=O	—	142–143.3	17, 83, 134
$C_{12}H_{20}O_2$	Me_2			Me			=O		H	150–152 (22)	76
	Semicarbazone									189–191	76
$C_{13}H_{16}O_3$	Me_2			—	=O	=O	Me	Me	—	109–110	18, 23–25, 83, 85–88, 134
$C_{13}H_{16}O_3$	Me_2			—	=O	Me	Me	=O	—	84	23, 88
$C_{13}H_{20}O_3$		Me_2	H, OH	—	=O		Me_2		—	130.5	60
$C_{13}H_{22}O_2$	H, Me			H	Me_2	=O			Me	Oil	70, 71
$C_{13}H_{22}O_2$	H, Me			H	Me_2		=O		Me	40–41	13–15
$C_{14}H_{18}O_2$	Me_2			—	$=CH_2$	=O	Me	Me	—	—	73

$C_{14}H_{18}O_3$	Me_2		Me_2	—	=O	H	Me	=O	—	112	89
$C_{14}H_{20}O_3$	Me_2			—	Me	=O	Me	Me	OH	—	144
$C_{14}H_{24}O_2$	H, Et	H, Me		H	=O	H, Me		H, Me	H	Oil	4
	Oxime									152.5–154.5	4
$C_{14}H_{24}O_2$	H, CH_2Pr^i		H, Me	H			=O		H	59.5–60.5[d]	80, 81
$C_{14}H_{24}O_2$	H, CH_2Pr^i		H, Me	H			=O		H	57.5–58[e]	80, 81
$C_{14}H_{24}O_2$	H, CH_2Pr^i		H, Me	H			=O		H	75.5–76.5[f]	81
$C_{16}H_{24}O_2$	Me, *g*			—	=O				—	138–142 (0.5)	55
$C_{16}H_{24}O_2$	Me_2			—	=O		H, *h*		—	—	59
$C_{16}H_{26}O_2$	Me, *i*			—	=O				—	124–125 (0.5)	55
	2,4-DNP									154–155	55
$C_{17}H_{18}N_2O_4$	=NH	H, NO_2	H, Ph	—	=O		Me_2		—	—	145
$C_{17}H_{22}O_4$							=O		H	125–126	78, 79
$C_{18}H_{20}N_2O_6$	=NH	H, NO_2	H, *k*	—	=O		Me_2		—	—	145
$C_{18}H_{26}O_4$	Me_2			—	=O	OH	C_7H_{15}	=O	—	88–89	28, 29
$C_{19}H_{23}N_3O_4$	=NH	H, NO_2	H, *l*	—	=O		Me_2		—	—	145
$C_{20}H_{30}O_3$	Me_2		H, Me	—	=O	=O	*m*	H	—	92–94	26
$C_{21}H_{36}O_2$	Me, *n*			—	=O				—	166–170 (0.4)	55
	2,4-DNP									105.5–106.5	55
$C_{23}H_{36}O_4$	Me_2			—	=O	*h,h*	OH	$COPr^i$	—	Oil	5–7
$C_{25}H_{36}O_4$	Me_2			—	=O	$COPr^i$	OH	*h,h*	—	—	6
$C_{25}H_{36}O_4$	Me_2			—	OH	$COPr^i$	=O	*h,h*	—	—	5
$C_{27}H_{44}O_4$ (Tocopurple)	Me, TMTD[o]			—	=O	OH	Me	=O	—	[p]	17, 19, 83
$C_{28}H_{38}O_4$	Me_2			—	H, *q*	=O	Me	Me	—	—	9
$C_{28}H_{46}O_3$ (Tocored)	Me, TMTD[o]			–	=O	=O	Me	Me	—	—	16, 18–22, 83, 84
$C_{28}H_{46}O_3$	Me, TMTD[o]			—	=O	Me	Me	=O	—	—	16, 146
$C_{29}H_{48}O_2$	Me, TMTD[o]			—	$=CH_2$	=O	Me	Me	—	—	72, 73
$C_{29}H_{50}O_3$	Me, TMTD[o]			—	Me, OH	=O	Me	Me	—	—	—
	O-Acetate									—	68

TABLE 2. (*Contd.*)

Mol. formula	Substituents[a] 2	3	4	4*a*	5	6	7	8	8*a*	mp or bp (mm)	Ref.
$C_{29}H_{50}O_3$	Me, TMTD[o]			—	Me	=O	Me	Me	OH	—	67–69
	O-Acetate									—	68
$C_{30}H_{52}O_3$	Me, TMTD[o]			—	Me	=O	Me	Me	OMe	—	10
$C_{31}H_{54}O_3$	Me, TMTD[o]			—	Me	=O	Me	Me	OEt	—	10, 64, 65, 68
$C_{32}H_{56}O_3$	Me, TMTD[o]			—	Me	=O	Me	Me	OPr^i	—	10
$C_{33}H_{58}O_3$	Me, TMTD[o]			—	Me	=O	Me	Me	OBu	—	10
$C_{37}H_{66}O_3$	Me, TMTD[o]			—	Me	=O	Me	Me	$O(CH_2)_7Me$	—	10
$C_{38}H_{50}O_6$	Compound (**23**) (see text)									135	27
$C_{38}H_{50}O_6$	Compound (**24**) (see text)									222	27
$C_{39}H_{54}O_{10}$	Compound (**45**) (see text)									221–222	61
$C_{43}H_{78}O_3$	Me, TMTD[o]			—	Me	=O	Me	Me	$O(CH_2)_{13}Me$	—	10
$C_{47}H_{86}O_3$	Me, TMTD[o]			—	Me	=O	Me	Me	$O(CH_2)_{17}Me$	—	10
$C_{58}H_{97}O_4$	Compound (**17**) (see text)									—	9

[a] The substituents at the position indicated are H_2 unless otherwise stated. [b] Cis isomer. [c] Trans isomer. [d] Structure (**70**) or (**71**).
[e] Structure (**71**) or (**70**). [f] Structure (**74**). [g] $CH_2CMe{=}CHCH_2Me$. [h] $CH_2CH{=}CMe_2$. [i] Isohexyl. [j] C_6H_3-2,3-diOMe. [k] C_6H_3-4-OH,3OMe.
[l] C_6H_4-4-NMe_2. [m] $CMe_2CH_2Bu^t$. [n] $CH_2CHMe(CH_2)_3CHMe(CH_2)_2Me$. [o] TMTD = $[(CH_2)_3CHMe]_3Me$. [p] Orange low-melting wax.

[q] OH Me (CH2)2 Me O Me Me

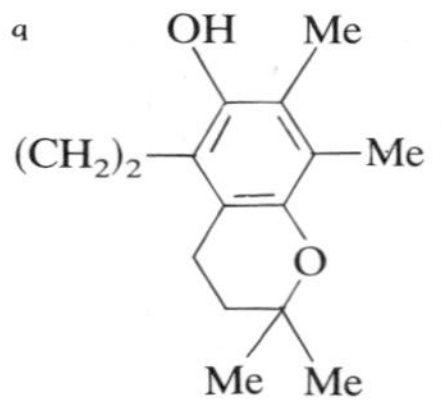

TABLE 3. OXO DERIVATIVES OF 2,3-DIHYDRO-4*H*-NAPHTHO[1,2-*b*]PYRANS

	Substituents[a]						
Mol. formula	2	3	5	6	7	mp	Ref.
$C_{13}H_{10}O_3$			=O	=O		181–182	101
$C_{14}H_{12}O_3$	H, Me		=O	=O		164	92
$C_{14}H_{12}O_3$		H, Me	=O	=O		148–148.5	96, 97, 137
$C_{15}H_{13}BrO_3$	Me_2	H, Br	=O	=O		137–138	2, 3, 45
$C_{15}H_{14}O_3$ (β-Lapachone)	Me_2		=O	=O		155–156	2, 3, 30–37, 41, 42, 90, 94, 95, 135, 136, 138, 139, 147
	6-Oxime					167	143
$C_{15}H_{14}O_4$	Me_2	H, OH	=O	=O		170–171[b]	45
	Acetate					153–154	45
$C_{15}H_{14}O_4$	Me_2	H, OH	=O	=O		204–206[c]	3, 45
$C_{15}H_{14}O_4$	Me_2		=O	=O	OH	173–173.5	99
						184–185	39
$C_{16}H_{16}O_2$	Me_2		$=CH_2$	=O		—	73
$C_{16}H_{16}O_3$	Me_2	H, Me	=O	=O		147–148	98
$C_{16}H_{16}O_4$	Me_2		=O	=O	OMe	162–163	99
$C_{31}H_{46}O_2$	Me, TMTD[d]		$=CH_2$	=O		—	73–75

[a] Substituents at positions 2, 3, and 4 are H_2, and at 7, 8, 9, and 10 are H unless otherwise indicated. None of the compounds reported in this group has the 4*a*, 10*a* or 10*b* positions available for substitution.

[b] (R) configuration; $[\alpha]_D^{20} + 32.5°$ in $CHCl_3$. [c] Racemic. [d] TMTD = $[(CH_2)_3CHMe]_3Me$.

TABLE 4. OXO DERIVATIVES OF 2,3-DIHYDRO-1*H*-NAPHTHO[2,1-*b*]PYRANS

	Substituents[a]													
Mol. formula	1	3	4*a*	5	6	6*a*	7	8	9	10	10*a*	10*b*	mp	Ref.
$C_{16}H_{16}O_5$		Me_2	—	OH	H	—	=O	Me	OH	=O	—	—	254–255	114
	Leucotetraacetate												188–189	114
$C_{16}H_{20}O_3$	=O		Me	H	H	—	H	H, Me		=O	H	Me	159–160	104
$C_{16}H_{20}O_3$	=O		Me	H	H	—		H, Me		=O	—	Me	136–137	104
$C_{16}H_{22}O_2$		H, Et	—			Me	=O			H	—	—	91–92	108, 148, 149
$C_{16}H_{24}O_2$		H, Et	—			Me	=O				H	—	—	108
$C_{16}H_{26}O_3$		H, Et	OH			Me	=O				H	H	—	108
$C_{17}H_{18}O_5$		Me_2	—	OMe	H	—	=O	Me	OH	=O	—	—	218–218.5	114
$C_{17}H_{24}O_2$		H, Et	—			Et	=O			H	—	—	—	108
$C_{17}H_{26}O_2$		H, Et	—			Et	=O				H	—	—	108
$C_{17}H_{28}O_3$		H, Et	OH			Et	=O				H	H	—	108
$C_{19}H_{30}O_4$		Me, CO_2H	Me		=O	H	Me_2				Me	H	184–185	103
	Methyl ester												155–156.5	103
$C_{19}H_{32}O_2$		Me_2	Me		=O	H	Me_2				Me	H	148–149	103

$C_{20}H_{30}D_4O_2$	Me, Et	Me		H	Me_2	D_2	=O	D_2	Me	H	91–92	107
$C_{20}H_{30}O_2$	Me, C≡CH	Me		H	Me_2		=O		Me	H	98–100	105
$C_{20}H_{32}O_2$	Me, CH=CH_2	Me		H	Me_2		=O		Me	H	—	107
$C_{20}H_{32}O_2$	Me, Et	Me		H	Me_2	H	H	=O	Me	H	135–136	106
$C_{20}H_{34}O_2$	Me, Et	Me		H	Me_2	=O			Me	H	70–71.5	106
$C_{20}H_{34}O_2$	Me, Et	Me		H	Me_2		=O		Me	H	—	106
$C_{20}H_{34}O_2$	Me, Et	Me		H	Me_2			=O	Me	H	80–81	106
$C_{23}H_{38}O_4$	Me, *b*	Me	=O	H	Me_2				Me	H	179–181	103
$C_{40}H_{58}O_4$	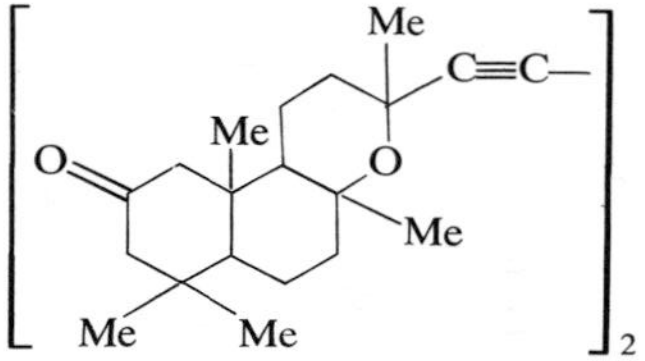										258–260	105

[a] The substituents at the positions indicated are H_2 unless otherwise stated.

[b] 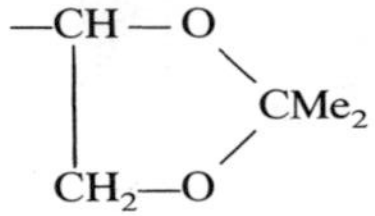

TABLE 5. 2,3-DIHYDRO-4*H*-NAPHTHO[2,3-*b*]PYRAN-5,10-DIONES

Mol. formula	Substituents 2	3	4	6	8	9	mp	Ref.
$C_{13}H_{10}O_3$							220–221	100, 101
$C_{14}H_{12}O_3$	Me						122.5	92
$C_{14}H_{12}O_3$		Me					—	96, 137
$C_{14}H_{12}O_6$ (Cryptosporin)	Me	OH	OH	OH			244	51
	Triacetate						—	51
	Tribenzoate						—	51
$C_{15}H_{12}O_4$	Me_2		=O				163–165	39
$C_{15}H_{14}O_3$ (α-Lapachone)	Me_2						118–119	2, 32–36, 38, 39 41–44, 95, 135, 136, 139, 147, 150
	10-Oxime						204	143
$C_{15}H_{14}O_4$	Me_2	OH					187	3
	Acetate						179.5	3
$C_{15}H_{14}O_4$	Me_2		OH				—	39, 44
	Acetate						139.5–140.5	39, 44
	Benzoate						139–142	39
$C_{15}H_{14}O_4$	Me_2					OH	120–122	38, 39, 99
	Acetate						176–180	38, 39
$C_{15}H_{14}O_5$	Me_2		OH			OH	130–131	38–40
	Diacetate						175–176	38, 39
	Dibenzoate						216–218	39
$C_{16}H_{16}O_3$	Me_2	Me					92–93	98
$C_{16}H_{16}O_4$	Me_2					OMe	168–170	38, 39
$C_{17}H_{16}O_6$	CH_2COMe			OH	OMe		146–151	49
							142–148[a]	47
	Acetate						160–161	47
$C_{17}H_{16}O_7$	CH_2COMe			OH		OH	—	—
	Diacetate						192–195[a]	47
$C_{17}H_{16}O_8$	CH_2COMe		OH	OH		OH	—	—
	Triacetate						163–165[a]	47
$C_{17}H_{18}O_6$	$CH_2CHOHMe$			OH	OMe		[a]	48
$C_{17}H_{18}O_7$	$CH_2CHOHMe$			OH	OMe	OH	—	—
	Triacetate						[b]	47
$C_{18}H_{18}O_6$	CH_2COMe			OMe	OMe		158–160	49
$C_{19}H_{21}NO_4$	Mor[c]	Me_2					153–155	109–111
$C_{21}H_{18}O_4$	Me, OMe		Ph				144–145	50, 112
$C_{21}H_{20}O_3$	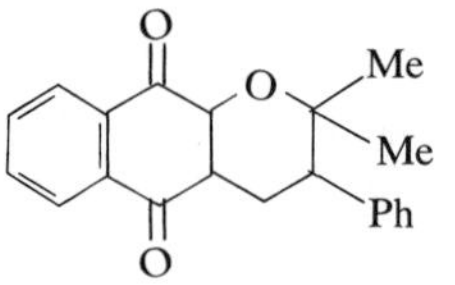 (structure: O, O, O, Me, Me, Ph)						91–92	102

[a] Mixed product. [b] Mixed product obtained as a glass. [c] Mor = *N*-morpholino.

TABLE 6. 2,3-DIHYDRO-4*H*-NAPHTHO[2,3-*b*]PYRAN-6,9-DIONES

Mol. formula	Substituents					mp	Ref.
	2	4	5	8	10		
$C_{14}H_{12}O_6$	Me		OH	OH	OH	—	113
$C_{16}H_{14}O_7$	CH_2COMe		OH	OH	OH	186–188	47
$C_{17}H_{14}O_8$	CH_2COMe	=O	OH	OMe	OH	207–209	47
$C_{17}H_{16}O_6$	CH_2COMe		OH[a]	OMe		175–178	49
	Acetate					194–196	47
$C_{17}H_{16}O_7$	CH_2COMe	OH	OH	OMe		—	46
$C_{17}H_{16}O_7$ (Deoxyerythrostominone)	CH_2COMe		OH	OMe	OH	148–150	46, 47
	Diacetate					192–195	47
$C_{17}H_{16}O_8$ (Erythrostominone)	CH_2COMe	OH	OH	OMe	OH	184–186	46–48
	Triacetate					163–165[b]	47
$C_{17}H_{18}O_6$	$CH_2CHOHMe$		OH[a]	OMe		137–141[b]	47
$C_{17}H_{18}O_7$	$CH_2CHOHMe$	OH	OH	OMe		—	46
$C_{17}H_{18}O_7$ (Deoxyerythrostominol)	$CH_2CHOHMe$		OH	OMe	OH	139–141	46–48
	Triacetate					[b,c]	47
$C_{17}H_{18}O_7$	$CH_2CHOHMe$		OH	OMe	OH	178–182[d]	48
$C_{17}H_{18}O_8$ (Epierythrostominol)	$CH_2CHOHMe$	OH	OH	OMe	OH	187–191	48
$C_{17}H_{18}O_8$	$CH_2CHOHMe$	OH	OH	OMe	OH	180–182[d]	48

[a] 5 or 10-OH or a mixture of these. [b] Mixed with further isomer. [c] Obtained as a glass.
[d] 2′-Epimer.

TABLE 7. Bz-ACYLCHROMANS AND RELATED COMPOUNDS

Mol. formula	Substituents						mp or bp (mm)	Ref.
	2	4	5	6	7	8		
$C_{11}H_{12}O_2$				Ac			45	118–121, 151
							100–102 (0.1)	
	Oxime						88	120
	Semicarbazone						261	118, 119
	Thiosemicarbazone						219	120
$C_{11}H_{12}O_3$				Ac	OH		93	123
	2,4-DNP						304d	123
	Semicarbazone						304d	123
$C_{11}H_{12}O_3$				Ac		OH	114–116	128, 141
							185–190 (0.2)	
	O-Acetate						107–108.5	141
$C_{11}H_{12}O_3$				OH	Ac		111–112	124–127
	2,4-DNP						251–252d	124
$C_{11}H_{12}O_3$					Ac	OH	86–87	129
							168–170 (0.2)	128
$C_{12}H_{14}O_2$		Me	[a]			[a]	110–112 (2.0)	131, 132
	Semicarbazone						185–186	131, 132
$C_{12}H_{14}O_2$				Ac		Me	58–59	152
$C_{12}H_{14}O_2$				COEt			37–38	118, 119
							198–200 (19)	
	Semicarbazone						242	118, 119
$C_{12}H_{14}O_3$				Ac		OMe	96–97	128
$C_{12}H_{14}O_3$				OMe	Ac		49–50	124
$C_{12}H_{14}O_3$				OMe		Ac	62–63.5	130
	2,4-DNP						155–156.5	130

$C_{13}H_{14}O_4$			OH	$COCH_2Ac$		143–145	124
$C_{13}H_{16}O_2$	Me_2		Ac			93–94	116, 153, 154
	2,4-DNP					247–249	116
$C_{13}H_{16}O_2$	Me_2				Ac	75	153
$C_{13}H_{16}O_2$			COPr			192–193 (18)	119
$C_{13}H_{16}O_4$	Me_2	OH	Ac	OH		229	52, 117, 140
	2,4-DNP					227.5	117
$C_{13}H_{16}O_4$	Me_2	OH		OH	Ac	150	52, 117, 142
	2,4-DNP					228.5	117
$C_{14}H_{18}O_4$	Me_2	OH	Ac	OMe		88	52, 140
	2,4-DNP					192	52
$C_{14}H_{18}O_4$ (Dihydroevodionol)	Me_2	OMe	Ac	OH		68–68.5	52, 155
	Acetate					84–85	52
	Nitro derivative					158.5	52
	Oxime					132	52
$C_{14}H_{18}O_4$	Me_2	OMe		OH	Ac	78	117, 155
$C_{15}H_{16}O_6$			OH	[b]		149–150	124–127
$C_{15}H_{20}O_4$	Me_2	OMe	Ac	OMe		91	52
	2,4-DNP					169	52
	Oxime					160–161	52
$C_{15}H_{20}O_4$	Me_2		Ac	OMe	OMe	—	133
	2,4-DNP					164–165	133
$C_{15}H_{20}O_4$	Me_2	OMe		OMe	Ac	76–77	156
$C_{16}H_{14}O_2$			COPh			365 (710)	115
						252–255 (14)	118, 119
$C_{16}H_{22}O_5$ (Dihydroevodione)	Me_2	OMe	Ac	OMe	OMe	63	122
	2,4-DNP					162	122

TABLE 7. (*Contd.*)

Mol. formula	Substituents 2	4	5	6	7	8	mp or bp (mm)	Ref.
$C_{17}H_{16}O_2$				$COCH_2Ph$			80.5–81 255–260 (15)	118, 119
$C_{18}H_{18}O_4$				[c]			103–104	115
$C_{18}H_{24}O_5$	Me_2		CH_2Ac	COEt		OMe	116.5–117	114
$C_{20}H_{20}O_4$	Me_2		OH		OH	COCH=CHPh	176–177	117
$C_{20}H_{22}O_4$	Me_2		OH	$CO(CH_2)_2Ph$	OH		171	117
$C_{20}H_{22}O_4$	Me_2		OH		OH	$CO(CH_2)_2Ph$	172	117, 142
$C_{20}H_{22}O_4$	Me_2		OH	Ac	OCH_2Ph		129.5–130	155
$C_{21}H_{22}O_4$	Me_2		OMe	COCH=CHPh	OH		114	52
$C_{21}H_{24}O_4$	Me_2		OMe	Ac	OCH_2Ph		Oil	155
$C_{21}H_{24}O_4$	Me_2		OMe		OCH_2Ph	Ac	105–105.5	155
$C_{21}H_{24}O_4$	Me_2		OMe	$CO(CH_2)_2Ph$	OH		88	52
$C_{21}H_{24}O_4$	Me_2		OH		OMe	$CO(CH_2)_2Ph$	158	117
$C_{21}H_{24}O_4$	Me_2		OMe		OH	$CO(CH_2)_2Ph$	104	117
$C_{22}H_{24}O_4$	Me_2		OMe	COCH=CHPh	OMe		104	52
$C_{22}H_{26}O_4$	Me_2		OMe	$CO(CH_2)_2Ph$	OMe		—	52
	Oxime						129.5	52
$C_{22}H_{26}O_4$	Me_2		OMe		OMe	$CO(CH_2)_2Ph$	74	117
$C_{23}H_{26}O_5$	Me_2		OMe	COCH=CHPh	OMe	OMe	114	122
$C_{25}H_{40}O_2$				$CO(CH_2)_{14}Me$			56–56.5 270–280 (1.0)	119
$C_{27}H_{28}O_4$	Me_2		OCH_2Ph		OH	$CO(CH_2)_2Ph$	112–113	117

[a] 5-Ac or 8-Ac. [b] $COCH_2COCO_2Et$. [c] COC_6H_3-2,4-diOMe.

VII. REFERENCES

1. E. Games, *Gazz. Chim. Ital.*, **12,** 337 (1882).
2. S. C. Hooker, *J. Chem. Soc.*, **61,** 611 (1892).
3. S. C. Hooker, *J. Chem. Soc.*, **69,** 1355 (1896).
4. V. Prelog, A. M. Gold, G. Talbot, and A. Zamojski, *Helv. Chim. Acta*, **45,** 4 (1962).
5. E. Byrne, D. M. Cahill, and P. V. R. Shannon, *Chem. Ind.* (London), 875 (1969).
6. E. Byrne, D. M. Cahill, and P. V. R. Shannon, *J. Chem. Soc.*, (*C*) 1637 (1970).
7. S. J. Shaw and P. V. R. Shannon, *Org. Mass Spectrom.*, **3,** 941 (1970).
8. K. Wada and T. Ishida, *J. Chem. Soc., Chem. Commun.*, 340 (1976).
9. A. S. Csallany, *Int. J. Vitam. Nutr. Res.*, **41,** 376 (1971); *Chem. Abstr.*, **75,** 140631 (1971).
10. C. T. Goodhue and H. A. Risley, *Biochemistry*, **4,** 854 (1965).
11. E. Demole and D. Berthet, *Helv. Chim. Acta*, **55,** 1866 (1972).
12. J. N. Schumacher, *Tob. Sci.*, **18,** 43 (1974); *Chem. Abstr.*, **81,** 75056 (1974).
13. T. Fujimori, R. Kasuga, H. Matsushita, H. Kaneko, and M. Noguchi, *Agr. Biol. Chem.* (Tokyo), **40,** 303 (1976).
14. R. A. Lloyd, C. W. Miller, D. L. Roberts, J. A. Giles, J. P. Dickerson, N. H. Nelson, C. E. Rix, and P. H. Ayres, *Tob. Sci.*, **20,** 125 (1976); *Chem. Abstr.*, **85,** 59847 (1976).
15. D. L. Roberts and J. N. Schumacher, U.S. Patent 3,217,716 (1965); *Chem. Abstr.*, **64,** 5466 (1966).
16. V. L. Frampton, W. A. Skinner, and P. S. Bailey, *J. Am. Chem. Soc.*, **76,** 282 (1954).
17. V. L. Frampton, W. A. Skinner, P. Cambour, and P. S. Bailey, *J. Am. Chem. Soc.*, **82,** 4632 (1960).
18. W. A. Skinner and R. M. Parkhurst, *Lipids*, **5,** 184 (1970).
19. J. Pokorny, Y. S. R. Sastry, T. T. Phan, and G. Janicek, *Nahrung*, **18,** 217 (1974); *Chem. Abstr.*, **81,** 62241 (1974).
20. A. N. Umanskaya, N. N. Safronova, and V. V. Klyuchkin, *Maslo-Zhir. Prom.*, 13 (1974); *Chem. Abstr.*, **81,** 36649 (1974).
21. M. Komoda and I. Harada, *J. Am. Oil Chem. Soc.*, **46,** 18 (1969).
22. M. Komoda, N. Onuki, and I. Harada, *Agr. Biol. Chem.* (Tokyo), **31,** 461 (1967).
23. W. A. Skinner, H. L. Johnson, M. Ellis, and R. M. Parkhurst, *J. Pharm. Sci.*, **60,** 643 (1971).
24. J. Schroeder, *Biochim. Biophys. Acta*, **343,** 173 (1974).
25. D. Glick, Y. Katsumata, and W. A. Skinner, *Proc. Soc. Exp. Biol. Med.*, **149**, 763 (1975).
26. J. R. Schaeffer and C. T. Goodhue, *J. Org. Chem.*, **36,** 2563 (1971).
27. C. G. Karanjgoakar, A. V. R. Rao, K. Venkataraman, S. S. Yemul, and K. J. Palmer, *Tetrahedron Lett.*, 4977 (1973).
28. A. Quilico, C. Cardani, and G. S. d'Alcontres, *Gazz. Chim. Ital.*, **83,** 754 (1953); *Chem. Abstr.*, **49,** 3058 (1955).
29. C. Cardani, G. Casnati, and B. Cavalleri, *Rend. Ist. Lombardo Sci., Pt. I. Classe Sci. Mat. e Nat.*, **91,** 624 (1957); *Chem. Abstr.*, **53,** 6218 (1959).
30. C. G. Casinovi, G. B. Marini-Bettolo, O. G. da Lima, M. H. D. Maia, and I. L. d'Albuquerque, *Ann. Chim.* (Rome), **52,** 1184 (1962); *Chem. Abstr.*, **59,** 7466 (1963).
31. C. G. Casinovi, G. B. Marini-Bettolo, O. G. da Lima, M. H. D. Maia, and I. L. d'Albuquerque, *Rend. Inst. Super. Sanità*, **26,** 5 (1963); *Chem. Abstr.*, **60,** 9593 (1964).
32. O. G. da Lima, I. L. d'Albuquerque, C. G. da Lima, and M. H. D. Maia, *Rev. Inst. Antibiot., Univ. Recife*, **4,** 3 (1962); *Chem. Abstr.*, **60,** 9099 (1964).
33. A. R. Burnett and R. H. Thompson, *J. Chem. Soc.* (*C*), 2100 (1967).
34. A. R. Burnett and R. H. Thompson, *J. Chem. Soc.* (*C*), 850 (1968).
35. G. Manners, L. Jurd, and K. Stevens, *J. Chem. Soc., Chem. Commun.*, 388 (1974).
36. G. D. Manners and L. Jurd, *Phytochemistry*, **15,** 225 (1976); *Chem. Abstr.*, **84,** 132684 (1976).

37. K. C. Joshi, L. Prakash, and P. Singh, *Phytochemistry*, **12,** 469 (1973); *Chem. Abstr.*, **78,** 94852 (1973).
38. H. Inouye, T. Okuda, and T. Hayashi, *Tetrahedron Lett.*, 3615 (1971).
39. H. Inouye, T. Okuda, and T. Hayashi, *Chem. Pharm. Bull.* (Tokyo), **23,** 384 (1975).
40. K. Inoue, S. Ueda, Y. Shiobara, and H. Inouye, *Tetrahedron Lett.*, 1795 (1976).
41. I. L. d'Albuquerque, *Rev. Inst. Antibiot., Univ. Recife*, **8,** 73 (1968); *Chem. Abstr.*, **74,** 34577 (1971).
42. C. F. da Santana, O. G. da Lima, I. L. d'Albuquerque, A. L. Lacerda, and D. G. Martins, *Rev. Inst. Antibiot., Univ. Recife*, **8,** 89 (1968); *Chem. Abstr.*, **74,** 40989 (1971).
43. I. L. d'Albuquerque, M. C. N. Maciel, A. R. P. Schuler, M. do C. M. DeAraujo, G. M. Maciel, M. da S. B. Cavalcanti, D. G. Martins, and A. L. Lacerda, *Rev. Inst. Antibiot., Univ. Fed. Pernamburo. Recife*, **12,** 31 (1972); *Chem. Abstr.*, **82,** 51459 (1975).
44. M. de L. D. Weinberg, O. R. Gottlieb, and G. G. De Oliveira, *Phytochemistry*, **15,** 570 (1976); *Chem. Abstr.*, **85,** 177197 (1976).
45. J. Mock, S. T. Murphy, E. Ritchie, and W. C. Taylor, *Austr. J. Chem.*, **26,** 1121 (1973).
46. B. E. Cross, M. N. Edinberry, and W. B. Turner, *J. Chem. Soc.* (*D*), 209 (1970).
47. B. E. Cross, M. N. Edinberry, and W. B. Turner, *J. Chem. Soc., Perkin Trans. I*, 380 (1972).
48. B. E. Cross and L. J. Zammitt, *J. Chem. Soc., Perkin Trans. I*, 2975 (1973).
49. B. E. Cross and L. J. Zammitt, *J. Chem. Soc., Perkin Trans. I*, 1936 (1975).
50. I. Chmielęwska and B. Jurecka, *Przemyst. Chem.*, **6,** 288 (1950); *Chem. Abstr.*, **46,** 8102 (1952).
51. A. Closse and H.-P. Sigg, *Helv. Chim. Acta*, **56,** 619 (1973).
52. F. N. Lahey, *Univ. Queensland Papers, Dept. Chem.*, **1,** 2 (1942); *Chem. Abstr.*, **37,** 3432 (1943).
53. I. N. Nazarov, S. N. Ananchenko, and I. V. Torgov, *Zh. Obshch. Khim.*, **26,** 819 (1956); *Chem. Abstr.*, **50,** 13843 (1956).
54. V. I. Gunar and S. I. Zav'yalov, *Izvest. Akad. Nauk SSSR, Otdel. Khim. Nauk*, 937 (1960); *Chem. Abstr.*, **54,** 24700 (1960).
55. V. I. Gunar, S. I. Zav'yalov, G. N. Pershin, S. N. Milovanova, N. S. Bogdanova, O. O. Makeeva, and A. I. Krotov, *Zh. Obshch. Khim.*, **30,** 3975 (1961); *Chem. Abstr.*, **57,** 12345 (1962).
56. J. W. Patterson and W. Reusch, *Synthesis*, 155 (1971).
57. U. Eder and G. Sauer, Ger. Offen. 2,006,372 (1971); *Chem. Abstr.*, **75,** 140698 (1971).
58. G. Sauer, U. Eder, and G. A. Hoyer, *Chem. Ber.*, **105,** 2358 (1972).
59. R. Verhé, L. De Buyck, and N. Schamp, *Bull. Soc. Chim. Belges*, **84,** 761 (1975).
60. K. Lucas, P. Weyerstahl, H. Marschall, and F. Nerdel, *Chem. Ber.*, **104,** 3607 (1971).
61. A. K. Kiang and K. H. Loke, *J. Chem. Soc.*, 480 (1956).
62. E. Guy and F. Winternitz, *Ann. Chim.* (Paris), **4,** 5 (1969).
63. P. D. Boyer, *J. Am. Chem. Soc.*, **73,** 733 (1951).
64. C. Martius and H. Eilingsfeld, *Biochem. Z.*, **328,** 507 (1957); *Chem. Abstr.*, **51,** 11655 (1957).
65. C. Martius and H. Eilingsfeld, *Justus Liebigs Ann. Chem.*, **607,** 159 (1957).
66. C. T. Goodhue and H. A. Risley, *Biochem. Biophys. Res. Commun.*, **17,** 549 (1964).
67. W. Dürckheimer and L. A. Cohen, *Biochem. Biophys. Res. Commun.*, **9,** 262 (1962).
68. W. Dürckheimer and L. A. Cohen, *J. Am. Chem. Soc.*, **86,** 4388 (1964).
69. M. F. Marcus and M. D. Hawley, *Biochim. Biophys. Acta*, **201,** 1 (1970).
70. V. A. Smit and A. V. Semenovskii, *Tetrahedron Lett.*, 3651 (1965).
71. M. Z. Krimer, V. A. Smit, and A. V. Semenovskii, *Izv. Akad. Nauk SSSR, Ser. Khim.*, 1573 (1967); *Chem. Abstr.*, **68,** 78426 (1968).
72. F. W. Knapp and A. L. Tappel, *J. Am. Oil. Chem. Soc.*, **38,** 151 (1961).
73. R. Azerad, *Bull. Soc. Chim. Fr.*, 2728 (1967).
74. R. E. Erickson, A. F. Wagner, and K. Folkers, *J. Am. Chem. Soc.*, **85,** 1535 (1963).

75. P. Mamont, R. Azerad, P. Cohen, M. Vilkas, and E. Lederer, *C.R. Acad. Sci. Paris*, **257**, 706 (1963).
76. A. J. Birch and J. S. Hill, *J. Chem. Soc.* (*C*), 419 (1966).
77. R. T. Blickenstaff and I. Y. C. Tao, *Tetrahedron*, **24,** 2495 (1968).
78. Y. Misaka, T. Mizutani, M. Sekido, and S. Uyeo, *Chem. Commun.*, 1258 (1967).
79. Y. Misaka, T. Mizutani, M. Sekido, and S. Uyeo, *J. Chem. Soc.* (*C*), 2954 (1968).
80. A. J. Birch, P. L. Macdonald, and V. H. Powell, *Tetrahedron Lett.*, 351 (1969).
81. A. J. Birch, P. L. Macdonald, and V. H. Powell, *J. Chem. Soc.* (*C*), 1469 (1970).
82. D. Becker and J. Kalo, *Tetrahedron Lett.*, 3725 (1971).
83. M. Komoda and I. Harada, *J. Am. Oil Chem. Soc.*, **47,** 249 (1970).
84. G. E. Inglett and H. A. Mattill, *J. Am. Chem. Soc.*, **77,** 6552 (1955).
85. L. I. Smith, W. B. Irwin, and H. E. Ungnade, *J. Am. Chem. Soc.*, **61,** 2424 (1939).
86. P. Karrer, H. Fritzsche, and R. Escher, *Helv. Chim. Acta*, **22,** 661 (1939).
87. W. John, E. Dietzel, and W. Emte, *Z. Physiol. Chem.*, **257,** 173 (1939); *Chem. Abstr.*, **33,** 2896 (1939).
88. W. John and W. Emte, *Z. Physiol. Chem.*, **268,** 85 (1941); *Chem. Abstr.*, **36,** 1607 (1942).
89. W. Baker, R. F. Curtis, J. F. W. McOmie, L. W. Olive, and V. Rogers, *J. Chem. Soc.*, 1007 (1958).
90. M. Gates and D. L. Moesta, *J. Am. Chem. Soc.*, **70,** 614 (1948).
91. S. C. Hooker, *J. Am. Chem. Soc.*, **58,** 1190 (1936).
92. L. F. Fieser, *J. Am. Chem. Soc.*, **49,** 857 (1927).
93. S. C. Hooker, *J. Am. Chem. Soc.*, **58,** 1181 (1936).
94. S. R. Gupta, K. K. Malik, and T. R. Seshadri, *Indian J. Chem.*, **7,** 457 (1969).
95. L. F. Fieser, *J. Am. Chem. Soc.*, **70,** 3232 (1948).
96. S. C. Hooker and A. Steyermark, *J. Am. Chem. Soc.*, **58,** 1198 (1936).
97. S. C. Hooker and A. Steyermark, *J. Am. Chem. Soc.*, **58,** 1207 (1936).
98. R. G. Cooke and T. C. Somers, *Austr. J. Sci. Res.*, **3A,** 466 (1950); *Chem. Abstr.*, **45,** 7086 (1951).
99. T. Matsumoto, C. Mayer, and C. H. Eugster, *Helv. Chim. Acta*, **52,** 808 (1969).
100. C. M. Moser and M. Paulshock, *J. Am. Chem. Soc.*, **72,** 5419 (1950).
101. H. Machatzke, W. R. Vaughan, C. L. Warren, and G. R. White, *J. Pharm. Sci.*, **56,** 86 (1967).
102. T. Otsuki, *Bull. Chem. Soc. Japan*, **47,** 3089 (1974); *Chem. Abstr.*, **82,** 170588 (1975).
103. B. Rodriguez and S. Valverde, *Tetrahedron*, **29,** 2837 (1973).
104. W. J. McGahren, G. A. Ellestad, G. O. Morton, and M. P. Kunstmann, *J. Org. Chem.*, **41,** 66 (1976).
105. R. Hodges and R. I. Reed, *Tetrahedron*, **10,** 71 (1960).
106. R. Hodges, *Tetrahedron*, **12,** 215 (1961).
107. P. K. Grant and R. Hodges, *Chem. Ind.* (London), 1300 (1960).
108. G. Saucy, Fr. Patent 1,531,486 (1968); *Chem. Abstr.*, **71,** 81178 (1969).
109. M. Von Strandtmann, M. P. Cohen, and J. Shavel, *Tetrahedron Lett.*, 3103 (1965).
110. M. Von Strandtmann, M. P. Cohen, and J. Shavel, U.S. Patent 3,518,273 (1970); *Chem. Abstr.*, **73,** 66553 (1970).
111. M. Von Strandtmann, M. P. Cohen, and J. Shavel, *J. Heterocycl. Chem.*, **7,** 1311 (1970).
112. H. E. Zaug, *J. Am. Chem. Soc.*, **71,** 1890 (1949).
113. R. E. Moore, H. Singh, and P. J. Scheuer, *Tetrahedron Lett.*, 4581 (1968).
114. J. R. Cannon, K. R. Joshi, I. A. McDonald, R. W. Retallack, A. F. Sierakowski, and L. C. H. Wong, *Tetrahedron Lett.*, 2795 (1975).
115. St. v. Kostanecki, V. Lampe, and C. Marschall, *Chem. Ber.*, **40,** 3660 (1907); *Chem. Abstr.*, **2,** 99 (1908).
116. K. Gollnick, E. Leppin, and G. O. Schenck, *Chem. Ber.*, **100,** 2462 (1967).
117. T. Backhouse and A. Robertson, *J. Chem. Soc.*, 1257 (1939).

118. G. Chatelus, *C.R. Acad. Sci. Paris,* **224,** 201 (1946).
119. G. Chatelus, *Ann. Chim.* (Paris), **4,** 505 (1949).
120. V. Hach, *Collect. Czech. Chem. Commun.*, **24,** 3136 (1959).
121. Belg. Patent 633,973 (1963), *Chem. Abstr.*, **61,** 673 (1964).
122. R. Huls and S. Brunelle, *Bull. Soc. Chim. Belges,* **68,** 325 (1959).
123. P. Naylor, G. R. Ramage, and F. Schofield, *J. Chem. Soc.*, 1190 (1958).
124. A. O. Fitton and G. R. Ramage, *J. Chem. Soc.*, 5426 (1963).
125. A. O. Fitton and G. R. Ramage, Brit. Patent 1,024,645 (1966); *Chem. Abstr.*, **64,** 17606 (1966).
126. A. H. Wragg and G. P. Ellis, Brit. Patent 1,023,373 (1966); *Chem. Abstr.*, **64,** 19566 (1966).
127. G. P. Ellis and A. H. Wragg, U.S. Patent 3,322,795 (1967); *Chem. Abstr.*, **68,** 49578 (1968).
128. P. S. Bramwell and A. O. Fitton, *J. Chem. Soc.*, 3882 (1965).
129. P. S. Bramwell, A. O. Fitton, and G. R. Ramage, Brit. Patent 1,116,562 (1968); *Chem. Abstr.*, **69,** 67355 (1968).
130. A. O. Fitton and G. R. Ramage, *J. Chem. Soc.*, 2481 (1964).
131. R. M. Lagidze, *Materialy Nauchn. Konf. Inst. Khim. Akad. Nauk Azerb., Arm. i Gruz. SSR, Akad. Nauk Arm. SSR, Inst. Organ. Khim., Erevan,* 252 (1957); *Chem. Abstr.*, **59,** 1543 (1963).
132. R. M. Lagidze and B. S. Potskhverashvili, *Soobsch. Akad. Nauk Gruz. SSR,* **19,** 685 (1957); *Chem. Abstr.*, **57,** 7081 (1962).
133. D. R. Taylor and J. A. Wright, *Phytochemistry,* **10,** 1665 (1971).
134. W. A. Skinner and R. M. Parkhurst, *J. Chromatogr.*, **13,** 69 (1964).
135. M. H. Simatupang and B. M. Hausen, *J. Chromatogr.*, **52,** 180 (1970).
136. T. A. Elwood, K. H. Dudley, J. M. Tesarek, P. F. Rogerson, and M. M. Bursey, *Org. Mass Spectrom.*, **3,** 841 (1970).
137. M. G. Ettlinger, *J. Am. Chem. Soc.*, **72,** 3090 (1950).
138. M. L. Josien, N. Fuson, J. M. Lebas, and T. M. Gregory, *J. Chem. Phys.*, **21,** 331 (1953).
139. A. R. Burnett and R. H. Thomson, *J. Chem. Soc. (C)*, 1261 (1967).
140. F. N. Lahey, *Univ. Queensland Papers, Dept. Chem.*, **1,** 18 (1942); *Chem. Abstr.*, **37,** 3343 (1943).
141. B. Willhalm, A. F. Thomas, and F. Gautschi, *Tetrahedron,* **20,** 1185 (1964).
142. R. A. Morton and Z. Sawires, *J. Chem. Soc.*, 1052 (1940).
143. S. C. Hooker and E. Wilson, *J. Chem. Soc.*, **65,** 717 (1894).
144. M. F. Marcus and M. D. Hawley, *J. Org. Chem.*, **35,** 2185 (1970).
145. L. N. Sokolova, A. S. Polyanskaya, and N. I. Aboskalova, *XXVII Gertsenovsk. Chteniya. Khimiya.*, 25 (1975); *Chem. Abstr.*, **84,** 17080 (1976).
146. W. John and W. Emte, *Z. Physiol. Chem.*, **261,** 24 (1939).
147. M. A. Floyd, D. A. Evans, and P. E. Howse, *J. Insect Physiol.*, **22,** 697 (1976); *Chem. Abstr.*, **85,** 105219 (1976).
148. G. Saucy, U.S. Publ. Pat. Appl. B 450,708 (1976); *Chem. Abstr.*, **84,** 180477 (1976).
149. G. Saucy, U.S. Publ. Pat. Appl. B 450,693 (1976); *Chem. Abstr.*, **84,** 180478 (1976).
150. M. G. Ettlinger, *J. Am. Chem. Soc.*, **72,** 3085 (1950).
151. P. Maitte, *C.R. Acad. Sci., Paris,* **234,** 1787 (1952).
152. G. Brancaccio, G. Lettieri, and R. Viterbo, *J. Heterocycl. Chem.*, **10,** 623 (1973).
153. G. R. Clemo and N. D. Ghatge, *J. Chem. Soc.*, 4347 (1955).
154. H. S. Bloch, U.S. Patent 3,551,456 (1970); *Chem. Abstr.*, **75,** 5708 (1971).
155. F. N. Lahey and R. V. Strick, *Austr. J. Chem.*, **26,** 2291 (1973).
156. A. C. Jain and M. K. Zutshi, *Tetrahedron Lett.*, 3179 (1971).

CHAPTER VIII

Chroman Carboxaldehydes

G. P. Ellis

Department of Chemistry, University of Wales Institute of Science and Technology, Cardiff, U.K.

I. INTRODUCTION

Chroman aldehydes are of interest mainly as synthetic intermediates (since a number of them may be readily synthesized) and as possible products formed when tocopherols act as antioxidants.

II. SYNTHESIS

1. From *o*-Allylphenols

o-Allylic phenols carrying a nuclear formyl substituent may be cyclized in good yields to chroman aldehydes under acidic conditions. Other substituents (usually electron releasing) may be present on the ring, for example, hydroxy,[1,2] methoxy,[3–5] or alkyl.[1,2,6,7] The allyl group is usually disubstituted at C-3 with similar[3–5,8–11] or different[1,7] alkyl substituents. An inorganic acid[1,4,6,10,11] or formic acid[3,5] may be used. 2,4-Dihydroxy-5-(3-methylbut-2-enyl)benzaldehyde (**1**) was converted in this way to 7-hydroxy-2,2-dimethylchroman-6-carboxaldehyde (**2**),[6] but when the benzaldehyde (**1**) was treated at room temperature with 3-chloroperoxybenzoic acid and a trace of toluene-4-sulfonic acid, the product was 3,7-dihydroxy-2,2-dimethylchroman-6-carboxaldehyde (**3**).[12] The 3-hydroxychroman however may be the minor product as in the oxidation of the benzaldehyde (**4**), where the main product was the epoxide (**5**).[8]

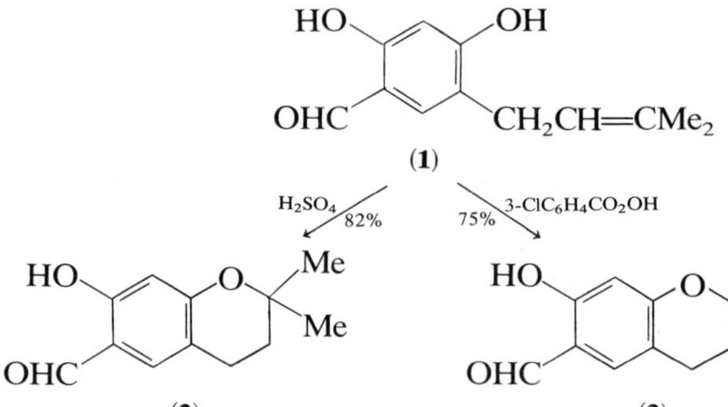

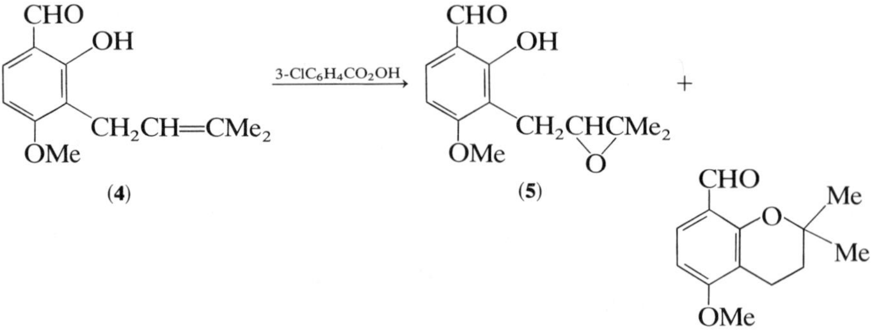

A double bond at the 3,4-position of an alkyl chain reacted with a hydroxyl group to form a pyran ring during the hydrolysis of the acetal (**6**).[13]

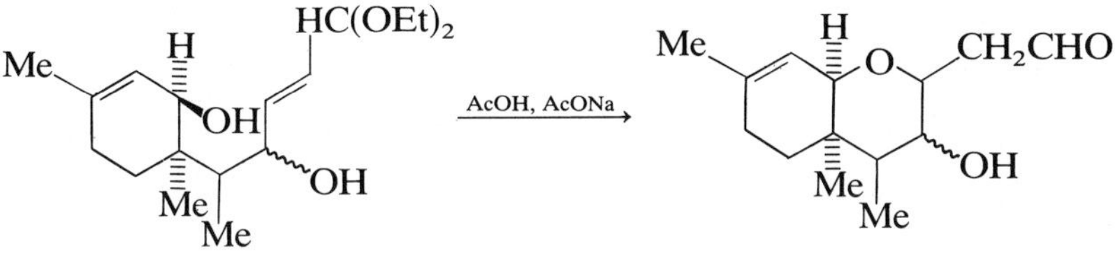

AcOH, AcONa

(6)

2. From 1,5-Diols

When the terpene sclareol (**7**) was oxidized with "chromic acid mixture," the two epimeric aldehydes (**8**) and (**9**) were among the products.[14–16]

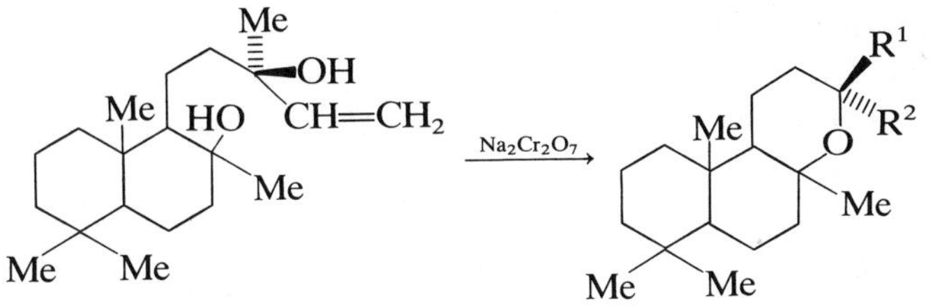

$Na_2Cr_2O_7$

(**7**)

(**8**) $R^1 = Me$, $R^2 = CH_2CHO$
(**9**) $R^1 = CH_2CHO$, $R^2 = Me$

3. From Other Chromans

A number of other functional groups are convertible into an aldehyde function without affecting the pyran ring. It is sometimes possible to formylate the benzene ring directly either by a Vilsmeier, Gattermann, of Duff reaction and to convert a carboxyl or carboxylate group into an aldehyde. Conversion of carbon–carbon double bonds carrying hydrogen atoms to aldehydes needs careful choice of reagent, and this reaction is more often used for structural elucidation than for synthesis.

A. *Direct Formylation*

Gattermann's original method has been used for the preparation of several chroman aldehydes[17,18] and this can give a good yield in the presence of aluminum chloride, for example, in the formylation of chroman to give the 6-aldehyde (**10**).[19] Adams's modification of this reaction has also been applied to chromans.[20–22]

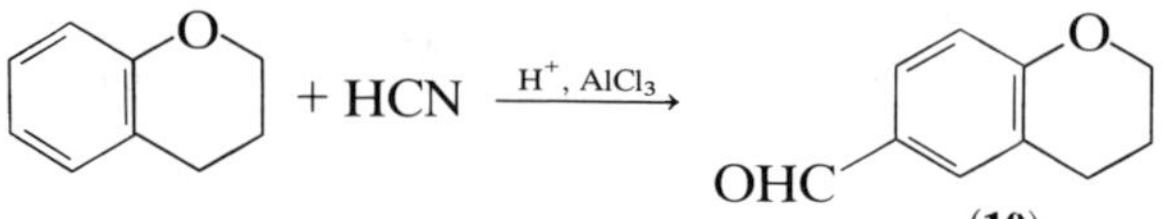

O + HCN $\xrightarrow{H^+, AlCl_3}$ O OHC (**10**)

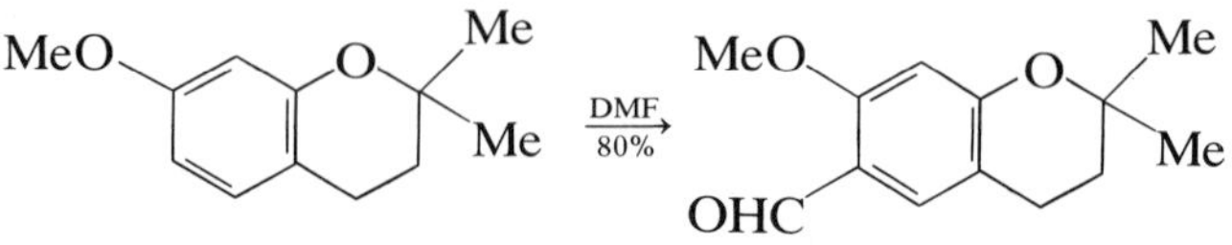

MeO O Me Me $\xrightarrow[80\%]{DMF}$ MeO O Me Me OHC (**11**)

2,2-Dimethyl-7-methoxychroman-6-carboxaldehyde (**11**) and analogs of this[23,24] have been prepared by the Vilsmeier reaction. The Duff reaction using hexamethylenetetramine appears to give lower yields.[25,26] An unexpected insertion of a formyl group occurred when γ-tocopherol (**12**) or δ-tocopherol (**13**) was heated at 180° for 1 hr with trimethylamine oxide.[27] α-Tocopherol (**14**) gave a mixture of products (Scheme 1).

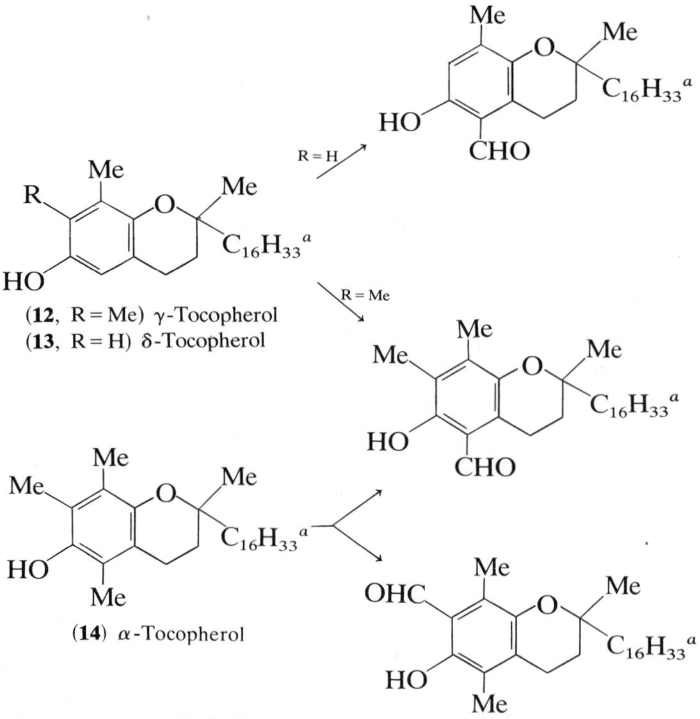

Me O Me $C_{16}H_{33}{}^a$ HO CHO

R = H

R Me O Me $C_{16}H_{33}{}^a$ HO

R = Me

(**12**, R = Me) γ-Tocopherol
(**13**, R = H) δ-Tocopherol

Me Me O Me $C_{16}H_{33}{}^a$ HO CHO

Me Me O Me $C_{16}H_{33}{}^a$ HO Me

(**14**) α-Tocopherol

OHC Me O Me $C_{16}H_{33}{}^a$ HO Me

[a] $[CH_2CH_2CH_2CHMe]_3Me$

Scheme 1. Oxidation of tocopherols with trimethylamine oxide.

B. *Ozonization of Alkenylchromans*

Treatment of suitable alkenyl side chains attached to chromans with ozone gives aldehydes. Although this reaction is used mostly for determining the position of a double bond,[28] it has occasionally found synthetic utility,[29,30] as in the synthesis of 6-acetyloxy-2,5,7,8-tetramethylchroman-2-acetaldehyde (**15**), which is an intermediate in the synthesis of certain tocopherols.[26] A double bond in the 2-pyrone ring of dihydroseselin (**16**) was similarly ozonized.[31]

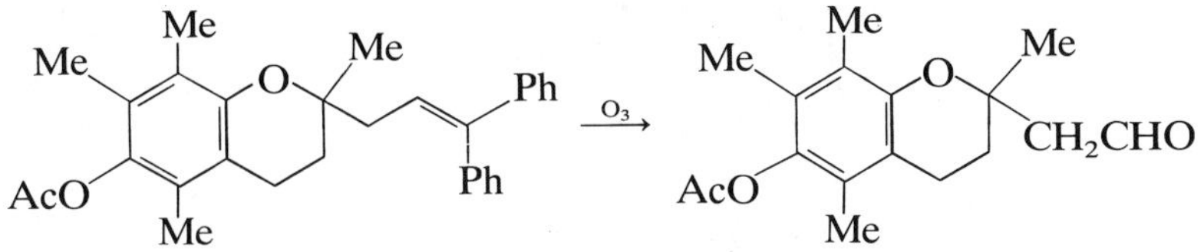

(**15**)

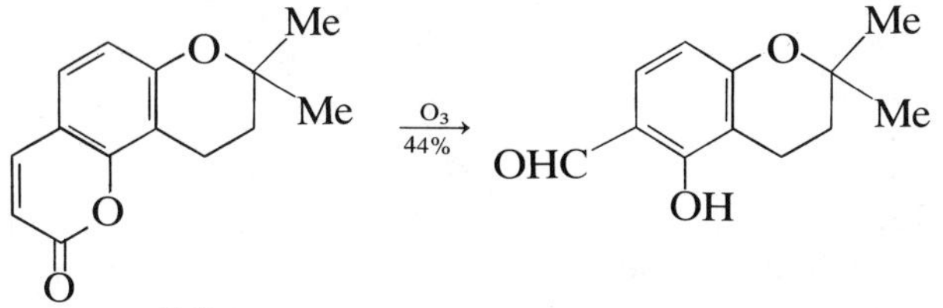

(**16**)

C. *Cleavage of* 1,2-*Diols*

1,2-Diols (usually obtained by the action of osmium tetroxide on a carbon–carbon double bond) are cleaved by periodic acid with the formation of an aldehyde. This reaction is commonly used in structural investigations.[32–36] For example, the structure of a stilbene (**17**) from the bark of *Lonchocarpus violaceus* was demonstrated by a series of reactions which led to the independently synthesized methyl 2,2-dimethyl-5-methoxychroman-7-carboxylate (**18**).

D. *Oxidation of Alkyl Groups*

A 5-methyl group is more readily oxidized by oxygen than one placed elsewhere in the chroman molecule. For example, 6-hydroxy-2,2,5,7,8-pentamethylchroman gave 6-hydroxy-2,2,7,8-tetramethylchroman-5-carboxaldehyde in very low yield[37] on bubbling oxygen through the chroman for 40 days. Less selectivity was shown by the oxidation of α-tocopherol by trimethylamine oxide as a small amount of the 7-aldehyde

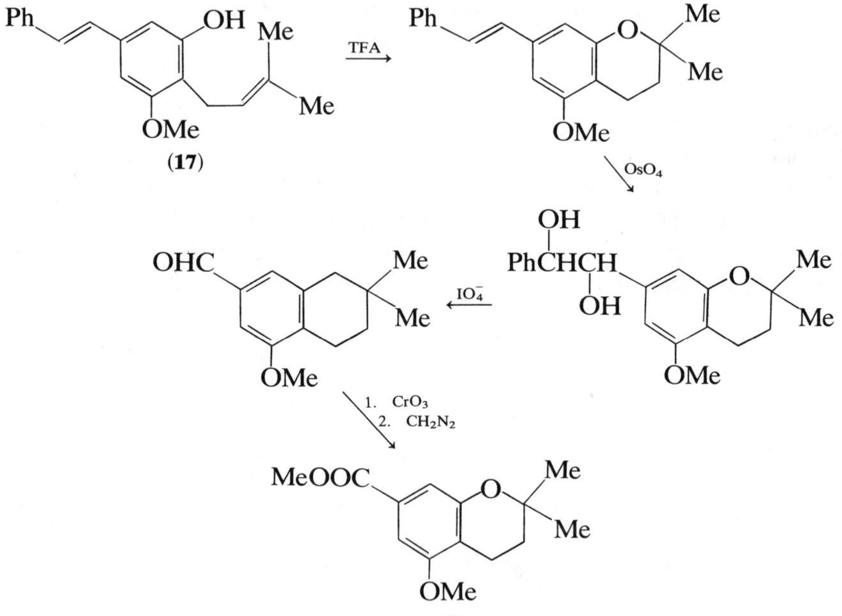

(17)

(18)

was isolated (Scheme 1).[27] An exocyclic methylene group is converted into an aldehyde function by the combined action of selenium dioxide and hydrogen peroxide at room temperature,[38] but this reaction is not of preparative value.

E. *Miscellaneous Methods*

Chroman aldehydes have been prepared by Rosenmund reduction of acyl halide,[39] diisobutylaluminum hydride reduction of a carboxylic ester,[39] and by hydrolysis of a benzoyloxymethyl function at the more reactive 5-position, for example, to give 6-hydroxy-2,2,7,8-tetramethylchroman-5-carboxaldehyde (**19**).[22]

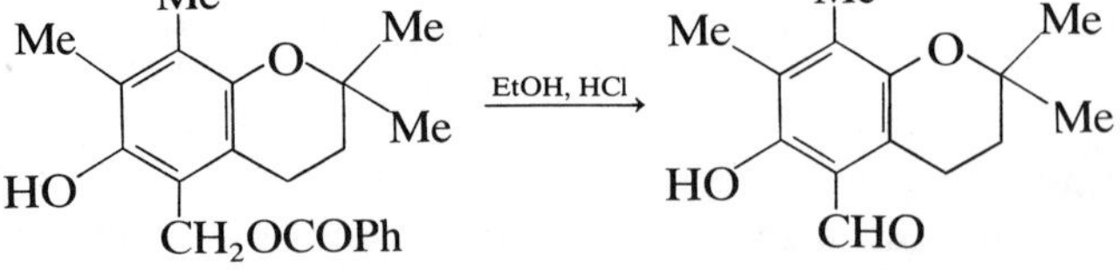

(**19**)

Photolytic decomposition of the nitrite ester of 5,7-dimethyl-2-(2-hydroxyethyl)chroman gave 4,7-dimethylchromanylacetaldehyde.[40]

III. PHYSICAL AND SPECTRAL PROPERTIES

Some chroman aldehydes of low molecular weight are liquids,[19,24,31] as are also a few of the perhydro analogs which may be regarded as terpene derivatives.[13] Chromans that have a biosynthetic origin are frequently disubstituted at C-2.[41] When the substituents are different from one another, the compounds are chiral, and their specific rotation is a guide to their optical purity.[7,33,39]

The absorption of different types of chroman aldehydes in the uv region has been recorded, for example, for nuclear bicyclic aldehydes[7,19,37] and side-chain aldehydes,[28,30,39] but these are not as characteristic as their absorption in the ir region. The carbonyl absorption appears at about 1730 cm^{-1} for nonhydrogen-bonded aldehydes[13,39], at 1740 cm^{-1} for aliphatic aldehydes[32], or at 1680 cm^{-1} for aryl aldehydes,[11] but most of the known chroman aldehydes have ortho-placed hydroxyl groups which may shift the carbonyl absorption frequency to values as low as 1634 cm^{-1}.[22,27]

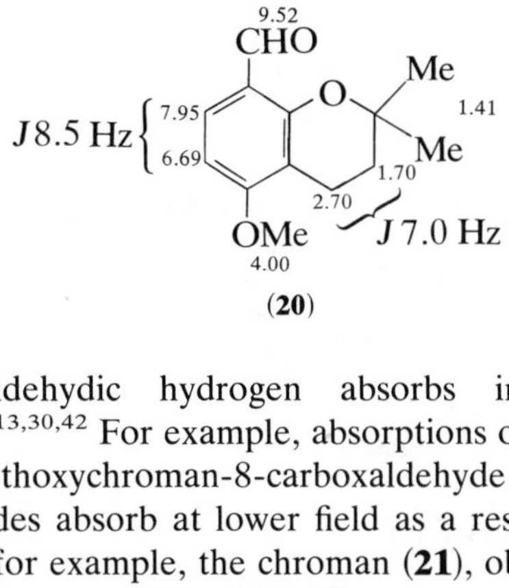

(20)

Typically, the aldehydic hydrogen absorbs in nmr spectra at $\delta 10 \pm 0.5$ ppm.[3,4,9–11,13,30,42] For example, absorptions of the hydrogen atoms of 2,2-dimethyl-5-methoxychroman-8-carboxaldehyde are shown in (**20**),[11] but *o*-hydroxyaldehydes absorb at lower field as a result of intramolecular hydrogen bonding,[27] for example, the chroman (**21**), obtained by cyclization of an antibiotic,[1] shows an aldehydic hydrogen signal at $\delta 12.71$. On the other hand, the aliphatic aldehyde (**22**) gave a triplet at $\delta 8.25$.[39]

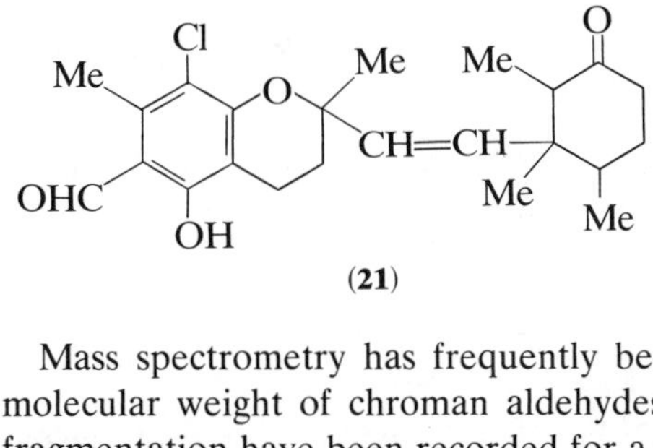

(21) 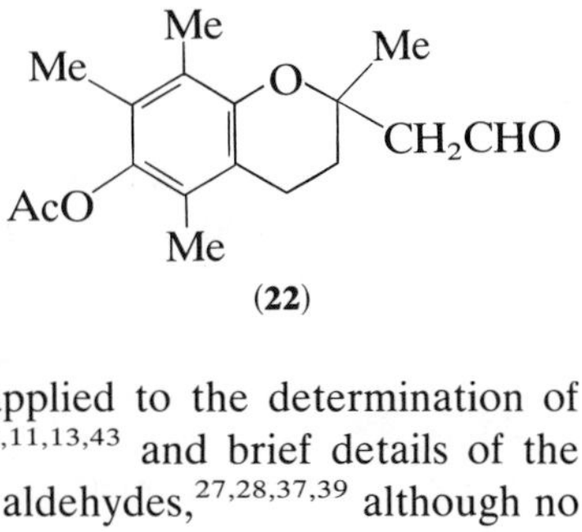 **(22)**

Mass spectrometry has frequently been applied to the determination of molecular weight of chroman aldehydes,[1,10,11,13,43] and brief details of the fragmentation have been recorded for a few aldehydes,[27,28,37,39] although no deliberate study appears to have been made.

IV. CHEMICAL PROPERTIES

1. Introduction

Examples of the various reactions of an aldehyde group and also of the cyclization of an *o*-hydroxybenzaldehyde are to be found in the literature of chromans.

2. Oxidation

The most common chemical reaction of chroman aldehydes is probably their oxidation to the carboxylic acid. Chromium trioxide,[34–36,42] silver oxide (or alkaline silver nitrate),[23,24,44,45] and potassium permanganate[23,30] have been used for this purpose; and yields can be quite good, for example, the oxidation of 2,2-dimethyl-7-methoxychroman-6-carboxaldehyde (**23**).[24] The aldehyde function may not be the most easily oxidized group in a molecule, and under mild conditions chromium trioxide in pyridine leaves the aldehyde group of (**24**) unchanged to give the ketoaldehyde (**25**), although further reaction by this reagent does give the carboxylic acid (**26**).[13]

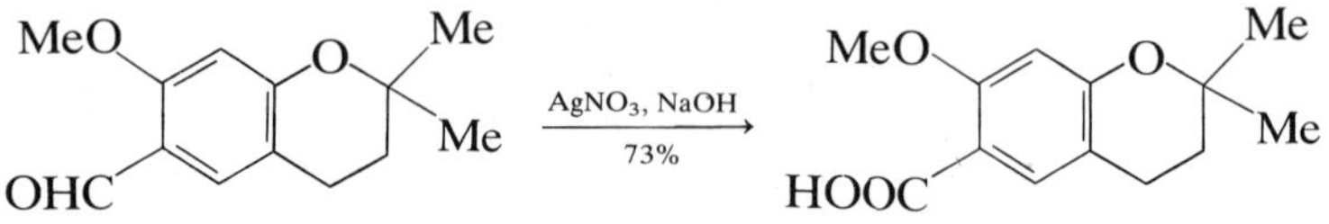

MeO, OHC, Me, Me — $\xrightarrow[73\%]{AgNO_3,\ NaOH}$ — MeO, HOOC, Me, Me

(**23**)

Me, H, CH_2CHO, OH, Me, Me + CrO_3 $\xrightarrow{Pyridine}$ Me, H, CH_2R, O, Me, Me

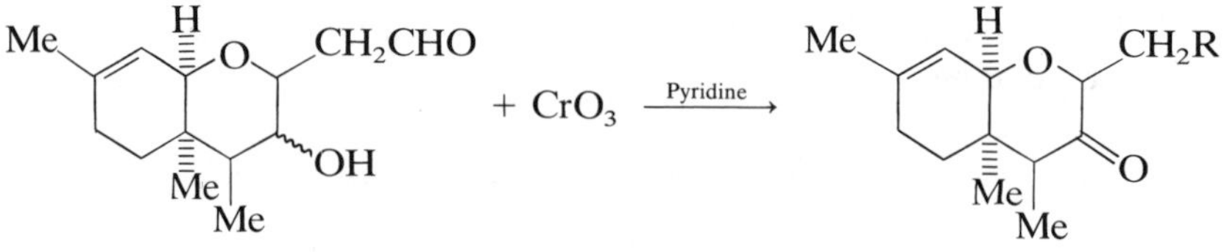

(**24**)

(**25**) R = CHO
(**26**) R = COOH

Alkaline hydrogen peroxide (Dakin reagent) converts an aldehyde into a phenol in good yield and is sometimes used in structural studies.[25,26]

3. Reduction

Mild catalytic hydrogenation of 2,2-dimethyl-7-hexyl-6-hydroxychroman-8-carboxaldehyde produced the corresponding primary alcohol.[9,46] An aromatic nitro group is more easily hydrogenated than an aldehyde group, as was demonstrated when 2,2-dimethyl-8-methoxy-5-nitrochroman-6-carboxaldehyde (**27**) was reduced under slight pressure to the aminoaldehyde (**28**).[23]

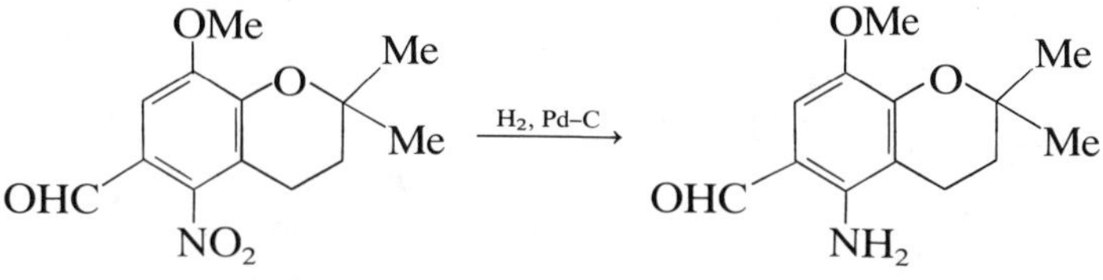

$$\xrightarrow{H_2,\ Pd\text{-}C}$$

(**27**) (**28**)

Reduction of an aldehyde with hydrazine hydrate and alkali gave the expected methyl derivative.[33]

4. Condensation Reactions

Conventional nitrogenous carbonyl reagents such as semicarbazide[17,19,21,24,33,34,47,48] and 2,4-dinitrophenylhydrazine[17,20,21,24,47] react normally with chroman aldehydes.

A Knoevenagel reaction of 2,2-dimethyl-3-hydroxy-5-methoxychroman-8-carboxaldehyde (**29**) gave the cinnamic acid (**30**).[8]

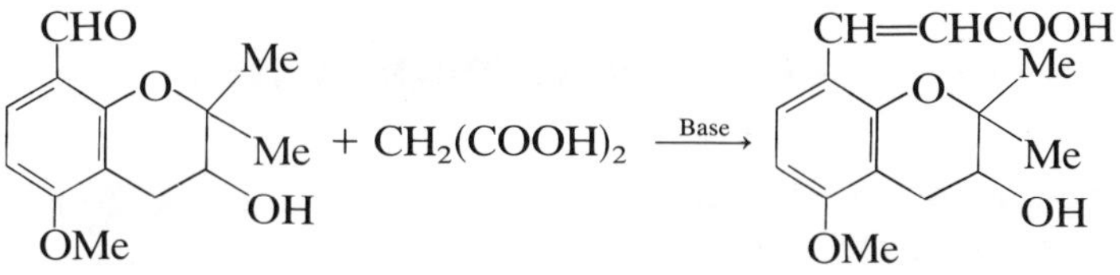

$$+ CH_2(COOH)_2 \xrightarrow{\text{Base}}$$

(**29**) (**30**)

5. Miscellaneous Reactions

An *o*-hydroxyaldehyde may be cyclized by a Perkin reaction to form a coumarin[3,25] such as (**31**).[26]

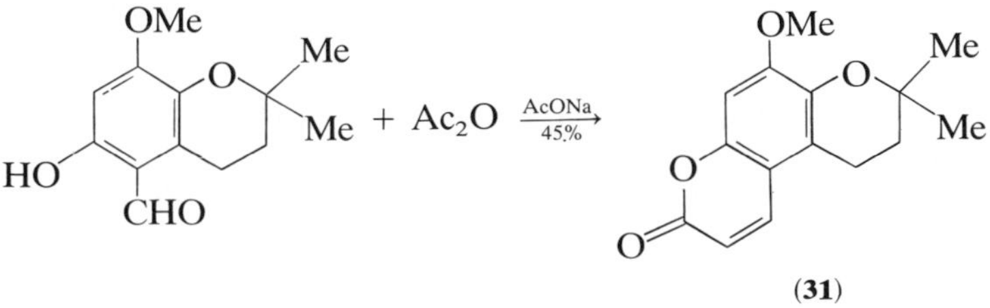

$$+ Ac_2O \xrightarrow[45\%]{AcONa}$$

(**31**)

Descent of a homologous series was demonstrated for a chroman-2-acetaldehyde (**32**) by reaction with piperidine in acid solution followed by ozonization to give 6-acetyloxy-2,5,7,8-tetramethylchroman-2-carboxaldehyde (**33**).[26]

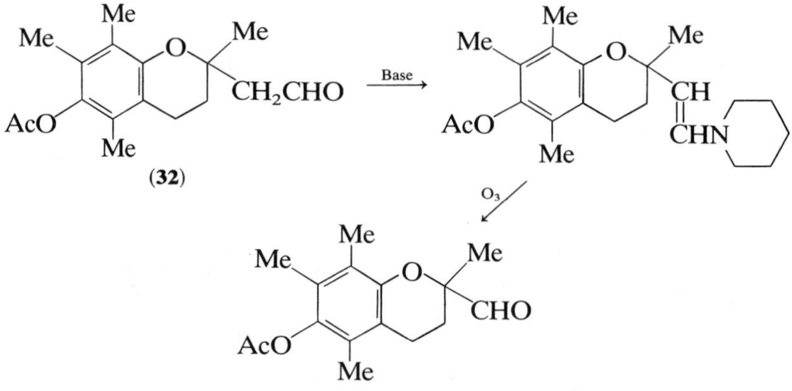

(**32**)

(**33**)

2,2-Dimethyl-7-methoxychroman-6-carboxaldehyde has been dehydrogenated to the 2*H*-chromene using 2,3-dichloro-5,6-dicyanobenzoquinone.[49] The expected secondary alcohol was obtained when 2,2-dimethyl-5-methoxychroman-8-carboxaldehyde was treated with butylmagnesium bromide.[11]

V. BIOLOGICAL PROPERTIES

Although chroman aldehydes have been shown to be formed when tocopherols and related compounds are oxidized,[27,37] the aldehydes do not seem to have any useful properties.[50]

VI. TABLES OF COMPOUN[illegible]S

TABLE 1. NUCLEAR CHROMAN ALDEHYDES AND THEIR DERIVATIVES

Mol. formula	Substituents 2	3	4	Other positions	mp or bp (mm)	Ref.
$C_{10}H_{10}O_2$			CHO		—	47
$C_{10}H_{10}O_2$				6-CHO	152–159 (9)	19
					160–163 (20)	24
	Semicarbazone				236d	24
$C_{12}H_{14}O_2$	Me_2			6-CHO	218–218.5d	19
					97–98 (0.2)	9
$C_{12}H_{14}O_3$	CHO			6-OH-7-Me	109–111	43
	Me					
$C_{12}H_{14}O_3$	Me_2			5-OH-6-CHO	70 (0.005)	31
$C_{12}H_{14}O_3$	Me_2			7-OH-6-CHO	105	6, 20, 51
$C_{12}H_{14}O_3$	Me_2			8-OH-7-CHO	33–34.5	25
					138 (0.5)	25
$C_{12}H_{14}O_4$	Me_2	OH		5-OH-6-CHO	Oil	6
$C_{12}H_{14}O_4$	Me_2	OH		7-OH-6-CHO	97.5–99	12
$C_{12}H_{14}O_4$	Me_2			5,7-$(OH)_2$-8-CHO	179–180	17
$C_{13}H_{15}NO_5$	Me_2			8-MeO-5-NO_2-6-CHO	161–163	23
$C_{13}H_{16}O_3$	Me_2			5-MeO-7-CHO	Oil	42
$C_{13}H_{16}O_3$	Me_2			7-MeO-6-CHO	76–77	24
	2,4-DNP				255	24
$C_{13}H_{16}O_3$	Me_2			5-MeO-8-CHO	108–109.5	5, 11
$C_{13}H_{16}O_3$	Me_2			8-MeO-6-CHO	64–65	4, 23
$C_{13}H_{16}O_3$	Me_2			8-MeO-7-CHO	44–45	25
					152–154 (1)	25
$C_{13}H_{16}O_4$	Me_2	OH		5-MeO-8-CHO	137–137.5	8
$C_{13}H_{16}O_4$	Me_2			7-OH-5-MeO-6-CHO	85–86	17
$C_{13}H_{16}O_4$	Me_2			5-OH-8-MeO-6-CHO	73–75	23
$C_{13}H_{16}O_4$	Me_2			6-OH-8-MeO-5-CHO	146–148	26

TABLE 1. (*Contd.*)

Mol. formula	Substituents 2	3	4	Other positions	mp or bp (mm)	Ref.
$C_{13}H_{16}O_4$	Me_2			7-OH-5-MeO-8-CHO	90	18
$C_{13}H_{16}O_4$	Me_2			7-OH-8-MeO-6-CHO	77–78	25
					157–158 (0.1)	25
$C_{13}H_{17}NO_3$	Me_2			5-NH_2-8-MeO-6-CHO	119–120	23
$C_{14}H_{18}O_3$	CHO			6-OH-5,7,8-Me_3	—	
	Me					
	O-Acetate				90–91	26
$C_{14}H_{18}O_3$	Me_2			6-OH-7,8-Me_2-5-CHO	116–118	37
					108–110	22
$C_{14}H_{18}O_4$	Me_2			5,7-$(MeO)_2$-6-CHO	81–82	17
$C_{14}H_{18}O_4$	Me_2			5,7-$(MeO)_2$-8-CHO	105–107	3
	Semicarbazone				217–218	17, 21
	2,4-DNP				242–243	17, 21
$C_{14}H_{24}O_2$	CHO, Me			5,5,8a-Me_3-perhydro	—	32
$C_{19}H_{28}O_3$	Me_2			6-OH-7-C_7H_{15}-8-CHO	102–103	2, 46
$C_{23}H_{29}ClO_4$	$C_{11}H_{17}O$[a], Me			8-Cl-5-OH-7-Me-6-CHO	97–99	1
$C_{23}H_{29}ClO_4$	$C_{11}H_{17}O$[a], Me			6-Cl-5-OH-7-Me-8-CHO	64–67	1
$C_{23}H_{31}ClO_4$	$C_{11}H_{19}O$[b], Me			8-Cl-5-OH-7-Me-6-CHO	139–146	1
$C_{23}H_{32}N_4O_3$	$C_8H_{13}N_4$[c], Me			5-OH-7-Me-6-CHO	123–145	1
$C_{23}H_{32}O_4$	$C_{11}H_{19}O$[b], Me			5-OH-7-Me-8-CHO	192–194	1, 7
$C_{23}H_{32}O_4$	$C_{11}H_{17}O$[b], Me			5-OH-7-Me-6-CHO	97–99	1, 7
$C_{23}H_{33}NO_4$	$C_{11}H_{20}NO$[d]			5-OH-7-Me-6-CHO	—	1
$C_{29}H_{48}O_3$	$C_{16}H_{33}$[e], Me			5,8-Me_2-6-OH-7-CHO	—	27
$C_{29}H_{48}O_3$	$C_{16}H_{33}$[e], Me			7,8-Me_2-6-OH-5-CHO	—	27

[a] Me, O, Me, Me

[b] Me, O, Me, Me

[c] N, N, N, N

[d] Me, NH, O, Me, Me

TABLE 2. SIDE-CHAIN CHROMAN ALDEHYDES AND THEIR DERIVATIVES

Mol. formula	Substituents				mp or bp (mm)	Ref.
	2	3	4	Other positions		
$C_{13}H_{16}O_2$	CH_2CHO		Me	7-Me	—	40
$C_{14}H_{20}O_3$	CH_2CHO	=O	Me	4a,7-Me_2-4a,5,6,8a-H_4	110 (0.5)	13
$C_{14}H_{22}O_3$	CH_2CHO	OH	Me	4a, 7-Me_2-4a,5,6,8a-H_4	Oil	13
$C_{15}H_{20}O_3$	CH_2CHO, Me			6-OH-5,7,8-Me_3	—	
	O-Acetate				63–64.5	52
$C_{16}H_{22}O_3$	$(CH_2)_2CHO$, Me			6-OH-5,7,8-Me_3	90.5–92.5	39
	O-Acetate				48–49	29
					145–146 (0.005)	29
$C_{16}H_{22}O_5$	$(CH_2)_2CHO$, Me			7,8-$(MeO)_2$-6-OH-5-Me	—	
	O-Acetate				—	30
$C_{17}H_{22}O_5$	CH_2CHO, Me			6-OH-5,7,8-Me_3	—	
	O-Acetate				87.5–90	39
$C_{20}H_{34}O_2$	Et, Me			5,8a-Me_2-5-CH=CH_2-6-CMe_2CHO perhydro	65.5–66.5	53
$C_{22}H_{26}O_3$	CH_2CHO, Me			6-$PhCH_2O$-5,7,8-Me_3	87.5–90	39
$C_{26}H_{22}O_3$	$C_{13}H_{21}O^a$, Me			6-OH-5,7,8-Me_3	—	
	O-Acetate				Resin	39

[a] $(CH_2CH{=}CMe)_2CH_2CH_2CHO$.

TABLE 3. DIHYDRONAPHTHOPYRAN ALDEHYDES AND THEIR DERIVATIVES

Mol. formula	Isomer	Substituents 2	3	5	Other positions	mp	Ref.
$C_{18}H_{20}O_3$	[1,2-*b*]	Me,$(CH_2)_2CHO$		Me	6-OH	—	
		O-acetate				92–93	28
$C_{19}H_{32}O_2$	[2,1-*b*]		CHO Me		4a,7,7,10a-Me_4 perhydro	Oil	33, 34
		Semicarbazone				225–227.5	33, 34
$C_{19}H_{32}O_3$	[2,1-*b*]		CHO Me		6-OH-4a,7,7,10a-Me_4 perhydro	120–122	33
$C_{20}H_{34}O_2$	[2,1-*b*]		CH_2CHO[a] Me		4a,7,7,10a-Me_4 perhydro	—	14, 16
$C_{20}H_{34}O_2$	[2,1-*b*]		CH_2CHO[a] Me		4a,7,7,10a-Me_4 perhydro	90–91	14–16
		Semicarbazone				263–264	14, 15
$C_{21}H_{34}O_2$	[2,1-b]		Et Me		2,3,4a,5,6,6a,7,10,10a,10b-H_{10}-4a,7,7,10a-Me_4-9-CHO	135–136	38
$C_{25}H_{30}O_3$	[1,2-*b*]	Me,$C_8H_{15}O$[b]		Me	3,4-H_2-6-OH	—	
		O-Acetate				—	28

[a] C-3 epimers.
[b] $(CH_2)_3CHMe(CH_2)_2CHO$.

VII. REFERENCES

1. D. C. Aldridge, A. Borrow, R. G. Foster, M. S. Large, H. Spencer, and W. B. Turner, *J. Chem. Soc., Perkin Trans. I,* 2136 (1972).
2. A. Quilico, C. Cardani, and I. Lucatelli, *Gazz. Chim. Ital.,* **83,** 1088 (1953).
3. P. W. Austin, T. R. Seshadri, M. S. Sood, and Vishwapaul, *Tetrahedron,* **24,** 3247 (1968).
4. J. A. Diment, E. Ritchie, and W. C. Taylor, *Austr. J. Chem.,* **20,** 565, (1967).
5. M. Murayama, E. Sato, T. Okuta, I. Morita, I. Dobaslic, and M. Maehara, *Chem. Pharm. Bull.,* **20,** 741 (1972).
6. W. Steck, *Can. J. Chem.,* **49,** 1197 (1971).
7. G. A. Ellstead, R. H. Evans, and M. P. Kunstmann, *Tetrahedron,* **25,** 1323 (1969).
8. M. Matsumato, K. Fujui, and M. Nanba, *Chem. Lett.,* 603 (1974).
9. D. W. Knight and G. Pattenden, *Chem. Commun.,* 635 (1976); *J. Chem. Soc., Perkin Trans. I,* 70 (1979).
10. K. Raj, P. P. Joshi, D. S. Bhakuni, R. S. Kapil, and S. P. Popli, *Indian J. Chem.,* **14B,** 332 (1976).
11. O. P. Malik, R. S. Kapil, and N. Anand, *Indian J. Chem.,* **14B,** 449 (1976).
12. W. Steck, *Can. J. Chem.,* **49,** 2297 (1971).
13. E. W. Colvin, S. Malchenko, R. A. Raphael, and J. S. Roberts, *J. Chem. Soc., Perkin Trans. I.* 1989 (1973).
14. V. E. Sibirtaeva, E. G. Chauser, and S. D. Kustova, *J. Gen. Chem. USSR,* **39,** 2297 (1969).
15. V. E. Sibirtaeva, L. I. Kaufman, and S. D. Kustova, *J. Gen. Chem. USSR,* **38,** 738 (1968).
16. P. F. Vlad and L. T. Xuan, *J. Gen. Chem. USSR.,* **45,** 1850 (1975).
17. A. Robertson and T. S. Subramanian, *J. Chem. Soc.,* 286 (1937).
18. A. Robertson and T. S. Subramanian, *J. Chem. Soc.,* 1545 (1937).
19. G. Baddeley, N. H. P. Smith, and M. A. Vickers, *J. Chem. Soc.,* 2455 (1956).
20. J. C. Bell, W. Bridge, A. Robertson, and T. S. Subramanian, *J. Chem. Soc.,* 1542 (1937).
21. F. N. Lahey, *Univ. Queensland Papers, Dep. Chem.,* **1,** No. 20, 2 (1942).
22. W. A. Skinner and R. M. Parkhurst, *J. Org. Chem.,* **31,** 1248 (1966).
23. A. Meidell, *Med. Norsk Farm. Selskap,* **27,** 101 (1965).
24. J. N. Chatterjea, *J. Indian Chem. Soc.,* **36,** 76 (1959); J. N. Chatterjea, K. D. Banerjii, and N. Prasad, *Ber.,* **96,** 2356 (1963).
25. A. Meidell, *Med. Norsk Farm. Selskap,* **28,** 1 (1966).
26. A. Meidell, *Med. Norsk Farm. Selskap,* **28,** 90 (1966).
27. Y. Ishikawa, *Agr. Biol. Chem.,* **38,** 2545 (1974).
28. R. Azerad and M. O. Cyrot-Pelletier, *Biochemie* (Paris), **55,** 591 (1973).
29. T. Ichikawa and T. Kato, Japan. Patent 11,064 ('47); *Chem. Abstr.,* **67,** 100003 (1967).
30. H. Morimoto, I. Imada, and G. Goto, *Justus Liebigs Ann. Chem.,* **729,** 171 (1969).
31. O. A. Stamm, H. Schmid, and J. Büchi, *Helv. Chim. Acta,* **41,** 2001 (1958).
32. C. J. W. Brooks and M. M. Campbell, *Phytochemistry,* **8,** 215 (1969).
33. B. Rodriguez and S. Valverde, *Tetrahedron,* **29,** 2837 (1973).
34. R. Hodges and R. I. Reed, *Tetrahedron,* **10,** 71 (1960).
35. C. von Corstenn-Lichterfelde, R. Rodriguez, and S. Valverde, *Experientia,* **31,** 757 (1975).
36. C. Marquez, B. Rodriguez Gonazlez, and S. Valverde-Lopez, *An. Quim.,* **71,** 603 (1975).
37. M. Fujimaki, K. Kanamaru, T. Kurata, and O. Igarashi, *Agr. Biol. Chem.,* **34,** 1781 (1970).
38. M. J. Francis, P. K. Grant, K. Show, and R. T. Weavers, *Tetrahedron,* **32,** 95 (1976).
39. J. W. Scott, F. T. Bizzaro, D. R. Parrish, and G. Saucy, *Helv. Chim. Acta,* **59,** 290 (1976).
40. D. J. Goldsmith and C. T. Helmes, *Synth. Commun.,* **3,** 231 (1973).
41. F. M. Dean, *Naturally Occurring Oxygen Ring Compounds,* Butterworth, London, 1963.

42. F. della Monache, F. Marletti, G. B. Marini-Bettolo, J. F. de Mello, and O. C. de Lima, *Lloydia*, **40,** 201 (1977).
43. J. W. Westley, R. H. Evans, T. Williams, and A. Stempel, *J. Org. Chem.*, **38,** 3431 (1973).
44. S. Bory and E. Lederer, *Croat. Chem. Acta*, **29,** 157 (1957).
45. B. Danieli, P. Manitto, F. Ranchetti, G. Russo, and G. Ferrari, *Experientia*, **28,** 249 (1972).
46. C. Cardani, G. Casnati, and B. Cavalleri, *Rend. Ist. Lombardo Sci. Pt. I, Cl. Sci. Mat. Nat.*, **91,** 624 (1957).
47. G. Fontaine, *Ann. Chim.* (Paris), **3,** 467 (1968).
48. T. Backhouse and A. Robertson, *J. Chem. Soc.*, 1257 (1939).
49. S. Y. Dike and J. R. Merchant, *Chem. Ind.*, 996 (1976).
50. W. A. Skinner and R. M. Parkhurst, *Lipids*, **5,** 184 (1970).
51. E. Spath and W. Mocnik, *Ber.*, **70B,** 2276 (1937).
52. H. Meyer, P. Schuedel, R. Rüegg, and O. Isler, *Helv. Chim. Acta*, **46,** 650 (1963).
53. P. K. Grant, L. N. Dixon, and J. M. Robertson, *Tetrahedron*, **26,** 1631 (1970).

CHAPTER IX

Chroman Carboxylic Acids and Their Derivatives

G. P. ELLIS

Department of Chemistry, University of Wales Institute of Science and Technology, Cardiff, U.K.

I. INTRODUCTION

Many chroman carboxylic acids have been prepared in the course of identification of other chromans, while others have been purposefully synthesized by various methods. Very few of these compounds exist in nature, but the terpenoidal gomeric acid and its epimer (see Section VII) have been isolated from a plant. Biological activity has been demonstrated in several chroman carboxylic acid derivatives, and these are discussed in Section VII.

II. SYNTHESIS OF CARBOXYLIC ACIDS AND ESTERS

1. Cyclization Reactions

A. *From Phenols and Alkenes*

Nuclear alkylation of a phenol by an alkene in the presence of a Lewis acid produces a chroman. When the phenol also has a carboxyl group, this method leads to a chroman carboxylic acid, for example, 2,2-dimethylchroman-6-carboxylic acid (**1**).[1] Several analogous syntheses in which the alkene contains a halogen or hydroxyl group have been described,[2,3] and in some instances a carboxyl[3] or a potential carboxyl group[4–7] is also present. The latter frequently takes the form of a lactone, as in the synthesis of 3-(6-hydroxy-2,5,7,8-tetramethylchroman-2-yl)propionic acid (**2**).[4,8]

EtOOC–C₆H₄–OH + $CH_2{=}CHCHMe_2$ $\xrightarrow{ZnCl_2}$ (**1**)

OH; EtOOC; Me; Me; O; HOOC; (**1**)

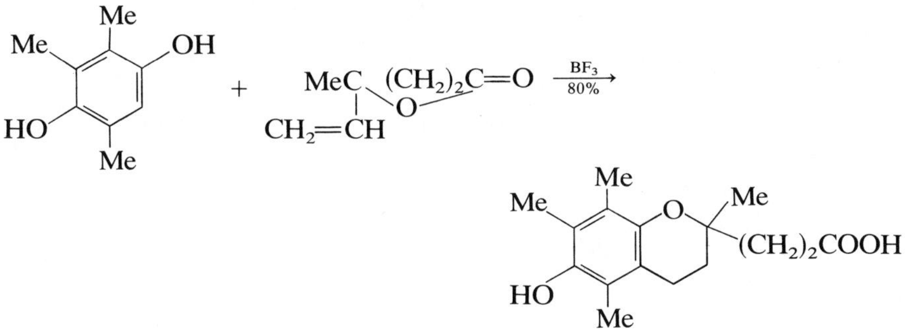

(2)

B. *Cyclization of o-Allylphenols*

Substituted *o*-allylphenols carrying a carboxyl group either on the benzene ring[9–16] or the side chain[17,18] are readily cyclized by acids or bases to the chromans. The syntheses of 2,2-dimethylchroman-6-carboxylic acid (**3**)[14] and methyl 6-acetyloxy-2,5,7,8-tetramethylchroman-2-ylacetate (**5**)[18] (which was synthesized in situ from the ketone (**4**) by a Wittig–Horner reaction) are typical examples of these methods. Fully saturated (terpenoid) naphtho[2,1-*b*]pyran-3-carboxylic acids have been prepared in this way.[19] A compound, (**6**) in which the positions of the double bond and the hydroxyl group are reversed was cyclized to perhydro-2,5,5,8a-tetramethylchroman-2-ylacetic acid (**7**).[20]

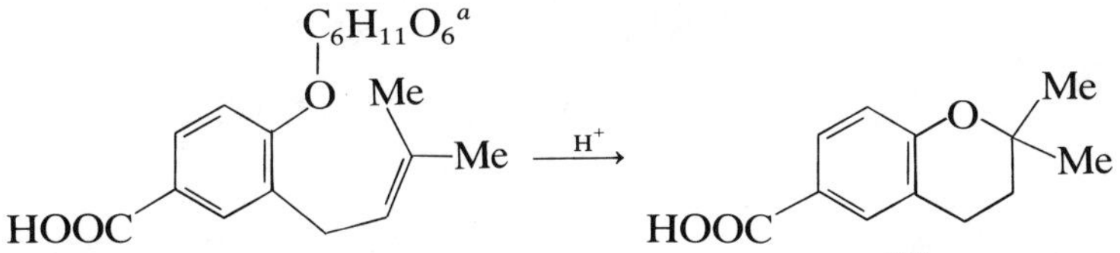

[a] D-Glucosyl

(3)

Photochemical cyclization of *o*-allylphenols such as ethyl 2-allyl-3-hydroxybenzoate (**8**) gives a mixture of chroman (**9**) and coumaran (**10**) in the ratio of 1 to 10, but the yield was low.[21]

C. *Cyclodehydration*

Chromans are formed when *o*-(3-hydroxypropyl)phenols are heated with acid,[17,22] as for example in the synthesis of 2,4,4,7-tetramethylchroman-2-carboxylic acid (**11**).[23] Demethylation is accompanied by hydrolysis of the carboxamide, and a nitrile is converted into carboxyl in a synthesis of 6-hydroxy-2,5,7,8-tetramethylchroman-2-carboxylic acid.[17]

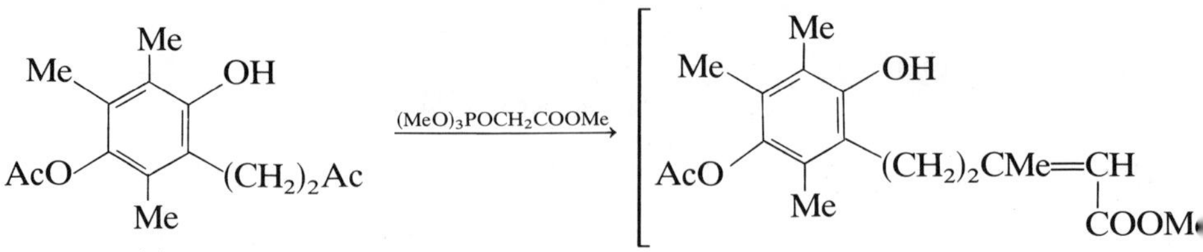

Me OH Me AcO $(CH_2)_2Ac$ Me (4) $\xrightarrow{(MeO)_3POCH_2COOMe}$ [Me OH Me AcO $(CH_2)_2CMe{=}CH$ Me COOMe]

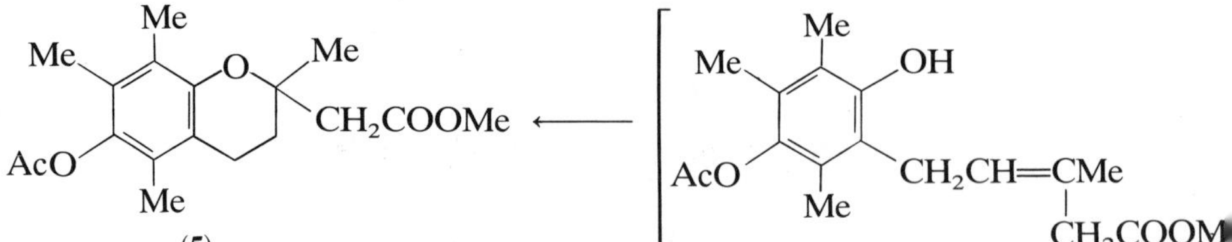

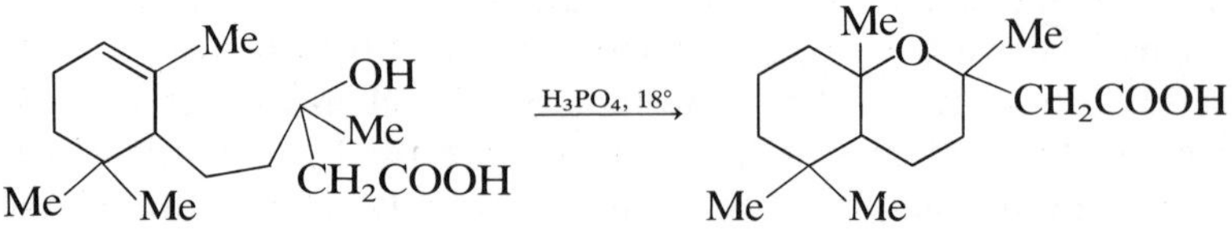

Me Me O Me AcO CH_2COOMe Me (5) ⟵ [Me OH Me AcO $CH_2CH{=}CMe$ Me CH_2COOMe]

Me OH Me CH_2COOH Me Me (6) $\xrightarrow{H_3PO_4,\ 18°}$ Me O Me CH_2COOH Me Me (7)

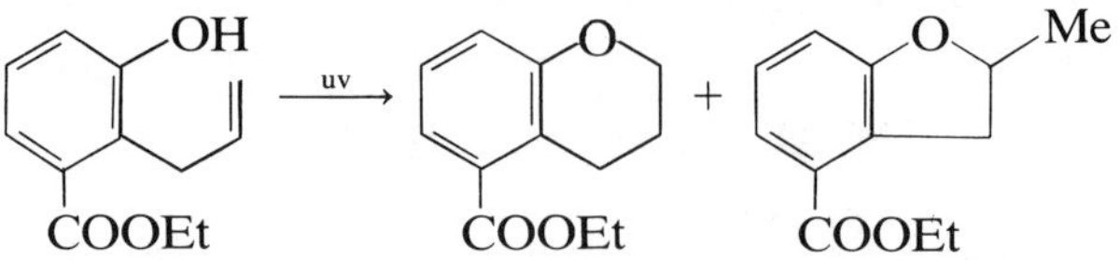

OH COOEt (8) $\xrightarrow{uv}$ O COOEt (9) + O Me COOEt (10)

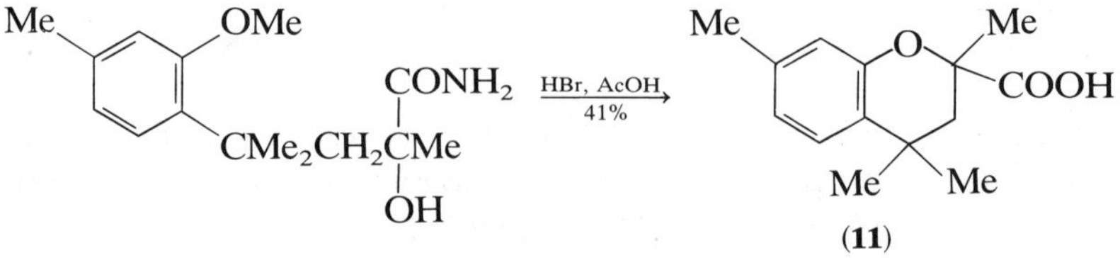

Me OMe $CONH_2$ CMe_2CH_2CMe OH $\xrightarrow[41\%]{HBr,\ AcOH}$ Me O Me COOH Me Me (11)

D. *Miscellaneous Methods*

A few chroman carboxylic acids have been synthesized from *o*-hydroxybenzyl alcohols, but there is uncertainty about the exact structure of the product of several of these reactions.[24,25] An alkenylphenol (**12**) was cyclized with methanolic hydrochloric acid in a biosynthetic study of the precursors of ubiquinone.[16]

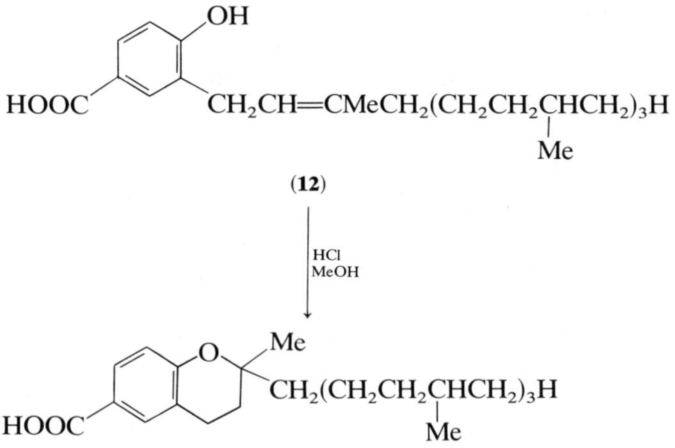

OH
HOOC $CH_2CH{=}CMeCH_2(CH_2CH_2CHCH_2)_3H$
Me

(**12**)

HCl
MeOH

Me
O
HOOC $CH_2(CH_2CH_2CHCH_2)_3H$
Me

Condensation of salicylaldehyde with levulinic acid and reduction of the resulting salicylidene derivative gave a keto acid (**13**) which spontaneously cyclized to 3-(2-hydroxychroman-2-yl)propionic acid (**14**).[26]

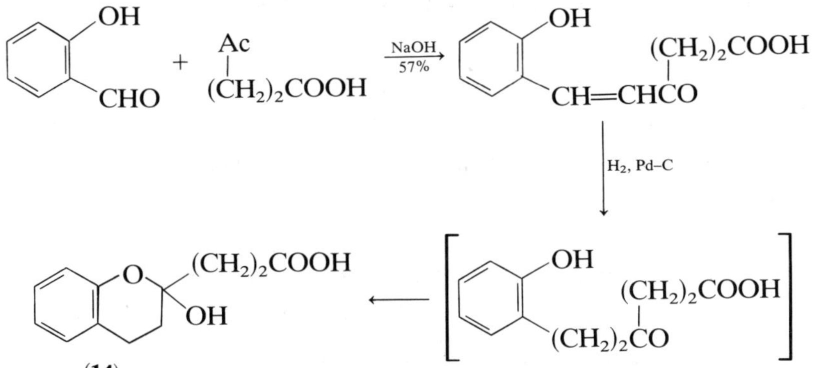

OH
CHO + Ac $(CH_2)_2COOH$ $\xrightarrow[57\%]{NaOH}$ OH $(CH_2)_2COOH$ CH=CHCO

H_2, Pd–C

O $(CH_2)_2COOH$ OH ⟵ [OH $(CH_2)_2COOH$ $(CH_2)_2CO$]

(**14**) (**13**)

Chroman-2-carboxylic acid has been obtained in low yield (<10%) by treatment of 2-bromo-1-tetralone successively with peroxybenzoic acid, hydrogen peroxide, and dilute sodium hydroxide (Scheme 1).[27] A much better method is now available from 4-oxo-4*H*-1-benzopyran-2-carboxylic acid (Section II, 2C).

When *o*-hydroxybenzyl alcohols are heated with a maleic acid ester, a chroman-2,3-dicarboxylic acid is obtained. The first-formed trans isomer of 6-*t*-butyl-8-methylchroman-2,3-dicarboxylic acid (**15**) was converted into the cis form via the anhydride.[24]

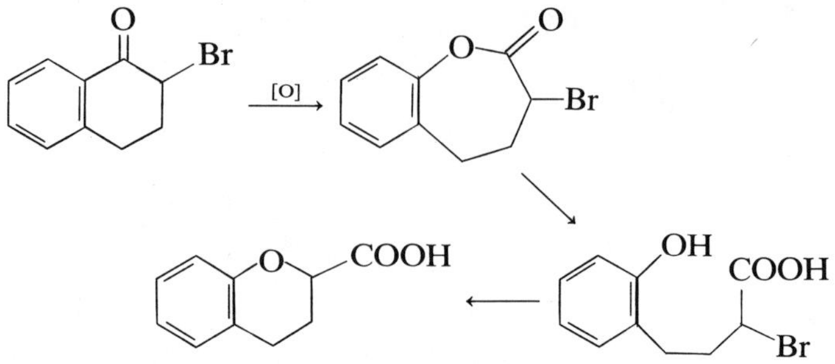

Scheme 1. A synthesis of chroman-2-carboxylic acid.

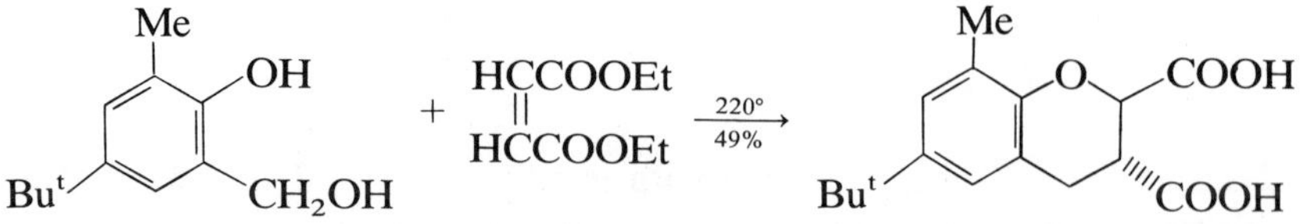

(**15**)

2. From Other Benzopyrans

A. *Oxidation of Aldehydes*

This important source of chroman carboxylic acids is discussed in Chapter VIII, Section IV, 2.

B. *Oxidation of Other Groups*

2-, 3-, or 4-Arylchromans are oxidized by potassium permanganate in moderate or low yield to the 2-, 3-, or 4-carboxylic acid.[28–34] For example, the formation of 2,4,4,7-tetramethylchroman-2-carboxylic acid (**17**) from the 2-arylchroman (**16**) was used in a proof of structure.[23] A 4*H*-chromen-3-yl group was similarly oxidized to a carboxylic acid.[35] Chromium trioxide has been used instead of permanganate.[36]

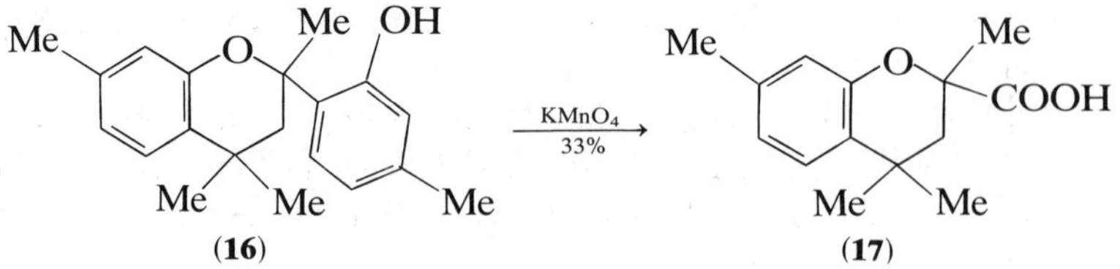

(**16**) (**17**)

Oxidation of alkyl and alkenyl groups to carboxyl is not well documented, but Baker, McOmie, and Wild[31] converted a 7-ethylchroman (**18**) to 2,4,4-trimethylchroman-2,7-dicarboxylic acid (**19**). A 6-acetyl group was oxidized to a carboxyl by the action of sodium hypochlorite,[37] and the mild oxidation of the benzylidene derivative (**20**) of dihydroevodionol[38] contributed to the determination of the structure of natural evodionol. Ozonization of a chroman containing a 2-$(CH_2)_2CH{=}CMe_2$ side chain gave the 2-$(CH_2)_2COOH$ derivative.[8]

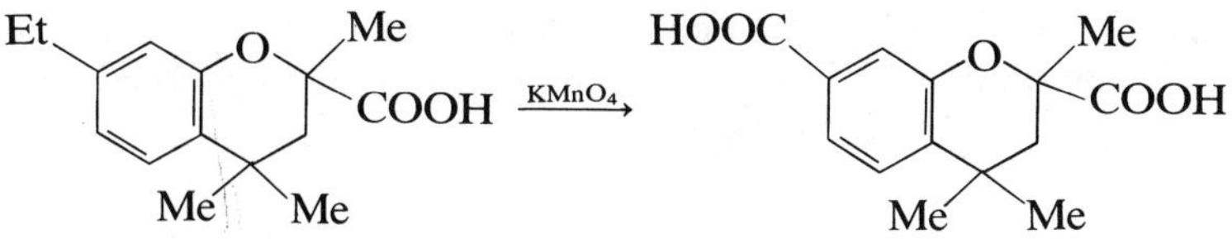

(**18**) $\xrightarrow{KMnO_4}$ (**19**)

A few instances of the conversion of primary alcohols to carboxylic acids occur in the chroman series.[39,40]

C. *Reduction of Chromenes, Chromanones, and Chromones*

A general discussion of these reductions has been published[41–43] and should be read in conjunction with this brief account. Catalytic hydrogenation of 2*H*- or 4*H*-chromencarboxylic acids or esters[44–54] in the presence of palladium,[45,50,51,53,55] platinum,[46–49,52] or nickel[54] usually proceeds without complication, but other changes sometimes accompany catalytic reduction. For example, a 3-hydroxyl or 3-acetyloxy group was lost when ethyl 3-acetyloxy-8-methoxy-2*H*-chromen-4-carboxylate was reduced under slight pressure in the presence of Adams's catalyst (platinum oxide);[48,49] however the same compound behaved normally when platinum was used.[48] Other examples of loss of hydroxyl or halogen are known.[49,56,57] Reduction of a conjugated dienic chromene gave ethyl chroman-2-ylacetate (**21**).[58]

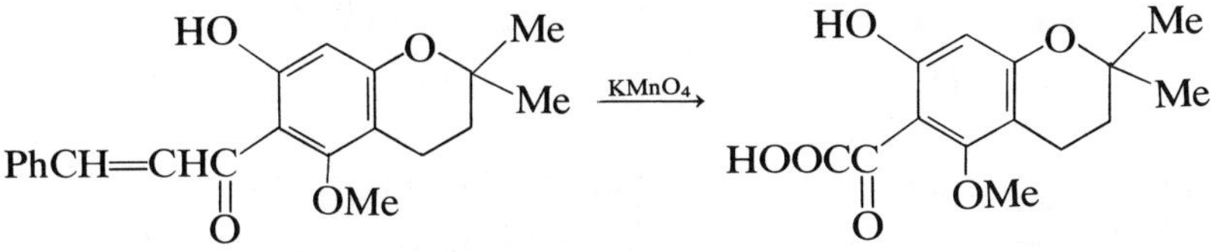

(**20**) $\xrightarrow{KMnO_4}$

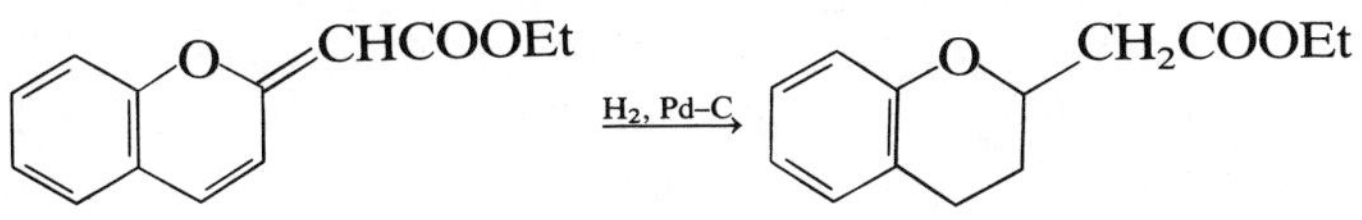

$\xrightarrow{H_2,\ Pd\text{–}C}$ (**21**)

4-Chromanones have been reduced to chromans by means of zinc amalgam in acid,[56,59,60] often in high yields.[61] For example, 6-chlorochroman-2-carboxylic acid (**22**) was efficiently prepared in this way. Another chemical method which has given a satisfactory yield is the reduction of the 4-toluenesulfonylhydrazide at room temperature with sodium cyanoborohydride.[62] This method maintains the stereochemistry of the bridgehead and is thus a valuable technique in conformational studies of the perhydrochromans such as methyl perhydrochroman-7-carboxylate (**23**), which was prepared by this method. Attempts to hydrogenate the 4-chromanone in the presence of ruthenium on charcoal[63] gave a mixture of stereoisomers of perhydro-4-chromanols, perhydrochromans, and hydrogenolytic products (such as 3-cyclohexylpropanol).[62]

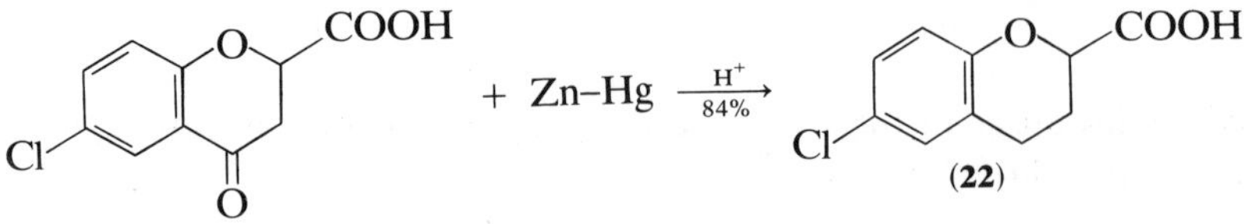

COOH O Cl O + Zn–Hg $\xrightarrow[84\%]{H^+}$ COOH O Cl

(**22**)

MeOOC O

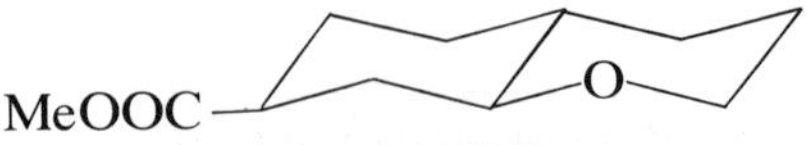

(**23**)

With more common noble metal catalysts, such as palladium–charcoal[45] or nickel,[64] a chromanol was obtained in good yield from ethyl 8-methoxy-3-oxochroman-4-carboxylate (**24**), for example.

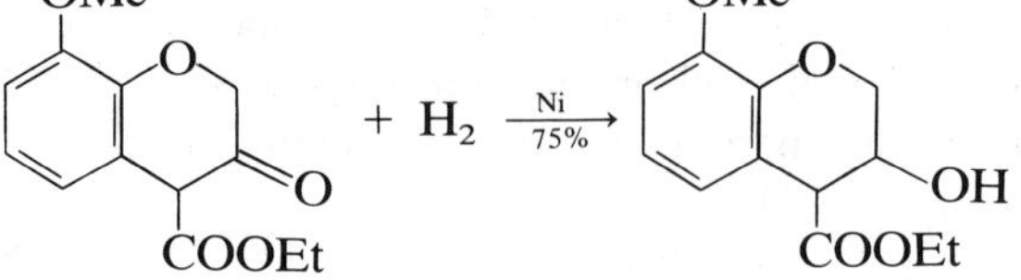

OMe O O COOEt + H_2 $\xrightarrow[75\%]{Ni}$ OMe O OH COOEt

(**24**)

Chromone carboxylic acids or esters need fairly drastic treatment for their complete reduction to chromans while retaining the carboxyl function. A Russian team[65] obtained good results by using palladium on barium sulfate in the reduction of ethyl 4-oxo-4*H*-1-benzopyran-2-carboxylate (**25**); the parent acid of (**25**) has been reduced in 87% yield to chroman-2-carboxylic acid (**26**) with hydrogen and palladium–charcoal at 70°.[66] The 6-chloro ester corresponding to (**25**) was reduced under pressure at 70° in the presence of palladium on charcoal.[61,67]

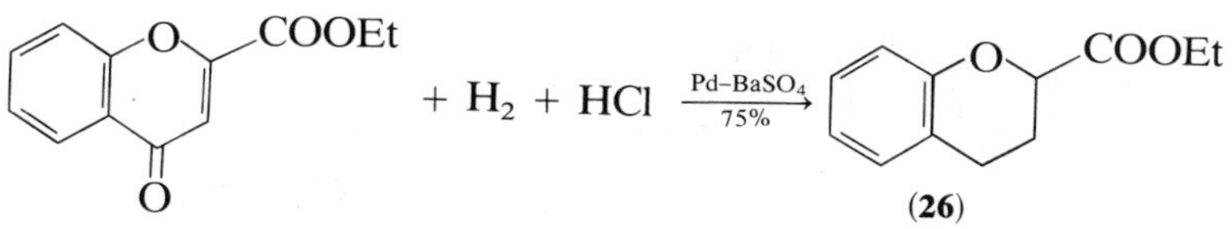

D. *Addition to Chromenes and Chromanones*

2*H*-Chromenes add on one mole of bromine to give a 3,4-dibromochroman.[15] But when *N*-bromosuccinimide and methanol are used, for example, methyl 3-bromo-4-methoxychroman-3-carboxylate (**27**) is formed.[68]

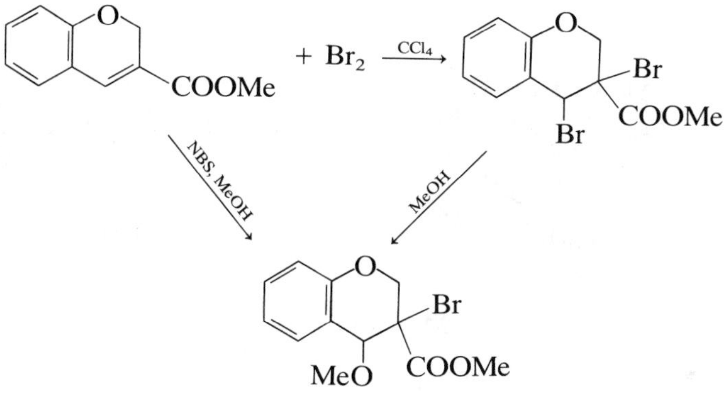

Heating a 4*H*-chromene with sulfuryl chloride results in the addition of two chlorine atoms across the 2,3-double bond.[69] A high yield of methyl 2,3,4,4-tetrachlorochroman-2-carboxylate was obtained in this way.

The reactivity of the carbonyl group of 4-chromanone was employed in a synthesis of 4-carboxychroman-4-ylacetic acid (**28**), an intermediate in the synthesis of a spiro[chroman-4,3′-furan]-2′,5′-dione (**29**).[70]

E. *Cleavage of a Ring*

Many naturally occurring compounds containing a pyran-2-one or dihydropyran-2-one ring have been cleaved by a base,[71] an oxidizing agent (such as permanganate),[72] or an alkali and a methylating agent in order to prevent the lactone from being formed again.[73–76] High yields of chroman carboxylic acids are sometimes obtained, for example, a constituent of the roots of *Poncirus trifoliata* was shown to have the structure (**30**) by mild reduction followed by ring opening and methylation.[74,75]

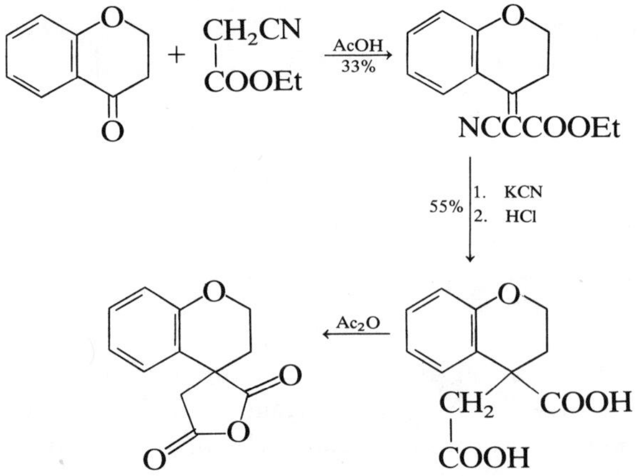

O + CH_2CN | COOEt —AcOH, 33%→ O NCCCOOEt

O

↓ 55% 1. KCN 2. HCl

O O O ←Ac_2O— O CH_2 COOH | COOH

(29) (28)

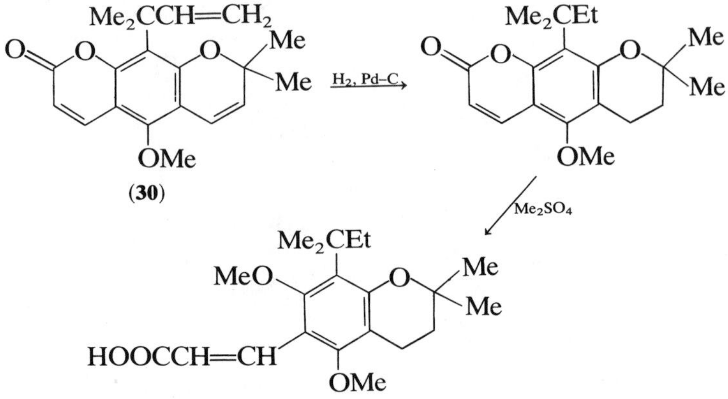

$Me_2CCH{=}CH_2$ O O O Me Me —H_2, Pd–C→ Me_2CEt O O O Me Me

OMe OMe

(30)

↙ Me_2SO_4

Me_2CEt MeO O Me Me HOOCCH=CH OMe

Chroman carboxylic acids have also been obtained in low yield by the opening of a pyran-4-one[77] or 2*H*-pyran[78] ring. Autoxidation of 3-ethyl-3,4a,7,7,10a-pentamethylperhydronaphtho[2,1-*b*]pyran-10-one ("dihydro-1-oxomanoyl oxide") in the presence of potassium *t*-butoxide gave the dicarboxylic acid (Scheme 2).[79]

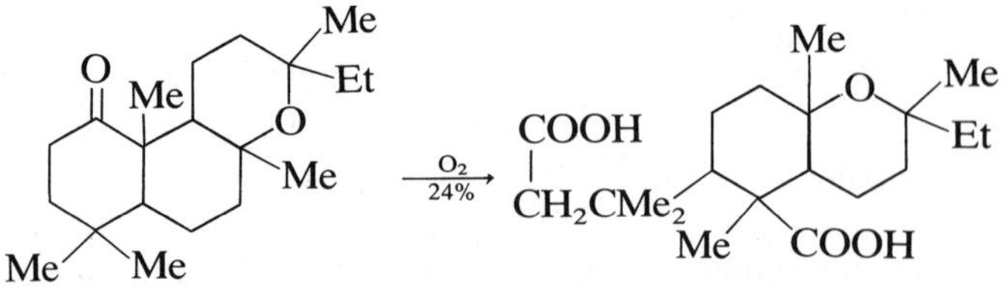

O Me Me Et O Me Me Me —O_2, 24%→ COOH | CH_2CMe_2 Me Me O Me Et Me COOH

Scheme 2. Autoxidation of a cyclohexanone ring.

F. *Rearrangement Reactions*

When 3-phenylbenzofuran-2-ones are treated with a base (sodium alkoxide or an alkylamine), rearrangement occurs and a 4-phenylchroman-4-carboxylic acid is formed, often in good yield.[80–87] For example, methyl 4-phenylchroman-4-carboxylate (**32**) is produced in high yield from the furanone (**31**).[80] Variation in the bromoethyl side chain of (**31**) and of the base leads to a differently substituted product such as the basic ester (**33**).[87]

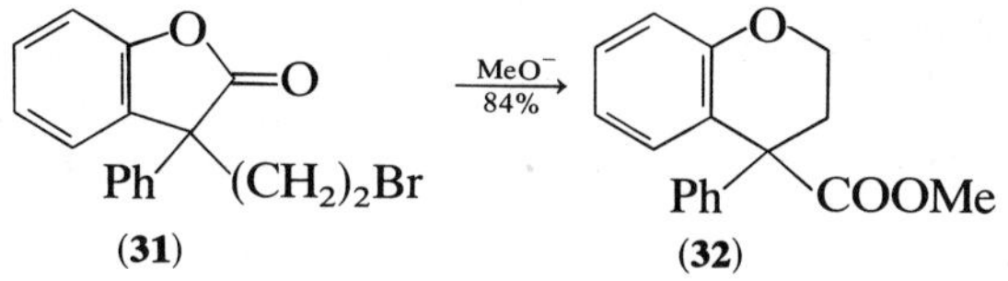

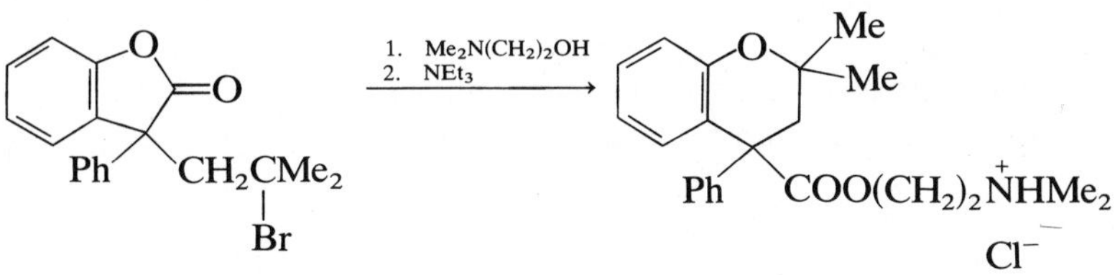

MeO^-, 84%: (**31**) [Ph, $(CH_2)_2Br$] → (**32**) [Ph, COOMe]

1. $Me_2N(CH_2)_2OH$ 2. NEt_3: [Ph, CH_2CMe_2Br] → (**33**) [Me, Me; Ph, $COO(CH_2)_2\overset{+}{N}HMe_2$ Cl^-]

3-Acyldihydrocumarins rearrange in anhydrous acidic solution, via ring opening, to a chroman-3-carboxylic acid ester under mild conditions and usually in good yield.[88,89] In this way, methyl 2-methoxychroman-3-carboxylate (**35**) was obtained from 3,4-dihydrocoumarin-3-carboxaldehyde (**34**).[88]

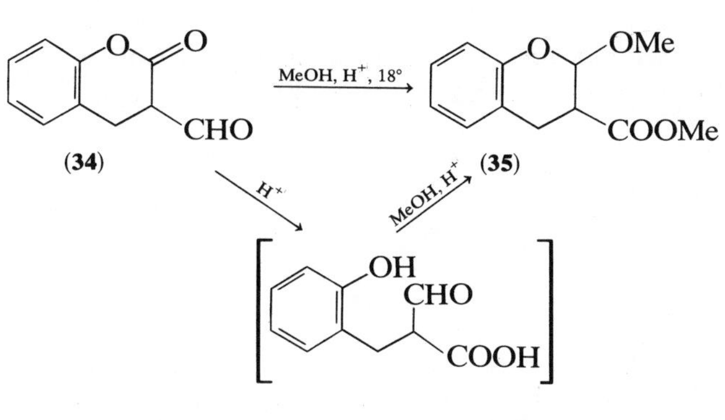

(**34**) [O, CHO] → MeOH, H^+, 18° → (**35**) [OMe, COOMe]

(**34**) → H^+ → [OH, CHO, COOH] → MeOH, H^+ → (**35**)

Ring opening of 4-(2-bromoethyl)-3,4-dihydrocoumarins under basic conditions leads to chromans which have a side-chain carboxyl or carboxamide group. Scheme 3 summarizes these reactions[90] which result from a nucleophilic attack on the lactone group and subsequent recyclization.

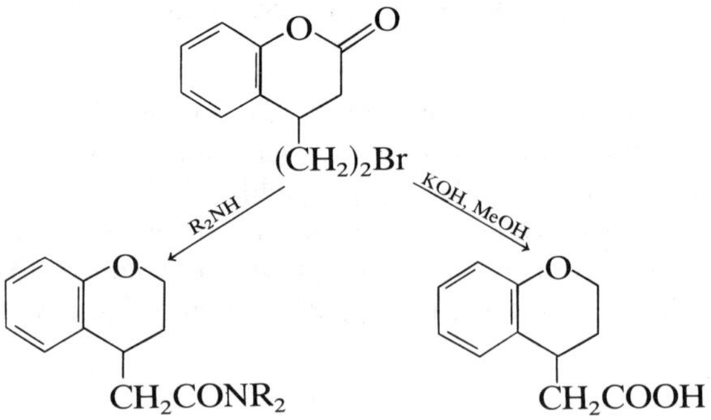

$(CH_2)_2Br$ — R_2NH → CH_2CONR_2; KOH, MeOH → CH_2COOH

Scheme 3. Chroman-4-ylacetic acids from a dihydrocoumarin.

A rearrangement is involved in the Willgerodt reaction of ketones to form carboxamides which may be hydrolyzed to the acids. Joshi and Kamat[91] converted 6-acetyl-2,2-dimethyl-7-methoxychroman (**36**) in this way to 2,2-dimethyl-7-methoxychroman-6-ylacetic acid (**37**), and a French team[92] recently prepared chroman-6-ylacetic acid by a similar method.

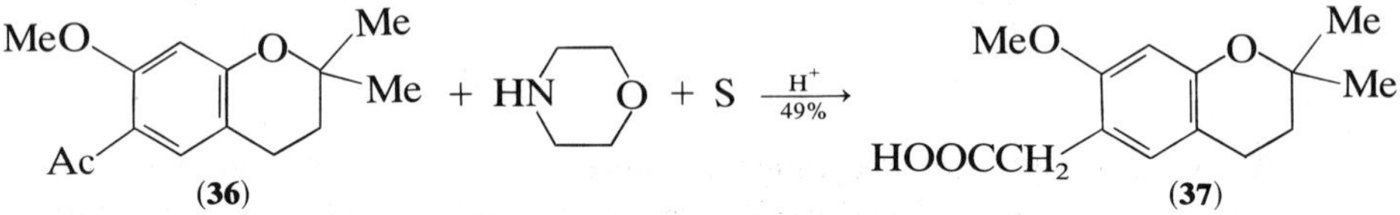

MeO, Me, Me, Ac (**36**) + HN O + S —H^+, 49%→ MeO, Me, Me, $HOOCCH_2$ (**37**)

G. *Carboxylation*

The Kolbe–Schmitt reaction when applied to 2,2-dimethyl-5-hydroxychroman (**38**) gave the 6-carboxylic acid (**39**) in low yield.[50,93]

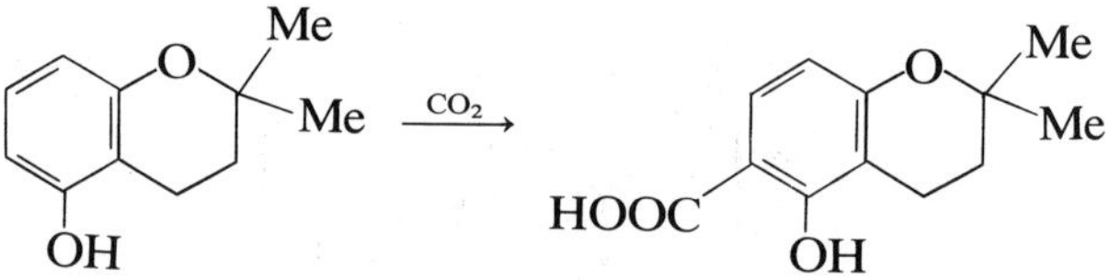

Me, Me, OH (**38**) —CO_2→ Me, Me, HOOC, OH (**39**)

Butyllithium and carbon dioxide converted 2,2-dimethylchroman into the 8-carboxylic acid in 39% yield,[94] while carboxylation of a terminal ethynyl group with a Grignard reagent and carbon dioxide gave a carboxylic acid (**40**) containing an additional carbon atom.[95]

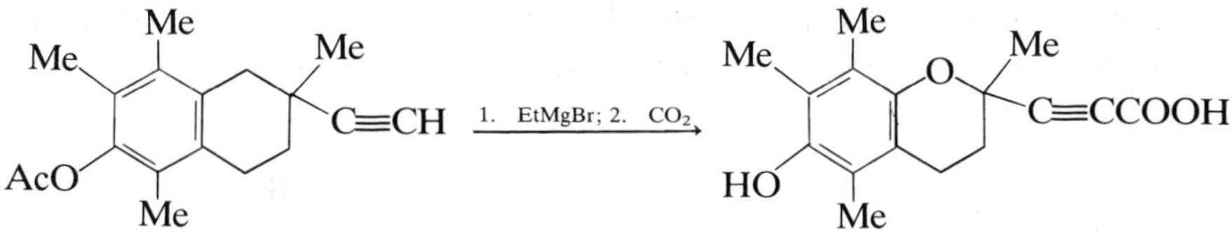

(40)

H. *Miscellaneous*

Many chroman carboxylic acids (and esters) have been obtained by methods which do not fall into the previously mentioned classes. Nevertheless, some of these may have a wider applicability than has so far been accorded to them. Others have been attempted on a very small number of compounds and have not given good yields but may be capable of being improved by the use of recently discovered reagents or modifications. Some of these methods are summarized in this section in the hope that they may attract further interest.

Friedel–Crafts acylation of chroman with the half-ester acid chloride of a dicarboxylic acid gave a good yield of the chroman-6-keto ester, which was then reduced by the Clemmensen reaction to the ω-carboxyalkyl compound.[96] The cyclic anhydride was also employed and both methods gave yields of about 60% of keto acid or keto ester.

Claisen condensation of 6-acetyl-7-hydroxychroman with diethyl oxalate was used by Naylor and Ramage[55] to prepare the diketo ester (**41**), which was then cyclized to the chromone (**42**).

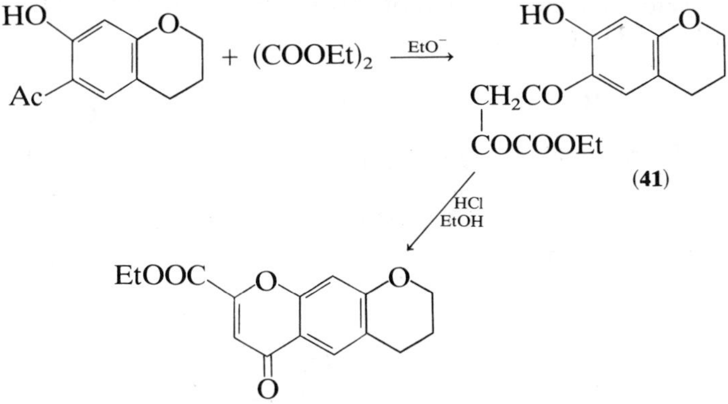

(41)

(42)

An unsaturated side-chain carboxylic acid was prepared from 2,2-dimethyl-3-hydroxy-5-methoxychroman-8-carboxaldehyde (**43**) by a Knoevenagel condensation (yield not stated).[97] Diethyl malonate reacted with the hemiacetal (**44**) to give the chromanyl malonic ester (**45**), which on hydrolysis and decarboxylation yielded 4,7-dimethylchroman-2-ylacetic acid (**46**).[98]

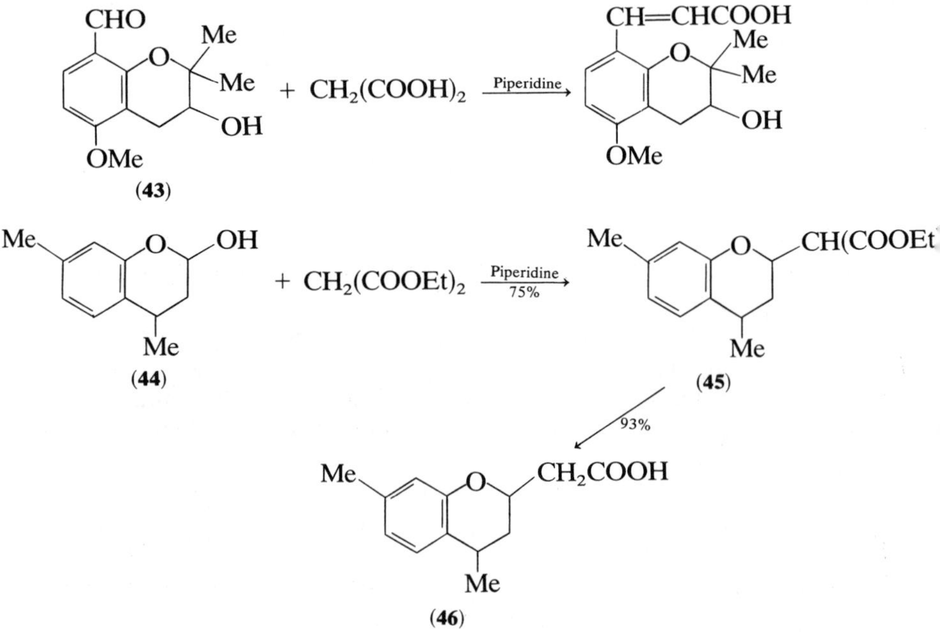
CHO
Me
Me
OH
OMe
(43)
+ $CH_2(COOH)_2$
Piperidine
CH=CHCOOH
Me
Me
OH
OMe
Me
OH
Me
(44)
+ $CH_2(COOEt)_2$
Piperidine
75%
Me
CH(COOEt
Me
(45)
93%
Me
CH_2COOH
Me
(46)

Synthesis of a chroman from a dihydropyran is described by Lebouc, Riobe, and Delaunay.[99] Dimethyl acetylenedicarboxylate was condensed with a 5,6-dihydro-2-vinylpyran (**47**) to form the quinonoid chroman in 40–68% yield. Dehydrogenation of this led to dimethyl 8-phenylchroman-5,6-dicarboxylate (**48**).

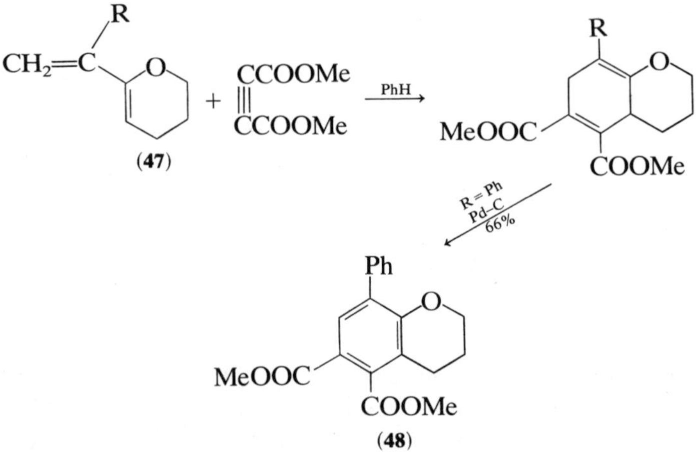
R
CH_2=C
(47)
+
CCOOMe
CCOOMe
PhH
R
MeOOC
COOMe
R = Ph
Pd–C
66%
Ph
MeOOC
COOMe
(48)

A partially reduced chromanylacetic ester (**51**) was synthesized in high yield from 5,5-dimethylcyclohexan-1,3-dione (dimedone, **49**) and methyl penta-2,4-dienoate (**50**) in the presence of dimethylsulfinyl ion.[100] A mechanism involving 5-(5,5-dimethyl-1,3-dioxocyclohex-2-yl)pent-2-enoic ester was proposed for this kind of reaction, which was also applied to the synthesis of methyl 5-oxo-5,6,7,8-tetrahydrochroman-2-ylacetate.

(**49**) + CHCOOMe=CH–CH=CH_2 (**50**) $\xrightarrow{MeSOCH_2^-}$ (**51**) (CH_2COOMe)

An acetyl group is easily introduced into a benzene ring of a chroman by Friedel–Crafts acylation. Using the haloform reaction, it is possible to convert this acyl group into a carboxyl; for instance, by tribromination of 6-acetyl-8-methylchroman (**52**),[101] 8-methylchroman-6-carboxylic acid (**53**) is obtained.

(**52**) + Br_2 $\xrightarrow[34\%]{1.\ OH^-;\ 2.\ H^+}$ (**53**)

Hydrolysis of a nitrile to a carboxylic acid is discussed in Section VI, 3.

III. SYNTHESIS OF ACYL HALIDES, CARBOXAMIDES, AND CARBAMATES

1. Acyl Halides

Chroman carboxylic acids have been converted into their acyl chlorides by the use of thionyl chloride,[23,66,80,93,102] phosphorus trichloride,[91] or oxalyl chloride.[18] One feature of interest is the simultaneous chlorination of one methyl group when ethyl 3-(6-acetyloxy-2,5,7,8-tetramethylchroman-2-yl)-propionate (**54**) was heated with thionyl chloride.

When the product (**55**) was treated with ammonia, only the chlorocarbonyl group reacted, and the carboxamide (**56**) was obtained.[5]

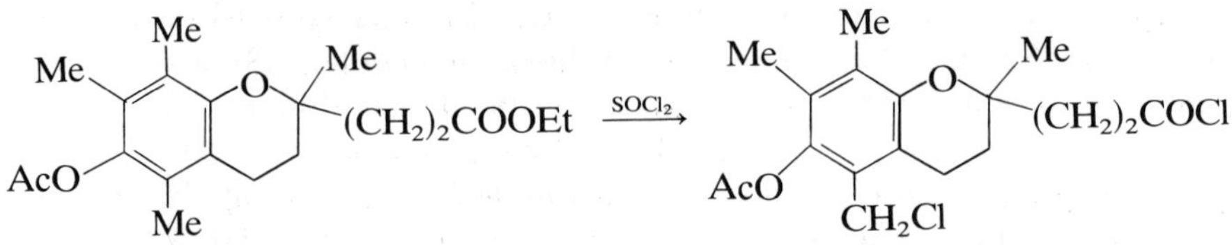

(54) $\xrightarrow{SOCl_2}$ (55)

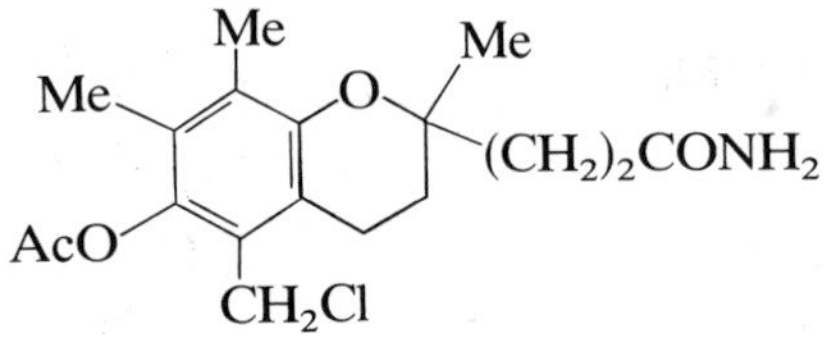

(56)

2. Carboxamides

Apart from the conventional method of reacting an acyl chloride with ammonia,[5,23,66,80,91] or an amine,[102] chroman *N*-arylcarboxamides have been obtained by heating an ester with the amine in xylene.[52,103] A benzofuran-2-one such as (**57**) rearranges when allowed to react at room temperature with a primary amine or ammonia to give a derivative of 4-phenylchroman-4-carboxamide (**58**) as one of the products.[84] With secondary amines, lower yields of the amides are obtained.[86]

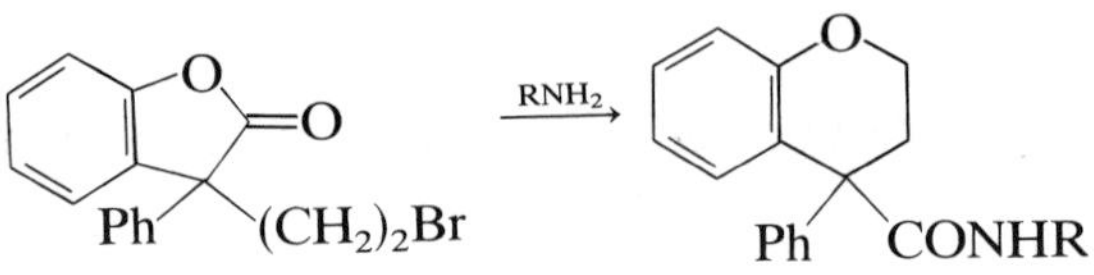

(57) $\xrightarrow{RNH_2}$ (58)

3. Carbamates

Two methods of preparing chroman carbamates are known. In the first method salicylaldehyde is heated in an autoclave with urethane, propene, or higher alkene and boron trifluoride, and an ethyl *N*-(2-methylchroman-4-yl)carbamate (**59**) is obtained in moderate or low yield.[104]

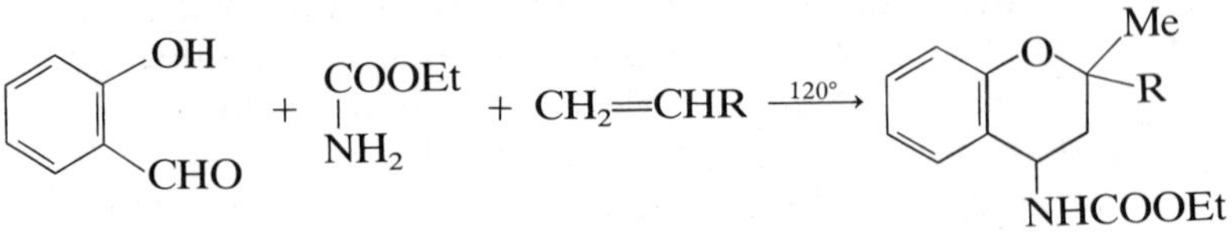

(59)

In the second method a 2*H*-chromen-3-ylcarbonyl chloride (**60**) is treated with sodium azide, and the corresponding acyl azide is formed. Boiling this with ethanol gives the carbamate with simultaneous migration of the double bond to yield ethyl chroman-3-ylidenecarbamate (**61**).[105]

(**60**) COCl $\xrightarrow{NaN_3}$ CON_3 → NCOOEt (**61**)

IV. SYNTHESIS OF CHROMAN CARBONITRILES

1. By Cyclization Reactions

Base-catalyzed cyanoethylation of a phenol which also contains an ortho-placed aldehyde group was originally[105–107] believed to give the 4-hydroxychroman-3-carbonitrile (**62**), but Augstein, Monro, Potter, and Scholfield[66] have shown spectroscopically and by synthesis from the free acid that the product of this reaction is 2*H*-1-benzopyran-3-carboxamide (**63**).

OH, CHO + CH_2=CHCN ⇸ CN, OH (**62**); → $CONH_2$ (**63**)

A cyanohydrin such as (**64**) is cyclized under acidic conditions to the nitrile (**65**) in modest yield.[108]

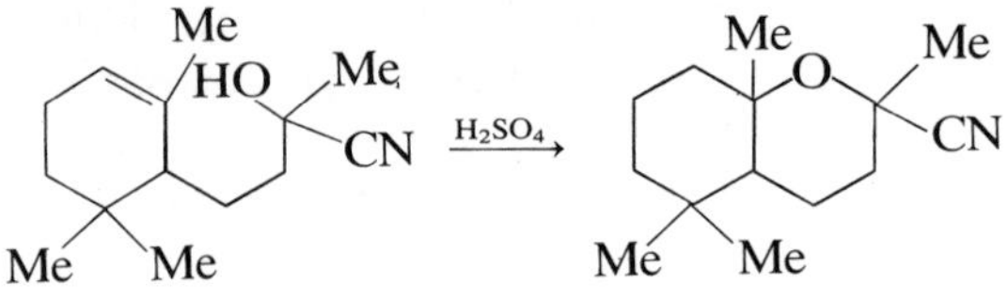

(64) $\xrightarrow{H_2SO_4}$ (65)

2. From Other Chromans

Conventional methods of preparing nitriles[109] have been applied to the synthesis of chroman carbonitriles, for example, dehydration of an oxime,[110,111] dehydration of a carboxamide,[91] displacement of an aliphatic halogen by cyanide ion,[112,113] addition of an alkali cyanide to a carbonyl group,[114] and condensation of malononitrile with a carbonyl compound.[115,116] The last-named method gives very high yields of dinitriles such as 6,8-dimethylchroman-4-ylidenemalononitrile (**67**) from the 4-chromanone (**66**)[116] using a Dean and Stark apparatus.

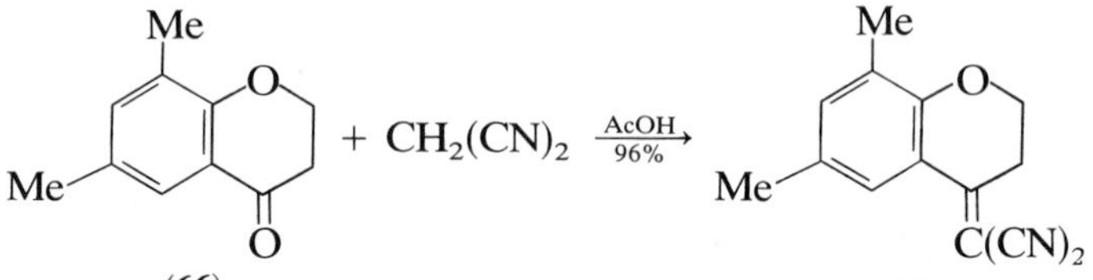

(66) + $CH_2(CN)_2$ $\xrightarrow[96\%]{AcOH}$ (67)

V. PHYSICAL AND SPECTRAL PROPERTIES

1. Physical Properties

All known chroman carboxylic acids are solids, with the exception of a few fully reduced compounds in which the carboxyl group is in a side chain, for example, perhydro-2,5,5,8*a*-tetramethylchroman-2-ylacetic acid,[20] and chromans in which the carboxyl is at the end of a C_8 side chain.[3] Many methyl esters are solids,[12,17,77,87,117] while others[63,80,88,90] and most higher esters,[70,80,96] but not all,[48,49,61,96] are oils. Simple amides are solids,[17,23,35,118] while *N*-substituted amides may be solids[77,84,87] or oils.[84,90] Few acyl chlorides have been characterized, but some of these are solids.[102] Chroman nitriles are usually solids,[105–107,110,115] but most of these contain other substituents such as hydroxyl which raise the melting point. Perhydro-2,5,5,8*a*-tetramethylchroman-2-carbonitrile is a liquid at ordinary temperatures.[108]

Optical isomers have been isolated for a few of the chiral carboxylic acids or esters,[17–19,32,40,56,80,95,97,119–124] while cis and/or trans forms of others have been described.[82,85] Conformational studies have been reported on some perhydrochroman carboxylic acid esters (see Section V, 2).

The physicochemical properties of 6-chlorochroman-2-carboxylic acid have been investigated because of the pharmacological properties of this acid and some of its derivatives (see Section VII). Its pK_a is 4.26, and its partition coefficients between octanol and water were determined spectrophotometrically.[67]

2. Spectral Properties

Although the majority of chroman carboxylic acids and their derivatives are identified by their elemental analysis and melting or boiling point, there are a number of compounds which have been isolated in small quantities, often by gas–liquid chromatography, and characterized only by their mass spectrum and their absorption patterns in the ir region and by nmr spectra.[16,21,58,63,114,125] Their chromatographic behavior (R_F or retention time) is sometimes also given. Such compounds are included in the tables at the end of this chapter even though no boiling or melting point is available.

Absorption of compounds in the uv region is less often quoted than previously and is of little value except to show whether compounds do or do not have conjugated double bonds.[14,15,74,100,115] However the following are typical absorptions of chroman carboxylic acids (in ethanol, log ε in brackets where available). 2,2-Dimethylchroman-6-carboxylic acid:[15] 261 (4.24), 278 *sh* (3.90), 289 (3.58) nm; 2,4,4-trimethyl-4*H*-1-naphtho[1,2-*b*]pyran-2-carboxylic acid:[36] 241, 265, 289, 310 nm. Its place has largely been taken by ir and nmr spectroscopy. Infrared absorption of esters,[6,12,15,36,59,60,65,89,114] acyl chloride,[102] and nitriles[91,114–116] are recorded for chromans with varying substitutions. Carboxylic acids similarly give characteristic absorptions.[12,15,18,53,74,125–127] For example, 2,2-dimethylchroman-6-carboxylic acid absorbs[15] at 3200–2280 (OH), 1670 (CO), 1608, 1588, 1500 (aromatic C–C). Optical rotatory dispersion of a few chroman carboxylic acids has been measured.[95]

Nuclear magnetic resonance spectroscopy has been used routinely in work on chroman carboxylic acids,[12,15,18,39,74,126,128] esters,[6,12,15,21,58,63,87,97,100,113,114,120,124,128,129] amides,[66] and nitriles[116] over the last decade or so. Some typical absorption peaks of a variety of chroman derivatives are listed in Table 1. A more detailed study has been made of the conformation of perhydrochroman-6- and 7-carboxylic esters.[62] This showed that the two rings are trans fused and that the ester group is equatorial in methyl perhydrochroman-7-carboxylate (**68**) but axial in the corresponding 6-ester (**69**).

TABLE 1. NMR SPECTRAL SIGNALS OF SOME CHROMAN CARBOXYLIC ESTERS AND NITRILE

Substituents			
Bz ring	Pyran ring	Assignments[a] (solvent)	Ref.
6-Ph	2-COOEt	6.90-7.74, m, 8H, Ar-H; 4.84, m, 1H, 2-H; 4.26, q, J7, 2H, OCH_2; 2.80, m, 2H, 4-CH_2; 2.24, m, 2H, 3-CH_2; 1.25, t, 3H, C-Me	128
6-COOMe	2,2-Me_2-3-OH	7.60–7.85, br, 2H, 5-H and 7-H; 6.80, d, J 8, 1H, 8-H; 3.93, s, OMe; 3.76–4.10, m, 1H, 3-H; 2.97, m, 2H, 4-CH_2; 2.40, s, 1H, OH; 1.38, s, 6H, 2Me	15
6-COOMe	2,2-Me_2-5-OH	11.1, s, 1H, OH; 7.48, d, J 8.5, 1H, 7-H; 6.18, d, J 8.5, 1H, 8-H; 3.80, s, 3H, OMe; 2.60, t, J 7, 2H, 4-CH_2; 1.72, t, J 7, 2H, 3-CH_2; 1.38, s, 6H, 2Me	12
6,8-Me_2	4-=$C(CN)_2$	8.11, s, 1H, 5-H; 7.36, s, 1H, 7-H; 4.38, t, J 6, 2H, CH_2; 3.07, t, J 6, 2H, CH_2; 2.28, s, 3H, Me; 2.17, s, 3H, Me	116
8-Ph-5,6-$(COOMe)_2$		7.45, s, 5H, Ph; 7.25, s, 1H, 7-H; 4.25, t, 2H, 2-H; 2.85, m, 2H, 4-CH_2; 2.05, m, 2H, 3-CH_2	99
5-MeO-8-CH=CHCOOMe	2,2-Me_2-3-OH	7.95, d, J 16, ArCH; 7.36, d, J 9, 1H, 7-H: 6.44, d, J 9, 1H, 6-H; 6.44, d, J 16, 1H, CHCO; 3.82, s, 3H, OMe; 3.80, 1H, 3-H; 3.76, s, 3H, COOMe; 3.00, dd, J 18 and 4.5, 1H, 4-H; 2.75, dd, J 18 and 5.5, 1H, 4-H; 2.32, s, 1H, OH; 1.37, s, 6H, 2Me	97

[a] br, broad; d, doublet; m, multiplet; q, quartet; t, triplet; coupling constants (J) are quoted in hertz.

Mass-spectrometric studies on chroman carboxylic acids and their derivatives are almost unknown, although the identity of some individual compounds has been ascertained with the help of molecular weight determinations in the mass spectrometer.[14,53,58,63,114] The fragmentation of molecules has been recorded for a few other compounds.[12,18,21,39,125] A combination of nmr, mass spectrometry and X-ray crystallography has been applied to the determination of the structure of aspulvinone (**69a**), a complex 2,2-dimethyl-7-hydroxychroman.[129a]

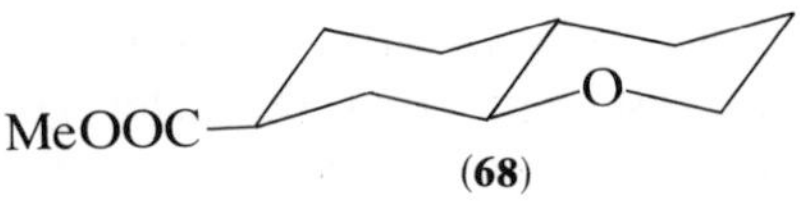

(**68**)

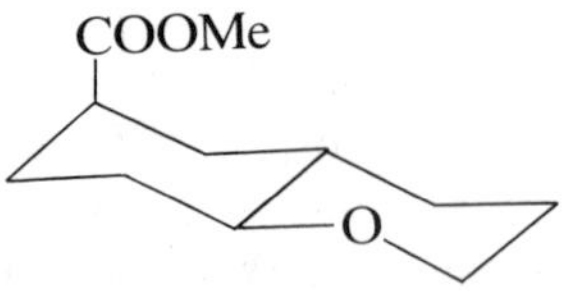

(**69**)

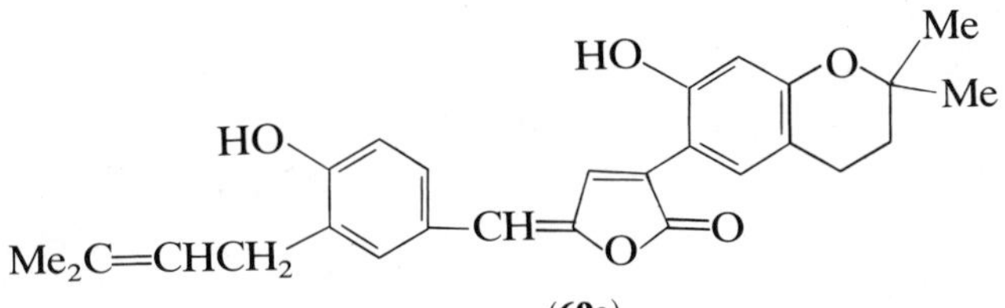

(**69**a)

VI. CHEMICAL PROPERTIES

1. Carboxylic Acids and Esters

There are many examples of the interconvertibility of carboxylic acids and esters in the chroman field. Conventional reaction conditions are employed in most of these, and so they will be referred to only briefly. Chroman carboxylic acids have been esterified by heating with the alcohol and an acidic catalyst such as concentrated sulfuric acid[6,20,61,111,128,130] or hydrogen chloride (Fischer–Speier method),[54,128] or by reaction with diazomethane.[3,7,15,39,40,77,78,90,119,124,131,132] A tocopherol derivative (**70**) was esterified very efficiently at room temperature by treatment with methyl iodide and sodium bicarbonate in dimethylformamide.[18]

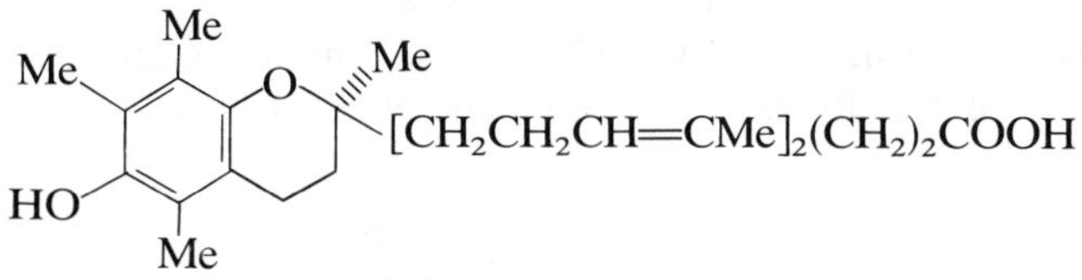

(**70**)

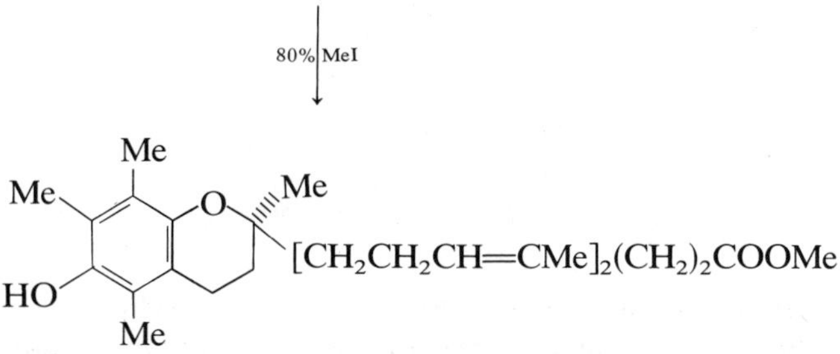

80% MeI

Esterification accompanied by methylation of one of the hydroxyl groups of 5,8-dihydroxy-2,2-dimethylchroman-6-carboxylic acid (**71**) occurred in rather low yield using dimethyl sulfate.[133]

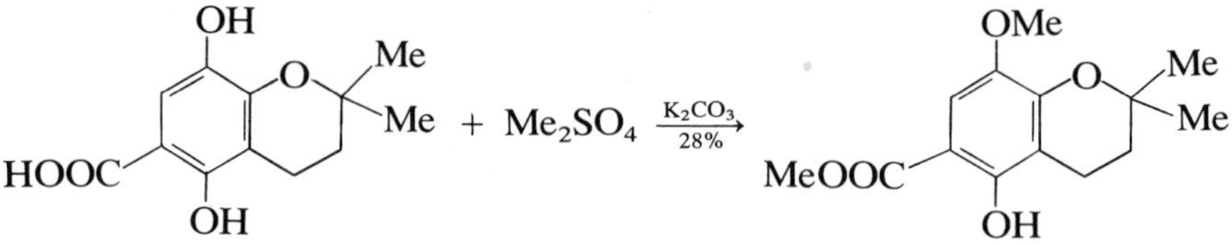

$$\text{(71)} + Me_2SO_4 \xrightarrow[28\%]{K_2CO_3}$$

Hydrolysis of esters of chroman carboxylic acids under both acidic and alkaline conditions have been described. Dilute hydrochloric acid[48,64] and alcoholic potassium hydroxide[77,80] or aqueous dioxan–potassium hydroxide[57] have given good results. But when ethyl 6,7-dimethoxy-3-hydroxychroman-4-carboxylate (ethyl hydroxynetorate, **72**) or its *O*-acetate was treated with ethanolic potassium hydroxide, dehydration also occurred to yield toxicaric acid (6,7-dimethoxy-2*H*-chromene-4-carboxylic acid, **73**).[45]

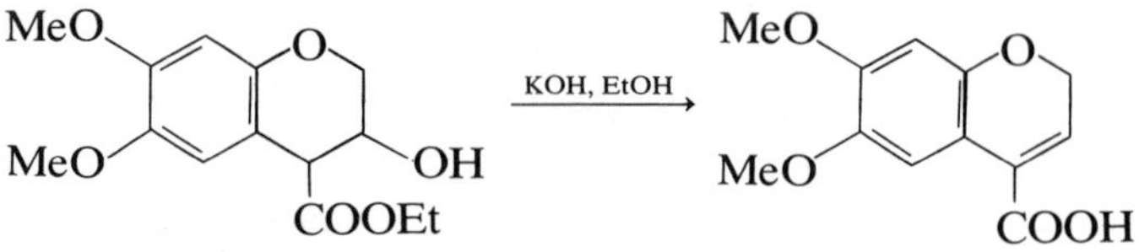

$$\text{(72)} \xrightarrow{\text{KOH, EtOH}} \text{(73)}$$

The influence of stereochemistry on the ease of hydrolysis of an ester was demonstrated by Giles, Schumacher, Mims, and Bernasek[122] who found that the methyl ester of manoylic acid (**74**) was saponified by ethanolic potash at 58° 38 times as rapidly as that of epimanoylic acid (**75**).

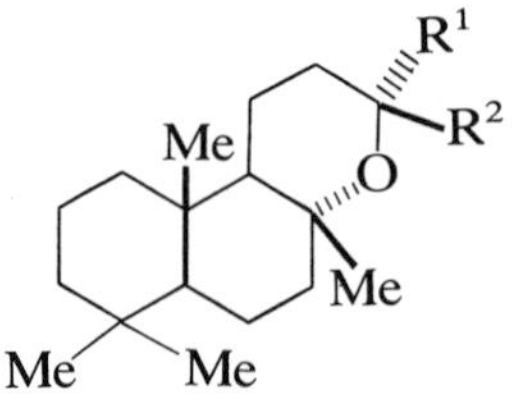

(**74**) R^1 = COOH, R^2 = Me
(**75**) R^1 = Me, R^2 = COOH

Decarboxylation occurs when most chroman carboxylic acids are heated under vacuum[77] or in the presence of copper chromite or quinoline, although decarbonylation (loss of carbon monoxide) sometimes occurs under these conditions.[23,29,30] Dicarboxylic acids can sometimes be selectively monodecarboxylated.[34] For instance, 2,4,4,-trimethylchroman-2,7-dicarboxylic acid (**76**) loses its 7-carboxyl first on heating with copper chromite and quinoline for 20 min.[29]

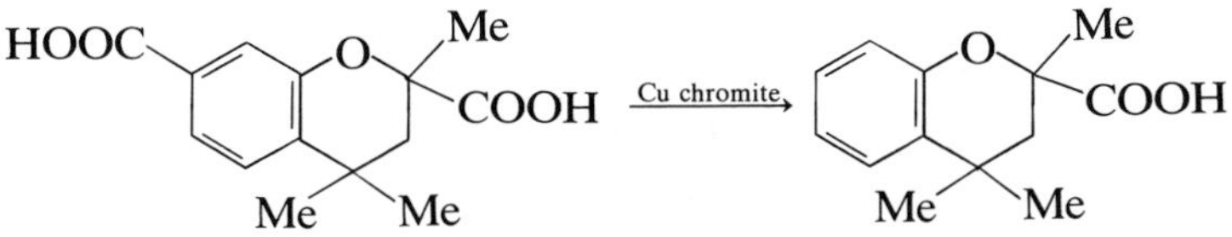

(76) $\xrightarrow{\text{Cu chromite}}$ (77)

α-Alkylcarboxylic acids such as (**77**) are particularly difficult to decarboxylate[28] but are decarbonylated by heating with concentrated sulfuric acid.[23]

Carboxylic acids have been reduced to the primary alcohol with lithium aluminum hydride,[20,23,39,98] and esters have been converted into aldehydes by means of diisobutylaluminum hydride.[18]

A carboxylic ester (**78**) was subjected to a Grignard reaction and gave a high yield of the expected tertiary alcohol (**79**).[95]

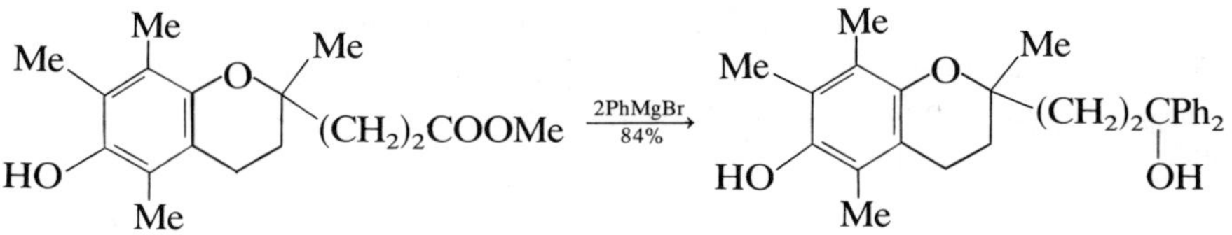

(78) $\xrightarrow[84\%]{\text{2PhMgBr}}$ (79)

Conversion of a carboxyl group into a ketone through the acyl chloride and its reaction with a cadmium alkyl bromide has been described.[93] A cyclic ketone (**81**) was formed when 3-(5,8-dimethoxy-2,2-dimethylchroman-6-yl)propionic acid (methyltetrahydrobraylinic acid, **80**) was heated briefly with sulfuric acid.[76]

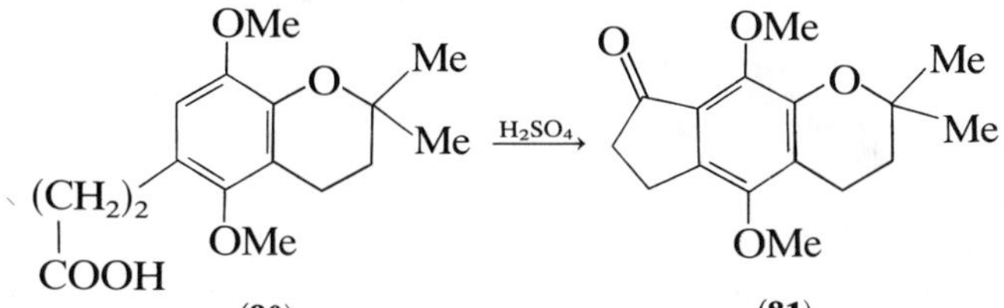

(80) $\xrightarrow{H_2SO_4}$ (81)

Chroman carboxylic acids or esters undergo degradation when oxidized with cerium(IV) sulfate,[4,123] the product being a 1,4-quinone from 6-hydroxychromans; thus 3-(6-hydroxy-2,5,7,8-tetramethylchroman-2-yl)-propionic acid (**82**) gives the quinonoid acid (**83**). Hydrogenolysis of ethyl chroman-2-carboxylate is claimed to occur when this ester is hydrogenated in the presence of Raney nickel and 25% aqueous sodium hydroxide with the formation of 4-(2-hydroxyphenyl)butanoic acid.[130]

(82) $\xrightarrow{Ce(SO_4)_3}$ (83)

An attempt to nitrate 8-methylchroman-6-carboxylic acid with fuming nitric acid at room temperature resulted in the displacement of the carboxyl group and the formation in 30% yield of 8-methyl-6-nitrochroman.[101]

2. Carboxamides

Little work has been reported on the reactions of chroman carboxamides. Zaugg, De Net, and Michaels[80] converted 4-phenylchroman-4-carboxamide (**84**) into 4-amino-4-phenylchroman (**85**) by a Hofmann degradation, but a similar reaction on 2,4,4-trimethylchroman-2-carboxamide (**86**) gave the alcohol (**87**) as an unexpected product.[35]

(84) $\xrightarrow{KOBr}$ (85)

(86) $\xrightarrow{NaOBr}$ (87)

Reduction of chroman-2-carboxamide with lithium aluminum hydride gave 2-aminomethylchroman.[66]

3. Carbonitriles

Chroman carbonitriles have been hydrolyzed to the carboxylic acid[108,111] and the carboxylate ester[112] and reduced to the primary amine.[107] The malononitrile derivatives (such as **88**) have been converted into tricyclic compounds (**89**) by heating with polyphosphoric acid,[116] although an earlier attempt to effect this kind of reaction had failed.[115]

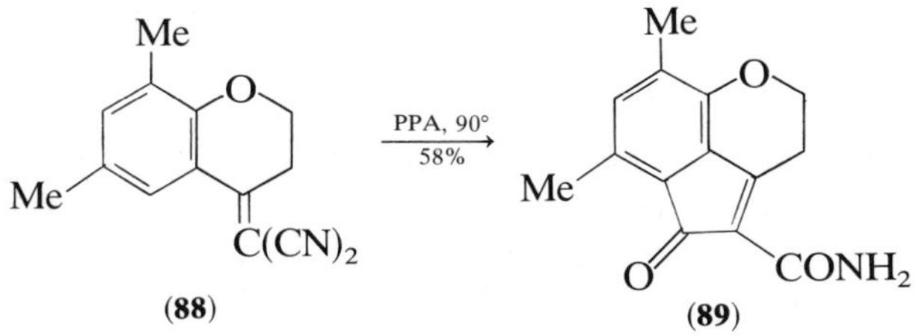

(88) $\xrightarrow[58\%]{\text{PPA, }90^\circ}$ (89)

VII. BIOLOGICAL SIGNIFICANCE

Analogs of tocopherols which contain a carboxyl group either at C-2 or in a chain attached to C-2 have been studied as antioxidants.[134,135] Some of these compounds are metabolites of the tocopherols,[136] while others have been studied as model compounds.[137–139] For example, the catabolism of α-tocopherol (**90**) in the rat is postulated to involve the carboxylic acids shown in Scheme 4.[136]

A new antioxidant was discovered in 1974 at the Hoffmann–LaRoche Laboratories in New Jersey. It was observed that, of the many compounds related to α- and γ-tocopherol which were screened for antioxidant activity, several chromans containing a carboxyl group attached directly or indirectly at C-2 showed a high level of activity.[17] Of these, 6-hydroxy-2,5,7,8-tetramethylchroman-2-carboxylic acid (Trolox C, TC, **91**) was chosen as the most promising compound. Some of its physical properties were determined,[140] and it was subjected to extensive testing as an antioxidant in various tests and in comparison with compounds which are well-established antioxidants. These studies showed that the carboxylic acid was an efficient antioxidant of

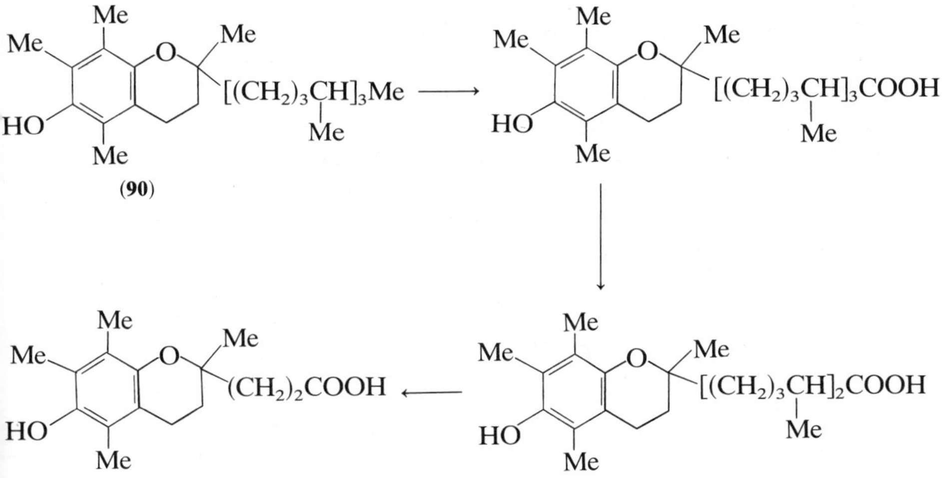

(90)

Scheme 4. Some steps in the catabolism of α-tocopherol.

corn oil, peanut oil, soybean oil, olive oil, palm oil, lard, and citrus oils.[140–142]

The carboxylic acid (**91**) has a synergistic effect on the antioxidant activities of ascorbic acid and ascorbyl palmitate. It was shown that this was due to the stabilizing effect of ascorbic acid on the chroman which is slowly decomposed by natural materials such as corn oil, cottonseed oil, and peanut oil.[140] The toxicity of the compound when administered orally to mice, rats, and rabbits is low, the lethal dose for 50% of the animals (LD_{50}) being greater than 1500 mg/kg.[140]

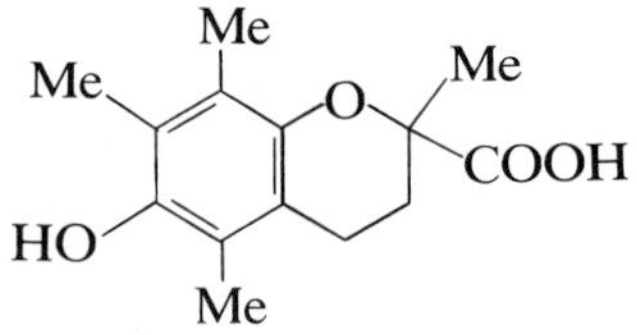

(**91**)

Olcott and Lin's theory of antioxidant action[117] when applied to these compounds postulates that the free-radical one-electron oxidation product of the antioxidant is the actual agent and that this combines with radicals produced during antioxidation. The observations made on the action of the chroman carboxylic acid, including its synergy with ascorbic acid, are in agreement with the theory.[141] The compound inhibits soybean lipoxidase but not prostaglandin biosynthesis in bovine microsomes.[143]

One of the most effective hypolipidemic drugs (used to lower abnormally high levels of lipids in blood) is ethyl 2-(4-chlorophenoxy)-2-methylpropionate (clofibrate, **92**).[144] An attempt to prepare related compounds that contained an asymmetric center showed that several chroman-2-carboxylic acids possessed promising activity.[61] Further studies showed that ethyl 6-chlorochroman-2-carboxylate (**93**) had a potency comparable with clofibrate,[145,146] but in some tests the parent acid (**94**) showed high activity.[147] Replacing chlorine by cyclohexyl gave a compound (**95**) which was comparable with clofibrate in lowering elevated levels of triglycerides in blood.[128] Clofibrate and the cyclohexyl (**95**) and phenyl (**96**) chromans reduced the level of cholesterol in blood serum of normal rats.[148]

Several chromanacetic acids and their esters were screened for anti-inflammatory activity, but none was found.[92]

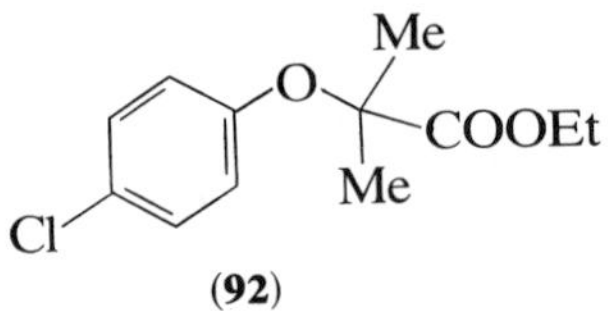

(**92**)

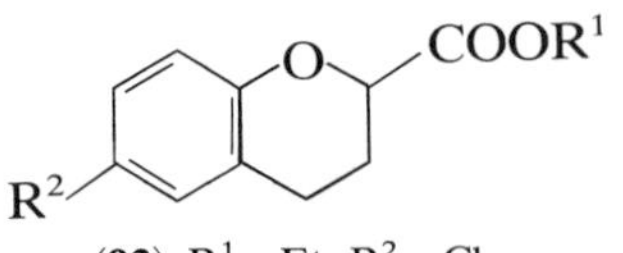

(**93**) $R^1 = Et$, $R^2 = Cl$
(**94**) $R^1 = H$, $R^2 = Cl$
(**95**) $R^1 = Et$, $R^2 = C_6H_{11}$
(**96**) $R^1 = Et$, $R^2 = Ph$

A compound which is claimed to prevent and reverse the genetic tropical disease sickle-cell anemia was described in a short paper as 4-(2,2-dimethyl-chroman-6-yl)butanoic acid (**97**).[149] Although this serves as a noncovalent inhibitor of polymerization of sickle deoxyhemoglobin, the techniques used to assess this property should be treated cautiously.[150]

(**97**)

Certain types of sulfonamides are known to be useful in the treatment of some forms of diabetes mellitus.[151] Sulfonamides based on chromans possess hypoglycemic activity, for example, derivatives of chroman-8-carboxamide (such as **98**)[152–155] or 6-chlorochroman-8-carboxamide (for example, **99**).[156]

(**98**) R = H or CONH–cyclohexyl

(**99**)

VIII. TABLES OF COMPOUNDS

TABLE 2. CHROMAN-2-CARBOXYLIC ACIDS AND THEIR DERIVATIVES

Mol. formula	Substituents 2	3	4	Other positions	mp or bp (mm)	Ref.
$C_{10}H_6Cl_4O_3$	Cl	Cl	Cl_2		—	
	Acyl chloride				—	69
	Methyl ester				113.5–114	69
$C_{10}H_9ClO_3$				6-Cl	152–153	61, 67
	Ethyl ester				111 (0.04)	61
$C_{10}H_{10}O_3$					98.5–100	26, 61, 66, 67
					93–96	61
	Ethyl ester				156–159 (6)	130
					124–125 (1)	65
					116–117 (0.5)	61
	Amide				125–126	66, 157
$C_{11}H_{12}O_4$	OMe				124.5–125	57
	Methyl ester				71–71.5	57
					142 (2)	57

TABLE 2. (*Contd.*)

Mol. formula	Substituents 2	3	4	Other positions	mp or bp (mm)	Ref.
$C_{11}H_{12}O_4$	Me			6-OH	189.5–192	17
$C_{12}H_{14}O_3$[a]				6,8-Me_2	99	158
$C_{13}H_{16}O_3$	Me		Me_2		172	29
					167–170	34, 35
	Amide				140–142	35
$C_{13}H_{16}O_4$	Me			7,8-Me_2	167.5–168.5	17
$C_{13}H_{16}O_4$				5,7,8-Me_3	208.5–210	17
$C_{14}H_{18}O_3$	Me		Me_2	7-Me	148–149	23, 28, 29
	Amide				148	23
$C_{14}H_{18}O_4$	Me			6-OH-5,7,8-Me_3	190–192	17
					189–195	140
	(2R)-Form				162–163	17
	(2S)-Form				161–162.5	17
	Methyl ester				158.5–161.5	17
	Ethyl ester				124–126	17
	O-Acetate				165.5–167	17
	Amide				220–220.5	17
$C_{14}H_{24}O_3$	Me			5,5,8a-Me_3	123	108
$C_{15}H_{20}O_3$	Me		Me_2	7-Et	123	31
$C_{15}H_{20}O_4$	Me			7-Bu^t	215–220	17
$C_{15}H_{20}O_4$	Et			5,7,8-Me_3	210.5–213	17
$C_{16}H_{14}O_3$				6-Ph	198–199	128
	Ethyl ester				49–50	128
					220–224 (0.2)	128
$C_{16}H_{14}O_4$				6-PhO	111–112	128
	Ethyl ester				190–195 (0.4)	128
$C_{16}H_{18}O_4$	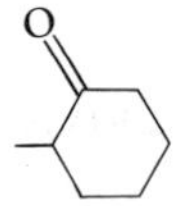				—	
	Methyl ester				110–111[b]	56
					109–110[b]	56
$C_{16}H_{20}O_3$				6-Cyclohexyl	168–169	128
	Ethyl ester				72–73	128
$C_{16}H_{20}O_4$	Me		Me_2	Bz-Ac-7-Me	169.5–170.5	127
$C_{16}H_{26}O_3$				6-Cyclohexyl perhydro	—	
	Ethyl ester				175–179 (0.2)	128
$C_{17}H_{24}O_4$	Me			5,7-Pr^i_2	187.5–190	17
$C_{18}H_{24}O_3$		Me[c]		6-Cyclohexyl-8-Me	158	24
$C_{21}H_{24}O_4$				6-$PhCH_2O$-5,7,8-Me_3	154.5–155	17

[a] Structure uncertain. [b] Stereoisomers. [c] May be 2-Me-3-COOH isomer.

TABLE 3. CHROMAN-3-CARBOXYLIC ACIDS AND THEIR DERIVATIVES

Mol. formula	Substituents 2	3	4	Other positions	mp or bp (mm)	Ref.
$C_{11}H_{11}BrO_4$		Br	OMe		—	
	Methyl ester				82	68
$C_{11}H_{12}O_4$	OMe				—	
	Methyl ester				103 (0.3)	88
$C_{11}H_{12}O_4$				7-MeO	149	44
	(+)-Form				146–149	32
$C_{13}H_{16}O_4$	Me, OEt				—	
	Ethyl ester				Oil	89
$C_{14}H_{18}O_4$	Me			6-OH-5,7,8-Me_3	210–212	54
	Ethyl ester				99	54
	O-Acetate				199	54
	Ethyl ester *O*-acetate				76–77	54
$C_{14}H_{18}O_5$	Me, OH			6-OH-5,7,8-Me_3	—	
	Ethyl ester *O*-acetate				136–137	54
$C_{18}H_{20}O_4$	Ph, OEt				—	
	Ethyl ester				96–98	89
$C_{18}H_{24}O_3$	Me[a]			6-Cyclohexyl	158	24

[a] May be 3-Me-2-COOH isomer.

TABLE 4. CHROMAN-4-CARBOXYLIC ACIDS AND THEIR DERIVATIVES

Mol. formula	Substituents 2	3	4	Other positions	mp or bp (mm)	Ref.
$C_{10}H_{10}O_3$					66	118
	Amide				194	118
$C_{11}H_{11}ClO_3$	Me			6-Cl	150–152	61
$C_{11}H_{12}O_4$				7-MeO	78	48
$C_{11}H_{12}O_4$				8-MeO	110	48
$C_{11}H_{12}O_5$		OH		7-MeO	—	
	Ethyl ester				97	48
$C_{11}H_{12}O_5$		OH		8-MeO	185–194	64
					197	48
	Ethyl ester				85	48, 64
	Ethyl ester *O*-acetate				94	48
$C_{12}H_{14}O_5$				6,7-$(MeO)_2$[a]	134	49, 52, 159
	Monohydrate				90–91	49
	Methyl ester				60	159
$C_{12}H_{14}O_6$		OH		6,7-$(MeO)_2$[b]	189	52
	Ethyl ester				106	49
	cis-DL-Form				161	102
	cis-DL-Form ethyl ester				108	45
	cis-DL-Form ethyl ester *O*-acetate				—	45
	trans-DL-Form				156–157	102
$C_{13}H_{16}O_3$	Me_2		Me		124–125	30
$C_{16}H_{14}O_3$			Ph		151–152	80, 81
	Acetic mixed anhydride				—	81
	Methyl ester				160–163 (1.4)	80
	Ethyl ester				175–178 (2.5)	80

	2-Dimethylaminoethyl ester, $HCl \cdot O.5H_2O$		214–215d	80
	2-Diethylaminoethyl ester, HCl		194–195	80
	2-Diisopropylaminoethyl ester, HCl		135–136	80
	2-Pyrrolidinoethyl ester, $HCl \cdot O.5H_2O$		205–206d	80
	2-Piperidinoethyl ester, $HCl \cdot O.5H_2O$		203–204	80
	2-(4-Methylpiperazino)ethyl ester, 2HCl		242–244d	80
	2-Dimethylaminopropyl ester, HCl		200–201	80
	2-Pyrrolidinopropyl ester, HCl		199–200	80
	2-Piperidinopropyl ester, HCl		218–220	80
	2-Hexamethyleneiminopropyl ester, HCl		223–224[c]	80
			182–184[c]	80
	2-(*N*-Methylpiperid-2-yl)ethyl ester		222–226 (0.1)	80
	(*N*-Methylpiperid-3-yl)methyl ester, HCl		233–234	80
	(*N*-Methylpiperid-4-yl)ester, HCl		215–217	80
	3α-Tropanyl ester		131–132	80
	Amide		180–182	80, 84
	N-Cyclopropylamide		Oil	84
	N-Cyclobutylamide		140–141	84
	N-*n*-Butylamide		Oil	84
	N-Cyclohexylamide		85–87	84
	N-Benzylamide		87–88	84
	Morpholide		127–128	86
	Piperidide		106–107	86
	Pyrrolidide		104–105	86
$C_{17}H_{15}BrO_3$	CH_2Br	Ph	—	
	cis Form		150–151	82
	cis Form methyl ester		106–107	82
	trans Form		217–218	82
	trans Form, methyl ester		111–112	82

TABLE 4. (*Contd.*)

Mol. formula	Substituents 2	3	4	Other positions	mp or bp (mm)	Ref.
$C_{17}H_{16}O_4$	CH_2OH		Ph		—	
	cis Form				176–177	85
	trans Form				189–190	85
$C_{18}H_{18}O_3$			Ph		195–196	87
	Methyl ester				108–109	87
	$Me_2NCH_2CH_2$ ester, HCl				162–163	87
	$Et_2NCH_2CH_2$ ester, HCl				196–197	87
	Amide				206–208	87
	4-Methylpiperazide				159–161	87
$C_{18}H_{17}ClO_3$			Ph	6-Cl	184–185	87
	Methyl ester				121–123	87
	Amide				184–185	87
	N,N-Dimethylamide				115–117	87
	Pyrrolidide				140–142	87
	4-Methylpiperazide				163–165	87
$C_{18}H_{18}O_4$			Ph	6-OH	—	
	Methyl ester				156–157	87
$C_{19}H_{20}O_4$			Ph	6-MeO	124–126	87
	N,N-Dimethylamide				85–86	87
	Pyrrolidide				109–110	87
	4-Methylpiperazide				120–121	87
$C_{24}H_{23}NO_3$	CH_2NHCH_2Ph		Ph		—	
	Methyl ester hydrochloride, trans form				229–230d	82

[a] Netoric acid. [b] Hydroxynetoric acid. [c] Diastereoisomers.

TABLE 5. CHROMAN-Bz-CARBOXYLIC ACIDS AND THEIR DERIVATIVES

Mol. formula	Substituents 2	3	4	Other positions	mp	Ref.
$C_{10}H_9BrO_3$				6-Br-8-COOH	167–168	155
$C_{10}H_9ClO_3$				5-Cl-8-COOH	161–163	155
$C_{10}H_9ClO_3$				6-Cl-8-COOH	159–160	153, 154
	Acyl chloride				96–97	153, 154
	N-[$(CH_2)_2$–C_6H_4–SO_2NH_2] amide				221–222	153, 154
	N-[$(CH_2)_2$–C_6H_4–$SO_2NHCONH$–C_6H_{10}–Me] amide				190–191	153, 154
	N-[$(CH_2)_2$–C_6H_4–$SO_2NHCONH$–C_6H_{10}–Me] amide, K salt				198–200	153, 154
	N-[$(CH_2)_2$–C_6H_4–Me] amide				97–99	156
	N-[$(CH_2)_2$–C_6H_4–Et] amide				58–59	156
$C_{10}H_{10}O_3$				5-COOH	—	
	Ethyl ester				—	21
$C_{10}H_{10}O_3$				6-COOH	—	
	Methyl ester				—	63
	Ethyl ester				—	21
$C_{10}H_{10}O_3$				7-COOH	—	
	Methyl ester				—	63

TABLE 5. (*Contd.*)

Mol. formula	Substituents 2	3	4	Other positions	mp	Ref.
$C_{10}H_{10}O_3$				8-COOH	—	
	N-[$(CH_2)_2$—C_6H_4—SO_2NH_2]amide				194–195	153, 154
	N-[$(CH_2)_2$—C_6H_4—$SO_2NHCONHBu^n$]amide				136–137	153, 154
	N-[$(CH_2)_2$—C_6H_4—$SO_2NHCONH$—C_6H_{11}]amide				198–199	153, 154
$C_{10}H_{10}O_4$			OH	6-COOH	—	
	Methyl ester				—	63
$C_{10}H_{10}O_4$			OH	7-COOH	—	
	Methyl ester				—	63
$C_{10}H_{16}O_3$				Perhydro 6-COOH[a]	—	
	Methyl ester				—	62
$C_{10}H_{16}O_3$				Perhydro 7-COOH[a]	—	
	Methyl ester				Oil	62, 63
$C_{10}H_{16}O_4$			OH	Perhydro 7-COOH	—	
	Methyl ester				Oil	63
	N-($C_{17}H_{20}N_3O_2S^b$)amide				118–120	152
$C_{11}H_{12}O_3$				5-Me-6-COOH	207	160
$C_{11}H_{12}O_3$				6-Me-8-COOH	—	
	N-[$(CH_2)_2$—C_6H_4—SO_2NH_2]amide				215–216	153, 154

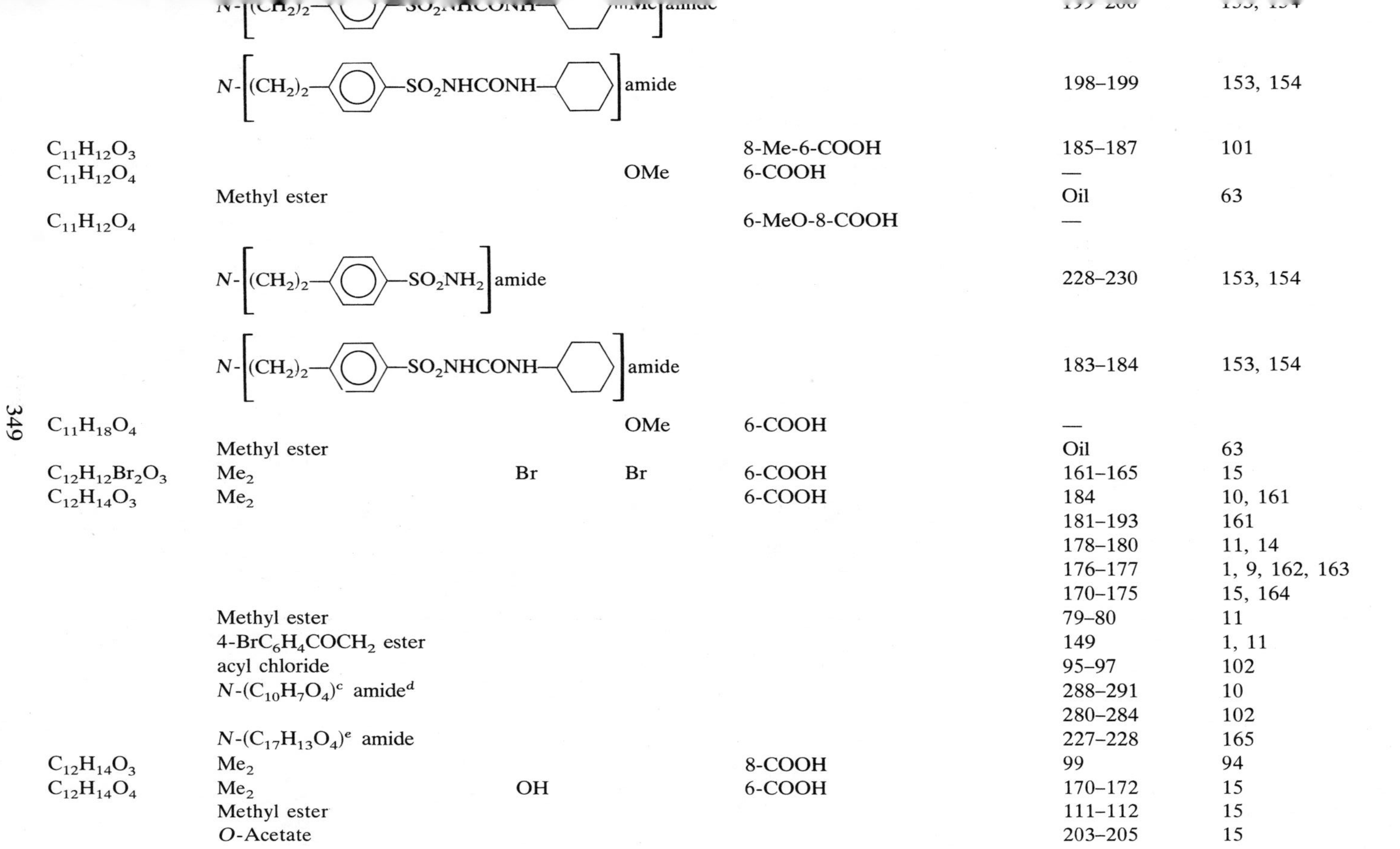

	N-[$(CH_2)_2$–⟨benzene⟩–$SO_2NHCONH$–⟨cyclohexyl⟩–Me] amide				[illegible]	[illegible]
	N-[$(CH_2)_2$–⟨benzene⟩–$SO_2NHCONH$–⟨cyclohexyl⟩] amide				198–199	153, 154
$C_{11}H_{12}O_3$				8-Me-6-COOH	185–187	101
$C_{11}H_{12}O_4$			OMe	6-COOH	—	
	Methyl ester				Oil	63
$C_{11}H_{12}O_4$				6-MeO-8-COOH	—	
	N-[$(CH_2)_2$–⟨benzene⟩–SO_2NH_2] amide				228–230	153, 154
	N-[$(CH_2)_2$–⟨benzene⟩–$SO_2NHCONH$–⟨cyclohexyl⟩] amide				183–184	153, 154
$C_{11}H_{18}O_4$			OMe	6-COOH	—	
	Methyl ester				Oil	63
$C_{12}H_{12}Br_2O_3$	Me_2	Br	Br	6-COOH	161–165	15
$C_{12}H_{14}O_3$	Me_2			6-COOH	184	10, 161
					181–193	161
					178–180	11, 14
					176–177	1, 9, 162, 163
					170–175	15, 164
	Methyl ester				79–80	11
	4-$BrC_6H_4COCH_2$ ester				149	1, 11
	acyl chloride				95–97	102
	N-$(C_{10}H_7O_4)^c$ amide[d]				288–291	10
					280–284	102
	N-$(C_{17}H_{13}O_4)^e$ amide				227–228	165
$C_{12}H_{14}O_3$	Me_2			8-COOH	99	94
$C_{12}H_{14}O_4$	Me_2	OH		6-COOH	170–172	15
	Methyl ester				111–112	15
	O-Acetate				203–205	15

TABLE 5. (*Contd.*)

Mol. formula	Substituents				mp	Ref.
	2	3	4	Other positions		
$C_{12}H_{14}O_4$	Me_2			5-OH-6-COOH[f]	173.5–174.5	46, 50, 93, 166–168
					165–167d	169
	Methyl ester				Oil	12
					57–58	169
	O-Acetate				165d	169
					151	93
$C_{12}H_{14}O_5$	Me_2	OH		7-OH-6-COOH	195	170
$C_{12}H_{14}O_5$	Me_2	OH	OH	6-COOH	214–216	15
	Di-*O*-acetate				202–207	15
$C_{12}H_{14}O_5$	Me_2			5,8-$(OH)_2$-6-COOH	179	133
$C_{13}H_{16}O_4$	Me_2			7-MeO-6-COOH	116	37, 111
	Methyl ester				112–113	111
$C_{13}H_{16}O_4$	Me_2			5-MeO-6-COOH	118	12, 170
	Methyl ester					12
$C_{13}H_{16}O_4$	Me_2			5-MeO-7-COOH	184–185	171
	Methyl ester				83	171
$C_{13}H_{16}O_4$	Me_2			8-MeO-6-COOH	212–214	172
$C_{13}H_{16}O_5$	Me_2	OH		5-MeO-6-COOH	145	170
$C_{13}H_{16}O_5$	Me_2			7-OH-5-MeO-6-COOH	94.5–95.5	173
$C_{13}H_{16}O_5$	Me_2			5-OH-8-MeO-6-COOH	203–204	77, 172
	Methyl ester				111–113	77, 133, 172

$C_{13}H_{17}NO_4$	Me_2		5-NH_2-8-MeO-6-COOH	—	
	Methyl ester			122	131
$C_{15}H_{24}O_3$	Bu^i	Me	$4a,5,6,8a$-H_4-7-COOH	—	
	Methyl ester			—	114
$C_{15}H_{24}O_3$	Bu^i	Me	$4a,5,8,8a$-H_4-7-COOH	—	
	Methyl ester			—	114
$C_{15}H_{26}O_4$	Bu^i	Me	7-OH-7-COOH perhydro	—	
	Methyl ester			—	114
$C_{17}H_{22}O_3$	Me_2		8-$CH_2CH{=}CMe_2$-6-COOH	120–121	14
$C_{17}H_{24}O_4$	Me_2		8-$(CH_2)_2CMe_2OH$-6-COOH	170–171	13, 14, 174
$C_{27}H_{44}O_3$	Me, $C_{16}H_{33}$[g]		6-COOH	—	
	Methyl ester			Oil	16

[a] More than one conformer.

[b] (structure: SO$_2$NH, N, N, SPr^i)

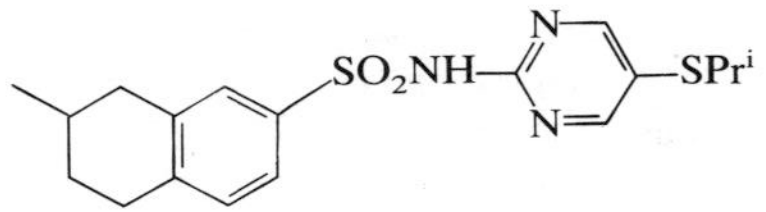

[c] (structure: HO, Me, O, O, OH)

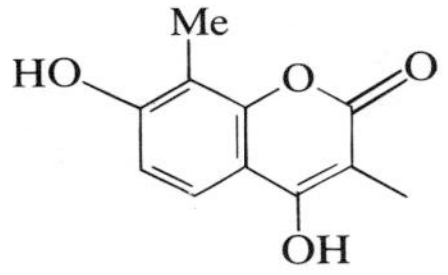

[d] Cyclonovobiocic acid.

[e] (structure: HO, Me, O, O, OCH$_2$Ph)

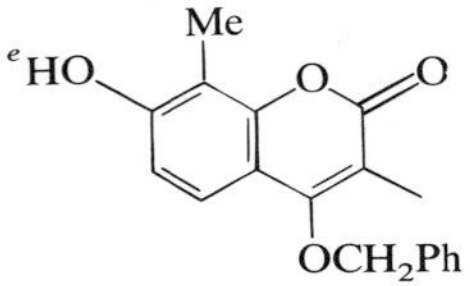

[f] β-Dihydrotubaic acid.

[g] $[(CH_2)_3CHMe]_3Me$.

TABLE 6. CHROMAN SIDE-CHAIN CARBOXYLIC ACIDS AND THEIR DERIVATIVES

Mol. formula	Substituents: 2	3	4	Other positions	mp or bp (mm)	Ref.
$C_{11}H_{10}O_3$				6-COCOOH	—	
	Ethyl ester				215 (15)	96
$C_{11}H_{12}O_3$				6-CH_2COOH	—	
	Methyl ester				Oil	92
$C_{11}H_{12}O_3$	CH_2COOH				—	
	Ethyl ester				Oil	58
$C_{11}H_{12}O_3$			CH_2COOH		91–92	90
	Methyl ester				Oil	90
	N-Dimethylamide				Oil	90
	N-Benzylamide				201–202	90
	N-Piperidide				98	90
	N-Cyclohexylamide				147–148	90
$C_{11}H_{14}O_4$	CH_2COOH			5,6,7,8-H_4-5-oxo	—	
	Methyl ester				53–54	100
$C_{12}H_{14}O_3$				6-CHMeCOOH	98	92
	Methyl ester				119–124 (0.2)	92
$C_{12}H_{14}O_3$	Me			6-CH_2COOH	—	175
$C_{12}H_{14}O_4$	$(CH_2)_2$COOH, OH				107	26
	Lactone				106	26
$C_{13}H_{12}O_6$				6-OH-7-$COCH_2$COCOOH	—	
	Ethyl ester				149–150	176, 177
$C_{13}H_{12}O_6$				7-OH-6-$COCH_2$COCOOH	—	
	Ethyl ester				132	55
$C_{13}H_{14}O_4$				6-$CO(CH_2)_2$COOH	118.5–119.5	96
	Methyl ester				63	96
					235 (17)	96
$C_{13}H_{14}O_4$				8-Ac-6-CH_2COOH	—	
	Methyl ester				154 (0.2)	92
[illegible]	Me			6-CH_2COOH	165	[illegible]

$C_{13}H_{16}O_3$	Me_2			8-CH_2COOH	181–182	178
	Methyl ester				87	178
$C_{13}H_{16}O_3$	Et			6-CH_2COOH	88–90	178a
$C_{13}H_{16}O_3$				6-$(CH_2)_3COOH$	57.5–58.5	96
$C_{13}H_{16}O_4$	Me_2			7-OH-6-CH_2COOH	111–113	96
	Lactone				141–142	96
$C_{13}H_{18}O_4$	CH_2COOH			7,7-Me_2-5,6,7,8-H_4-5-oxo	—	
	Methyl ester				118–119 (0.2)	100
$C_{14}H_{14}O_6$				6-MeO-8-$COCH_2COCOOH$	—	
	Ethyl ester				106–108	22
$C_{14}H_{16}O_4$				6-$CO(CH_2)_3COOH$	93–94	96
	Ethyl ester				44–45	96
	Semicarbazone				255	96
$C_{14}H_{16}O_4$				8-Ac-6-CHMeCOOH	—	
	Methyl ester				156–160 (0.2)	92
$C_{14}H_{18}O_3$				6-$(CH_2)_4COOH$	88–88.5	96
					238 (13)	96
	Amide				103–104	96
$C_{14}H_{18}O_3$	Et			6-CHMeCOOH	66–67	178
$C_{14}H_{18}O_4$	Me_2			7-MeO-6-CH_2COOH	130	91, 111
	Amide				126	91
$C_{14}H_{18}O_4$	Me_2			8-MeO-5-CH_2COOH	—	
	Methyl ester				51.5–52	112
$C_{14}H_{18}O_4$	CH_2COOH			6-OH-7,8-Me_2	131–132.5	17
$C_{14}H_{20}O_4$	CH_2COOH	=O	Me	4*a*,7-Me_2-4*a*,5,6,8*a*-H_4	—	126
	Methyl ester				72.5–73.5	126
$C_{14}H_{22}O_4$	CH_2COOH	OH	Me	4*a*,7-Me_2-4*a*,5,6,8*a*-H_4	—	126
	Methyl ester				89.5–90.5	126
$C_{15}H_{18}O_4$	Me_2			7-MeO-6-CH=CHCOOH	141–142	179
$C_{15}H_{18}O_4$	Me_2			6-C(COOH)=$CHCH_2OH$	—	
	Lactone				143–144	113
$C_{15}H_{18}O_5$	Me_2			7-MeO-6-$CH_2COCOOH$	127–130	111
$C_{15}H_{18}O_5$	Me_2	OH		5-Me-8-CH=CHCOOH	187.5–188.5	97
	Methyl ester				121–122	97
$C_{15}H_{18}O_6$	Me_2			5,7-$(MeO)_2$-6-COCOOH	169d	72

TABLE 6. (*Contd.*)

Mol. formula	Substituents 2	3	4	Other positions	mp or bp (mm)	Ref.
$C_{15}H_{18}O_6$	Me_2	OH	OMe	5-OH-6-CH=CHCOOH	—	
	Methyl ester				117–118	55, 71
$C_{15}H_{20}O_3$	Me_2			$6\text{-}(CH_2)_3COOH$	—	149
$C_{15}H_{20}O_4$	Me, CH_2COOH			6-OH-5,6,7-Me_3	172–174	17, 18
	(2S)-Form				145.5–148.5[h]	18
					127–129[h]	18
	(2S)-α-Methylbenzylamine salt				164–166.5	18
	(2R)-Form				—	18
	(2S)-Form, *O*-acetate				125–126.5	18
$C_{15}H_{20}O_4$	Me_2			7-MeO-6-$(CH_2)_2COOH$	99–100	179
$C_{15}H_{20}O_5$	Me_2			7,8-$(MeO)_2$-5-CH_2COOH	166–168	59
$C_{15}H_{26}O_3$	Me			5,5,8a-Me_3-perhydro	145–150 (0.2)	20
	Methyl ester				85 (0.01)	20
$C_{16}H_{17}NO_4$	Me_2			6-C(CN)=C(OMe)COOH	—	
	Ethyl ester				63–63.5	113
$C_{16}H_{18}O_4$	Me, C≡CCOOH			6-OH-5,7,8-Me_3	—	
	(±)-Form				195–196	95
	(2R)-(−)-Form				205–207	95
	(2S)-(+)-Form				205–207	95
	(2R)-(−)-Quinine salt				149–150	95
	(2S)-(+)-Quinine salt				—	95
$C_{16}H_{18}O_5$	Me_2			6-$C_5H_5O_4$[i]	160–161	113
$C_{16}H_{20}O_4$	Me, CH=CHCOOH			6-OH-5,7,8-Me_3	—	
	(2R)-(+)-Form				121–122	95
	(2S)-(−)-Form				121.5–122.5	95
	(2R)-(+)-Methyl ester *O*-acetate				108.5–109.5	95
	(2S)-(−)-Methyl ester *O*-acetate				108–109	95
$C_{16}H_{20}O_5$	Me_2			5,7-$(MeO)_2$-8-CH=CHCOOH	178d	73
$C_{16}H_{20}O_5$	Me_2			5,8-$(MeO)_2$-6-CH=CHCOOH	155	76

$C_{16}H_{20}O_5$	Me_2	6-C(COOH)=C(OMe)CH_2OH		
	Lactone		142–143	113
$C_{16}H_{21}ClO_4$	$(CH_2)_2COOH$, Me	6-OH-5-CH_2Cl-7,8-Me_2	—	
	O-Acetate		—	5
	Amide *O*-acetate		153–155	5
	Methyl ester *O*-acetate		99–100	5
$C_{16}H_{22}O_4$	$(CH_2)_2COOH$, Me	6-OH-5,7,8-Me_3	173	4, 5, 8, 17
	O-Acetate		159.5–160.5	4, 180, 181
	Methyl ester		94	45, 95
	(2R)-Form		151–162	95
	(2S)-Form		161–162	123
	Ethyl ester *O*-acetate		—	5
	Amide *O*-acetate		153–155	5
$C_{16}H_{22}O_5$	Me_2	5,7-$(MeO)_2$-6-$(CH_2)_2COOH$	108.5	73
$C_{16}H_{22}O_5$	Me_2	5,7-$(MeO)_2$-8-$(CH_2)_2COOH$	155–157	53
$C_{16}H_{22}O_5$	Me_2	5,8-$(MeO)_2$-6-$(CH_2)_2COOH$	102	76
$C_{16}H_{22}O_5$	Me_2	8-MeO-6-EtCO-5-CH_2COOH	—	
	Methyl ester		116.5–117	112
$C_{16}H_{22}O_6$	Me, $(CH_2)_2COOH$	6-OH-7,8-$(MeO)_2$-5-Me	Oil	7, 125
	O-Acetate		Oil	7, 125
	Methyl ester *O*-acetate		158–160 (0.05)	7
$C_{17}H_{18}O_4$		8-$C_6H_7O^j$-6-CH_2COOH	141	92
$C_{17}H_{22}O_4$		8-$C_6H_{11}O^k$-6-CH_2COOH	108	92
$C_{17}H_{22}O_6$	Me_2	6-Ac-7,8-$(MeO)_2$-5-CH_2COOH	147–148	59
$C_{18}H_{20}O_4$		8-$C_6H_7O^j$-6-CHMeCOOH	—	92
$C_{19}H_{28}O_4$	Me, $C_6H_{11}O_2{}^l$	6-OH-5,7,8-Me_3	—	134
	Methyl ester		145 (0.001)	3, 134
	O-Acetate		119–123	3
	Methyl ester *O*-acetate		158 (0.03)	3
$C_{20}H_{30}O_5$	Me_2	7-OH-5-MeO-8-CH(C_5H_{11})COOH	—	
	Methyl ester		93–95	60
$C_{21}H_{28}O_5$	Me_2	5,7-$(MeO)_2$-8-Me_2C=$CHCH_2$-6-CH=CHCOOH	169–171	182
$C_{21}H_{30}O_5$	Me_2	5,7-$(MeO)_2$-8-$CEtMe_2$-6-CH=CHCOOH	168–169	74

TABLE 6. (*Contd.*)

Mol. formula	Substituents: 2	3	4	Other positions	mp or bp (mm)	Ref.
$C_{21}H_{32}O_5$	Me_2			5,7-$(MeO)_2$-8-$CEtMe_2$-6-$(CH_2)_2COOH$	126–127	74
$C_{22}H_{16}O_4$	Me, CH_2COOH			6-$PhCH_2O$-5,7,8-Me_3	—	
	Methyl ester				51–52	18
$C_{24}H_{26}O_6$	Me_2			6-$C_{13}H_{13}O_5{}^m$	—	
	Lactone				168.5–169	113
$C_{24}H_{38}O_4$	Me, $C_{11}H_{21}O_2{}^n$			6-OH-5,7,8-Me_3	193–195 (0.001)	3, 134
	Methyl ester				178–179 (0.001)	3, 134, 136
	O-Acetate				214–218 (0.1)	3
$C_{26}H_{38}O_4$	Me, $C_{13}H_{21}O_2{}^o$			6-OH-5,7,8-Me_3	97–101	18
	Methyl ester				64–66	18
	Ethyl ester				205–210 (0.05)	18
	Methyl ester *O*-Acetate				44–46	18
$C_{27}H_{28}O_7$	Me_2			7-OH-6-$C_{16}H_{15}O_5{}^p$	—	
	Lactone				234–235	164
$C_{27}H_{30}O_6$	Me_2			6-$C_{16}H_{17}O_5{}^q$	—	
	Lactone				243–245	164
$C_{27}H_{30}O_6$	Me_2			6-$C_{16}H_{17}O_5{}^r$	—	
	Lactone				187–189	164
$C_{27}H_{30}O_7$	Me_2			7-OH-6-$C_{16}H_{17}O_5{}^q$	—	
	Lactone				257–259	164
$C_{27}H_{30}O_7$	Me_2			7-OH-6-$C_{16}H_{17}O_5{}^r$	—	
	Lactone				257–258	129a, 164
$C_{27}H_{30}O_8$	Me_2			7-OH-6-$C_{16}H_{17}O_6{}^s$	—	
	Lactone				234–235	164
$C_{28}H_{36}O_6$	Me_2			6-$C_{17}H_{19}O_5{}^t$	—	
	Lactone				133–135	113
$C_{29}H_{32}O_7$	Me_2			7-MeO-6-$C_{17}H_{17}O_5{}^u$	—	
	Lactone				108–110	164

$C_{29}H_{34}O_6$	Me_2	$6\text{-}C_{18}H_{21}O_5$[v]	—	
	Lactone		106–108	113
$C_{29}H_{34}O_7$	Me_2	$7\text{-MeO-}6\text{-}C_{17}H_{19}O_5$[w]	—	
	Lactone		157–162	164
$C_{29}H_{34}O_7$	Me_2	$7\text{-MeO-}6\text{-}C_{17}H_{19}O_5$[x]	194–198	164
$C_{29}H_{48}O_4$	Me, $C_{16}H_{31}O_2$[y]	6-OH-5,7,8-Me_3	—	
	Methyl ester		181–183 (0.005)	3
	Methyl ester *O*-acetate		180 (0.001)	3
	O-Acetate		223–227 (0.001)	3, 136
$C_{29}H_{49}NO_3$	Me, $C_{16}H_{33}$[z]	6-$CH_2CH(NH_2)COOH$	185–187	2
$C_{30}H_{50}O_3$	$(CH_2)_7COOH$, C_8H_{17}[aa]	6-Bu^t-8-Me	—	
	Methyl ester		204–205 (0.04)	25

[a] COOH; $NMeC_{18}H_{37}$

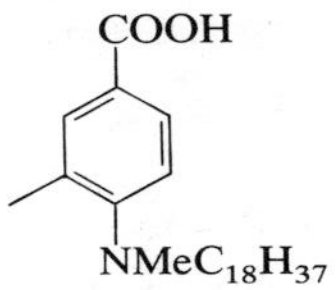

[b] SO_2OH; $NMeC_{18}H_{37}$

[c] O; Cl

[d] OMe; Me

[e] NO_2; OMe

[f] NH_2; OMe

[g] NHCOPh; OMe

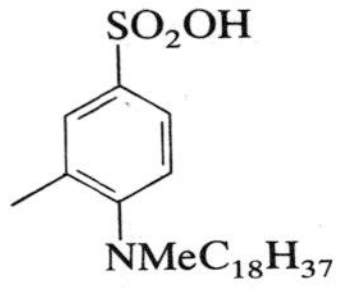

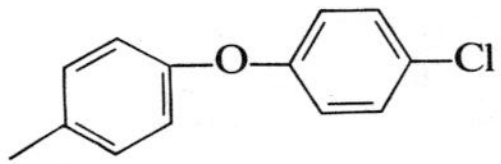

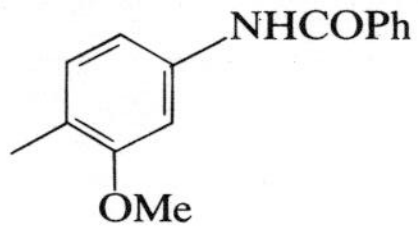

[h] Polymorphic.

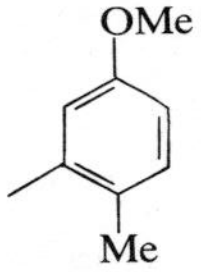

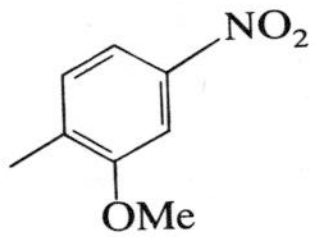

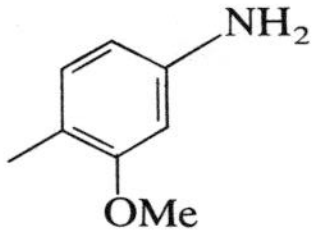

[i] OMe; O; O; OH

[j] O

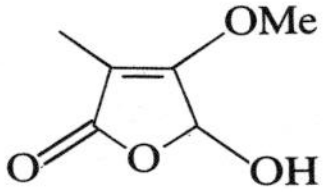

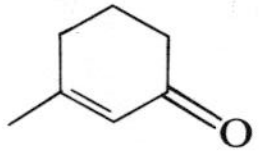

[k] OH

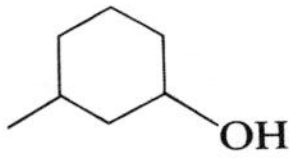

[l] $(CH_2)_3CHMeCOOH$.

TABLE 6. (*Contd.*)

Mol. formula	Substituents				mp or bp (mm)	Ref.
	2	3	4	Other positions		

Footnotes to Table 6 continued.

[m] CH=CC(OMe)=CC_6H_4—4—OMe. (with OH on the first C and COOH on CC_6H_4)

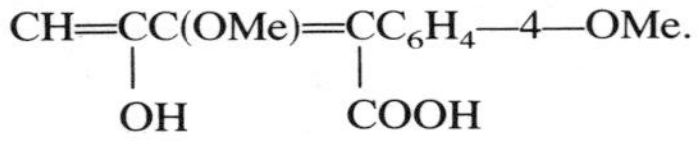

[n] $[(CH_2)_3CHMe]_2COOH$.

[o] $[(CH_2)_3CH{=}CMe]_2(CH_2)_2COOH$.

[p] OH; CH=CC=C; HO; COOH; O; Pr^i

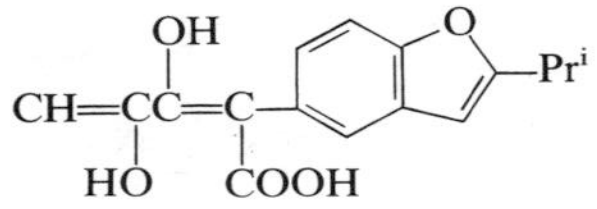

[q] OH; CH=CC=C; HO; COOH; O; Me; Me

[r] OH; CH=CC=C; HO; COOH; $CH_2CH{=}CMe_2$; OH

[s] OH; CH=CC=C; HO; COOH; OH; CH_2 CH—CMe_2; O

[t] OMe; C=CC=CH; OH; COOH; O; Me; Me

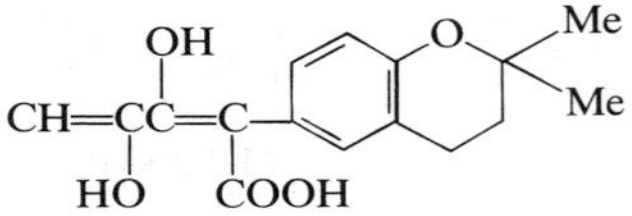

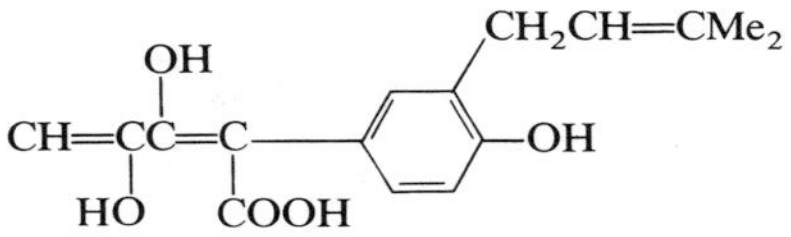

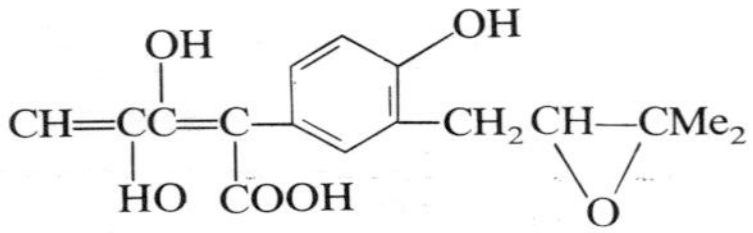

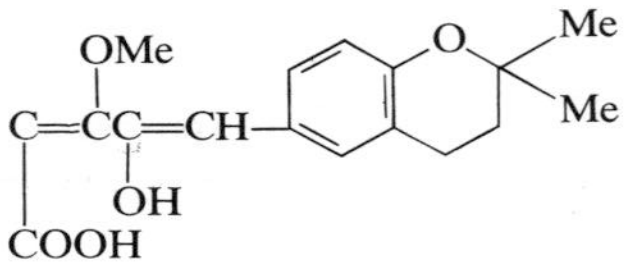

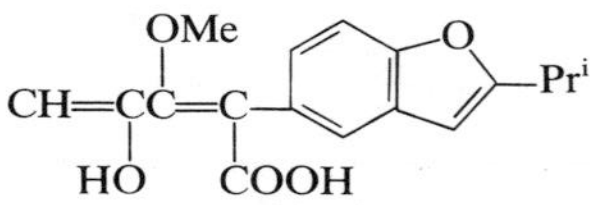

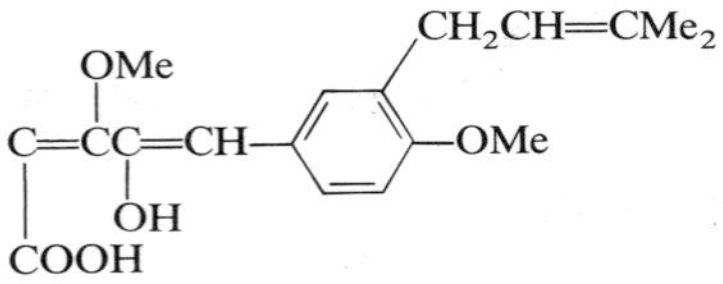

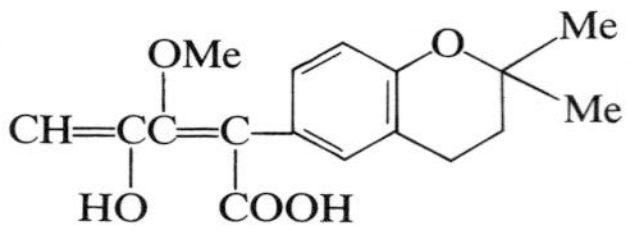

[u] OMe; CH=CC=C; HO; COOH; O; Pr^i

[v] OMe; C=CC=CH; OH; COOH; $CH_2CH{=}CMe_2$; OMe

[w] OMe; CH=CC=C; HO; COOH; O; Me; Me

[x] OMe; $CH_2CC{=}C$; O; COOH; O; Me; Me

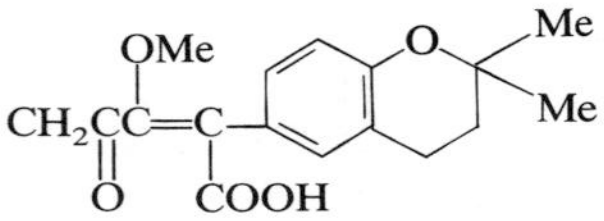

[y] $[(CH_2)_3CHMe]_3COOH$. [z] May be a mixture of this and the 2-C_8H_{17}-3-$(CH_2)_7COOMe$ isomer.

[aa] $[(CH_2)_3CHMe]_3Me$.

TABLE 7. DIHYDRONAPHTHO[1,2-*b*]PYRAN CARBOXYLIC ACIDS AND THEIR DERIVATIVES

Mol. formula	Substituents					mp	Ref.
	2	4	5	6	Other positions		
$C_{16}H_{16}O_4$	Me_2		COOH	OH		—	
	Methyl ester					104	51
$C_{16}H_{20}O_4$	Me_2		COOH	OH	7,8,9,10-H_4	—	
	Methyl ester					87	51
$C_{17}H_{18}O_3$	COOH, Me	Me_2			2,3-H_2	—	
	Monohydrate					195–198	36
	Methyl ester monohydrate					184	36
$C_{20}H_{22}O_5$	Me, $(CH_2)_2COOH$		Me	OH	3,4-H_2	—	
	Methyl ester *O*-acetate					104.5–106	6

TABLE 8. DIHYDRONAPHTHO[2,1-*b*]PYRAN CARBOXYLIC ACIDS AND THEIR DERIVATIVES

Mol. formula	Substituents: 2	3	Other positions	mp or bp (mm)	Ref.
$C_{15}H_{14}O_3$			8-CH_2COOH	165–166	178
$C_{16}H_{16}O_3$			8-CHMeCOOH	162–164	178a
$C_{17}H_{18}O_4$	COOH	Me, OEt	3,4-H_2	—	
	Ethyl ester			101	89
$C_{19}H_{30}O_4$		COOH, Me	Perhydro-6-oxo-4*a*,7,7,10*a*-Me_4	184–185	124
	Methyl ester			155–156.5	124
$C_{19}H_{32}O_3$		COOH[a], Me	Perhydro-4*a*,7,7,10*a*-Me_4	97–98	40, 119, 120, 124
	Methyl ester			85–86	119, 122, 124
$C_{19}H_{32}O_3$		COOH[a], Me	Perhydro-4*a*,7,7,10*a*-Me_4	152–154	121, 122
	Methyl ester			99–100	122
$C_{19}H_{32}O_4$		COOH, Me	Perhydro-6-OH-4*a*,7,7,10*a*-Me_4	—	
	Methyl ester			112–113	124
	Methyl ester *O*-acetate			Oil	124
$C_{20}H_{34}O_3$		CH_2COOH[b], Me	Perhydro-4*a*,7,7,10*a*-Me_4	130–131[a]	19
				127–129[c]	39
				122.5–124	40
				116–118[a]	19, 183
				113–115[d]	39
	Methyl ester			109–111	19, 132
				86–87[a]	40
				64–65	19
				Oil	39
				115–116[a]	19, 39, 40
				106–107	39
	N,*N*-Dimethylamide			102–103	19
				185 (0.4)	19, 183
$C_{21}H_{32}O_5$	Pr^iCO		Perhydro-4*a*-CHO-7,-10*a*-Me_2-7-COOH	179–180	184
	Methyl ester			102–104	184
$C_{22}H_{20}O_4$	COOH	Ph, OEt	3,4-H_2	—	
	Ethyl ester			110–113	89

[a] C-3 Epimers. [b] (3S) Epimer. Also called (13S) 8α,13-epoxylabdan-15-oic acid. [c] [illegible] [d] [illegible]

TABLE 9. CHROMAN POLYCARBOXYLIC ACIDS AND THEIR DERIVATIVES

Mol. formula	Substituents 2	3	4	Other positions	mp	Ref.
$C_{11}H_{10}O_6$	OH, COOH				—	
	Diethyl ester				60–61	88
$C_{12}H_{12}O_5$	COOH	COOH		6-Me	247.5	24
$C_{12}H_{12}O_5$			CH_2COOH, COOH		199–200	70
$C_{13}H_{11}NO_5$	CH(CN), COOH (CH bonded to COOH)				—	
	Diethyl ester				92.5–93[a]	65
					82–83.5[a]	65
$C_{13}H_{11}NO_5$	COOH	CHCOOH (CH bonded to CN)			—	
	Diethyl ester				87.5–88.5	65
$C_{13}H_{14}O_5$	COOH	COOH		5,7-Me_2	220–235	24
$C_{14}H_{16}O_5$	COOH, Me		Me_2	6-COOH	266	29
$C_{14}H_{16}O_5$	COOH, Me		Me_2	7-COOH	261	29, 31
$C_{14}H_{16}O_5$	$CH(COOH)_2$		Me	7-Me	—	98
	Diethyl ester				—	98
$C_{14}H_{16}O_5$	COOH, Me		Me_2	8-COOH	198	34
$C_{16}H_{20}O_5$	COOH	COOH		6-Bu^t-8-Me	—	
	Cis				245	24
	Trans				245	24
	Anhydride				—	24
$C_{17}H_{14}O_5$				8-Ph-5,6-$(COOH)_2$	—	
	Dimethyl ester				105	99
$C_{17}H_{16}O_5$				4*a*,7-dihydro-8-Ph-5,6-$(COOH)_2$	—	
	Dimethyl ester				110	99

TABLE 9. (*Contd.*)

Mol. formula	Substituents 2	3	4	Other positions	mp	Ref.
$C_{18}H_{22}O_9$	Me_2			7-MeO-6,8-$(COOH)_2$-5-$OCMe_2COOH$	—	
	6-Methyl ester				200d	78
$C_{19}H_{20}O_5$				4*a*,7-dihydro-8-Ph$(CH_2)_2$-5,6-$(COOH)_2$	—	
	Dimethyl ester				90	99
$C_{20}H_{34}O_5$	Me, Et			5,8*a*-Me_2-6-CMe_2CH_2COOH-5-COOH perhydro	148–149	79
$C_{20}H_{34}O_5$	Me, Et			5,8*a*-Me_2-5-CH_2COOH-6-CMe_2COOH perhydro	—	
	Dimethyl ester				54–56	129
$C_{27}H_{30}O_{11}$	Me_2			5-$OCMe_2COOH$-6-COOH-7-MeO-8-$C_{10}H_9O_4$[b]	205d	78
	Dimethyl ester				133.5–134	78

[a] Crystalline modifications. [b] COCO–(3,4-dimethoxyphenyl: ring bearing OMe, OMe)

TABLE 10. CHROMAN CARBAMATES

Mol. formula	Substituents 2	3	4	6	8	mp or bp (mm)	Ref.
$C_{10}H_9NO_3$		=NCOOH				—	
	Ethyl ester					237–238	105
$C_{11}H_{13}NO_3$	Me		NHCOOH			—	
	Ethyl ester					105	104
$C_{11}H_{13}NO_3$	$CH_2NHCOOH$					—	
	Amidine					216–220	157
$C_{12}H_{15}NO_3$	Me_2		NHCOOH			—	
	Ethyl ester					97	104
$C_{13}H_{13}Cl_2NO_3$	Me, $CH{=}CH_2$		NHCOOH	Cl	Cl	—	
	Ethyl ester					170–180 (0.12)	104
$C_{13}H_{15}NO_3$	Me, $CH{=}CH_2$		NHCOOH			—	
	Ethyl ester					75–76	104
$C_{13}H_{17}NO_3$	Me, Et		NHCOOH			—	
	Ethyl ester					68	104
$C_{14}H_{17}NO_3$	Me, $CMe{=}CH_2$		NHCOOH			—	
	Ethyl ester					—	104
$C_{14}H_{17}NO_4$	Me, $CH{=}CH_2$		NHCOOH		OMe	—	
	Ethyl ester					144	104

TABLE 11. CHROMAN CARBONITRILES

Mol. formula	Substituents 2	3	4	Other positions	mp or bp (mm)	Ref.
$C_{10}H_9NO_2$		CN	OH[a]		—	
$C_{12}H_8N_2O$			$=C(CN)_2$		148–150	115
$C_{13}H_{10}N_2O$			$=C(CN)_2$	6-Me	103–106	115
$C_{13}H_{10}N_2O_4S$			$=C(CN)_2$	6-Me-8-SO_3H	—	
	S-Benzylthiouronium salt				234–235.5	115
$C_{13}H_{15}NO$	Me_2			6-CH_2CN	55	113
$C_{14}H_{12}N_2O$			$=C(CN)_2$	5,7-Me_2	137–139	116
$C_{14}H_{17}NO_2$	Me_2			7-MeO-6-CH_2CN	130	91
					127–128	111
$C_{14}H_{17}NO_3$	Me_2			5,7-$(MeO)_2$-6-CN	126–127	110
$C_{14}H_{23}NO$	CN, Me			5,5,8*a*-Me_3-perhydro	86–88 (0.15)	108
$C_{15}H_{25}NO_2$	Bu^i			7-OH-7-CN-perhydro	—	114
$C_{23}H_{25}NO_5$	Me_2			7-MeO-6-COCH(CN)-C_6H_4-OMe	131	111
$C_{23}H_{25}NO_5$	Me_2			7-MeO-6-CH(CN)CO-C_6H_3(OMe)-OMe	129–130	111

[a] See Section IV, 1.

TABLE 12. MISCELLANEOUS CHROMAN DERIVATIVES

Mol. formula	Structure	mp or bp (mm)	Ref.
$C_{15}H_{14}O_6S$	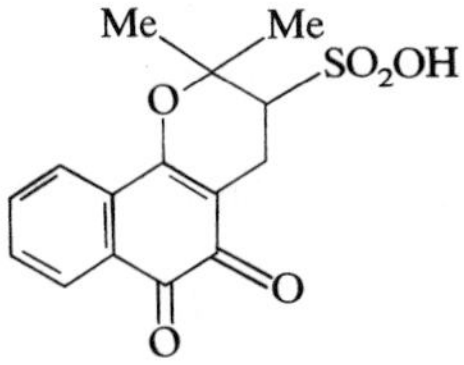	91	185
	Sodium salt	206–207d	185
	Sulfonyl chloride	Ca. 190	185
	Sulfonamide	188–189d	185
	Ethyl ester	162–163	185
$C_{19}H_{32}O_3$[a]		149	186
$C_{20}H_{32}O_4$	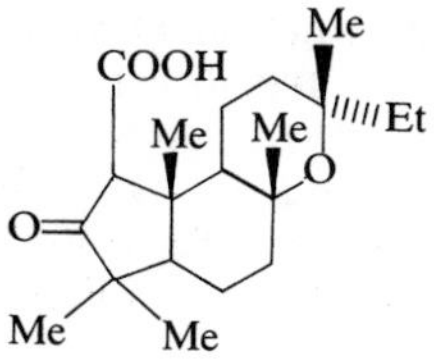	—	
	Methyl ester	80 (0.02)	129
$C_{20}H_{34}O_4$	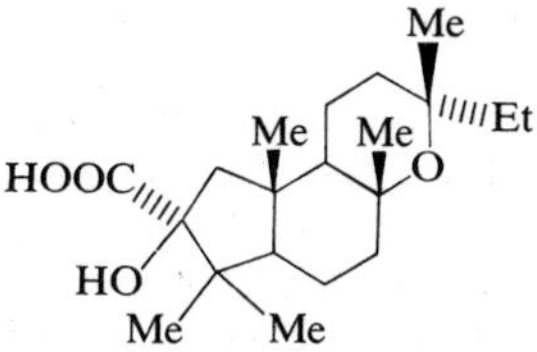	175–176	79
$C_{30}H_{52}N_2O_3S$	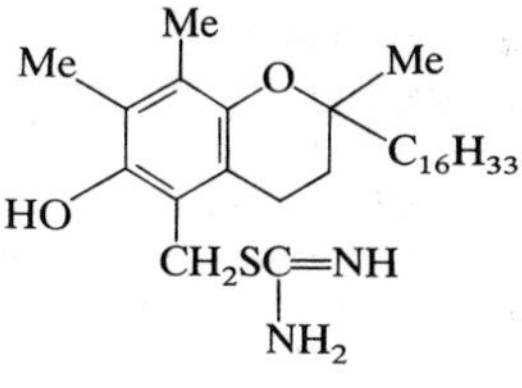	—	
	O-Acetate hydrochloride	130–135d	137

[a] Structure unknown.

IX. REFERENCES

1. W. M. Lauer and O. Moe, *J. Am. Chem. Soc.*, **65,** 289 (1943).
2. A. A. Svishcuk and E. D. Basalkevich, *Ukr. Khim. Zh.* (*Russ. Ed.*), **40,** 770 (1974).
3. J. Weichet, L. Blaha, B. Kakac, and J. Hodrova, *Collect. Czech. Chem. Commun.*, **31,** 2434 (1966).
4. J. Weichet, L. Blaha, and B. Kakac, *Collect. Czech. Chem. Commun.*, **24,** 1689 (1959); *Chem. Listy*, **52,** 722 (1958).
5. K. Murase, J. Matsumoto, K. Tamazawa, K. Takahashi, and M. Murakami, *Rep. Yamanouchi Res. Lab.*, No. 2, 66 (1974).
6. U. Gloor, J. Würsch, H. Mayer, O. Isler, and O. Wiss, *Helv. Chim. Acta*, **49,** 2582 (1966).
7. L. Blaha and J. Weichet, *Collect. Czech. Chem. Commun.*, **30,** 2068 (1965).
8. J. Green and D. McHale, Brit. Patent 947,885 (1964); *Chem. Abstr.*, **60,** 11991 (1964).
9. H. Hoeksema, J. L. Johnson, and J. W. Hinman, *J. Am. Chem. Soc.*, **77,** 6710 (1955).
10. J. W. Hinman, E. L. Caron, and H. Hoeksema, *J. Am. Chem. Soc.*, **79,** 3789 (1957).
11. E. A. Kaczka, C. H. Shunk, J. W. Richter, F. J. Wolf, M. M. Gasser, and K. Folkers, *J. Am. Chem. Soc.*, **78,** 4125 (1956).
12. F. della Monache, G. della Monache, G. B. Marini-Bettolo, M. M. de Albuquerque, J. F. de Mello, and O. G. de Lima, *Gazz. Chim. Ital.*, **106,** 935 (1976).
13. K. Nishikawa and Y. Hirata, *Tetrahedron Lett.*, 2591 (1967).
14. K. Nishikawa, M. Miyamura, and Y. Hirata, *Tetrahedron*, **24,** 2723 (1968).
15. K. Shima, S. Hisada, and I. Inagaki, *Yakugaku Zasshi*, **92,** 1410 (1972).
16. G. D. Daves, H. W. Moore, D. E. Schwab, R. K. Olsen, J. J. Wilezynski, and K. Folkers, *J. Org. Chem.*, **32,** 1414 (1967).
17. J. W. Scott, W. M. Cort, H. Harley, D. R. Parrish, and G. Saucy, *J. Am. Oil Chem. Soc.*, **51,** 200 (1974).
18. J. W. Scott, F. T. Bizzaro, D. R. Parrish, and G. Saucy, *Helv. Chim. Acta*, **59,** 290 (1976).
19. M. Belardini, G. Scudari, and L. Mangoni, *Gazz. Chim. Ital.*, **94,** 829, (1964).
20. A. Caliezi, E. Lederer, and A. Schinz, *Helv. Chim. Acta*, **34,** 879 (1951).
21. S. Houry, S. Gersch, and A. Shani, *Isr. J. Chem.*, **11,** 805 (1973).
22. A. O. Fitton and G. R. Ramage, *J. Chem. Soc.*, 2481 (1964).
23. W. Baker, R. F. Curtis, and J. F. W. McOmie, *J. Chem. Soc.*, 76 (1951).
24. K. Hultzsch, *J. Prakt. Chem.*, **158,** 275 (1941).
25. G. R. Sprengling, *J. Am. Chem. Soc.*, **74,** 2937 (1952).
26. G. Mixich and A. Zinke, *Monatsb. Chem.*, **96,** 220 (1965).
27. G. Baddeley and J. R. Cooke, *J. Chem. Soc.*, 2797 (1958).
28. W. Baker and D. M. Bealy, *J. Chem. Soc.*, 1103 (1940).
29. W. Baker, R. F. Curtis, and J. F. W. McOmie, *J. Chem. Soc.*, 1774 (1952).
30. W. Baker, A. J. Floyd, J. F. W. McOmie, G. Pope, A. S. Wearing, and J. H. Wild, *J. Chem. Soc.*, 2010 (1956).
31. W. Baker, J. F. W. McOmie, and J. H. Wild, *J. Chem. Soc.*, 3060 (1957).
32. A. J. Birch, B. Moore, S. K. Mukerjee, and C. W. L. Bevan, *Tetrahedron Lett.*, 673 (1962).
33. C. W. L. Bevan, A. J. Birch, B. Moore, and S. K. Mukerjee, *J. Chem. Soc.*, 5991 (1964).
34. N. Sugiyama and K. Taya, *Nippon Kagaku Zasshi*, **80,** 673 (1959).
35. W. Webster and D. P. Young, *J. Chem. Soc.*, 4785 (1956).
36. Y. Wang, K. Yamada, and N. Sugitama, *Nippon Kagaku Zasshi*, **86,** 954 (1965).
37. L. F. Bjeldanes and T. A. Geissman, *Phytochemistry*, **8,** 1293 (1969).
38. E. E. Schweizer and D. Meeder-Nycz, in *Chromenes, Chromanones, and Chromones*, G. P. Ellis, Ed., John Wiley, New York, 1977, p. 35.

39. A. G. Gonzalez, B. M. Fraga, M. G. Hernandez, F. Larruga, and J. G. Luis, *Phytochemistry*, **14,** 2655 (1975).
40. P. F. Vlad and L. T. Xuan, *J. Gen. Chem. USSR*, **45,** 1850 (1975).
41. E. E. Schweizer and D. Meeder-Nycz, in *Chromenes, Chromanones, and Chromones*, G. P. Ellis, Ed., John Wiley, New York, 1977, p. 70.
42. I. M. Lockhart, in *Chromenes, Chromanones, and Chromones*, G. P. Ellis, Ed., John Wiley, New York, 1977, p. 301.
43. G. P. Ellis, Ed., *Chromenes, Chromanones, and Chromones*, John Wiley, New York, 1977, p. 970.
44. A. McGookin, A. Robertson, and W. B. Whalley, *J. Chem. Soc.*, 787 (1940).
45. M. Miyano and M. Matsui, *Bull. Chem. Soc. Japan*, **31,** 267, 271 (1958).
46. J. Nickl, *Chem. Ber.*, **91,** 1372 (1958).
47. J. Nickl, *Chem. Ber.*, **92,** 1989, (1959).
48. R. W. H. O'Donnell, F. P. Reed, and A. Robertson, *J. Chem. Soc.*, 419 (1936).
49. A. Robertson and G. L. Busby, *J. Chem. Soc.*, 212 (1936).
50. O. A. Stamm, H. Schmid, and J. Büchi, *Helv. Chim. Acta*, **41,** 2006 (1958).
51. H. Schildknecht, F. Straub, and V. Scheidel, *Justus Liebigs Ann. Chem.*, 1295 (1976).
52. E. P. Clark, *J. Am. Chem. Soc.*, **54,** 2537 (1937).
53. D. B. de Correa, O. R. Gottlieb, and A. P. de Padua, *Phytochemistry*, **14,** 2059 (1975).
54. L. I. Smith and R. B. Carlin, *J. Am. Chem. Soc.*, **64,** 435 (1942).
55. P. Naylor and G. R. Ramage, *J. Chem. Soc.*, 1956 (1960).
56. E. K. Orlova, I. D. Tsvetkova, V. S. Troitskaya, V. G. Vinokurov, and V. A. Zagorevskii, *Chem. Heterocycl. Comp.*, **5,** 321 (1969).
57. V. A. Zagorevskii, I. D. Tsvetkova, E. K. Orlova, and S. L. Portnova, *Chem. Heterocycl. Comp.*, **3,** 627 (1967).
58. M. Pailer and O. Vostrowsky, *Monatsh. Chem.*, **102,** 951 (1971).
59. D. H. R. Barton and J. B. Hendrikson, *J. Chem. Soc.*, 1028 (1956); *Chem. Ind.*, 682 (1955).
60. G. H. Stout and K. D. Sears, *J. Org. Chem.*, **33,** 4185 (1968).
61. D. T. Witiak, E. S. Stratford, R. Nazareth, G. Wagner, and D. R. Feller, *J. Med. Chem.*, **14,** 758 (1971).
62. J. A. Hirsch and G. Schwartzkopf, *J. Org. Chem.*, **39,** 2040 (1974).
63. J. A. Hirsch and G. Schwartzkopf, *J. Org. Chem.*, **38,** 3534 (1973).
64. H. A. Offe and H. Jatzkewitz, *Chem. Ber.*, **80,** 469 (1947).
65. V. A. Zagorevskii, I. D. Tsvetkova, E. K. Orlova, and S. L. Protnova, *Chem. Heterocycl. Comp.*, **5,** 316 (1969).
66. J. Augstein, A. M. Monroe, G. W. H. Potter, and P. Scholfield, *J. Med. Chem.*, **11,** 844 (1968).
67. R. I. Nazareth, T.-D. Sokoloski, D. T. Witiak, and A. T. Hopper, *J. Pharm. Sci.*, **63,** 203 (1974).
68. D. Anker, J. Anarieux, M. Barau-Marszak, and D. Molho, *C.R. Acad. Sci.* (Paris), **274C,** 650 (1972).
69. I. D. Tsvetkova, E. K. Orlova, and V. A. Zagorevskii, *Chem. Heterocycl. Comp.*, **4,** 825 (1968).
70. L. M. Rice, B. S. Seth, and T. H. Zalucky, *J. Heterocycl. Chem.*, **8,** 155 (1971).
71. H. D. Schroeder, W. Bencze, O. Halpern, and H. Schmid, *Chem. Ber.*, **92,** 2338 (1959).
72. F. N. Lahey, *Univ. Queensland Papers, Dep. Chem.*, **1,** No. 20, 2 (1942).
73. A. Robertson and T. S. Subramanian, *J. Chem. Soc.*, 1545 (1937).
74. T. Tomimatsu, H. Hasegawa, and K. Tori, *Tetrahedron*, **30,** 939 (1974).
75. T. Tomimatsu, H. Hashimoto, T. Shinga, and K. Tori, *Chem. Commun.*, 168 (1969).
76. F. A. L. Anet, G. K. Hughes, and E. Ritchie, *Austr. J. Sci. Res.*, **2A,** 608 (1949).
77. J. S. P. Schwarz, A. I. Cohen, W. D. Ollis, E. A. Kaezka, and L. M. Jackson, *Tetrahedron*, **20,** 1317 (1964).

78. M. L. Wolfrom, W. D. Harris, G. F. Johnson, J. E. Mehan, S. M. Moffett, and B. Wilde, *J. Am. Chem. Soc.*, **68,** 406 (1946).
79. P. K. Grant and N. R. Hill, *Austr. J. Chem.*, **17,** 66 (1964).
80. H. E. Zaugg, R. W. De Net, and R. J. Michaels, *J. Org. Chem.*, **26,** 4821 (1961).
81. H. E. Zaugg, R. W. De Net, and R. J. Michaels, *J. Org. Chem.*, **26,** 4828 (1961).
82. H. E. Zaugg, R. W. De Net, and E. T. Kimura, *J. Med. Pharm. Chem.*, **5,** 430 (1962).
83. H. E. Zaugg and R. J. Michaels, *Tetrahedron*, **18,** 893 (1962).
84. H. E. Zaugg, R. W. De Net, and R. J. Michaels, *J. Org. Chem.*, **28,** 1795 (1963).
85. H. E. Zaugg and R. J. Michaels, *J. Org. Chem.*, **28,** 1801 (1963).
86. H. E. Zaugg, F. E. Chadda, and R. J. Michaels, *J. Am. Chem. Soc.*, **84,** 4567 (1962).
87. H. E. Zaugg, J. E. Leonard, R. W. De Net, and D. L. Arendson, *J. Heterocycl. Chem.*, **11,** 797 (1974).
88. F. Korte and K. H. Büchel, *Chem. Ber.*, **93,** 1025 (1960).
89. H. Wamkoff, G. Schorn, and F. Korte, *Chem. Ber.*, **100,** 1296 (1967).
90. J. A. Vide and M. Gut, *J. Org. Chem.*, **33,** 1202 (1968).
91. B. S. Joshi and V. N. Kamat, *J. Chem. Soc., Perkin Trans. I*, 907 (1973).
92. J. Maillard, M. Langlois, P. Delounay, T. V. Van, J. P. Meingan, M. Rapin, R. Morrin, C. Manuel, and C. Mazmanian, *Eur. J. Med. Chem.*, **12,** 161 (1977).
93. R. Huls, *Bull. Classe Sci. Acad. Roy. Belg.*, **39,** 1064 (1953).
94. M. Hallet and R. Huls, *Bull. Soc. Chim. Belg.*, **61,** 33 (1952).
95. H. Mayer, P. Schudel, R. Rüegg, and O. Isler, *Helv. Chim. Acta*, **46,** 650 (1963).
96. G. Chatelus, *Ann. Chim.* (Paris), **4,** 505 (1949); *C.R. Acad. Sci.* (Paris), **224,** 201 (1946).
97. T. Matsumoto, K. Fukui, and M. Nanbu, *Chem. Lett.*, 603 (1974).
98. D. J. Goldsmith and C. T. Helmes, *Synth. Commun.*, **3,** 231 (1973).
99. A. Lebouc, O. Riobe, and J. Delaunay, *C.R. Acad. Sci.* (Paris), **282C,** 357 (1976).
100. S. Danishefsky, G. Koppel, and R. Levine, *Tetrahedron Lett.*, 2257 (1968).
101. G. Brancaccio, G. Lettieri, and R. Viterbo, *J. Heterocycl. Chem.*, **10,** 623 (1973).
102. C. F. Spencer, J. O. Rodin, E. Walton, F. W. Holly, and K. Folkers, *J. Am. Chem. Soc.*, **80,** 140 (1958).
103. N. S. Corby, Brit. Patent 819,827 (1959); *Chem. Abstr.*, **54,** 7738 (1960); U.S. Patent 2,953,576 (1960); *Chem. Abstr.*, **55,** 2322 (1961).
104. R. Marten and G. Müller, *Chem. Ber.*, **97,** 682 (1964).
105. J. W. Clark-Lewis, A. H. Ilsley, and E. J. McGarry, *Austr., J. Chem.*, **29,** 2741 (1976).
106. H. V. Taylor and M. L. Tomlinson, *J. Chem. Soc.*, 2724 (1950).
107. G. B. Bachman and H. A. Levine, *J. Am. Chem. Soc.*, **70,** 599 (1948).
108. M. Stoll, L. Ruzicka, and C. F. Seidd, *Helv. Chim. Acta*, **33,** 1245 (1950).
109. G. P. Ellis and I. L. Thomas, *Progr. Med. Chem.*, **10,** 245 (1974).
110. T. Backhouse and A. Robertson, *J. Chem. Soc.*, 1257 (1939).
111. J. N. Chatterjea, K. D. Banerji, and N. Prasad, *Chem. Ber.*, **96,** 2356 (1963).
112. J. R. Cannon, K. R. Joshi, I. A. McDonald, R. W. Retallack, A. F. Sierakowski, and L. C. H. Wong, *Tetrahedron Lett.*, 2795, (1975).
113. D. W. Knight and G. Pattenden, *Chem. Commun.*, 635 (1976); *J. Chem. Soc., Perkin Trans. I*, 70 (1979).
114. A. J. Birch, P. L. McDonald, and V. H. Powell, *J. Chem. Soc.* (*C*), 1469 (1970).
115. E. Campaigne and C. D. Blanton, *J. Heterocycl. Chem.*, **7,** 1179 (1970).
116. S. W. Schneller, D. R. Moore, and M. A. Smith, *J. Heterocycl. Chem.*, **13,** 123 (1976).
117. H. Olcott and J. S. Lin, *65th Meeting of Am. Oil Chem. Soc., Dallas, Texas*, April 1975, Paper No. 10.
118. G. Fontaine, *Ann. Chim.* (Paris), **3,** 179 (1968).
119. R. Hodges and R. I. Reed, *Tetrahedron*, **10,** 71 (1960).
120. C. Marquez, B. Rodriguez Gonzalez, and S. Valverde-Lopez, *An. Quim.*, **71,** 603 (1975).
121. C. von Carstenn-Lichterfelde, R. Rodriguez, and S. Valverde, *Experientia*, **31,** 757 (1975).

122. J. A. Giles, J. N. Schumacher, S. S. Mims, and E. Bernasek, *Tetrahedron*, **18,** 169 (1962).
123. H. Mayer, P. Schudel, R. Rüegg, and O. Isler, *Helv. Chim. Acta*, **47,** 229 (1964).
124. B. Rodriguez and S. Valverde, *Tetrahedron*, **29,** 2837 (1973).
125. H. Morimoto, I. Imada, and G. Goto, *Justus Liebigs Ann. Chem.*, **729,** 171 (1969).
126. E. W. Colvin, S. Malchenko, R. A. Raphael, and J. S. Roberts, *J. Chem. Soc., Perkin Trans. I*, 1989 (1973).
127. G. E. Svadovskaya, N. E. Kologrivova, and L. A. Kheifits, *Khim. Geterosikl. Soedin. Sb.*, 191 (1970).
128. D. T. Witiak, W. P. Heilman, S. K. Sankarappa, R. C. Cavestri, and H. A. I. Newman, *J. Med. Chem.*, **18,** 934, (1975).
129. P. K. Grant, L. N. Dixon, and J. M. Robertson, *Tetrahedron*, **26,** 1631 (1970).
129a. M. J. Begley, D. R. Gedge, D. W. Knight, and G. Pattenden, *J. Chem. Soc., Perkin Trans. I*, 77 (1979).
130. S. Mitsui, A. Kasahara, T. Oika, and K. Hanya, *Nippon Kagaku Zasshi*, **83,** 581 (1962).
131. B. Danieli, P. Manito, F. Ronchetti, G. Russo, and G. Ferrari, *Experientia*, **28,** 249 (1972).
132. S. Bory and E. Lederer, *Croat. Chem. Acta*, **29,** 157 (1957).
133. S. F. Dyke, W. D. Ollis, M. Sainsbury, and J. S. P. Schwarz, *Tetrahedron*, 20, 1331 (1964).
134. Z. Placer and J. Weichet, *Nahrung*, **12,** 749 (1968).
135. Z. Placer and J. Weichet, *Nahrung*, **12,** 491 (1968).
136. J. Proll, H. Schmandke, and R. Maune, *Int. Z. Vitaminforsch.*, **39,** 299 (1969).
137. H. Schmandke, *Int. Z. Vitaminforsch.*, **35,** 346 (1965); *J. Chromatogr.*, **14,** 123 (1964).
138. M. Nishikimi and L. J. Machlin, *Arch. Biochem. Biophys.*, **170,** 680 (1975).
139. K. Fukuzawa, Y. Suzuki, and M. Uchiyama, *Biochem. Pharmacol.*, **20,** 279 (1971).
140. W. M. Cort, J. W. Scott, M. Aranjo, W. J. Mergens, M. A. Cannalonga, M. Osadea, H. Harley, D. R. Parrish, and W. R. Pool, *J. Am. Oil Chem. Soc.*, **52,** 174 (1975).
141. W. M. Cort, J. W. Scott, and J. H. Harley, *Food Technol.*, **29,** (11) 46 (1975).
142. S. C. Chang, *Shih Pin Kung Yeh* (Hsinchu, Taiwan), **8,** 33 (1976).
143. R. V. Panganamala, J. S. Miller, E. T. Gweba, H. M. Sharma, and D. G. Cornwell, *Prostaglandins*, **14,** 261 (1977).
144. E. C. Witte, *Progr. Med. Chem.*, **11,** 136 (1975).
145. N. J. Lewis, D. R. Feller, G. K. Pooikian, and D. T. Witiak, *J. Med. Chem.*, **17,** 41 (1974).
146. H. A. I. Newman, W. P. Heilman, and D. T. Witiak, *Lipids*, **8,** 378 (1973).
147. T. F. Whayne and D. T. Witiak, *J. Med. Chem.*, **16,** 228 (1973).
148. A. P. Goldberg, W. S. Mellor, D. T. Witiak, and D. R. Feller, *Atherosclerosis*, **27,** 15 (1977).
149. D. E. U. Ekong, J. I. Okogun, V. U. Enyenihi, V. Balogh-Nair, K. Nakanishi, and C. Natta, *Nature*, **258,** 743 (1975).
150. W. N. Poillon and J. F. Bertles, *Biochem. Biophys. Res. Commun.*, **75,** 636 (1977).
151. J. D. H. Slater, *Progr. Med. Chem.*, **1,** 187 (1961).
152. M. Huebner, R. Heerdt, F. H. Schmidt, M. Thiel, and R. Weyer, Brit. patent 1,346,705 (1973); *Chem. Abstr.*, **79,** 5361 (1973).
153. W. Andersen, H. Christensen, F. G. Gronvald, and B. F. Lundt, Brit. Patent 1,314,325 (1973); *Chem. Abstr.*, **79,** 53172 (1973).
154. H. Christensen, B. F. Lundt, F. G. Gronvald, and W. Andersen, Brit. Patent 1,314,324; U.S. Patent 3,803,176; *Chem. Abstr.*, **81,** 4956 (1974).
155. H. Christensen and W. Andersen, Brit. Patent 1,358,684; Ger. Offen. 2,127,236 (1971); *Chem. Abstr.*, **76,** 99520 (1972).
156. V. Hitzel, R. Weyer, and E. Bosies, Ger. Offen. 2,600,513 (1977); *Chem. Abstr.*, **87,** 167760 (1978).
157. J. Augstein, A. M. Monro, and T. I. Wrigley, Brit. Patent, 1,004,468 (1965); *Chem. Abstr.*, **63,** 18036 (1965).

158. E. Adler, H. von Euler, and G. Gie, *Ark. Kemi, Minerol. Geol.*, **16A,** No. 12 (1943).
159. S. Takei, S. Miyajima, and M. Ono, *Ber.*, **65B,** 279 (1932); L. E. Smith and F. B. La Forge, *J. Am. Chem. Soc.*, **52,** 4595 (1930); F. B. La Forge, H. L. Haller, and L. E. Smith, *J. Am. Chem. Soc.*, **53,** 4400 (1931); A. Budenadt and W. McCartenay, *Justus Liebigs Ann. Chem.*, **494,** 17 (1932); A. Budenadt and G. Hilgetag, *Justus Liebigs Ann. Chem.*, **495,** 172 (1932).
160. G. Sarker and D. Nasipuri, *J. Indian Chem. Soc.*, **45,** 200 (1968).
161. H. Hoeksma, E. L. Carron, and J. W. Hinman, U.S. Patent 2,929,821 (1960); *Chem. Abstr.*, **55,** 5534 (1961).
162. J. W. Hinman, H. Hoeksma, E. L. Carron, and W. G. Jackson, *J. Am. Chem. Soc.*, **78,** 1072 (1956).
163. C. A. Bunton, G. W. Kenner, M. J. T. Robinson, and B. R. Webster, *Tetrahedron*, **19,** 1001 (1963).
164. N. Ojima, S. Takenaka, and S. Seto, *Phytochemistry*, **14,** 573 (1975); *ibid.*, **12,** 2527 (1973).
165. F. Hoffmann–LaRoche AG, Belg. Patent 609,859; *Chem. Abstr.*, **58,** 1528 (1964).
166. J. Baudrenghien, J. Jadol, and R. Huls, *Bull. Soc. Roy. Sci. Liège*, **18,** 52, (1949).
167. S. Takei, S. Miyajima, and M. Ono, *Ber.*, **66,** 1826 (1933).
168. H. L. Haller, *J. Am. Chem. Soc.*, **53,** 733 (1931).
169. F. Dallacker and H. van Wersch, *Chem. Ber.*, **108,** 561 (1975).
170. B. F. Burrows, N. Finch, W. D. Ollis, and I. O. Sutherland, *Proc. Chem. Soc.*, 150 (1959).
171. F. Della Monache, G. delle Monache, G. B. Marini-Bettolo, M. M. F. de Albuquerque, J. F. de Mello, and O. G. de Lima, *Gazz. Chim. Ital.*, **106,** 935 (1976).
172. A. Meidell, *Med. Norsk Farm. Selskap*, **27,** 101 (1965).
173. A. J. East, W. D. Ollis, and R. E. Wheeler, *J. Chem. Soc.* (*C*), 365 (1969).
174. K. Nishikawa, M. Miyamura, and Y. Hirata, *Tetrahedron Lett.*, 2597 (1967).
175. G. F. Tamagnone, G. Mascellani, C. Casalini, and M. Fantuz, *Boll. Chim. Farm.*, **116,** 81 (1977).
176. A. H. Wragg and G. P. Ellis, Brit. Patent 1,023,373 (1966); U.S. Patent 3,222,795; *Chem. Abstr.*, **64,** 19567 (1966).
177. A. O. Fitton and G. R. Ramage, *J. Chem. Soc.*, 5426 (1963).
178. G. R. Clemo and N. D. Ghatge, *J. Chem. Soc.*, 4347 (1955).
178a. G. F. Tamagnone and F. de Marchi, Ger Offen. 2,526,089; *Chem. Abstr.*, **84,** 135469 (1976).
179. J. C. Bell and A. Robertson, *J. Chem. Soc.*, 1828 (1936).
180. J. Green and D. McHale, Brit. Patent 949,715 (1964); *Chem. Abstr.*, **60,** 13227 (1964).
181. T. Ichikawa and T. Kato, *Bull. Chem. Soc. Japan*, **41,** 1224 (1968).
182. E. V. Lassak and J. T. Pinhay, *Austr. J. Chem.*, **22,** 2175 (1969).
183. H. Audier, S. Bory, and M. Fetizon, *Bull. Soc. Chim. Fr.*, 1381 (1964).
184. W. Herz and V. Baburao, *J. Org. Chem.*, **36,** 3271 (1971).
185. L. F. Fieser, *J. Am. Chem. Soc.*, **70,** 3232 (1948).
186. D. H. McLean and S. N. Slater, *J. Soc. Chem. Ind.*, **64,** 28 (1945).

CHAPTER X

Spirochromans

S. SMOLINSKI

Institute of Chemistry, Jagiellonian University, Kraków, Poland

I. INTRODUCTION

Significant developments have been made in the chemistry of spirans containing chroman, but no thorough review of these compounds has been published. The chroman system may be spiroannulated at the C-2, C-3, or C-4 atoms of the dihydropyran ring or, if the benzene ring of the chroman is reduced, at C-5, C-6, C-7, or C-8. The benzene ring of the chroman part may be fused to another benzene or to a naphthalene system. The chroman is spiroannulated to another system which may or may not contain one or more hetero atoms. Compounds in which a ring is fused to the dihydropyran ring are excluded from this review, as also are those in which a hetero ring is fused to the benzene ring. The spirochromans containing the chroman component in the form of chromanone, dihydrocoumarin moiety, or hemiacetal grouping are omitted.

Generally the spirans are not a new class of compounds in the chemical sense, but they retain the chemical properties of their parent moieties, the only distinction lying in the geometric aspects. Because of the lack of detailed stereochemical data, this review reports mainly on the synthetic aspects. There are however some papers describing interesting behavior of spirans that are sterically hindered or not very stable and that change their structure when excited. They seem to provide a valuable basis for the synthesis of physiologically active compounds since they represent fixed structures of at least two components oriented more or less orthogonally to each other in which the whole space is filled. One effect is remarkable: a spiro junction located near or at the center of a molecule serves as an antichromophoric group and extinguishes the color of polycyclic systems when they do not bear any chromophoric substituents.

II. 2,2′-SPIROBICHROMAN AND ITS DERIVATIVES

1. 2,2′-Spirobichroman

A. *Preparation*

Borsche first prepared 2,2′-spirobichroman (**3**) by the catalytic hydrogenation of the disalicylideneacetone (**1**) to 1,5-di-*o*-hydroxyphenylpentane-3-one (**2**), which on vacuum distillation loses a molecule of water and yields spiran (**3**).[1] The success of this reaction was due to the facile condensation of *o*-hydroxyaromatic aldehydes with acetone and other ketones.

Mora and Szeki[2] have found that compound (**2**) was easily converted to 2,2′-spirobichroman (**3**) after refluxing with fused sodium acetate in acetic anhydride or with a solution of hydroxylamine hydrochloride in ethanol, or after shaking compound (**2**) in ethanol with palladium–charcoal and hydrogen at 60–70°. Spiran (**3**) has also been prepared by self-condensation of 3,4-dihydrocoumarin (**5**) with the aid of sodium hydride.[3] The product (**6**) lost a molecule of carbon dioxide on hydrolysis to form the final product (**3**).

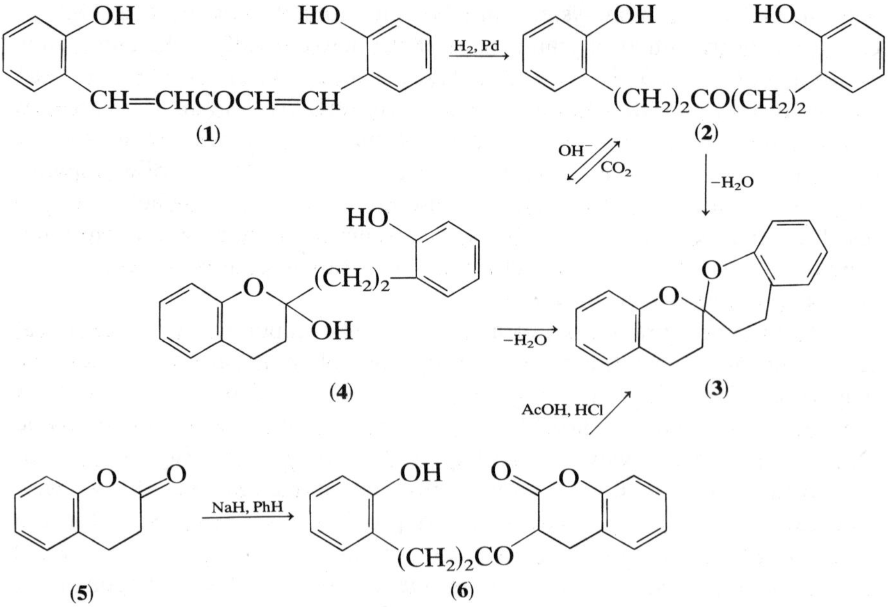

(**1**) (**2**) (**3**) (**4**) (**5**) (**6**)

B. *Properties*

2,2′-Spirobichroman (**3**) forms needles, mp 110°, when crystallized from methanol.[1] It has been reported that spiran (**3**) has no absorption in the uv region.[3]

2. Alkyl-2,2′-spirobichromans

A. *Preparation*

2-Propenylphenol (**7**) when treated with 2 moles DDQ in boiling benzene gave spiran (**8**) which, on catalytic hydrogenation, was reduced to 3-methyl-2,2′-spirobichroman (**9**).[4]

The mechanistic details for this reaction have not yet been elucidated.[5,6] The first step seems to be the formation of a charge-transfer complex between the strong electron acceptor DDQ and the donor phenol.[7] On the other hand Becker[8] has pointed out the importance of the one-electron process in the oxidation of phenols with DDQ in methanol.

Studies of the condensation of *m*- and *p*-cresol with acetone have a long

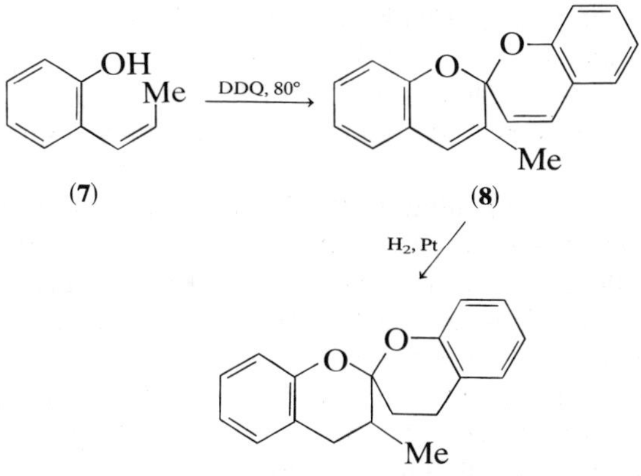

(7) (8) (9)

history.[9–11] The reaction was carried out in the presence of hydrogen chloride[9] or cold, concentrated sulfuric acid.[10,11] The molecular formulas and the spirochroman structures of (**10**) and (**11**) were ascribed ultimately to the compounds after preparing the *m*-cresol product from *m*-cresol and phorone in much better yield than from *m*-cresol and acetone.[12] The condensation was afterwards reinvestigated by Baker and Besly,[13] who confirmed the empirical formulas and the spirochroman structure. Both isomers were prepared[10,11] in yields of about 10%, and purified samples exhibited dimorphism: the spiran (**10**) melted at 128° (after solidification, 136°), and the spiran (**11**) melted at 136° (after solidification, 144°).[13] Italian chemists[14] prepared the spiran (**10**) in about 34% yield by condensation in the presence of acetic acid and hydrochloric acid; the spiran (**11**) was formed in about 10% yield.

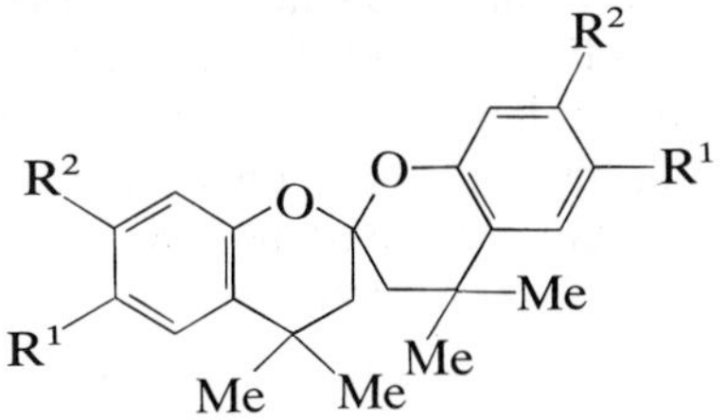

(**10**) $R^1 = H, R^2 = Me$
(**11**) $R^1 = Me, R^2 = H$
(**12**) 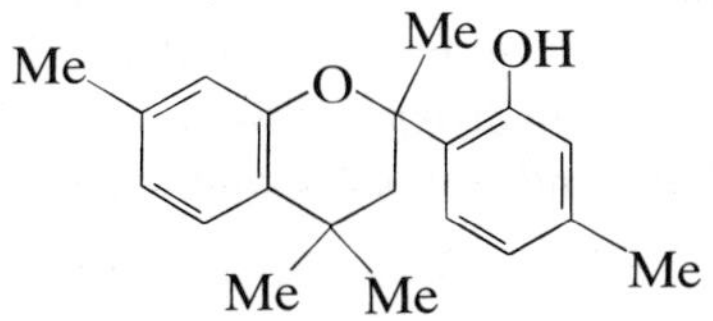

Hydrogenolysis of the *m*-cresol product with acetone gave a mixture which was fractionally distilled to give spiran (**10**) in about 33% yield and 2′-hydroxy-2,4,4,7,4′-pentamethylflavan (**12**) in 38% yield.[15] These results are in agreement with the suggestion originally advanced by Baker and Besly[13] that the spirochroman may have been formed from the flavan.[16] Chazan and Ourisson[17] were successful in preparing the flavan (**12**) and spiran (**10**) by the condensation of acetone with *m*-cresol in the presence of hydrogen chloride or phosphorus oxychloride. The three final products of the condensation of *m*-cresol with acetone in the presence of hydrochloric acid at room temperature were probably stereoisomers of spiran (**10**), the flavan (**12**), and 2,2-bis(2-hydroxy-4-methylphenyl)propane.[18] However in the presence of sulfuric acid at room temperature only compound (**10**) was obtained.[18]

Baker and Besly[13] degraded both the *m*- and *p*-cresol products to phoronic anhydride by oxidation with potassium permanganate in acetic acid and thus proved the partial structure of spirans (**10**) and (**11**), respectively.

Structure determination of spiran (**10**) was facilitated by destructive ozonolysis followed by oxidation with hydrogen peroxide to yield 4,4-dihydroxy-2,2,6,6-tetramethylheptanedioic acid (**13**) which in acidic medium was converted to 4,4,4′,4′-tetramethyl-2,2′-spirobibutyrolactone (**14**) and in neutral or alkaline medium to 4-oxo-2,2,6,6-tetramethylheptanedioic acid (**15**).[19]

$$(\mathbf{10}) \xrightarrow{O_3,\ H_2O_2} \left[HOOC\text{-}CMe_2CH_2C(OH)_2CH_2CMe_2\text{-}COOH \right] (\mathbf{13}) \xrightarrow{H^+} (\mathbf{14})$$

$$(\mathbf{13}) \xrightarrow{-H_2O} HOOC\text{-}CMe_2CH_2C(=O)CH_2CMe_2\text{-}COOH\ (\mathbf{15})$$

The condensation of acetone with meta-alkylated phenols in the presence of hydrogen chloride or concentrated sulfuric acid[16,20] was extended to include *m*-ethylphenol, which was also condensed with phorone (derived from acetone by self-condensation). In the acetone condensation, 7,4′-diethyl-2′-hydroxy-2,4,4-trimethylflavan (**16**) was obtained as an intermediate and was subsequently converted to spiran (**17**).

$$3Me_2CO + 2m\text{-}EtC_6H_4OH \xrightarrow{HCl,\ 40°} (\mathbf{16}) \xrightarrow{H_2SO_4} (\mathbf{17})$$

$$3Me_2CO + 2m\text{-}EtC_6H_4OH \xrightarrow{H_2SO_4\ 25°} (\mathbf{17})$$

$$Me_2C{=}CHCOCH{=}CMe_2 + 2m\text{-}EtC_6H_4OH \xrightarrow{HCl,\ 25°} (\mathbf{17})$$

(18) $\xrightarrow{H_2SO_4,\ 190°}$ (19)

2′-Hydroxy-2,4,4,6,7,4′,5′-heptamethylflavan (**18**), when melted and heated to 190° for 30 sec with sulfuric acid, gave the octamethyl spiran (**19**) and 3,4-dimethylphenol.[16]

B. *Properties*

The spiran (**9**) was obtained as an oil after preparative thin-layer chromatography on silica gel with a mixture of *n*-hexane–benzene (1:1). The uv spectrum in 95% ethanol shows maxima at 272 (ε 3800) and 279 (ε 3650).[4] In the nmr spectrum,[4] the protons of the methyl group appear as a doublet (δ1.16, J6.5 Hz).[4] Its mass spectrum exhibits fragments of *m/e* 266, 251, 159, 145, 131, and 107.

The R_f value of spiran (**10**) is 0.80 using a chromatographic method in a thin layer of alumina and developing with methanol and benzene (1:9).[23] Spiran (**10**) is soluble in ethanol, chloroform, and ether.[14]

Spiran (**17**) from acetone forms colorless needles[16] or rhombohedra,[20] mp 114°, whereas spiran (**17**) from phorone crystallizes as plates, mp 146°.[20] The compound exhibits dimorphism. The tetranitro derivative of both forms gives identical yellow crystals from ethanol.[20]

3. Other 2,2′-Spirobichromans

A. *Preparation*

Hydroxyquinol or its triacetate (**20**) when allowed to react with a warm mixture of acetone, acetic acid, and hydrochloric acid gives spiran (**21**) in about 88% yield.[13,21] The assignment was based on permanganate oxidation of spiran (**21**) to yield phoronic anhydride, thus proving (**21**) to be a derivative of spiran (**3**) and to possess four hydroxyl groups.[13]

The reaction has a general character and can be used for preparing 6,6′ and 7,7′-dihydroxy-4,4,4′,4′-tetramethyl-2,2′-spirobichromans by condensing acetone with hydroquinone and resorcinol, respectively, in the presence of a mineral acid at ambient temperature for an extended

(20) + Me_2CO $\xrightarrow{\text{AcOH, HCl}}$ (21)

period.[21,22] In general the yields obtained from hydroquinone are not very satisfactory, and therefore the use of hydroxyhydroquinone is preferred. Various derivatives of hydroquinone and hydroxyhydroquinone may be used, and any suitable ketone or even unsaturated aliphatic ketones may be employed.[21]

B. *Properties*

The spiran (**21**) forms colorless glistening plates[13,21] that dissolve in aqueous alkalis with production of a green color which soon changes to bright red.[13] Boiling spiran (**21**) with methanol and dimethyl sulfate in the presence of aqueous sodium hydroxide gives the 6,7,6′,7′-tetramethoxy derivative. Refluxing spiran (**21**) with acetic anhydride and anhydrous sodium acetate converts the spiran (**21**) to its tetraacetoxy derivative. Treatment of (**21**) with bromine in acetic acid gives a red product, probably a dibromodiquinone, which after reduction and acetylation forms 8,8′- (and less of 5,5′-) dibromo-6,7,6′,7′-tetraacetoxy-4,4,4′,4′-tetramethyl-2,2′-spirobichroman.[13] When treated with oxidizing agents, such as lead peroxide, silver oxide, or nitric acid, spiran (**21**) gives a red solution due to the formation of an unstable double *o*-quinone.[13] Compound (**21**) gave a bright green ferric chloride reaction.

4. Uses

Many representatives of 6,6′- or 7,7′-dihydroxy and 6,7,6′,7′-tetrahydroxy derivatives of 4,4,4′,4′-tetramethyl-2,2′-spirobichroman have useful properties. They are used as stabilizers for photographic magenta azomethine color emulsions of desirable spectral properties and high fastness to light, temperature, and humidity.[24–27] Polyesters containing the 2,2′-spirobichroman-7,7′-diyl nucleus were claimed to have good film-forming properties and good thermal characteristics useful for photographic silver halide as well as for multilayer elements used in color photography.[22] Hydrocarbons and edible fats and oils are stabilized by addition of 0.001–0.1 wt-% spiran (**21**).[21] It is also reported that epoxy resins of high thermal stability are produced by reacting epichlorhydrin with 7,7′-dihydroxy-4,4,4′,4′-tetramethyl-2,2′-spirobichroman in the presence of alkaline hydroxides.[28] Spiran (**21**) exhibits antioxidative activity in isotactic polypropylene.[29]

Japanese chemists prepared[30] 6,6′-dichloro-, 6,6′-dimethyl-, and 6,6′,8,8′-tetramethyl derivatives of 2,2′-spirobichroman (**3**) by treatment of the dihydrocoumarin (**22**, R = Cl or Me, R′ = H or Me) with boron tribromide. When R = R′ = Me, they obtained the pentanone (**24**) which gives spiran (**23**, R = R′ = Me) by reaction with ethanol–sulfuric acid.

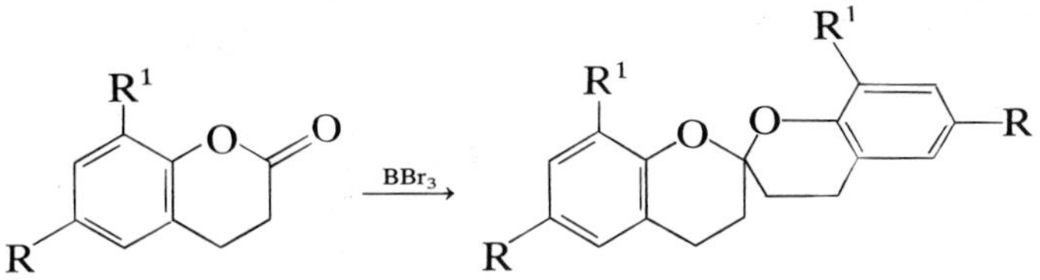

R^1, O, O, R, BBr_3, R^1, O, O, R, R^1, R

(**22**) (**23**)

Me, OH, HO, Me, $(CH_2)_2CO(CH_2)_2$, Me, Me

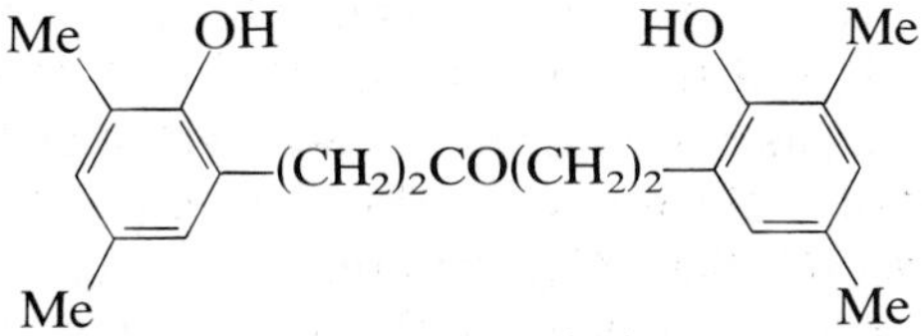

(**24**)

III. CHROMANS SPIROANNULATED AT C-2 WITH HOMOCYCLIC SYSTEMS

1. Spiro[chroman-2,1′-cyclohexanes] and Spiro[chroman-2,1′-cyclopentan]-6-ols

In an attempt to prepare chroman derivatives with tocopherol-like structure, Karrer and Kehrer[31] condensed derivatives of the allyl bromide (**25**) with trimethylhydroquinone (TMHQ) to give 5,7,8-trimethyl-2,2-pentamethylene-6-hydroxychroman (**26**), which was isolated as a crystalline allophanate in 36% yield. The free spiran (**26**) is a viscous oil with intense reducing properties. When the Mannich base 2-dimethylaminomethylcyclohexanone was allowed to stand for 16 hr with methyl iodide, a high

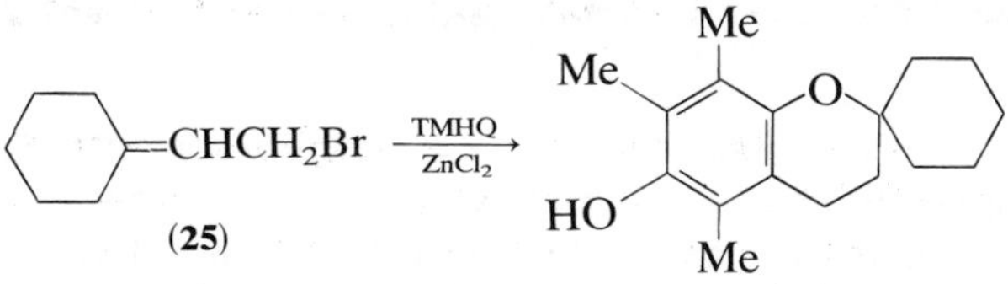

$=CHCH_2Br$, TMHQ, $ZnCl_2$, Me, Me, O, HO, Me

(**25**) (**26**)

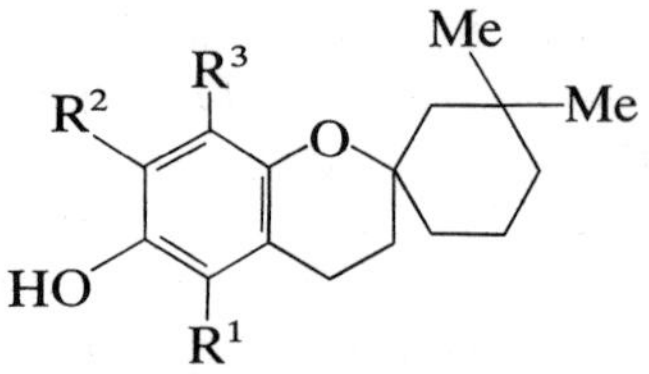

(**27**) $R^1 = R^2 = R^3 = H$
(**28**) $R^1 = R^2 = R^3 = Me$
(**29**) $R^1 = R^3 = H, R^2 = t\text{-}Bu$
(**30**) $R^1 = R^3 = H, R^2 = CMe_2CH_2CMe_3$

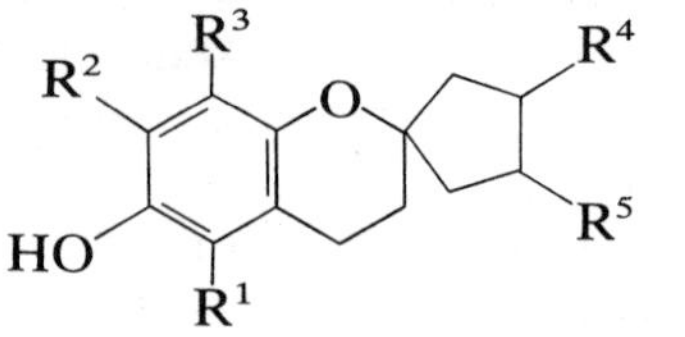

(**31**) $R^1 = t\text{-}Bu, R^2 = R^3 = R^5 = H, R^4 = CHMe_2$

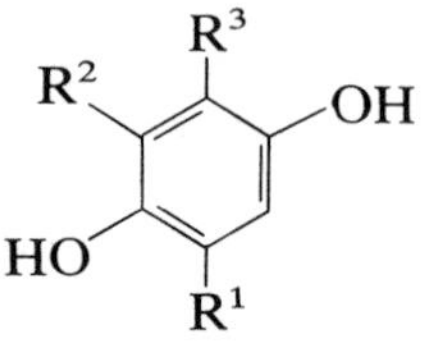

(**32**) 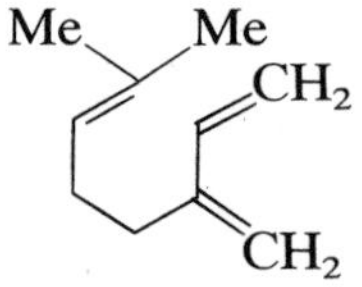 (**33**)

yield of 5,6,7,8-tetrahydrospiro[chroman-2,1′-cyclohexan]-2′-one was obtained and was converted into several related spiro compounds.[31a] A quinone methide is probably an intermediate in this synthesis.

According to the patent literature,[25,32] the method used for preparation of the tricyclic spirans (**27**) through (**31**) is to condense an appropriately substituted quinol (**32**) with a C_{10}-terpene, for example, myrcene (**33**), in the presence of zinc chloride or boron trifluoride etherate. Spirans of type (**31**) may be isolated together with spirans (**27**) through (**30**). Stern and coworkers[33] have synthesized spiro[chroman-2,1′-cyclohexan]-6-ols (**27–30**) in 7–31% yield via the acid-catalyzed condensation of alkylated hydroquinones with myrcene (**33**). The structures were assigned on the basis of pmr and mass spectral data. The formation of structures (**27**) through (**30**) is rationalized in Scheme 1.[33]

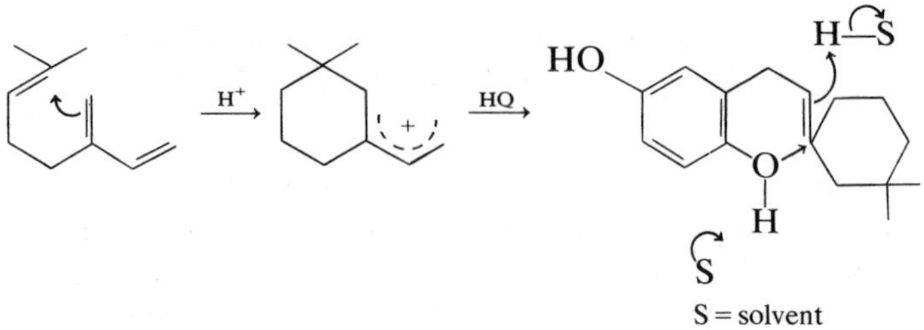

S = solvent

Scheme 1. Synthesis of spiro[chroman-2,1′-cyclohexan]-6-ols.

2. Spiro[chroman-2,1′-cyclohexa-3′,5′-dien and -4′-en]-2′-ones

Although *o*-quinone methides often give spiro compounds,[35] 3′,5′,6,8-tetramethylspiro[chroman-2,1′-cyclohexa-2′,4′-diene]-6′-one (**35**) has been made by another route, namely, by alkaline potassium ferricyanide oxidation of *o,o*′-dihydroxybimesityl (**34**).[34] Compound (**34**) can be recovered when reducing spiran (**35**). Spirans of type (**35**) are very stable compounds;[34] but though they resist polymerization,[34] they are converted into a trimer by reaction with *o*-quinone methide.[35] Spirochromans of type (**35**) have also been synthesized by reaction of a 2-chloromethylphenol (such as 2-chloromethyl-4-methoxy-3,5,6-trimethylphenol) with aqueous sodium bicarbonate.[35a]

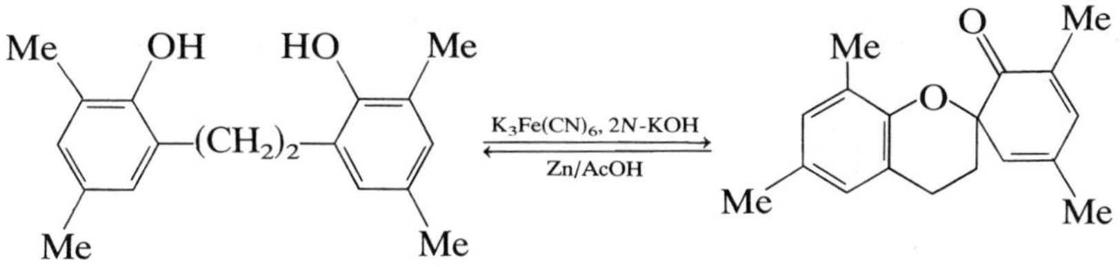

$K_3Fe(CN)_6$, 2*N*-KOH ⇄ Zn/AcOH

(**34**) (**35**)

A compound related to that obtained by Fries and Brandes[34] is 3′,5′,6,8-tetra-*t*-butylspiro[chroman-2,1′-cyclohexa-2′,4′-diene]-6′-one (**37**), which was formed by silver oxide oxidation of 2,4-di-*t*-butyl-6-methylphenol (**36**).[36] Its infrared spectrum reveals a strong carbonyl band at 1681 cm^{-1} but no hydroxyl band at 3571–3333 cm^{-1}. As before,[34] a ketoether of a quinol was reduced with zinc dust in boiling acetic acid solution to give a biphenol (**38**). Monohydric phenols play an important role in retarding the autoxidation of benzaldehyde,[37] and in a kinetically similar way they retard the autoxidation of hydrocarbons.[36] The mechanism proposed (Scheme 2) involves the abstraction of hydrogen atoms from phenol molecules and the dimerization of the radicals so produced with broken oxidation chains. The formation of a substituted hydroxybenzyl radical was realized by hydrogen removal from methyl and not from the hydroxyl group of the phenol.[36]

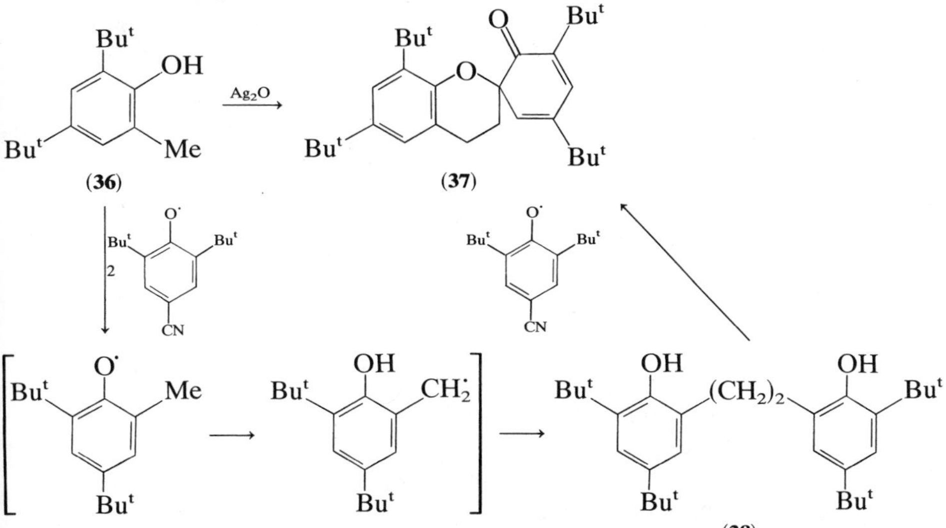

Scheme 2. Oxidation of 2,4-di-*t*-butyl-6-methylphenol.

Müller and co-workers[38] have prepared the inner spirocyclic quinonoid ether (**37**) in 60% yield by dehydrogenation of sterically hindered bisphenols obtained via short-lived aroxyl radicals using 4-cyano-2,6-di-*t*-butylphenol radical. During the dehydrogenation it is possible to record the existence of a new monovalent aroxyl radical with the aid of epr technique. The formation of spiran (**37**) can be formulated as in Scheme 2.

Cook and Butler[39] have shown that lead dioxide or alkaline potassium ferricyanide oxidation of 2,4-di-*t*-butyl-6-ethylphenol (**35**) does not lead to the isolation of the expected *o*-quinone methide (**43**) but produces two ketochromans, (**44**) and (**45**), and an unidentified trimer.

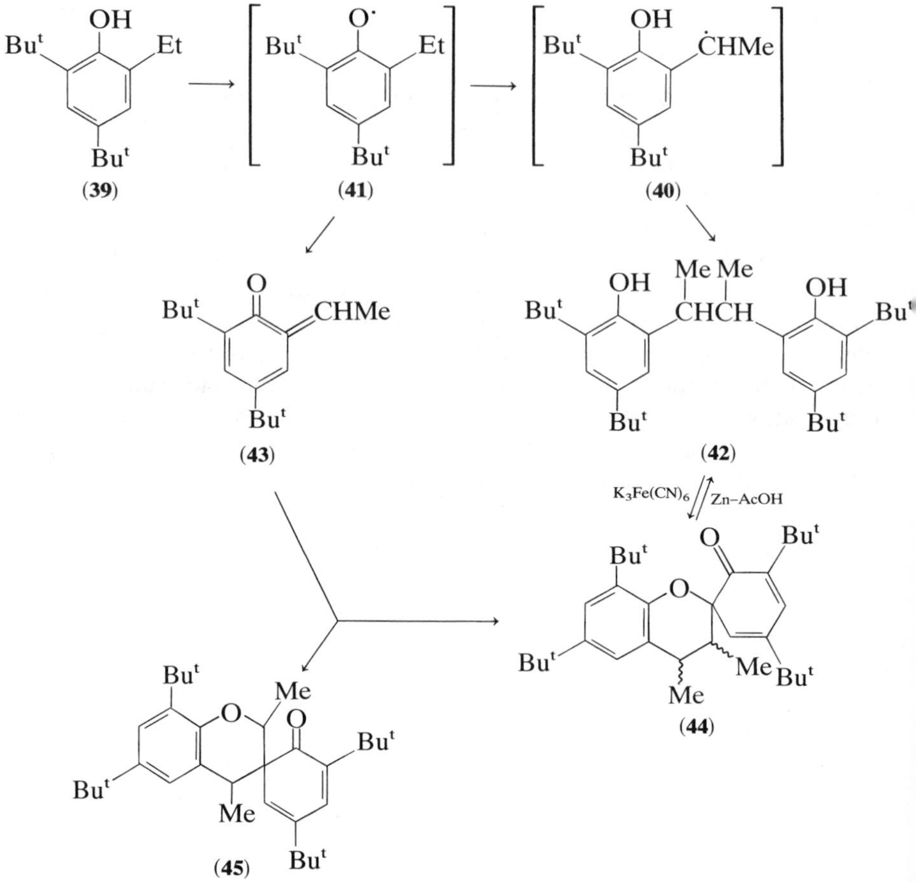

(**39**) (**41**) (**40**) (**43**) (**42**) $K_3Fe(CN)_6$ Zn–AcOH (**44**) (**45**)

The formation of both dimers occurs via the phenoxy radical (**41**), which most likely disproportionates to the phenol (**39**) and *o*-quinone methide (**43**). The latter reacts with itself in concerted fashion to form two compounds: one spiroannulated at C-2 (**44**) and the other at C-3 (**45**). An additional possible pathway to the spiran (**44**) exists via the benzyl radical (**40**).

Oxidation of the substituted dihydroxydiphenylethane (**42**) gives (**44**), which upon treatment with zinc and acetic acid forms compound (**42**). However spiran (**45**) cannot be obtained by this route and is unreactive toward zinc and acetic acid. This Diels–Alder reaction was the first one to give more than one product.

In a unique reaction, Bolon[40] demonstrated that when the trimer (**47**) is treated with hydrogen iodide, the double bond is reduced and the carbonyl is deconjugated with simultaneous cleavage of the aryloxy bond in the 4 position of the cyclohexanone ring to form spiran (**48**). Since the analogous trimer (**47**, R = Me) was obtained and did not undergo hydrogen iodide reduction, the reason for this deconjugation is most likely the steric requirements of the *t*-butyl group. Numerous attempts to use acid or base catalysts to conjugate the double bond and the carbonyl failed.[40] The reduced yellow trimer (**48**) was obtained in 73% yield. The carbonyl shifted from 1700 to 1720 cm^{-1} in the ir spectrum, and the conjugated carbonyl in the uv spectrum at 297 nm disappeared. A strong hydroxyl band in the ir spectrum was recorded. The mass spectrum[40] exhibits the molecular ion at *m/e* 530, the base peak at *m/e* 177, and the most abundant ion at *m/e* 354.

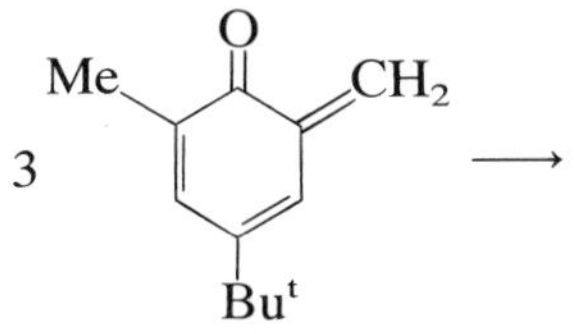

3 (**46**) → (**47**) —HI→ (**48**)

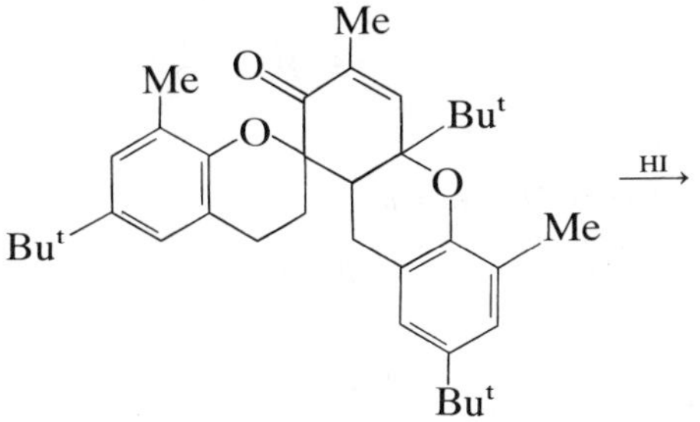

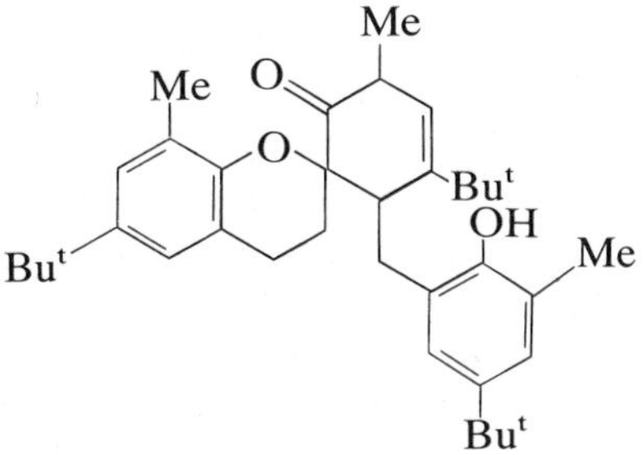

3. Spiro[chroman-2,1′-cyclohexa-2′,5′-dien]-4′-ones

Barton and co-workers[41] reported the synthesis in 11% yield of the spirodienone (**50a**) by alkaline potassium ferricyanide oxidation of 2,4′-dihydroxybibenzyl (**49a**). Spiran (**50a**) showed ir bands characteristic of cyclohexa-2,5-dienones at 1680, 1640, and 1615 cm^{-1}. It also showed a fourth weaker band at 1700 cm^{-1}. The uv spectra of cyclohexa-2,5-dienones usually show a high-intensity (*K*) band at 230–265 nm (log ε 3.9–4.3) along with a low-intensity band above 300 nm. Spiran (**50a**) exhibits high-intensity bands at 267 nm (log ε 3.3). The pmr spectrum reveals signals at δ7.2–6.7 (6H, *m*), 6.18 (2H, *d*, $J = 11$ Hz, α-dienone protons), 2.85 (2H, *t*), and 2.05 (2H, *t*, $J = 7$ Hz). In the mass spectrum of spiran (**50a**), the molecular ion is the base peak; there is also present a prominent M-1 ion.[42]

NaOH, $K_3Fe(CN)_6$

(**49a**) $R^1 = R^2 = H$
(**49b**) $R^1 = OMe, R^2 = H$
(**49c**) $R^1 = R^2 = OMe$

Ac_2O, H_2SO_4, 6h, RT, in the dark

(**50a**) $R^1 = R^2 = H$
(**50b**) $R^1 = OMe, R^2 = H$
(**50c**) $R^1 = R^2 = OMe$

(**51**)

(**52**) R = Ac

Spirodienone (**50a**) was converted in high yield into 2-acetoxydibenz[*b*, *f*]dihydrooxepin (**51**) by dienone–phenol rearrangement when treated with acetic anhydride containing a trace of sulfuric acid.[42] This rearrangement involves migration of aralkyl rather than of aryloxy groups[42] to give a small amount of compound (**52**).

In the same way as previously, ferricyanide oxidation of 1-(2-hydroxyphenyl)-2-(4-hydroxy-3-methoxyphenyl)ethane (**49b**) gave the spirodienone (**50b**) in less than 1% yield.[43]

This alkaline oxidation is thought to generate aryloxyl radicals[43] which may, according to Becker's mechanism,[44] disproportionate to form quinone methides. The uv absorption of the spiran (**50b**) in cyclohexane occurs at λ_{max} 248.5 (ε 7260), 271.5 (5400) and 281.6 (4760) nm. The infrared spectrum in KBr shows bands at 1673 and 1637 cm^{-1}. The pmr spectrum exhibits signals at 7.4–6.7 (6H, *m*), 5.85 (1H, *d*, $J = 3$ Hz, α-dienone proton), 3.66 (3H, *s*), 2.95 (2H, *m*), and 2.15 (2H, *m*).

The spiran (**50c**) was obtained by oxidation of bisphenol (**49c**) with alkaline ferricyanide in 13% yield.[45] It undergoes ring opening when reacted with methanolic hydrogen chloride to give (**53**), (**54a**), and (**54b**); but with methanol and a trace of sulfuric acid, the resulting phenol was (**53**) only. On treatment of (**50c**) with acetic anhydride and a trace of sulfuric acid, a polymeric material was obtained.

(**50c**)

MeOH, HCl

(**53**)

(**54a**) R = H
(**54b**) R = Me

These results demonstrate that under acidic conditions neither oxygen nor alkyl migration occurs. Dienone (**50c**) is less favored sterically to undergo ring expansion to form the seven-membered ring system of the oxepin type (**51**). The uv, ir, pmr, and mass spectra of (**50c**) have been published.[45]

4. Spiro[chroman-2,1′-naphthalene]

Russian chemists[46,47] carried out the microbial reduction of (±)-3-methoxy-8,14-seco-D-homoestra-1,3,5(10),9(11)-tetraene-14,17a-dione (**55**) to ketol (**56**) and diol (**57**) by yeasts, particularly *Sacharomyces Carlsbergensis* or bakers' yeast. The unstable compounds (**56**) and (**57**) rearranged into spirans (**58**) and (**59**), respectively. The alcohol (**59**) is an oil, and full spectral data of it have been published. The usual acetylation of the alcohol (**59**) gave the acetate (**60**).

(**55**)

(**56**) (**57**)

(**58**)

(**59**) R = OH
(**60**) R = OAc

On the chemical route the same group of chemists[47] also obtained 9,14-oxido compounds that possessed a carbonyl function at C-5 in the perhydrochroman system. Thus it may be possible to use 8,14-secosteroids in the synthesis of (±)-D-homoestrone via 9,14-spiroethers.

5. Spiro[chroman-2,17′-$\Delta^{5'}$-androstene] Derivative

Karrer and Kehrer[31] obtained spiran (**62**) by conversion of Δ^5-17-ethynyl-3,17-dihydroxyandrostene acetate into bromide (**61**) in 96.5% yield and condensation with trimethylquinol to give spiran (**62**) in 17% yield [cf. their synthesis of spiran (**26**)]. Spiran (**62**), also called 1′,2′,3′,4′,5′,8′,9′,10′,11′, 12′,13′,14′,15′,16′-tetradecahydro-5,7,8,10′,13′-pentamethylspiro[chroman-2,17′-(17-cyclopenta[*a*]phenanthrene)], has strong reducing properties but is neither an estrogen nor an androgen.

6. Uses

Spiro[cyclohexane-1,2′-chromans] such as compounds (**27**) through (**30**) are used as stabilizers in color photographic emulsions.[25,32] The most active compounds contain R^1 = H, Me; R^2 = Me, *t*-Bu, or $CMe_2CH_2CMe_3$.

(62) (61)

Spiran (**37**) was formed from phenolic inhibitors during the autoxidation of cumene.[36]

IV. CHROMANS SPIROANNULATED AT C-2 WITH HETEROCYCLIC SYSTEMS

1. With Monoheterocyclic Systems

A. *Spiro*[*chroman*-2,2′(5′*H*)-furan]-5′-ones

Mixich and Zincke[48] have obtained spiran (**66a**) by the hydrogenation of 5-salicylidenelevulinic acid (**63**) to 3-(2-hydroxychroman-2-yl)propanoic acid (**65**), which after treatment with dry hydrogen chloride gave spiran (**66a**) in 96% yield. This compound was also prepared by Bagli and Immer[49] by dehydration of 5-salicyllevulinic acid (**64a**) with acetic anhydride. They recorded its ir and nmr spectra. 5-Salicylidenelevulinic acid (**63**) and ketones (**67**) in the presence of hydrogen chloride undergo a nucleophilic addition of the activated methylene groups to (**63**) in a Michael-type reaction. The adducts (**68**) cannot be isolated since they cyclize to chroman-2-ol derivatives (**69**). After the second ring closure, spirochromans (**70**) are obtained; their structure has been proved by the presence of two characteristic uv absorption bands in the range of 270–290 nm.[48]

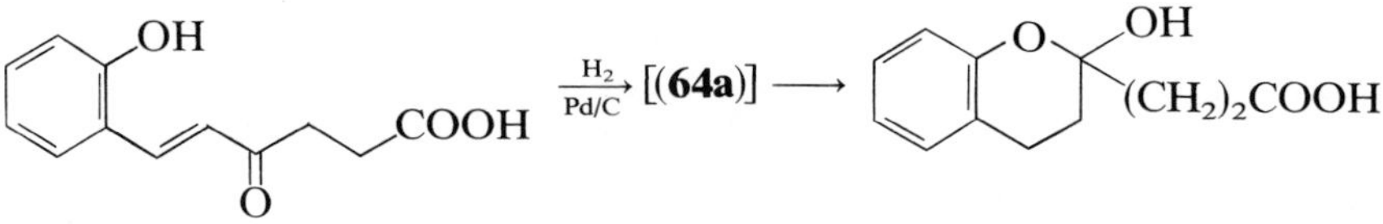

(63) (65)

(64a) R = H
(64b) R = OMe

(66a) R = H
(66b) R = OMe

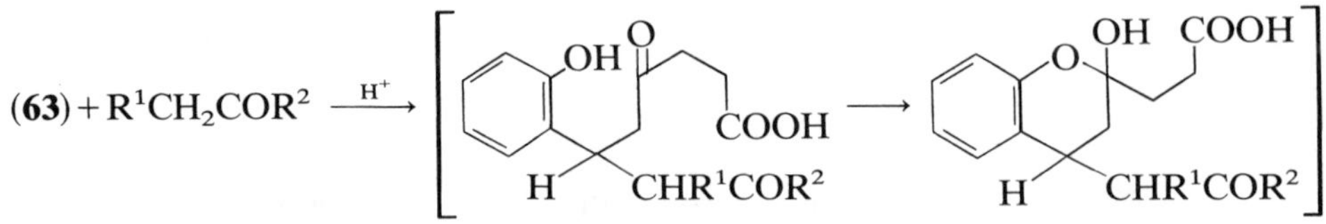

(67) (68) (69)

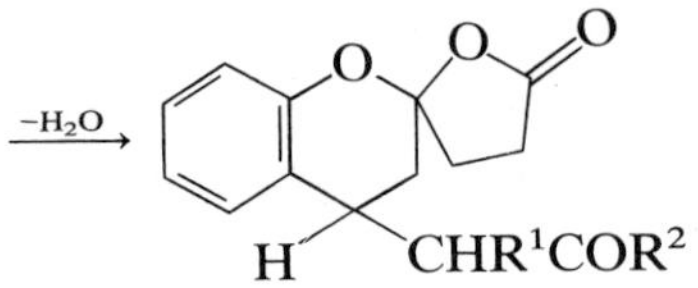

(70)

B. *Spiro[chroman-2,2′-[1.3]dithiolan]*

Corey and Beames[50] developed a new technique for the protection of lactones and esters against nucleophilic attack using bis(dimethylaluminum)-1,2-ethanedithiolate (**72**) and 3,4-dihydrocoumarin (**71**) to give 1,3-dithiolan derivatives (**73**) in 81% yield.

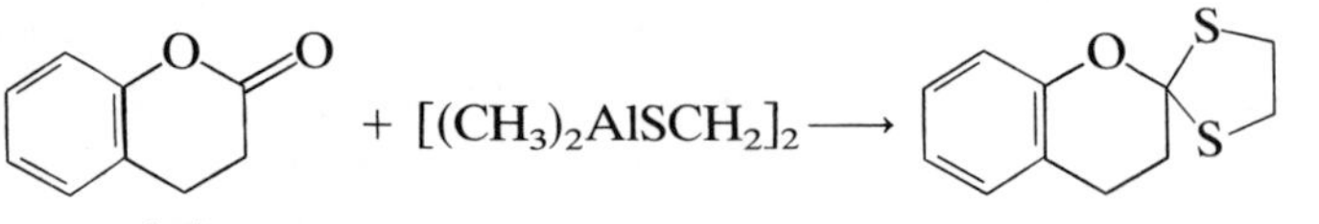

$$(71) + [(CH_3)_2AlSCH_2]_2 \longrightarrow (73)$$

(71) (72) (73)

Compounds like (**73**) were obtained via the corresponding keten thioacetals under mild conditions in high yield using reagent (**72**), which is readily prepared. They are more stable to weakly acidic reagents than the labile oxygen analog. The reconversion into lactone or esters can also be achieved under mild conditions. Infrared, pmr, and high-resolution mass spectral data were in complete consonance with structure (**73**).[50]

C. *Spiro[chroman-2,2'-[2H]pyran]*

Macrocyclic ring carbonate (**74**) under the influence of methanolic potassium carbonate and subsequent acidification can be quantitatively transformed via a transannular reaction into the cyclic ether (**75**).[51] The ir spectrum displayed weak aromatic absorption at 1605 cm^{-1}, whereas the uv spectrum confirmed the presence of an isolated aromatic ring (288 nm, ε 2850). The nmr spectrum has signals in agreement with structure (**75**).

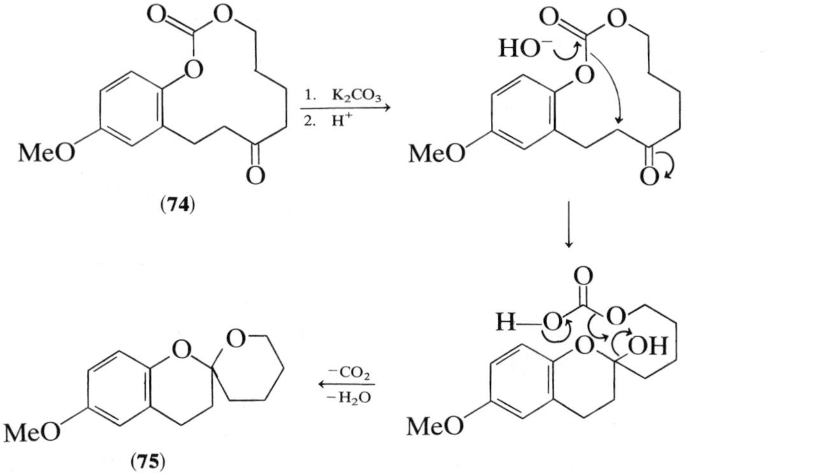

MeO (**74**) —1. K_2CO_3, 2. H^+→ MeO … HO⁻ … → H—O … OH … —$-CO_2$, $-H_2O$→ MeO (**75**)

D. *Derivatives of Spiro[chroman-2,4'(1'H)-pyrimidine]-2'(3'H)-one or -thione*

In a series of papers, Zigeuner and his co-workers[52–58] described the transformation of the Mannich base derivatives of 2-oxo- or 2-thiono-6-(β-dialkylaminoethylidene)hexahydropyrimidine (**76**, X = O or S) into spirans (**77**, X = O or S) by heating with 2,4-xylenol. The product (**77**, X = S)[52,54] was also obtained under similar conditions using the Mannich bases (**78**), (**79**), or (**80**).[54,55]

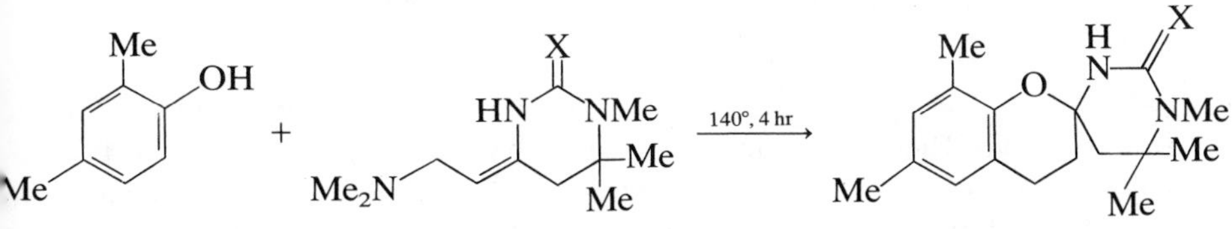

Me, OH, Me + X, HN, NMe, Me, Me, Me_2N (**76**) —140°, 4 hr→ Me, O, H, N, X, NMe, Me, Me, Me

(**77**) X = O or S

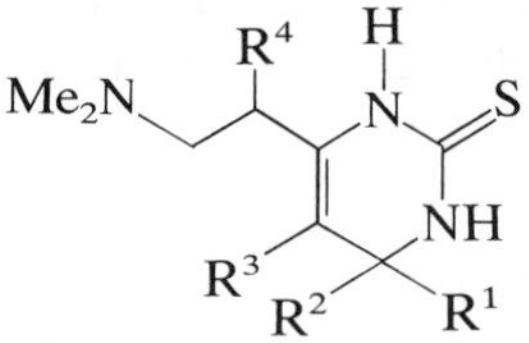

(**78**)
R^1–R^4 = H, Me or Et

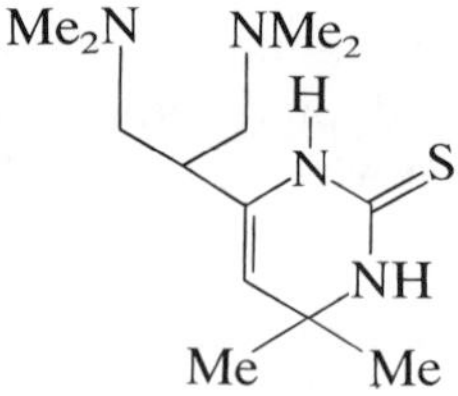

(**79**)

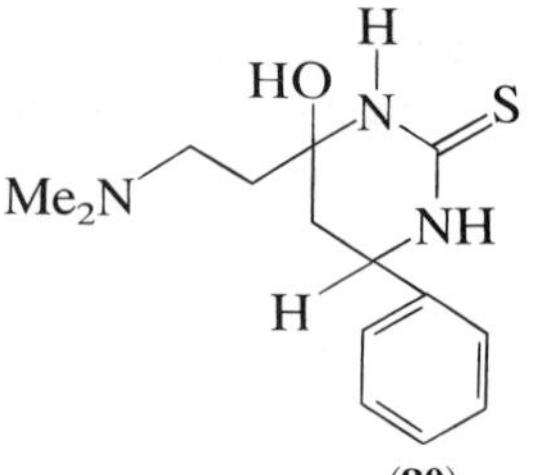

(**80**)

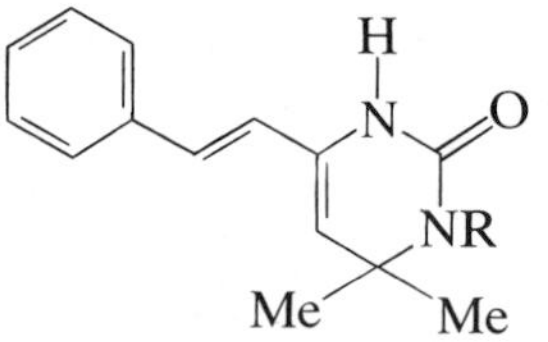

(**81**)
R = Me, H

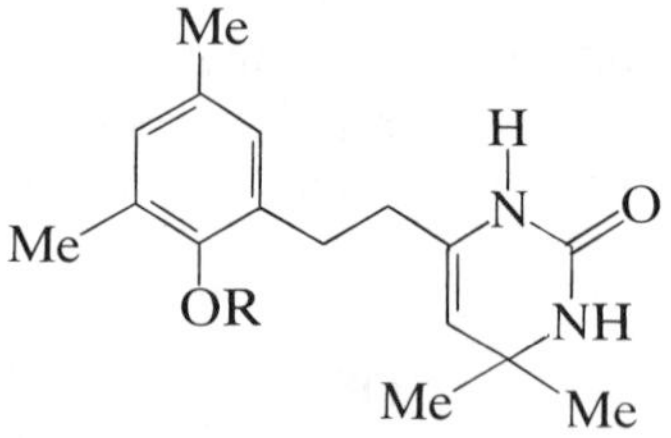

(**82**)
R = Me, H

The *C*-alkylation of the phenol and ring closure could be achieved by reacting (**81**) with 2,4-xylenol to give a 4-phenyl derivative of the spiran (**77**).[56] The formation of the spiran is the result of a Michael addition of styryl derivative (**81**) with the reactive phenol. Spirans such as (**77**) have also been obtained by heating in an autoclave an acetic acid solution of the ether (**82**) with pyridine hydrochloride[57] to effect cleavage of the ether bond. The last procedure was developed to prove the structures of ketones and thiones by synthesis.[53,57] The thiones are readily converted into ketones by treatment with hydrogen peroxide in alkaline medium.[57] Zigeuner, Korsatko and Fuchsgruber[58] modified the reaction for preparing spirans (**85**) and (**86**). These were obtained by heating 3,4-dihydro-4,4,6-trimethyl-2(1*H*)-pyrimidinone (**84**) with salicylaldehyde (**83**, R = H) or 2-hydroxy-3,5-dimethylbenzaldehyde (**83**, R = Me) and also on reaction in alkaline medium.[58]

2. Chromans, Benzochromans, and Dibenzochromans Spiroannulated at C-2 with Condensed Heterocycles

A. *Spiro[chroman-2,1′(2′H)-xanthen]-2′-ones*

The most facile reaction of the very reactive *o*-quinone methide is its trimerization. The trimer, an α,β-unsaturated ketone, (**89a**), may be prepared from salicylaldehyde in four stages or from saligenin by methylation to (**87**).[59] After subsequent pyrolysis, and especially at high concentrations

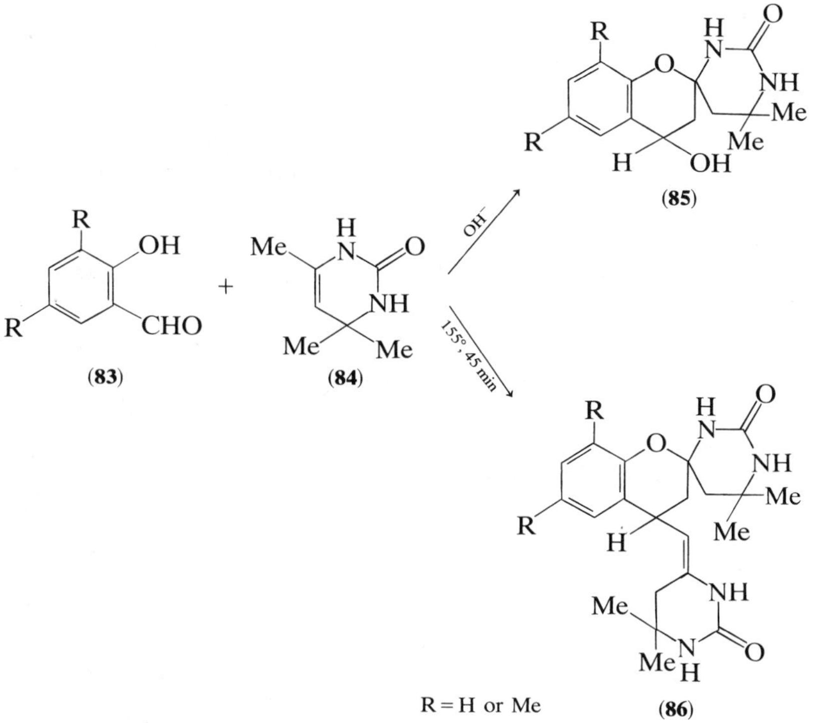

(83) (84) (85) (86)

R = H or Me

of substrate, the trimer was produced.[59] Its structure was proved by partial degradation and by comparison of spectral data from known xanthones. Spiran (**89a**) absorbs at 1684 cm^{-1} (5.94 μ) in the ir due to the presence of an α,β-unsaturated carbonyl group. It consumes two mole-equivalents of hydroxylamine: one for conjugative addition and the other for normal oxime condensation to give (**90**).[59] Quantitative hydrogenation confirmed the presence of one double bond since the product shows ir absorption at 1727 cm^{-1} (5.79 μ) characteristic of a saturated ketone (**91**).[59] The carbonyl group may be reduced with sodium borohydride or lithium aluminum hydride to yield the corresponding alcohol (**92**). The trimer (**89a**) seems to be the product of two successive Diels–Alder reactions.[59] From the degradation product it was possible to formulate the trimer as in (**89a**) and not as in (**93**).

It was reported[60] in 1907 that substituted *o*-quinone methides smoothly dimerized and trimerized; the trimer of 3,5-dimethylquinone-2-methide has been extensively investigated.[59] Because no less than three structures have been proposed for the trimer, Merijan, Shoulders, and Gardner[61] reexamined the properties of earlier reported substituted trimers using as representatives the spirans (**89b**) and (**89c**). They obtained evidence that both spirans are related and have a ring system identical with the parent spiran (**89a**). Spiran (**89b**) and spiran (**89a**) were prepared by pyrolysis[61] and were subsequently reduced under similar conditions as shown in Scheme 3.

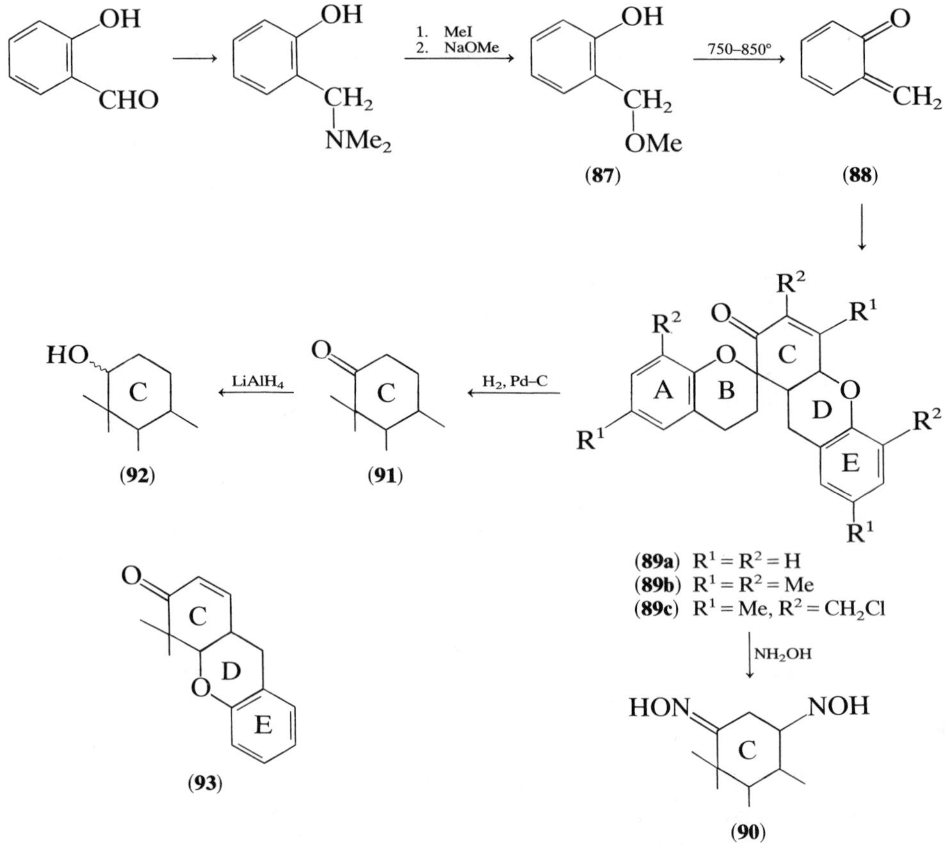
1. MeI
2. NaOMe
750–850°
(87)
(88)
LiAlH4
H2, Pd–C
(92)
(91)
(89a) R1 = R2 = H
(89b) R1 = R2 = Me
(89c) R1 = Me, R2 = CH2Cl
NH2OH
(93)
(90)

The pmr data tend to support the assignment since the lone olefinic proton of (**89b**) resonates at 6.37δ as a quartet, and the small coupling ($J = 1.3$ Hz) indicates the absence of protons on adjacent carbon atoms. The small splitting is due to coupling with the methyl group adjacent to the carbonyl and resonating at 1.79δ as a doublet. The angular methyl group is at 1.63δ while three of the aromatic methyls are at 2.17δ and one at 2.13δ, probably because of steric repulsion by the angular methyl group. The pmr spectrum gives no evidence of proton on carbon atoms adjacent to ethereal oxygen. On hydrogenation of the spiran (**89c**) the product (**94**) was obtained and was identical with a sample obtained from spiran (**89b**).[61]

The structure of the unsubstituted *o*-quinone methide trimer (**89a**) and the methylated trimer (**89b**) was decided chiefly on the analytical evidence for the former[59] and spectroscopic data for the latter.[61] Westra and his co-workers[62] afforded a detailed analysis of the spiran (**98**) produced by the oxidation of the phenol (**97**) using the pmr single and multiple resonance techniques. The pmr and ir data favor structure (**98a**) rather than (**98b**) because of (a) the presence of four typical acetal bands in the ir spectrum, (b) the low-field part of the nmr spectrum does not contain a bridgehead

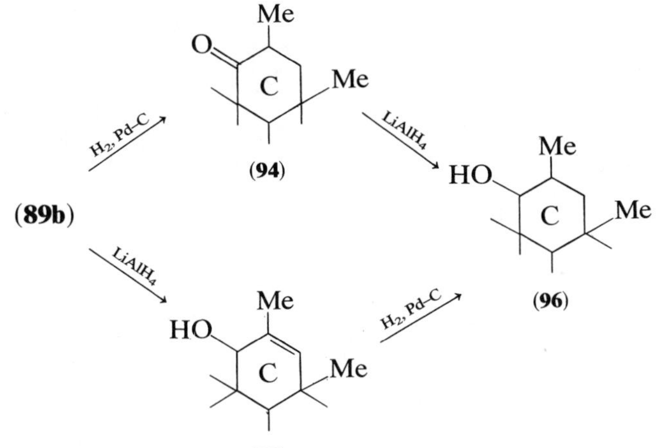

Me
O
C
Me
(**94**)
H_2, Pd–C
$LiAlH_4$
HO
Me
C
Me
(**96**)
(**89b**)
$LiAlH_4$
H_2, Pd–C
HO
Me
C
Me
(**95**)

Scheme 3. Syntheses of spiro[chroman-2,1′(2′*H*)-xanthen]-2′-ones.

proton, and (c) the existence of long-range spin–spin interactions of aromatic or vinylic protons with methoxy or aliphatic protons.[62]

The spiran (**98a**) was also synthesized by Bolon,[40] who used basic potassium ferricyanide or silver oxide for oxidation of the phenol (**97**) in pentane or benzene. The silver oxide oxidation of 4-phenyl-2,6-xylenol in benzene for 3 hr and 4-*t*-butyl-2,6-dimethylphenol for 18 hr respectively led to the spirans (**99**) and (**100**).[40]

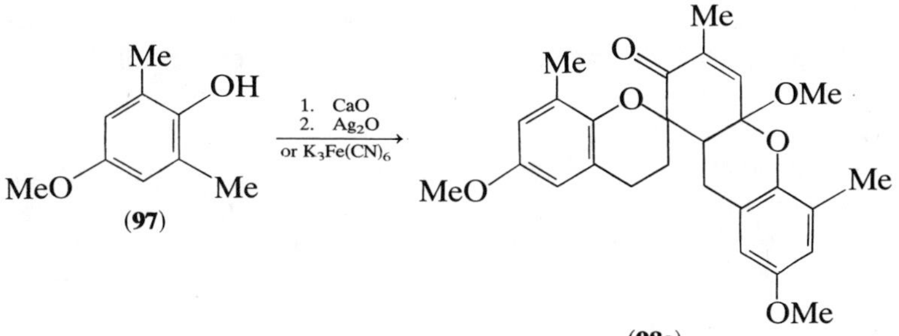

Me
OH
MeO
Me
(**97**)
1. CaO
2. Ag_2O
or $K_3Fe(CN)_6$
Me
Me
O
O
OMe
O
MeO
Me
OMe
(**98a**)

Me
Me
O
O
OMe
MeO
O
Me
OMe
(**98b**)

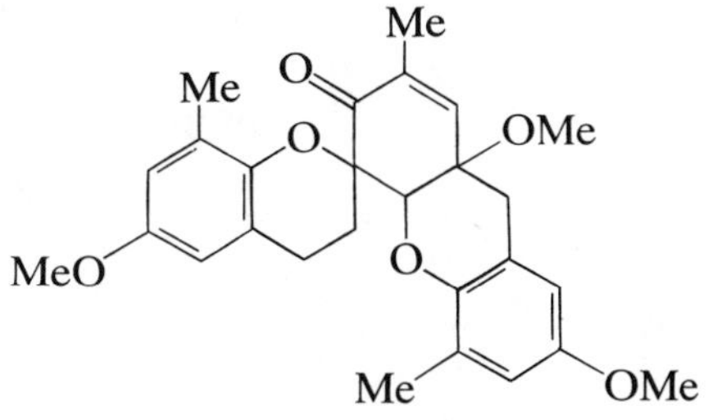

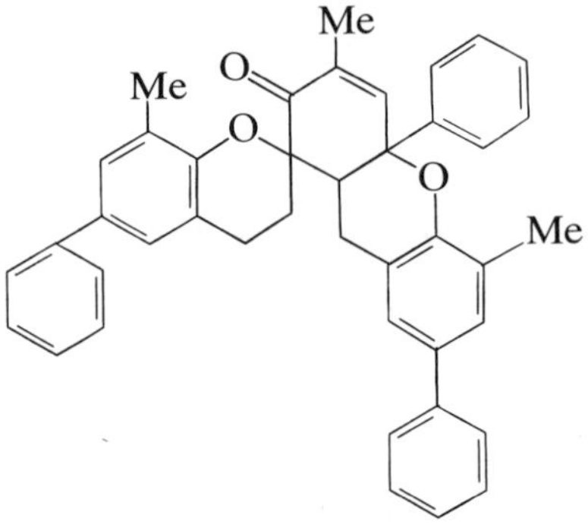

(**99**)

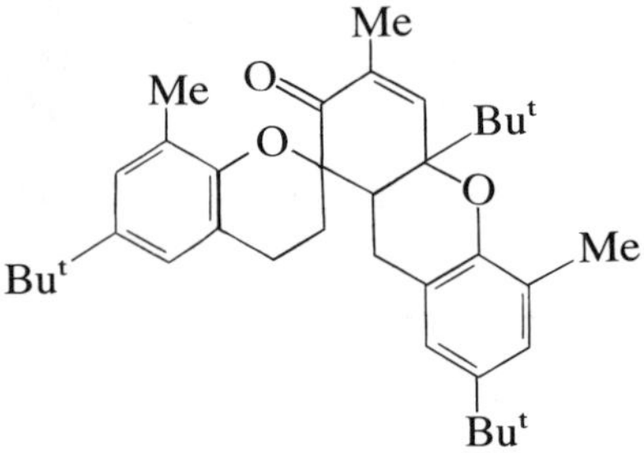

(**100**)

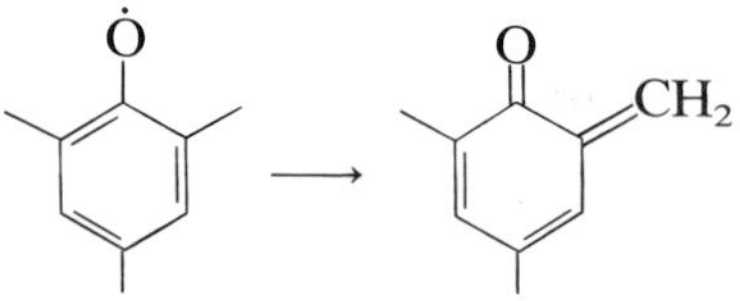

(**101**) (**102**)

The phenols mentioned are blocked in the para position to prevent oxidative coupling and *p*-quinone methide formation. They form phenoxy radicals (**101**) in solution which change into *o*-quinone methide (**102**), and this reacts with itself to give the trimers. Spiran (**100**) exhibits in the double-resonance pmr spectrum the signal of the methyl group on the cyclohexenone ring coupled with the vinyl proton. There was also a solvent effect on one of the aromatic methyls. In carbon tetrachloride, the two aromatic methyls resonate at 2.04 and 2.07; but in deuterochloroform, the 2.07 signal shifts to 2.16.

Bolon[40] found that silver oxide was superior to all other oxidizing agents, and the trimer formation is favored over other self-condensations.

Because of the steric hindrance of the 4-*t*-butyl group, the spiran (**100**) is unreactive and the only reaction which occurred was a reduction with hydrogen iodide described in Section III, 2. Many other reducing agents failed to cause reduction. Only lithium aluminum hydride reduced the carbonyl group to the corresponding alcohol.[40]

Cook and Butler[39] have isolated an unidentified trimeric compound (see Section III, 2) based on 2,4-di-*t*-butyl-6-ethylphenol and have described its physical and spectroscopic properties.

B. *Spiro*[*chroman*-2,4′(4′*aH*)-[3*H*]-*xanthen*]-3′-*one*

Chauhan, Dean, McDonald, and Robinson[35] were successful in preparing the spiran (**105**) by heating the dimer (**103**) in mesitylene or cyclo-octadiene followed by partial thermal dissociation into quinone methide (**104**) and subsequent coupling to give the final product (**105**). Dimer (**103**) does not easily undergo conversion into trimer (**105**), whereas isomers (**106**) and (**107**) are totally converted into spiran (**89b**). The isomer (**105**) has a signal at 4.35 (doublet, $J = 2$ Hz) which can be attributed to its isolated angular proton adjacent to ether oxygen. The small splitting probably reflects long-range coupling with a methylenic proton. Isomer (**89b**) does not possess such a signal.[35]

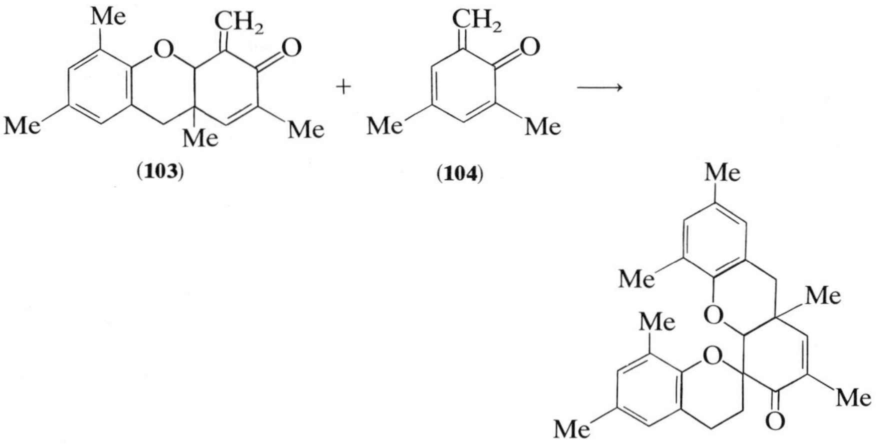

(**103**) + (**104**) → (**105**)

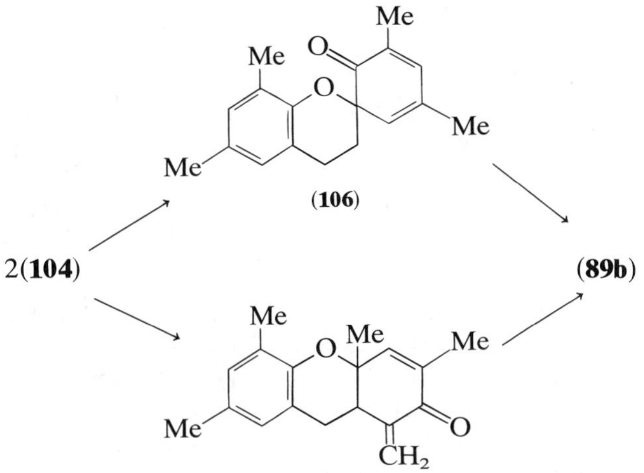

2(**104**) → (**106**) → (**89b**)

2(**104**) → (**107**) → (**89b**)

C. 4-*Substituted Spiro*[*chroman-2-heterocycles*]

a. 4-SUBSTITUTED SPIRO[CHROMAN-2,9′-[9*H*]XANTHENE]. Russian chemists[63] have obtained spiran (**111**) in 85% yield by saturation of an ethereal solution of 9-methylxanthen-9-ol (**108**) and salicylaldehyde with hydrogen chloride. Additionally they obtained spiran (**110**) which exhibited photochromism in benzene solution. It can be converted into spiran (**111**) by saturation with hydrogen chloride in ether and addition of (**108**) to (**109**). The phenetol solution of (**111**) when boiled did not display photochromism. The finding is an extension of the observation made by Irving who did not isolate any nonthermochromic spirans.[64] Spiran (**111**) is not stable toward strong concentrated acids.

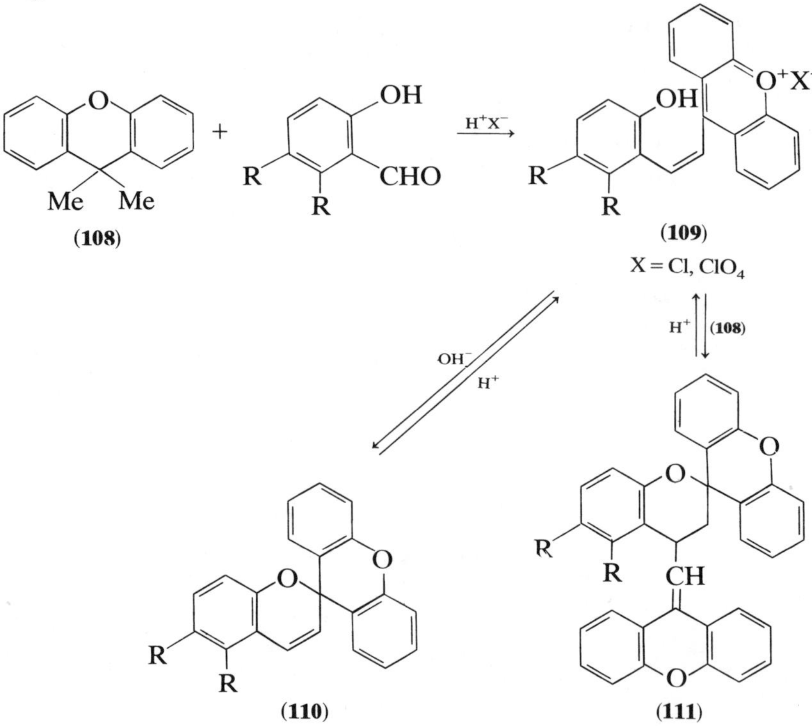

(**108**) (**109**) $X = Cl, ClO_4$ (**110**) (**111**) R = H or benzo

b. 4-SUBSTITUTED SPIRO[CHROMAN-2,2′-INDOLINES]. Condensation of salicylaldehyde with the Fischer base 1,3,3-trimethyl-2-methyleneindoline (**112**) has been reported[65] to give spiran (**113**). However, Koelsch, and Workman[66] found that the reaction actually yields two products: spiropyran (**113**) and a secondary product also having a spiropyran structure. Hinnen, Audic, and Gautron[67] proved that the secondary product exists as the spirodihydropyran (**114**). This assignment was made on the basis of the spectral data of (**114**) as well as its mode of formation. The absence of a signal at 5.6, characteristic of the proton at C-3 in benzo[*b*]pyran, supports (**114**). Also the diminution of the ir band at 780 cm^{-1}, characteristic of the α-pyran structure, and the appearance of the 805 cm^{-1} band attributed to an ethylenic grouping are in accordance with the suggested structure. The substitution at C-4 rather than at C-3 was also discussed. Two competing

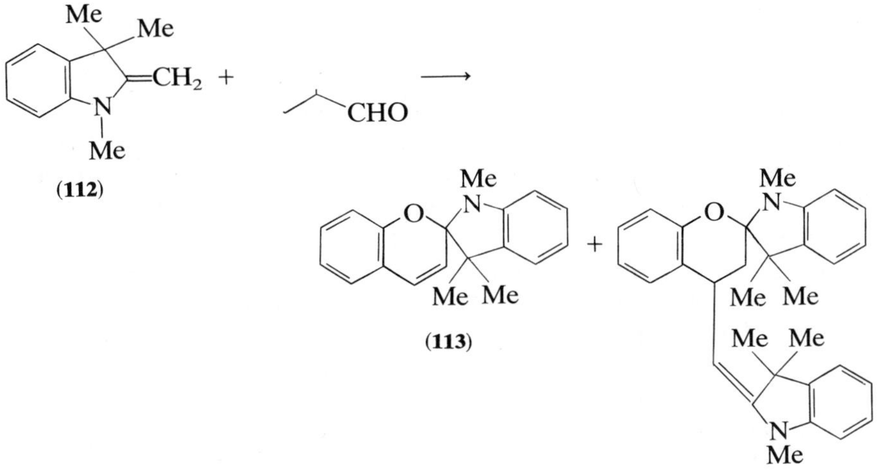

(112)

(113)

(114)

routes for the conversion of an initially formed phenol to spiropyrans (**115**) and (**116**) have been suggested (Scheme 4). The substitution was favored over the Michael condensation.[67]

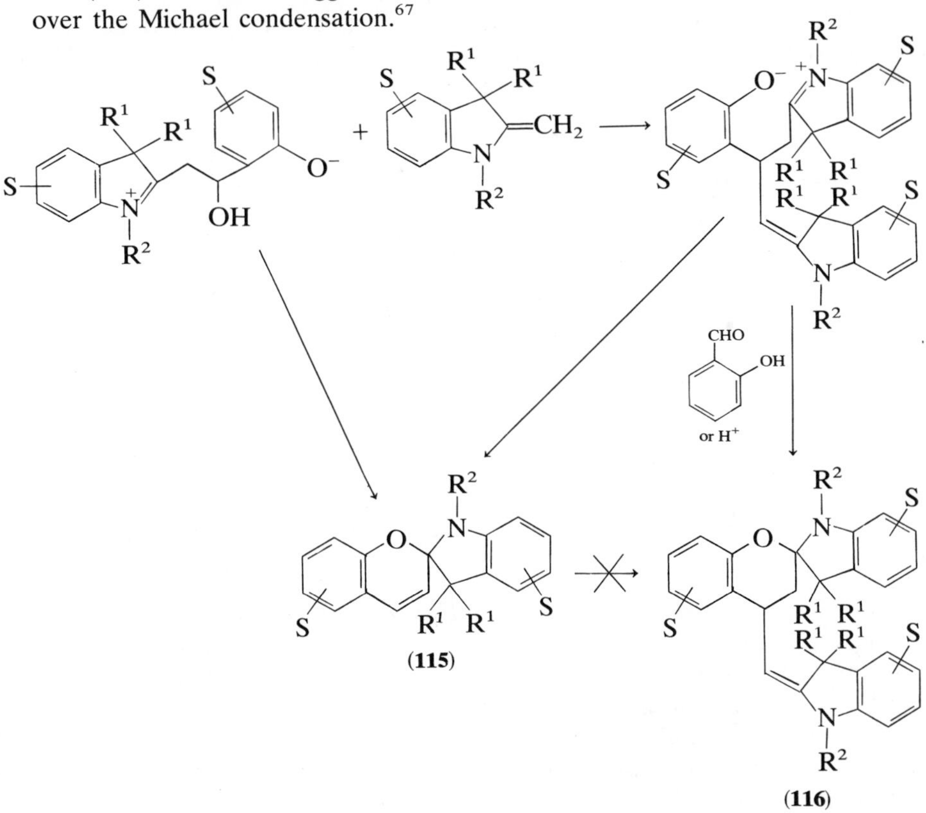

(115)

(116)

Scheme 4. Synthesis of spiro[chroman-2,2′-indolines].

Balny, Hinnen, and Mosse[68] found the characteristic fluorescence $\lambda_{exc.}^{EtOH}$ 282 nm, λ_{emi}^{EtOH} 370 nm shown by spiran (**116**, S = H). Spiran (**115**) exhibited a similar fluorescence, diminishing on recrystallization and evidently due to contamination with spiran (**116**). Spiran (**115**) substituted with 6′-CN, 5-Cl-6′-OMe, 6′,8′-Br_2, 6′-OMe, and 3′-D shows analogous fluorescence in alcohol, $\lambda_{exc.}$ 285 nm, $\lambda_{emi.}$ 365 nm, indicating the presence of dihydropyran (**116**) impurities. Since all fluorescent spirans examined were the *N*-methyl compounds and the analogous *N*-phenyl compounds showed no fluorescence, it was suggested that the phenyl group orients the condensation reaction rather toward the spiropyran form than the dihydropyran.[68]

Spiran (**114**) forms colorless crystals which melt to a pink liquid. The compound gave a violet solution in hot diphenyl ether.[66]

In the patent literature[69] there is mentioned a dihydro compound related to spiran (**113**), namely, 6-chloro-8-methoxyspiro[chroman-2,2′-[2*H*]indole] and claimed as a pressure-sensitive record sheet material.

c. 4-Substituted Spiro[benzothiazole-2(3*H*),2′-chromans]. Spirans (**118a**) and (**118b**) described in the patent literature[70] are claimed to be useful in thermographic recording material. They were prepared by adding piperidine to a mixture of 2,3-dimethylbenzothiazolium *p*-tosylate (**117**, X = 4-$MeC_6H_4SO_3$) and salicylaldehyde or *o*-vanillin, respectively, in the molar ratio of 2:1 at room temperature.

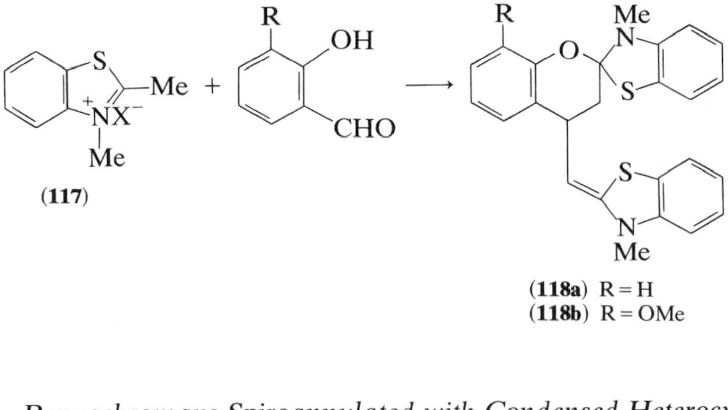

(**117**)

(**118a**) R = H
(**118b**) R = OMe

D. *Benzochromans Spiroannulated with Condensed Heterocycles*

a. Spiro[naphthalene-1(2*H*),3′-[3*H*]naphtho[2,1-*b*]pyran]-2-one. As early as 1919, Pummerer and Cherbuliez[71] demonstrated that 1,2-naphthoquinone-1-methide (**120**), generated from compound (**119**) by heating in xylene or benzene,[72] dimerizes to spiran (**121**), which after reduction gives dinaphthol (**122**).

Spiran (**121**) was also prepared by pyrolysis of 1-methoxymethyl-2-naphthol (**123**).[59] Dimer (**121**) dissociates into the quinone methide (**120**) at temperatures at which the product (**124**) of trapping (**120**) by styrene is stable,[35] and this shows that dimerization is less favored than addition to a simple alkene derivative.

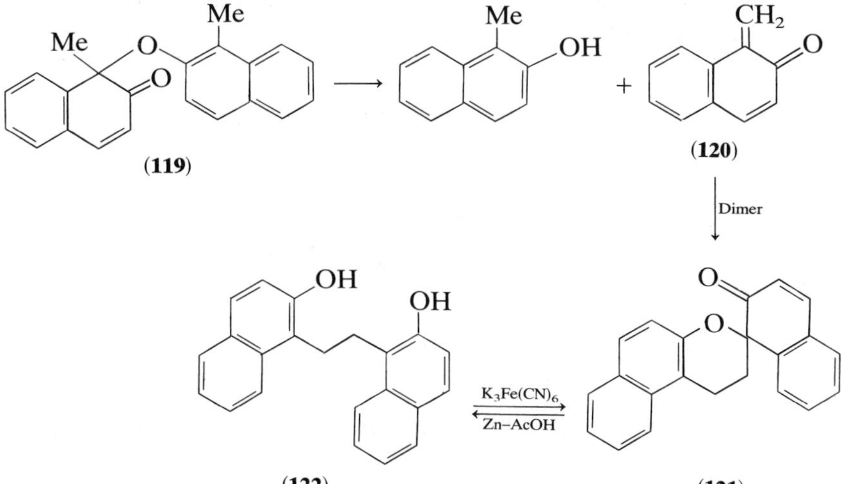

(119) (120) Dimer (122) $K_3Fe(CN)_6$ Zn–AcOH (121)

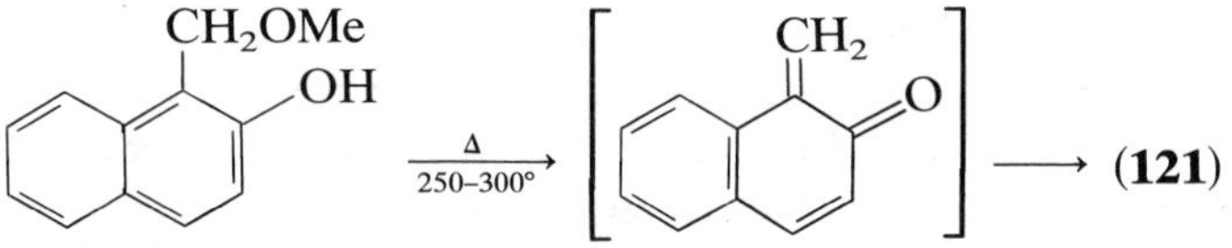

(123) Δ 250–300° → (121)

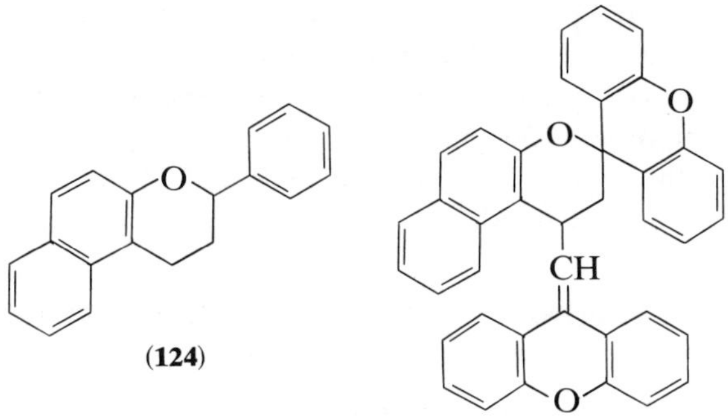

(124) (125)

b. Spiro[3*H*-naphtho[2,1-*b*]pyran-3,9′-[9*H*]xanthene] Derivatives. The spiran (**125**) was prepared by the method used for (**111**).[63] In contrast to spiran (**111**), spiran (**125**) is not thermostable when boiled for 20 min in phenetol. Treated with an excess of hydrogen chloride, it decomposes with formation of the corresponding photochromic spiropyran, which may be converted into (**125**) as described earlier for spiran (**110**).

c. SPIRO[BENZOTHIAZOLE-2(3*H*),3′-[3*H*]NAPHTHO[2,1-*b*]PYRAN DERIVATIVES. The quaternary salt (**117**) and 2-hydroxy-1-naphthaldehyde in acetone were stirred in piperidine to give the spiran (**126**), which is used in thermographic recording materials.[70]

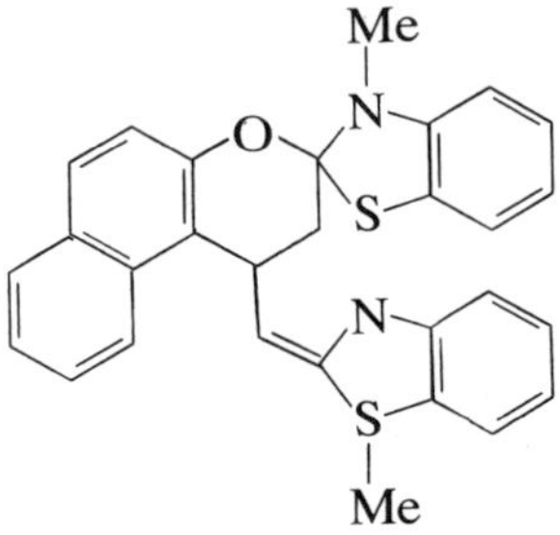

(**126**)

E. *Dibenzochroman Spiroannulated with Condensed Heterocycles*

a. SPIRO[PHENANTHRENE-9(10*H*),2′-[2*H*]PHENANTHRO[9,10-*b*]PYRAN]-10-ONE DERIVATIVE. Dimer (**130**) of 10-methylene-9(10*H*)-phenanthrone (**129**) was synthesized[73] by condensation of 9-phenanthrol (**127**) with formaldehyde and dimethylamine under mild conditions to form the Mannich base (**128**), which is very unstable and during purification easily loses the amine function to give spiran (**130**). The same result can be achieved when the unstable methiodide of (**128**) is treated as before. Reaction of 9-phenanthrol with formaldehyde leads directly to (**130**) due to the instability of the 10-methylol compound. Spiran (**130**) was reduced with lithium aluminum hydride to the corresponding alcohol. Replacement of the dimethylamine by a catalytic quantity of pyridine and refluxing for 10 hr afforded the spiran (**130**) in 40% yield; the absence of base diminishes the yield to 36%. Although compound (**130**) has some intense absorptions at 250, 275, 295, and 306 nm, the less intense peaks at 340 and 360 nm suggest that it has only one 9,10-double bond.[73]

V. CHROMANS SPIROANNULATED AT C-3

1. 3,3′-Spirobichromans

Sodium 2-methoxyphenoxide suspended in boiling xylene and treated with 1,3-dibromo-2,2-bis(bromomethyl)propane for 80 hr gave spiran (**132**).[74,75] The reaction failed when the substituent in the phenol was an electron-withdrawing group located at the ortho or para position. Spiran (**132**) was converted by the action of pyridine hydrochloride to the 8,8′-dihydroxy derivative which was characterized as the bis-3,5-dinitrobenzoate.

(127) (128)

(130) (129)

The unexpected survival of the pyran ring may be due to the protecting action of hydroxyl groups at C-8.[74,75]

The presence of hypochromism (1:1.35 in n-propanol and 1:1.38 in THF) suggests that the dominant conformation of (**132**) is helical and chiral with a twofold axis of symmetry.[75] Similar treatment of 4-hydroxydiphenyl with compound (**131**) for 160 hr afforded spiran (**133**).[76] The dominant conformation with helical structure was also selected for this compound.[76] Spectroscopic data of these spirans have been published.[74–76].

(131)

(**132**) R^1 = MeO, R^2 = H
(**133**) R^1 = H, R^2 = Ph

2. Spiro[chroman-3,1′-cyclopropan]-4-ols

As has been reported in an earlier volume in this Series (Volume 31, Chapter III), spiro[chroman-3,1′-cyclopropan]-4-ols of type (**134**, R^3 = H) have been prepared by reduction of the corresponding 4-chromanone with sodium borohydride.[77] Reaction of the 4-chromanone with phenylmagnesium bromide afforded 4-phenyl analogs (**134**, R^3 = Ph). Stereoisomers were separated and their configurations assigned from nmr spectral data. The action of acetic acid on the spirochromanols effected a stereoselective rearrangement to 3-substituted 2*H*-chromenes of type (**135**).[77]

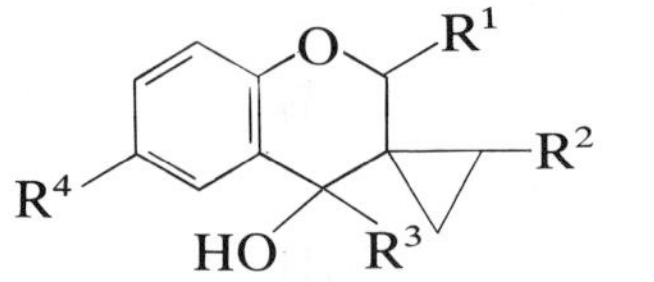

(134) 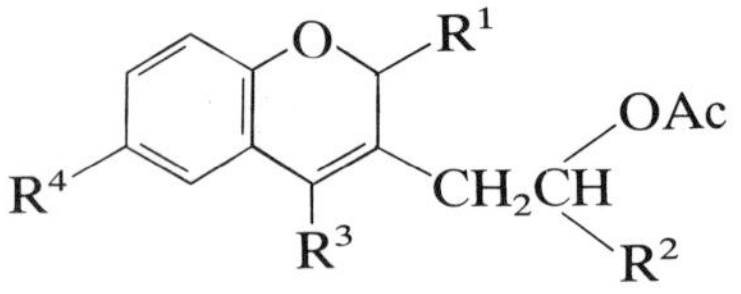 (135)

3. Spiro[chroman-3,2′-oxiran]-4-ols

A series of papers[78–81] on selective reduction of some oxidoketones and analogues (**136**) to cis and trans oxidoalcohols (**137**) was published in connection with studies on the synthesis of the brasilin skeleton,[78] (±)hematoxylin,[79] some hematoxylin derivatives,[80] and homoisoflavonoids.[81] Such reduction occurs smoothly with sodium borohydride, which attacks only the carbonyl group and not the epoxide ring, to give a pair of stereoisomeric racemic alcohols. They can be reoxidized in pyridine with chromium trioxide.

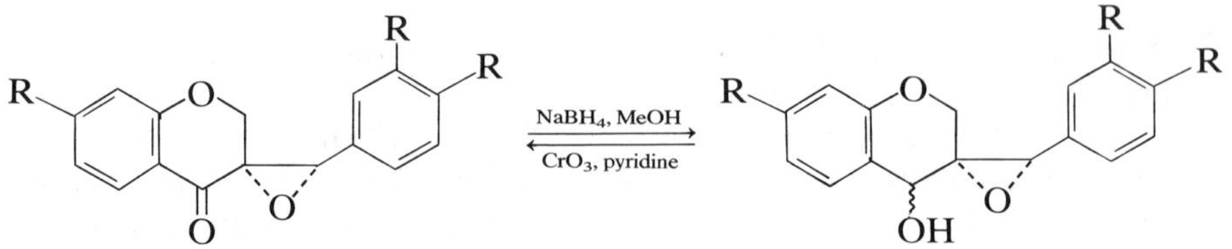

$NaBH_4$, MeOH / CrO_3, pyridine

(136) (137)

R = H, OMe, or OCH_2Ph

The mixtures of stereoisomeric epoxy alcohols can be separated by fractional crystallization. In the cis isomer, the 4-OH and the epoxy oxygen are closer to each other than in the trans isomer. The stereochemistry of the products was determined on the basis of solubility of the racemates; the cis isomer is poorly soluble because of intramolecular hydrogen bond formation whereas the trans form is easily soluble due to intermolecular hydrogen bond formation supporting solvation. This feature is evident in the respective ir spectra[81]: the trans forms absorb at about 3490 cm^{-1}, in contrast to cis forms absorbing at 3250 cm^{-1}. The nmr spectra of both cis and trans isomers exhibit signals for α-C-H at 4.5 ppm, the same as for (**136**). The chemical shifts of 4-H and 4-OH are nearly the same for cis and trans forms, but the coupling constants differ: 8.5 Hz in the cis and 4.5 Hz in the trans form. The nonequivalent 2-H protons resonate at 4.1–4.0, but they do not reveal any significant mutual splitting.[81]

4. Spiro[chroman-3,4′-imidazolidine]-2′,5′-dione

Spiro[chroman-3,4′-imidazolidine]-2′,5′-dione (**139**) was synthesized for comparative biological studies[82] from 3-chromanone (**138**) by the Bucherer reaction[83] in a yield of 18.6%. The compound has some interest as possible agent for the treatment of epilepsy in human subjects.[82]

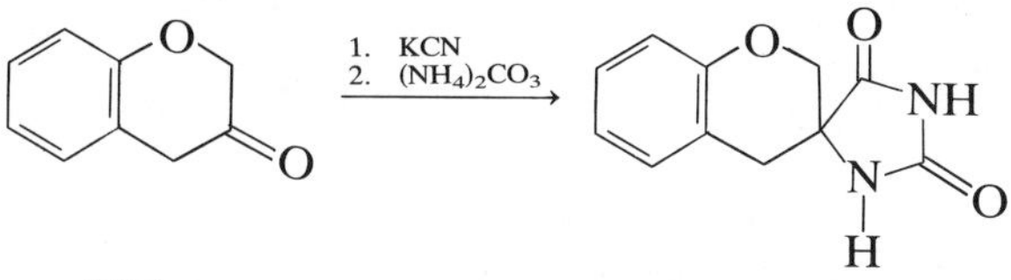

1. KCN
2. $(NH_4)_2CO_3$

(138) (139)

5. Hexahydrospiro[chroman-3,2′-oxiran] Derivative

4,7-Dimethyl-4*a*-hydroxymethylhexahydrospiro[chroman-3,2′-oxiran]-2,4-dicarboxaldehyde (**141**), a relative of diacetoxyscirpenol, the naturally occurring ester of sesquiterpene alcohol and an important mycotoxin from *Fusarium* sp., was found to be noncytotoxic and nonlarvicidal compared with the parent ester (**142**). These results are in accord with the suggestion that toxicity is associated with the presence of the protected 12,13-epoxy group on the trichothecane system. Spiran (**141**) is the ring-C seco product in which the 12,13-epoxy group was saved, but it became accessible to rearside nucleophilic attack which leads to detoxication of the molecule.[84,85]

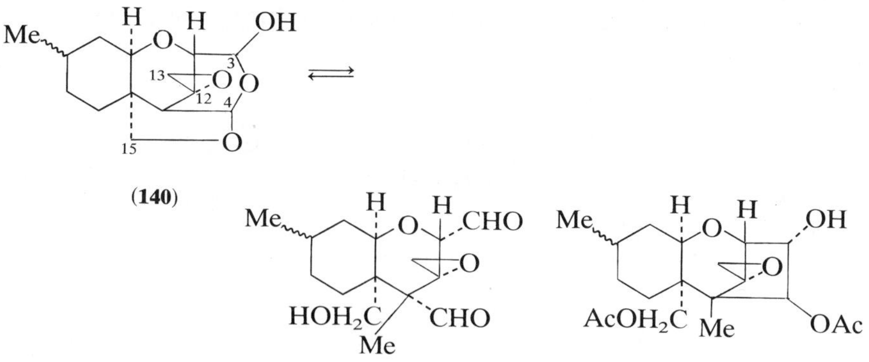

(140)

(141) (142)

VI. CHROMANS SPIROANNULATED AT C-4

1. 4,4′-Spirobichroman Derivative

A 26% yield of the spiran (**143**) is obtained when resorcinol, mesityl oxide, and ferric chloride are refluxed in benzene for 5 hr.[86] The spiran was an unexpected product, but it bears a structural resemblance to acronycine, one of a group of alkaloids exhibiting broad spectrum antitumor activity isolated from the bark of an Australian tree. It was therefore used in the studies on structure–activity relationships. It was identified from ir, nmr, and mass spectral data.[86]

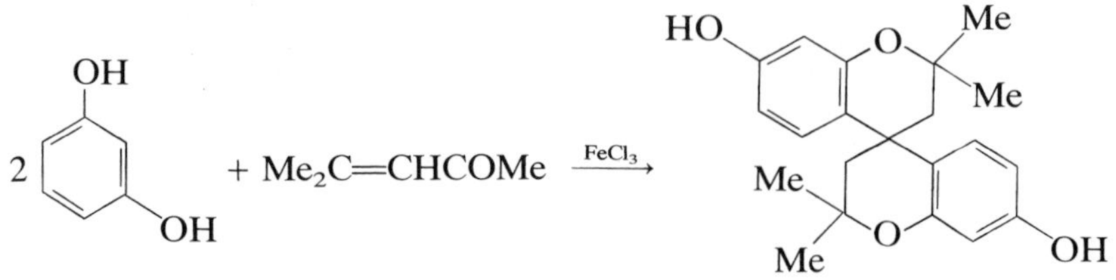

$$2\ \text{(resorcinol)} + Me_2C{=}CHCOMe \xrightarrow{FeCl_3}$$

(143)

2. Spirans Derived from 4-Chromanone

A. *Spiro[chroman-4,2′-[1,3]dioxolan Derivative*

When *p*-toluenesulfonic acid, 3-bromo-4-chromanone, and ethylene glycol in dry benzene were refluxed under a Soxhlet extractor filled with molecular sieves, the ketal (**144**) was formed in 95% yield.[87] The bromine atom of the ketal (**144**) is very difficult to displace with tertiary amines; the only product is the α,β-unsaturated dioxolan resulting from elimination of hydrogen bromide.[87]

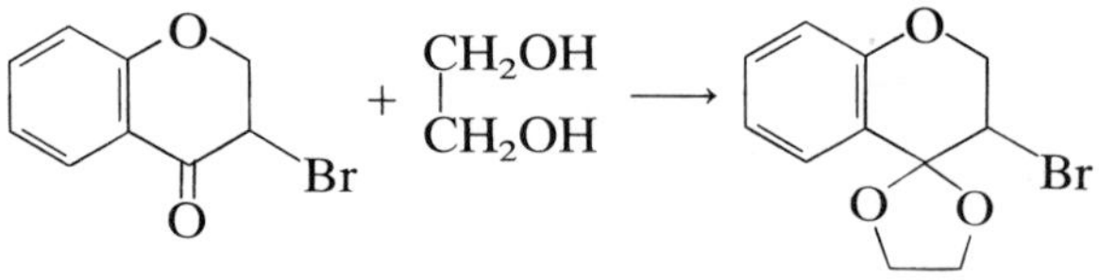

$$+ \begin{array}{c} CH_2OH \\ | \\ CH_2OH \end{array} \longrightarrow$$

(144)

B. *Spiro[chroman-4,3′-pyrrolidine] and Spiro[chroman-4,3′-furan] Derivatives*

4-Chromanone was converted by a modified Cope procedure[88] into the corresponding cyanoalkylidene ester (**145**) in 33% yield. Addition of potassium cyanide followed by hydrolysis of the dried salt with concentrated hydrochloric acid resulted in the formation of the acid (**146**) in 55% yield. This cyclized to its anhydride (**147**) and on treatment with 3-dimethylaminopropylamine yielded the imide (**148**), which was reduced with lithium aluminum hydride to give the azaspirochroman (**149**). This compound inhibited the growth of KB tissue culture at 4–6 mg/ml.[89]

$NCCH_2COOEt$ → NCCCOOEt (145); 1. KCN 2. HCl → (146) COOH COOH; Ac_2O, 3h →

(147) $Me_2N(CH_2)_3NH_2$, 180–200° → (148) $N(CH_2)_3NMe_2$; $LiAlH_4$ → (149) $N(CH_2)_3NMe_2$

C. *Spiro[chroman-4,2′-[1,3]dithiolan] Derivative*

Studies leading to the synthesis of DL-6-chlorochroman-2-carboxylic acid (**150**), a hypocholesterolemic and antilipidemic agent, indicated that the dithioketal ester (**152**) on reductive desulfurization in ethanol with Raney nickel afforded the ester (**153**), a result of dechlorination as well as desulfurization.[90] The nmr signal for the H_x in spiran (**152**) reveals four peaks: at $\delta 4.83$, $J_{AX} = 7.5$ Hz, $J_{BX} = 5.5$ Hz.

(150) (151) (152) (153) Raney Ni

D. *Spiro[chroman-4,4′-imidazolidine]-2′,5′-dione*

Hydantoin (**154**) and its derivatives possessing the aromatic ring substituted with one or more halogen atoms or alkyl groups, as well as their alkali metal salts and the 3′-alkyl derivatives, are strong anticonvulsants with low toxicity in therapeutic doses.[91] They were produced from the corresponding chromanones, potassium cyanide, and ammonium carbonate under carbon dioxide pressure at elevated temperature in yields of 50–60%. By treatment of (**154**) with an alkali metal hydroxide or alkoxide, the corresponding alkali metal salts were obtained. Standard alkylating agents in alcoholic solution give the 3′-alkyl derivatives.[91]

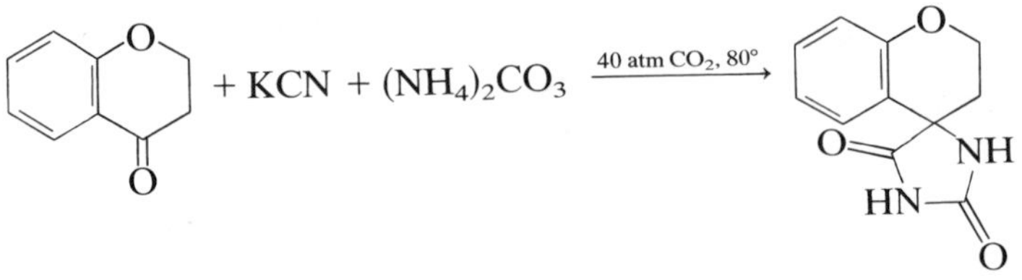

+ KCN + $(NH_4)_2CO_3$ $\xrightarrow{\text{40 atm } CO_2\text{, 80°}}$

(**154**)

3. Spiro[chroman-4,2′(5′*H*)-cyclopenta[*b*]pyrans]

Kasturi, Srinivasan, and Damodaran[92] found that sodium borohydride reduction of the secodione (**155**) in methanol yields two spiro by-products whose structures were established as (**156**) and (**157**).

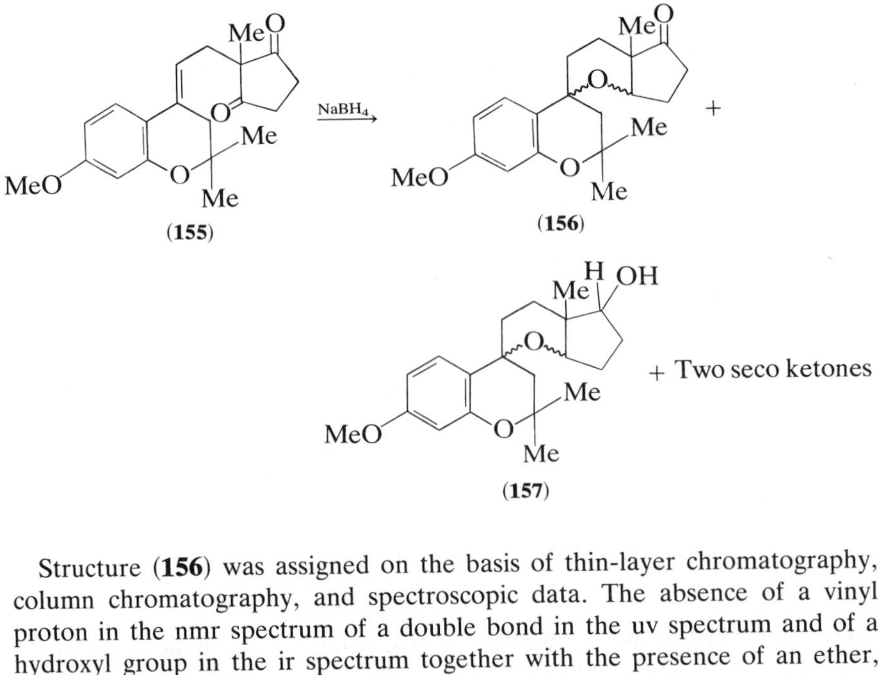

(**155**) $\xrightarrow{NaBH_4}$ (**156**) + (**157**) + Two seco ketones

Structure (**156**) was assigned on the basis of thin-layer chromatography, column chromatography, and spectroscopic data. The absence of a vinyl proton in the nmr spectrum of a double bond in the uv spectrum and of a hydroxyl group in the ir spectrum together with the presence of an ether, which was supported by the nmr signal for the proton adjacent to an ether oxygen centered at 4.13 (1H, $W_H = 6$ Hz), formed the evidence. The formation of spiran (**156**) could be the result of an acid-catalyzed attack of the hydroxyl on the double bond. Analogous rearrangements have been reported by Russian chemists[46] (see Section III, 4). The stereochemistry at C-9 and C-14 was not established from the available data.

4. Spirorotenoids

Crombie and his associates[93–96] described in a series of papers the structure of the unnatural (±)-isorotenolone C as a spiroketal (**159**). Compound (**159**) is obtained when isorotenone (**158**) is treated in alkaline methanol with hydrogen peroxide.[97] Spiran (**159**) is structurally related to spirans (**163**) and (**166**). The latter are obtained by treating rotenol (**162**) and isorotenol (**165**) respectively with manganese dioxide and alkaline ferricyanide.[93] Using lead tetraacetate in glacial acetic acid, spiran (**166**) was formed in addition to acetoxy-substituted material.[95] Isorotenol (**165**) and isoderritol (**161**) are produced when spiran (**159**) is treated with zinc and alkali. The reaction could be explained by attack as shown in (**168**) to yield the β-hydroxy compound (**169**).[94] Spiran (**159**) was reduced to the diol (**160**) with potassium borohydride. The reaction may be reversed by oxidation of this diol (**160**) with manganese dioxide. Spirans (**163**) and (**164**) are similarly interconverted.[95] Spirans (**163**) and (**166**) are presumably formed in one-electron transfer reactions via a mono- or a diradical (**170**).[95]

The formation of (±)-isorotenolone C (**159**) from (−)-isorotenone (**158**) may occur as attack by an OOH radical on the α-unsaturated ketonic ion (**171**) which is available in an alkaline solution of (**158**), giving an epoxide (**172**), which is subsequently opened intramolecularly.[94] Consequently the hydroxyl group may be cis to the keto group in the spiro system.

The ir spectrum of (±)-isorotenolone C exhibits an intermolecularly bonded hydroxyl at 3378 cm^{-1} and in dilute solution a free (3609 cm^{-1}) and a bonded (3578 cm^{-1}) hydroxyl group. The ketone frequency for (±)-isorotenolone C at 1706 cm^{-1} in comparison with that of isorotenone (1680 cm^{-1}) and rotenone (1674 cm^{-1}) indicates that it is in a five-membered environment and that the ketone is not enolizable. There exists a close similarity between the uv spectra of spiran (**159**), unchanged by acid or alkali, and of spiran (**166**) ($\nu = 1709$ cm^{-1}). Spiran (**159**) shows no hemiketal properties. Its 12-keto grouping is intact and can be reduced with borohydride.

Spiran (**163**) gave on crystallization only one diastereoisomer; spiran (**166**) is a racemate and was obtained crystalline.[95] Spiran (**163**) as well as spiran (**166**) give rotenol and isorotenol, respectively, without formation of derritol and isoderritol, upon treatment with zinc and alkali.[95]

High-resolution pmr spectroscopy allowed immediate insight into stereochemical details. For spiran (**166**) and (±)-isorotenolone C acetate

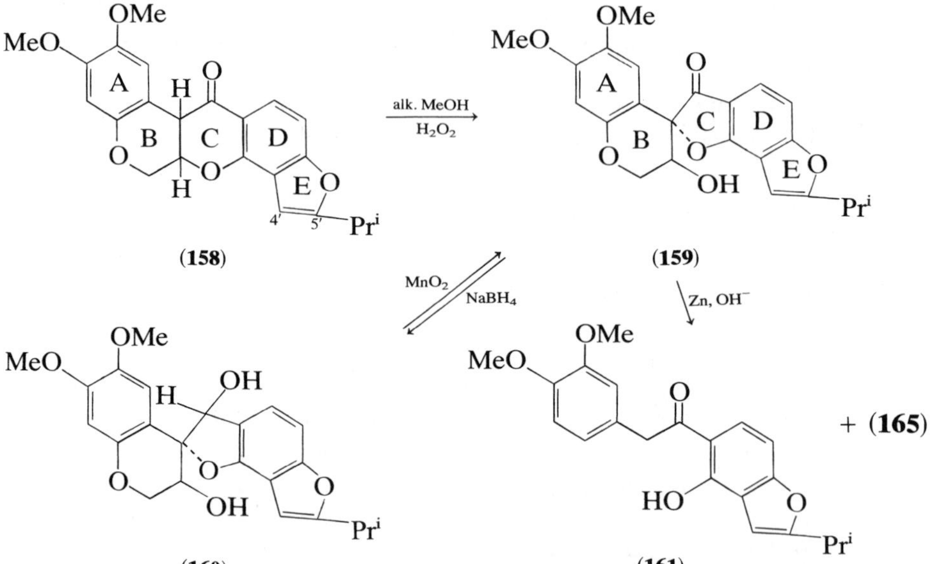

OMe
MeO
A B C D E
H H O
4′ 5′ Pr^i

(**158**)

alk. MeOH
H_2O_2

(**159**)

MnO_2
$NaBH_4$

Zn, OH^-

(**160**)

(**161**)

+ (**165**)

MnO_2, OH^- or
$K_3Fe(CN)_6$

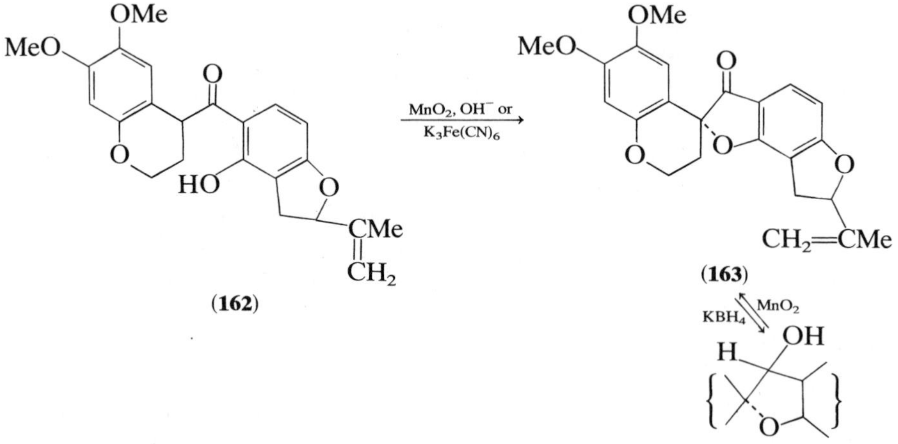

(**162**)

$CH_2{=}CMe$

(**163**)

KBH_4
MnO_2

(**164**)

MnO_2, $K_3Fe(CN)_6$
or $Pb(OAc)_4$

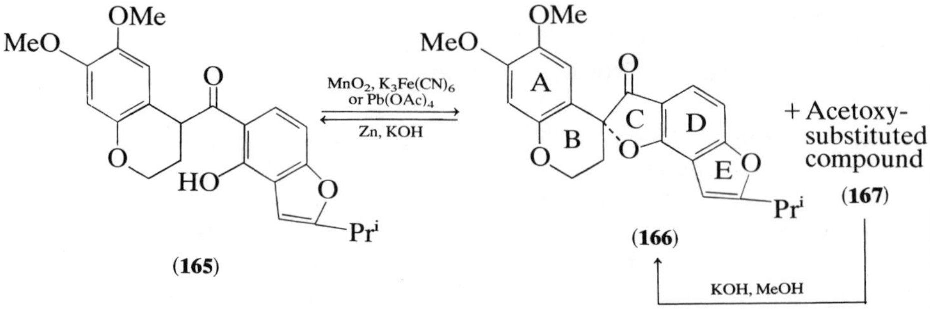

Zn, KOH

(**165**)

(**166**)

+ Acetoxy-
substituted
compound
(**167**)

KOH, MeOH

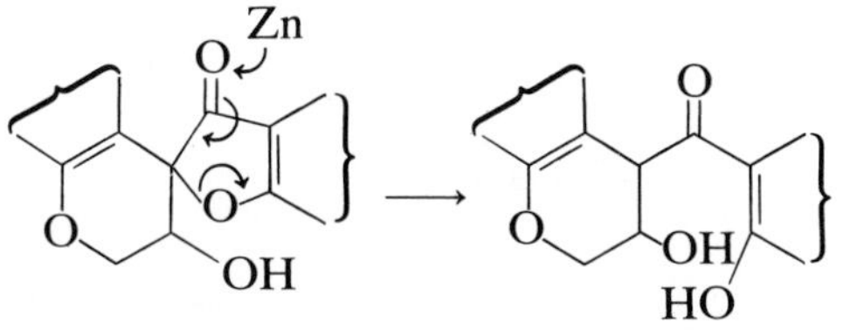

(168) (169)

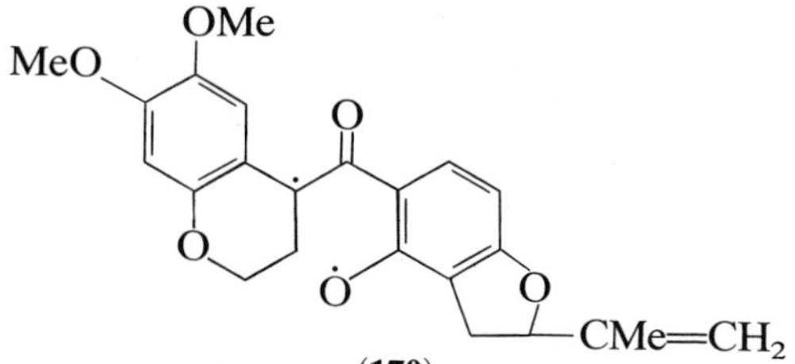

(170)

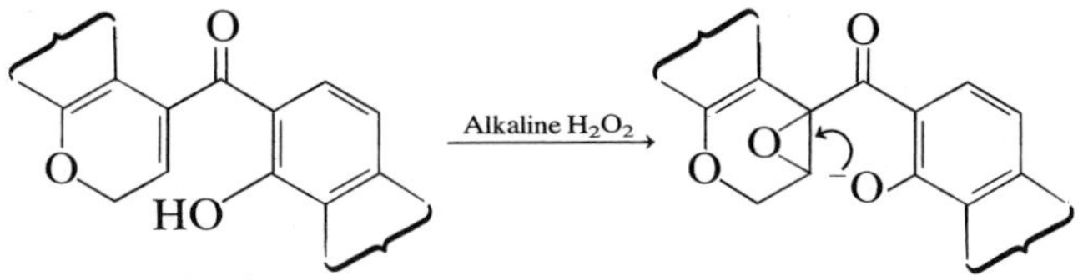

(171) (172)

(**159** acetate), the main quantitative results[96] are shown in Table 1. The 2,3-methoxyl groups of rotenone resonate in the region of 3.90–3.50 ppm, but sometimes when asymmetric shielding is operative, the signals are separated by up to 0.23 ppm. Such separation occurs when the structure forms an unnatural 12*a*-spiro-B/C system.[96]

TABLE 1. NMR SPECTRA OF SPIRANS **159** ACETATE AND **166** (CHEMICAL SHIFT IN δ ppm)

		Protons of				
Spiran	OMe	Ring A	Ring D	Ring E	Me of ring E	OAc
(**166**)	3.54	(5′) 6.28	(4) 6.67		1.76	
	3.72	(8′) 6.45	(5) 7.62			
			J 8.5			
(**159**)	3.60	(5′) 6.68	(4) 7.28	(8) 6.70	1.38	1.91
	3.83	(8′) 6.39	(5) 7.58		1.26	
			J 8.2			

VII. TABLES OF COMPOUNDS

TABLE 2. 2,2′-SPIROBICHROMANS

Mol. formula	Substituents	Yield (%)	mp	Ref.
$C_{17}H_{12}N_4O_{10}$	6,6′,8,8′-$(NO_2)_4$	64	180–182	2
$C_{17}H_{14}Cl_2O_2$	6,6′-Cl_2	—	—	30
$C_{17}H_{16}O_2$	—	72	110	1–3
$C_{19}H_{20}O_2$	6,6′-Me_2	—	—	30
$C_{19}H_{20}O_4$	8,8′-$(MeO)_2$	90	138	2
$C_{21}H_{22}Br_2O_6$	8,8′-(or 5,5′)-Br_2-6,7,6′,7′-$(OH)_4$-4,4,4′,4′-Me_4	—	—	
	Tetraacetate		227–230	13
$C_{21}H_{24}O_2$	6,8,6′,8′-Me_4	—	—	30
$C_{21}H_{24}O_4$	6,6′-$(OH)_2$-4,4,4′,4′-Me_4	—	210–212	21
$C_{21}H_{24}O_6$	6,7,6′,7′-$(OH)_4$-4,4,4′,4′-Me_4	—	270d	13
			265–267	21
	Tetraacetate		205–206	13
$C_{23}H_{28}O_2$	4,4,7,4′,4′,7′-Me_6	—	90–91[a]	18
			135–136[a]	18
			130–131[a]	18
$C_{23}H_{28}O_4$	6,6′-$(OH)_2$-4,4,7,4′,4′,7′-Me_6	24	203–206	24–27
	With EtOH of crystallization		145–149	24
$C_{24}H_{30}O_6$	4,4′-Et_2-6,7,6′,7′-$(OH)_4$-3,4,4′-Me_3	—	214–215	21
$C_{25}H_{28}N_4O_{10}$	7,7′-Et_2-4,4,4′,4′-Me_4-$(NO_2)_4$		246–248	20
$C_{25}H_{32}O_2$	7,7′-Et_2-4,4,4′,4′-Me_4	—	114	16, 20
			146	20
$C_{25}H_{32}O_2$	4,4,6,7,4′,4′,6′,7′-Me_8	—	199–200	16
$C_{25}H_{32}O_6$	6,7,6′,7′-$(MeO)_4$-4,4,4′,4′-Me_4	—	214–216	13
$C_{33}H_{48}O_6$	3-Bu-4,4′-Me_2-4,4′-$(C_5H_{11})_2$-6,7,6′,7′-$(OH)_4$	—	204–205	21
$C_{35}H_{36}O_4$	6,6′-$(OH)_2$-7,7′-$(4\text{-}MeC_6H_4)_2$-4,4,4′,4′-Me_4	18	293–296	26

[a] Stereoisomers.

TABLE 3. SPIRO[CHROMAN-2,1′-CYCLOHEXANES]

Mol. formula	Substituents	Yield (%)[a]	mp or bp (mm)	Ref.
$C_{14}H_{20}O_2$	5,6,7,8-H_4-2′-one	83	138–140 (2)	31a
$C_{14}H_{22}O_2$	5,6,7,8-H_4-2′-ol	—	130–135 (1)	31a
$C_{14}H_{22}O_2$	4*a*,5,6,7,8,8*a*-H_6-2′-one	—	160–162 (3)	31a
	Semicarbazone	—	235–236	31a
$C_{14}H_{24}O_2$	4*a*,5,6,7,8,8*a*-H_6-2′-ol	—	67–68	31a
$C_{15}H_{26}O_2$	4*a*,5,6,7,8,8*a*-H_6-2′-Me-2′-ol	—	145–150 (1)	31a
$C_{16}H_{22}O_2$	3′,3′-Me_2-6-OH	10 (15)	113–115	33
$C_{19}H_{28}O_2$	5,7,8,3′,3′-Me_5-6-OH	31 (80)	100–102	33
$C_{20}H_{30}O_2$	7-Bu^t-3′,3′-Me_2-6-OH	8 (15–20)	145–147	32–34
$C_{24}H_{38}O_2$	3′,3′-Me_2-6-OH-7-C_8H_{17}	7 (15–20)	151–153	32, 33

[a] Estimated overall yields, including filtrate residues, are given in parentheses.

TABLE 4. SPIRO[CHROMAN-2,1′-CYCLOHEXEN]-2′-ONE AND SPIRO[CHROMAN-2,1′-CYCLOHEXA-2′,5′-DIEN]-4′-ONES

Mol. formula	Structure	Yield (%)	mp	Ref.
$C_{14}H_{12}O_2$	(**50a**)	11	135–136	41
$C_{15}H_{14}O_3$	(**50b**)	1	144–146	43
$C_{16}H_{16}O_4$	(**50c**)	13	154–155	45
$C_{36}H_{50}O_3$	(**48**)	73	198–200	40

TABLE 5. SPIRO[CHROMAN-2,1′-CYCLOHEXA-3′,5′-DIEN]-2′-ONES

Mol. formula	Substituents	Yield (%)	mp	Ref.
$C_{18}H_{14}Br_6O_2$	8,3′-$(CH_2Br)_2$-6,5′-Me_2-5,7,4′,6′-Br_4	—	194	34
$C_{18}H_{16}Br_4O_2$	5,7,4′,6′-Br_4-6,8,3′,5′-Me_4	—	168	34
$C_{18}H_{20}O_2$	6,8,3′,5′-Me_4	—	123	34
$C_{22}H_{25}O_4$	6,5′-$(MeO)_2$-5,7,8,3′,4′,6′-Me_6	—	137–138	35a
$C_{24}H_{22}Br_3NO_2$	4′-PhNH-6,8,3′,5′-Me_4-5,7,6′-Br_3		206d	34
$C_{24}H_{23}Br_3N_2O_2$	4′-PhNHNH-5,7,6′-Br_3-6,8,3′,5′-Me_4	—	193	34
	N,N′-Diacetate	—	221	34
$C_{26}H_{38}O_4$	6,5′-$(Pr^iO)_2$-5,7,8,3′,4′,6′-Me_6	—	133	35a
$C_{30}H_{44}O_2$	3′,5′,6,8-Bu^t_4	—	151–152	36
$C_{32}H_{48}O_2$	3,4-Me_2-6,8,3′,5′-Bu^t_4	21	179–181	39

TABLE 6. SPIRO[CHROMAN-2,2′(5′*H*)-FURAN]-5′-ONES

Mol. formula	Substituents	Yield (%)	mp	Ref.
$C_{12}H_{12}O_3$	—	96	106	48
		100	102–104	49
$C_{13}H_{14}O_4$	6-MeO	56	103–104	49
$C_{15}H_{16}O_4$	4-CH_2COMe	56	155	48
$C_{16}H_{16}O_6$	4-CH(COOH)COMe	—	207–213d	48
$C_{16}H_{18}O_4$	4-CH_2COEt	16.5	123	48
$C_{20}H_{18}O_4$	4-CH_2COPh	80	183	48

TABLE 7. SPIRO[CHROMAN-2,4′(1′*H*)-PYRIMIDIN]-2′(3′*H*)-ONES AND -THIONES

Mol. formula	2′-One or -thione	Substituents	Yield (%)	mp	Ref.
$C_{14}H_{18}N_2O_3$	One	6′,6′-Me_2-4-OH	43	202	58
$C_{15}H_{20}N_2OS$	Thione	6,8,5′-Me_3	21	225	54
$C_{16}H_{22}N_2OS$	Thione	6,8,6′,6′-Me_4	94	247	53–55
$C_{16}H_{22}N_2O_2$	One	6,8,6′,6′-Me_4	44	212	53, 54, 57
$C_{16}H_{22}N_2O_3$	One	4-OH-6,8,6′,6′-Me_4	9.7	218	58
$C_{17}H_{24}N_2O_2$	One	6,8,1′,6′,6′-Me_5	—	221	52, 53
$C_{20}H_{22}N_2OS$	Thione	6,8-Me_2-6′-Ph	7.8	224	55
$C_{21}H_{28}N_4O_3$	One	6′,6′-Me_2-4-CH= [4,4-dimethyl-2-oxo-hexahydropyrimidin-6-ylidene: ring N(H), C=O, NH, C(Me)(Me), CH_2]	55	340	58
$C_{22}H_{26}N_2OS$	Thione	4-Ph-6,8,6′,6′-Me_4	22	215	57
$C_{22}H_{26}N_2O_2$	One	4-Ph-6,8,6′,6′-Me_4	64	247	56, 57
$C_{23}H_{28}N_2O_2$	One	6,8,1′,6′,6′-Me_5-4-Ph	40	276	56
$C_{23}H_{29}N_3OS$	Thione	6,8-Me_2-5-CH_2NMe_2-6′-Ph	80	143	55
$C_{23}H_{32}N_4O_3$	One	6,8,6′,6′-Me_4-4-CH= [4,4-dimethyl-2-oxo-hexahydropyrimidin-6-ylidene: ring N(H), C=O, NH, C(Me)(Me), CH_2]	41	352	58

TABLE 8. SPIRO[CHROMAN-2,1′(2′*H*)-XANTHEN]-2′-ONES

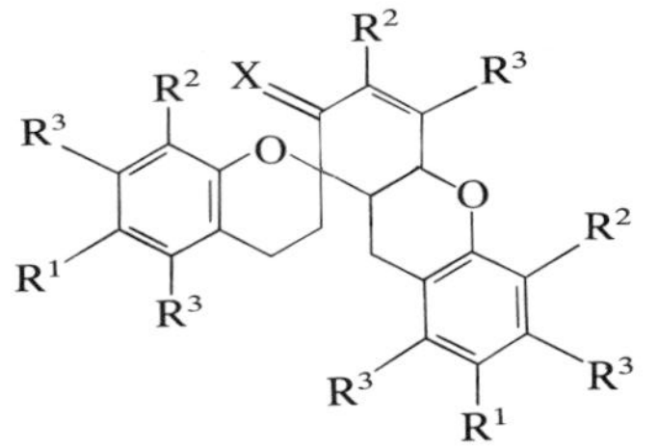

(**A**)

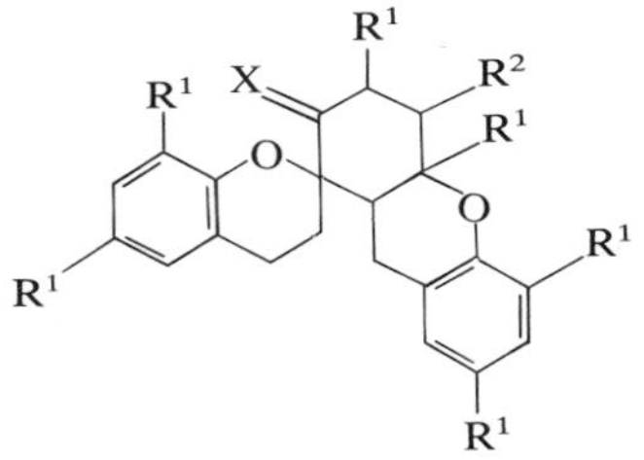

(**B**)

Mol. formula	A or B	Substituents R[1]	R[2]	R[3]	X	Yield (%)	mp	Ref.
$C_{21}H_{18}O_3$	A				O	80–85	190–192	59
$C_{21}H_{20}O_3$	B				O	90	223–224	59
$C_{21}H_{21}NO_3$	B				NOH	—	229–230d	59
$C_{21}H_{22}N_2O_4$	B		NHOH		NOH	—	232–233	59
$C_{21}H_{22}O_3$	B				H, OH	81	124–126	59
$C_{27}H_{30}O_3$	A	Me	Me		O[a]	60	199–201	61
$C_{27}H_{30}O_6$	A	MeO	Me		O	85	178–178.5	62
						20	172–175	40
						50	167–169	40
$C_{27}H_{32}O_3$	A	Me	Me		H, OH	70	145–146	61
$C_{27}H_{32}O_3$	B	Me			O[a]	64	180–180.5	61
$C_{27}H_{34}O_3$	B	Me			H, OH	33	133.5–135	61
$C_{33}H_{42}O_6$	A	MeO	Me	Me	O	66	190	35a
$C_{36}H_{48}O_3$	A	Bu^t	Me		O	72	148–149	40
$C_{36}H_{50}O_3$	A	Bu^t	Me		H, OH	35	87–90	40
$C_{42}H_{36}O_3$	A	Ph	Me		O	88	198–200	40

[a] This compound does not form carbonyl derivatives.

TABLE 9. 4-SUBSTITUTED SPIRO[CHROMAN-2,2′-INDOLINES]

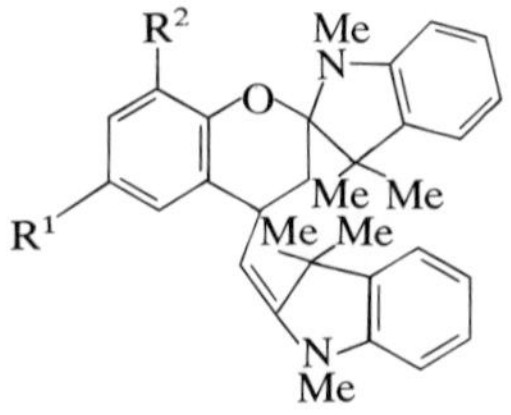

Mol. formula	Substituents R^1	R^2	mp	Ref.
$C_{31}H_{33}BrN_2O$	Br		199–200	66
$C_{31}H_{34}N_2O$			209–210	66–68
$C_{32}H_{36}N_2O_2$		MeO	233–234	66

TABLE 10. SPIRO[BENZOTHIAZOLE-2(3*H*),2′-CHROMANS] AND SPIRO[BENZOTHIAZOLE-2(3*H*),3′-[3*H*]NAPHTHO[2,1-*b*]PYRAN]

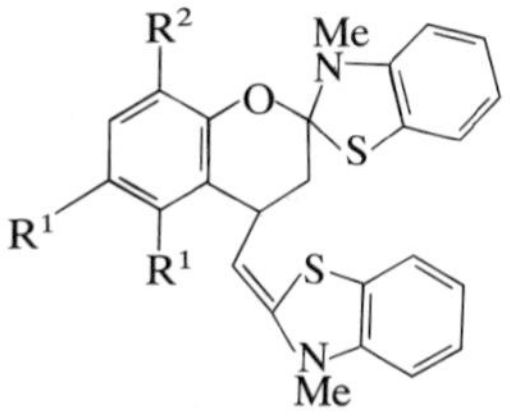

Mol. formula	Substituents R^1	R^2	Yield (%)	mp	Ref.
$C_{25}H_{22}N_2OS_2$			59	165–170	70
$C_{26}H_{24}N_2O_2S_2$		MeO	74	170–175	70
$C_{30}H_{24}N_2OS_2$	Benzo		62	172–176	70

TABLE 11. SPIRO[NAPHTHALENE-1(2*H*),3′-[3*H*]NAPHTHO[2,1-*b*]PYRAN]-2-ONES AND BENZO ANALOGS

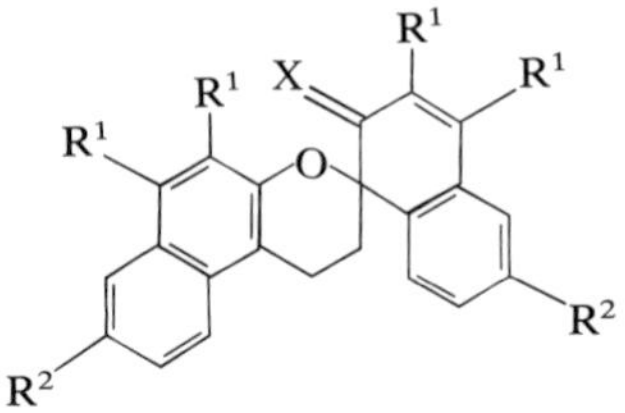

Mol. formula	Substituents R¹	R²	X	Yield (%)	mp	Ref.
$C_{22}H_{14}Br_2O_2$		Br	O	—	177	71
	Phenylhydrazone				237–238	71
$C_{22}H_{16}O_2$			O	80	143	59, 71, 72
	Phenylhydrazone				233	59, 71, 72
$C_{30}H_{20}O_2$	Benzo		O	43	251–252	73
$C_{30}H_{22}O_2$	Benzo		H, OH	90	249–250	73

TABLE 12. 4-SUBSTITUTED SPIRO[CHROMAN-2,9′-[9*H*]XANTHENE] AND ITS 5,6-BENZO ANALOG

Mol. formula	Structure	Yield (%)	mp	Ref.
$C_{35}H_{24}O_3$	(**111**, R = H)	85	180–181	63
$C_{39}H_{26}O_3$	(**111**, R = benzo)	37	199.7–200.7	63

TABLE 13. MISCELLANEOUS 2-SPIROCHROMANS

Mol. formula	Formula	Yield (%)	mp	Ref.
$C_{11}H_{12}OS_2$	(**73**)	—	—	50
$C_{14}H_{18}O_3$	(**75**)	—	—	51
$C_{20}H_{28}O_3$	(**59**)	—	Oil	46, 47
	O-Acetate	—	145–148	46, 47
$C_{20}H_{30}O_2$	(**31**, $R^1 = Bu^t$, $R^2 = R^3 = R^5 = H$, $R^4 = Pr^i$)	—	—	2
$C_{27}H_{30}O_3$	(**105**)	30	165	35
$C_{30}H_{20}O_2$	(**130**)	40	—	73
$C_{30}H_{48}O_3$	(**62**)	17	226–228	39
$C_{48}H_{72}O_3$	Structure unknown	11	181–183	39

TABLE 14. 3,3′-SPIROBICHROMANS (**132**) AND (**133**)

Mol. formula	Substituents R[1]	R[2]	Yield (%)	mp	Ref.
$C_{17}H_{16}O_4$	OH		—	—	74, 75
	3,5-Dinitrobenzoate			142d	74, 75
$C_{19}H_{20}O_4$	MeO		10	190–191	74, 75
$C_{29}H_{24}O_2$		Ph	10	164–165	76

TABLE 15. SPIRO[CHROMAN-3,2′-OXIRAN]-4-OLS[a]

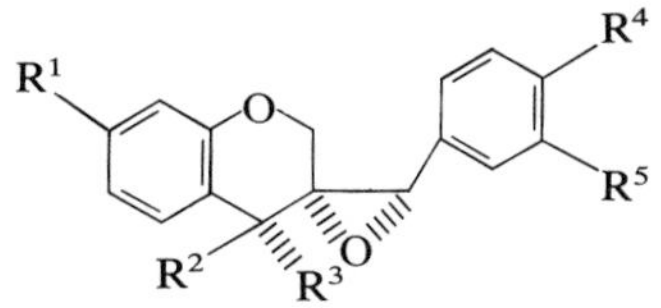

Mol. formula	Isomer	Substituents R[1]	R[2]	R[3]	R[4]	Yield (%)	mp	Ref.
$C_{17}H_{16}O_3$	Trans		OH		Me	—	110–111	81
$C_{17}H_{16}O_3$	Cis			OH	Me	—	143–144	81
$C_{17}H_{16}O_4$	Trans		OH		MeO	—	149–150	81
$C_{17}H_{16}O_4$	Cis			OH	MeO	—	143–144	81
$C_{18}H_{18}O_4$	Cis	MeO		OH	Me	—	110–111	81

[a] Other compounds of this type are described and tabulated in Volume 31, Chapter III, of this series.

TABLE 16. MISCELLANEOUS 3-ANNULATED SPIROCHROMANS

Mol. formula	Structure	Yield (%)	mp	Ref.
$C_{11}H_{10}N_2O_3$	(**139**)	18.6	288–289	82
$C_{15}H_{22}O_5$	(**141**)		135–145	84
$C_{32}H_{48}O_2$	(**45**)	15	129–130	39

TABLE 17. 4,4′-SPIROBICHROMAN-7,7′-DIOL

Mol. formula	Structural formula	mp	Ref.
$C_{21}H_{24}O_4$	(**143**)	199–200	86

TABLE 18. SPIRO[CHROMAN-4,3′-FURAN] AND SPIRO[CHROMAN-4,3′-PYRROLIDINE] DERIVATIVES

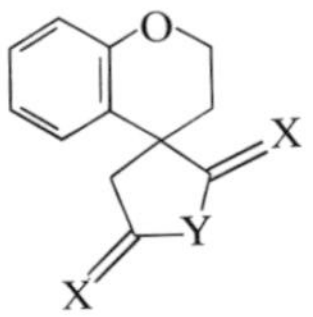

Mol. formula	X	Y	mp or bp (mm)	Yield (%)	Ref.
$C_{12}H_{10}O_4$	O	O	129–132 (0.06)	72	89
$C_{17}H_{22}N_2O_3$	O	$N(CH_2)_3NMe_2$	156–162 (0.05)	72	89
$C_{17}H_{26}N_2O$	H_2	$N(CH_2)_3NMe_2$	120–125 (0.05)	91	89
	di-HCl		255–257	—	89

TABLE 19. SPIRO[CHROMAN-4,2′-DITHIOLAN]

Mol. formula	Substituents	mp	Ref.
$C_{12}H_{11}ClO_3S_2$	6-Cl-2-COOH	—	
	Ethyl ester	111.5–113	90

TABLE 20. SPIRO[CHROMAN-4,4′-IMIDAZOLIDINE]-2′,5′-DIONES[91]

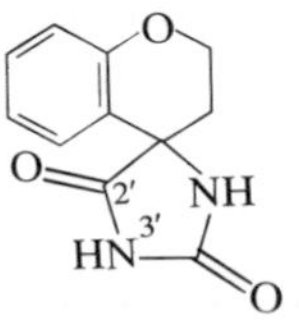

Mol. formula	Substituents	mp
$C_{11}H_9BrN_2O_3$	6-Br	264–269
$C_{11}H_9ClN_2O_3$	6-Cl	70–73
	Sodium salt	267–270
$C_{11}H_9ClN_2O_3$	8-Cl	231–235
$C_{11}H_{10}N_2O_3$	—	236–242
$C_{12}H_{11}ClN_2O_3$	6-Cl-7-Me	232–237
$C_{12}H_{11}ClN_2O_3$	8-Cl-5-Me	262–265
$C_{12}H_{12}N_2O_3$	6-Me	242–246
$C_{13}H_{13}ClN_2O_3$	6-Cl-3′-Et	131–134
$C_{13}H_{14}N_2O_3$	6-Et	211–215
$C_{15}H_{17}ClN_2O_3$	6-Cl-3′-Bu	155–159

TABLE 21. SPIRO[CHROMAN-4,2′(5′H)-CYCLOPENTA[*b*]PYRAN]-5′-ONE AND -5′-OL DERIVATIVES

Mol. formula	Formula	Yield (%)	mp or bp (mm)	Ref.
$C_{20}H_{26}O_4$[a]	(**156**)	40	139–140	92
$C_{20}H_{28}O_4$	(**157**)	3	142 (0.0023)	92

[a] Also called 3-methoxy-6-oxa-7,7-dimethyl-9,14-epoxy-8,14-secoestra-1,3,5(10)-trien-17-one.

TABLE 22. SPIROROTENOIDS: SPIRO[BENZO[1,2-*b*:3,4-*b*′]DIFURAN-2(3H),4′-CHROMAN]-3-ONES

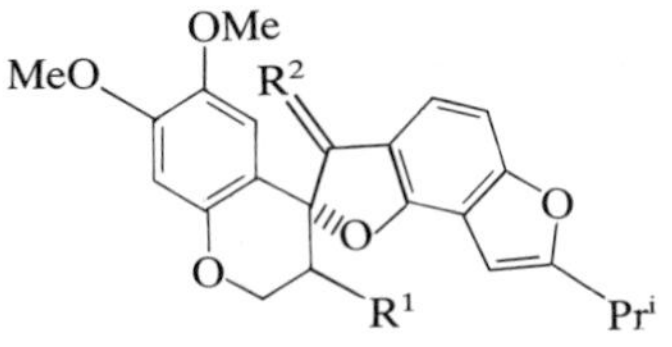

OMe MeO R² O O R¹ Prⁱ (**A**)

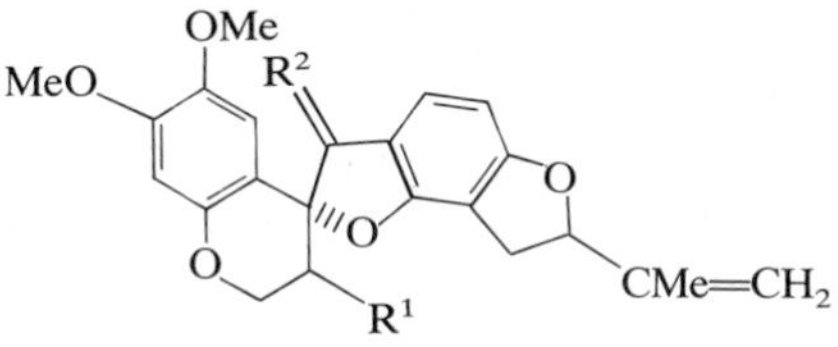

OMe MeO R² O O R¹ CMe=CH₂ (**B**)

Mol. formula	Substituents: A or B	R¹	R²	mp	Ref.
$C_{23}H_{21}ClO_6$	A	Cl	O	171	94
$C_{23}H_{22}O_6$	A		O	126[a]	94, 95, 97
$C_{23}H_{22}O_6$	B		O	75–80[b]	95
				119–121[c]	95
$C_{23}H_{22}O_7$	A	OH	O	212	93, 94, 97
	Acetate			145–146	95, 97
	Tosylate			187	95
$C_{23}H_{24}O_6$	B		H, OH	—	95
$C_{23}H_{24}O_7$	A	OH	H, OH	197–198	95
$C_{24}H_{24}O_7$	A	MeO	O	157	95

[a] (±)-Form.
[b] Two 5′β-diastereoisomers of isorotenolone C.
[c] Single diastereoisomer, $[\alpha]_D -88°$.

TABLE 23. MISCELLANEOUS 4-ANNULATED SPIROCHROMANS

Mol. formula	Structure	bp (mm)	Ref.
$C_{11}H_{11}BrO_3$	(**144**)	130 (0.02)	87

VIII. REFERENCES

1. W. Borsche, *Ber.*, **45,** 46 (1912).
2. P. T. Mora and T. Szeki, *J. Am. Chem. Soc.*, **72,** 3009 (1950).
3. J. N. Chatterjea, *J. Indian Chem. Soc.*, **35,** 47 (1958).
4. G. Cardillo, R. Cricchio, and L. Merlini, *Tetrahedron*, **27,** 1875 (1971).
5. H. D. Becker, *J. Org. Chem.*, **34,** 1211 (1969).
6. D. L. Coffen, and P. E. Garret, *Tetrahedron Lett.*, 2043 (1969).
7. W. Brown, A. B. Turner, and A. S. Wood, *Chem. Commun.*, 876 (1969).
8. H. D. Becker, *J. Org. Chem.*, **30,** 982 (1965); *ibid.*, **34,** 1198 (1969); *ibid.*, **34,** 1203 (1969).
9. T. Zincke and W. Gaebel, *Justus Liebigs Ann. Chem.*, **388,** 299 (1912).
10. J. B. Niederl, *J. Am. Chem. Soc.*, **50,** 2230 (1928).
11. J. B. Niederl and R. Casty, *Monatsh. Chem.*, **51,** 1038 (1929).
12. J. B. Niederl, *Z. Angew. Chem.*, **44,** 467 (1931).
13. W. Baker and D. M. Besly, *J. Chem. Soc.*, 195 (1939).
14. F. Mattu, R. Pirisi, L. Sancio, B. Follesa, M. R. Manca–Mura, and R. Perra, *Rend. Seminario Fac. Sci. Univ. Cagliari*, **35,** 155 (1965).
15. J. S. Dolskaya, G. E. Svadkovskaya, and L. A. Heificz, *Tr. Vses. Nauchn.-Issled. Inst. Sint. Nat. Dushistykh Veshchestv.*, **6,** 50 (1963).
16. W. Baker, J. F. W. McOmie, and J. H. Wild, *J. Chem. Soc.*, 3060 (1957).
17. J. B. Chazan and G. Ourisson, *Bull. Soc. Chim. Fr.*, 1384 (1968).
18. G. E. Svadkovskaya, N. E. Kologrivova, and L. A. Heificz, *Khim. Geterotsikl. Soedin.*, **2,** 191 (1970).
19. W. J. Moskviczev and L. A. Heificz, *Zh. Vses. Khim. Obshchest.*, **20,** 114 (1975).
20. J. B. Niederl and R. H. Nagel, *J. Am. Chem. Soc.*, **62,** 324 (1940).
21. R. B. Thompson, U.S. Patent 2,746,871 (1956); *Chem. Abstr.*, **50,** 15060 (1956).
22. F. L. Hamb and J. C. Wilson, U.S. Patent 3,859,097 (1975); *Chem. Abstr.*, **82,** 140984 (1975).
23. G. E. Svadkovskaya and T. A. Denisovich, *Tr. Vses. Nauchn.-Issled. Inst. Sint. Nat. Dushistykh Veshchestv.*, **6,** 122 (1963).
24. A. Arai, M. Jamada, and J. Oishi, Ger. Patent 2,165,371 (1972); *Chem. Abstr.*, **78,** 36244 (1973); U.S. Patent 3,830,778; Brit. Patent 1,381,499.
25. J. Oishi, M. Jamada, and H. Amano, Ger. Patent 2,420,066 (1974); *Chem. Abstr.*, **82,** 162977 (1975).
26. A. Arai, M. Jamada, and J. Oishi, Ger. Patent 2,166,076 (1973); *Chem. Abstr.*, **79,** 116089 (1973).
27. J. Oishi, M. Jamada, and M. Okazaki, Ger. Patent 2,510,538 (1975); *Chem. Abstr.*, **84,** 143038 (1976). Japan K Patent 75 122935 (1976).
28. J. Wiesner, Czech. Patent 111,355 (1964); *Chem. Abstr.* **63,** 5850 (1965).
29. J. Pospisil, L. Kotulak, and L. Taimr, *Eur. Polym. J.*, **7,** 255 (1971); *Chem. Abstr.*, **75,** 6800 (1971).
30. T. Tanaka and T. Tomimatsu, *Hukusokan Kagaku Toronkai Koen Joshishu*, **8,** 194 (1975); *Chem. Abstr.*, **85,** 32895 (1976).
31. P. Karrer and F. Kehrer, *Helv. Chim. Acta*, **25,** 29 (1942).

31a. E. Jena, V. G. Nikade, J. P. Sheth, and S. C. Bhattacharyya, *Indian J. Chem.*, **15B,** 867 (1977).

32. M. H. Stern and G. J. Lestina, Ger. Patent 2,005,301 (1970); *Chem. Abstr.*, **74,** 70252 (1971); Fr. Patent 2,034,008 (1971); U.S. Patent 3,574,627 (1971).
33. M. H. Stern, T. H. Regan, D. P. Maier, C. D. Robeson, and J. G. Thweatt, *J. Org. Chem.*, **38,** 1264 (1973).
34. K. Fries and E. Brandes, *Justus Liebigs Ann. Chem.*, **542,** 48 (1939).
35. M. S. Chauhan, F. M. Dean, S. McDonald, and M. S. Robinson, *J. Chem. Soc., Perkin Trans. I*, 359 (1973).

35a. F. M. Dean and M. Matkin, *J. Chem. Soc., Perkin Trans. I*, 2289 (1977).
36. R. F. Moore and W. A. Waters, *J. Chem. Soc.*, 243 (1954).
37. W. A. Waters and C. Wickham–Jones, *J. Chem. Soc.*, 812 (1951); *ibid.*, 2420 (1952).
38. E. Müller, R. Mayer, B. Narr, A. Rieker, and K. Scheffler, *Justus Liebigs Ann. Chem.*, **645,** 25 (1961).
39. C. D. Cook and L. C. Butler, *J. Org. Chem.*, **34,** 227 (1969).
40. D. A. Bolon, *J. Org. Chem.*, **35,** 715 (1970).
41. D. H. R. Barton, Y. L. Chow, A. Cox, and G. W. Kirby, *J. Chem. Soc.*, 3571 (1965).
42. A. M. Choudhury, K. Schofield, and R. S. Ward, *J. Chem. Soc. (C)*, 2543 (1970).
43. K. Schofield, R. S. Ward, and A. M. Choudhury, *J. Chem. Soc. (C)*, 2834 (1971).
44. H. D. Becker, *J. Org. Chem.*, **30,** 982 (1965); *ibid.*, **34,** 1198, 1203 (1969).
45. A. M. Choudhury, *J. Chem. Soc., Perkin Trans. I*, 132 (1974).
46. L. M. Kogan, V. E. Gulaya, and I. V. Torgov, *Tetrahedron Lett.*, 4673 (1967).
47. L. M. Kogan, V. E. Gulaya, B. Lacoume, and I. V. Torgov, *Izv. Akad. Nauk SSSR, Ser. Khim.*, 1356 (1970).
48. G. Mixich and A. Zinke, *Monatsh. Chem.*, **96,** 220 (1965).
49. J. F. Bagli and H. Immer, *J. Org. Chem.*, **35,** 3499 (1970).
50. E. J. Corey and D. J. Beames, *J. Am. Chem. Soc.*, **95,** 5829 (1973).
51. H. Immer and J. F. Bagli, *J. Org. Chem.*, **33,** 2457 (1968).
52. G. Zigeuner, W. Adam, and W. Galatik, *Monatsh. Chem.*, **97,** 52 (1966).
53. G. Zigeuner and R. Swoboda, *Monatsh. Chem.*, **97,** 1422 (1966).
54. G. Zigeuner, A. Frank, H. Dujmovits, and W. Adam, *Monatsh. Chem.*, **101,** 1415 (1970).
55. G. Zigeuner, A. Frank, and W. Adam, *Monatsh. Chem.*, **101,** 1788 (1970).
56. G. Zigeuner, G. Duesberg, E. Fuchs, and F. Paltauf, *Monatsh. Chem.*, **101,** 1794 (1970).
57. W. Korsatko, C. Knopp, A. Fuchsgruber, and G. Zigeuner, *Monatsh. Chem.*, **107,** 745 (1976).
58. G. Zigeuner, W. Korsatko, and A. Fuchsgruber, *Monatsh. Chem.*, **107,** 1355 (1976).
59. S. B. Cavitt, H. R. Sarrafizadeh, and P. D. Gardner, *J. Org. Chem.*, **27,** 1211 (1962).
60. K. Fries and K. Kann, *Justus Liebigs Ann. Chem.*, **353,** 339 (1907).
61. A. Merijan, B. A. Shoulders, and P. D. Gardner, *J. Org. Chem.*, **28,** 2148 (1963).
62. J. G. Westra, W. G. B. Huysmans, W. J. Mijs, H. A. Gaur, J. Vriend, and J. Smidt, *Rec. Trav. Chim. Pays-Bas*, **87,** 1121 (1968).
63. N. D. Dmitrieva, V. M. Zolin, and Y. E. Gerasimenko, *Zh. Org. Khim.*, **10,** 1505 (1974).
64. F. Irving, *J. Chem. Soc.*, 1093 (1929).
65. R. Wizinger and H. Wenning, *Helv. Chim. Acta*, **23,** 247 (1940).
66. C. F. Koelsch and W. R. Workman, *J. Am. Chem. Soc.*, **74,** 6288 (1952).
67. A. Hinnen, C. Audic, and R. Gautron, *Bull. Soc. Chim. Fr.*, 2066 (1968).
68. C. Balny, A. Hinnen, and M. Mossé, *Tetrahedron Lett.*, 5097 (1968).
69. B. W. Brockett, J. W. Stutz, and D. J. Kay, S. Afr. Patent 6,806,851 (1969); *Chem. Abstr.*, **72,** 14031 (1970).
70. A. Samat, R. Guglielmetti, and J. Metzger, Ger. Patent 2,522,877 (1975); *Chem. Abstr.*, **84,** 123407 (1976); Fr. Patent 2,272,082 (1976).
71. R. Pummerer and E. Cherbuliez, *Ber.*, **52B,** 1392 (1919).
72. L. I. Smith and J. W. Horner Jr, *J. Am. Chem. Soc.*, **60,** 676 (1938).
73. P. D. Gardner and H. R. Sarrafizadeh, *J. Org. Chem.*, **25,** 641 (1960).
74. S. Smoliński, *Tetrahedron Lett.*, 457 (1965).
75. S. Smoliński, *Tetrahedron*, **22,** 199 (1966).
76. S. Smoliński, A. Płatkowska, and W. Kamiński, *Tetrahedron*, **25,** 2155 (1969).
77. P. Bennett, J. A. Donelly, D. C. Meaney, and P. O'Boyle, *J. Chem. Soc., Perkin Trans. I*, 688 (1973).
78. O. Dann and H. Hofmann, *Chem. Ber.*, **96,** 320 (1963).
79. O. Dann and H. Hofmann, *Chem. Ber.*, **98,** 1498 (1965).

80. K. Drescher, H. Hofmann, and K. H. Frömming, *Chem. Ber.*, **101,** 2494 (1968).
81. J. N. Chatterjea, S. C. Shaw, and J. N. Singh, *J. Indian Chem. Soc.*, **52,** 210 (1975).
82. J. A. Faust, L. H. Jules, L. Yee, and M. Sahyun, *J. Am. Pharm. Assoc. Sci. Ed.*, **46,** 118 (1957).
83. H. T. Bucherer and W. Steiner, *J. Prakt. Chem.*, **140,** 291 (1934).
84. J. F. Grove and P. H. Mortimer, *Biochem. Pharmacol.*, **18,** 1473 (1969).
85. J. F. Grove and M. Hosken, *Biochem. Pharmacol.*, **24,** 959 (1975).
86. K. J. Liska, *J. Med. Chem.*, **15,** 1177 (1972).
87. E. U. Kahlert and F. Zymalkowski, *Arch. Pharm.* (Weinheim), **308,** 946 (1975).
88. A. C. Cope, C. M. Hofmann, G. Wyckoff, and E. Hardenbergh, *J. Am. Chem. Soc.*, **63,** 3452 (1941).
89. L. M. Rice, B. S. Sheth, and T. B. Zalucky, *J. Heterocycl. Chem.*, **8,** 155 (1971).
90. D. T. Witiak, E. S. Stratford, R. Nazareth, G. Wagner, and D. R. Feller, *J. Med. Chem.*, **14,** 758 (1971).
91. H. Arnold, E. Kuhas, and N. Brock, Ger. Patent 1,135,915 (1962); *Chem. Abstr.*, **58,** 3439 (1963).
92. T. R. Kasturi, A. Srinivasan, and K. N. Damodaran, *Indian J. Chem.*, **8,** 877 (1970).
93. L. Crombie, P. J. Godin, K. S. Siddalingaiah, and D. A. Whiting, *Proc. Chem. Soc.*, 19 (1961).
94. L. Crombie and P. J. Godin, *J. Chem. Soc.*, 2861 (1961).
95. L. Crombie, P. J. Godin, D. A. Whiting, and K. S. Siddalingaiah, *J. Chem. Soc.*, 2876 (1961).
96. L. Crombie and J. W. Lown, *J. Chem. Soc.*, 775 (1962).
97. F. B. La Forge and H. L. Haller, *J. Am. Chem. Soc.*, **56,** 1620 (1934).

Author Index

Numbers in parentheses are reference numbers and indicate that the author's work is referred to although his name is not mentioned in the text. Numbers in *italics* show the pages on which the complete references are listed.

Subject Index